109.95

MEANS
GRAPHIC
CONSTRUCTION
STANDARDS

PUBLISHER

E. Norman Peterson, Jr.

EDITOR-IN-CHIEF

William D. Mahoney

SENIOR EDITOR

F. William Horsley

CONTRIBUTING EDITORS

Allan Cleveland
Robert Crosscup
Donald Denzer
Dean Entwistle
Roger Grant
Melville Mossman
John Moylan
Jeannene Murphy
Kornelis Smit
Arthur Thornley
James Treichler
Edward B. Wetherill
Rory Woolsey

ILLUSTRATIONS BY

Carl W. Linde
John Brennan
Harold L. Jordan

MEANS GRAPHIC CONSTRUCTION STANDARDS

COPYRIGHT
1986

R. S. MEANS COMPANY, INC.
CONSTRUCTION CONSULTANTS & PUBLISHERS
100 CONSTRUCTION PLAZA
KINGSTON, MA 02364-0800
(617) 585-7880

11   12   13   14   15   16   17   18   19   20

Printed in the United States of America

ISBN 0-911950-79-6
Library of Congress Catalog Card Number 16-181938

TABLE OF CONTENTS

# FOREWORD

Throughout North America, the language of construction has many variations, though its methods and procedures are basically the same. This book is designed to bridge the gap between construction terminology as it is written and spoken, and actual construction practices and methods. It does so by defining and illustrating construction materials in relationship to one another as an assembly or part of a system. The graphics are clearly labeled and can be easily understood.

For more than forty years, R.S. Means Company, Inc. has been compiling data on construction products and costs, and publishing this information in annually updated books. *Means Construction Graphic Standards* offers detailed graphic information about those materials and techniques priced in our annual publications.

R.S. Means' cost books are presented in two standard formats, namely: the Construction Specifications Institute's MASTERFORMAT, and the "Uniformat." The MASTERFORMAT is the standard for construction specifications and is broken down into 16 major divisions as listed below:

Division 1 General Requirements
          2 Site Work
          3 Concrete
          4 Masonry
          5 Metals
          6 Wood & Plastics
          7 Moisture - Thermal Control
          8 Doors, Windows & Glass
          9 Finishes
       10 Specialties
       11 Equipment
       12 Furnishings
       13 Special Construction
       14 Conveying Systems
       15 Mechanical
       16 Electrical

Each of these divisions is further broken down into major subdivisions. For example, Division 3, Concrete, contains the following listings: 3.1 Formwork, 3.2 Reinforcing Steel, 3.3 Cast-in-Place Concrete, 3.4 Precast Concrete, and 3.5

Cementitious Decks. Each subdivision is further divided into "Major Classifications." These Major Classifications are comprised of activity descriptions, called "line numbers." Each activity is identifiable by a unique line number. Each line number shows a description and usually includes an installation crew and a man-hour or daily output figure for a specific unit, material, labor and/or equipment. A typical line number example as a cost for follows:

Major MASTERFORMAT subdivisions = 03.1
(2 digits plus decimal
point plus last digit)

3.1 Formwork		CREW	DAILY OUTPUT	UNIT	BARE COSTS			TOTAL INCL O&P
					MAT.	INST.	TOTAL	
25 FORMS IN PLACE, COLUMNS								
650	24" x 24" plywood columns, 1 use	C-1	190	S.F.C.A.	1.49	3.33	4.82	6.40

Major classification with
UCI subdivision = 25

Complete line number = 03.1-25-650

Item line number = 650

Uniformat is an organizational system that groups building components and procedures according to their placement and sequential order in the construction project. The assemblies in the Uniformat divisions are comprised of one or several divisions of the MASTERFORMAT. While the MASTERFORMAT is commonly used to specify construction projects, the Uniformat is used in this book because its format is compatible with illustrations of complete systems. Following is a list of Uniformat divisions.

Division 1 Foundations
         2 Substructures
         3 Superstructure
         4 Exterior Closure
         5 Roofing
         6 Interior Construction
         7 Conveying
         8 Mechanical
         9 Electrical
        10 General Conditions
        11 Special Conditions
        12 Site Work

In *Means Construction Graphic Standards*, the assemblies of each Uniformat division are graphically illustrated and captions provided to clearly identify each assembly item. Text is included with each system to further clarify the graphics and to explain the construction sequence. A man-hour chart shows the average time required to install both the pictured components and some substitute items.

In this book, special emphasis is given to mechanical and electrical materials and procedures. Not only do these two activities represent about 35% of a construction project, but they also have the widest margin for misunderstanding, and benefit most from a clear, graphic explanation.

Most graphics shown in *Means Construction Graphic Standards* are originals, but some of the charts and illustrations have been reproduced with permission from trade or institute publications as listed below. We would like to express our appreciation for the use of these helpful materials.

American Institute of Steel Construction, Inc.

Concrete Reinforcing Steel Institute

Precast Concrete Institute

Construction Specifications Institute

GUIDES

INTRODUCTION

HOW THIS BOOK IS ARRANGED

There are three basic elements to the description of the assemblies in this book. The first is a drawing showing each component of the assembly. There may be additional drawings illustrating other configurations and details. The accompanying text describes the assembly, explains its uses, and how it is constructed. Also, each assembly has a man-hour table listing the components and the man-hours required for their installation.

How to Use the Man-hour Tables

The labor or man-hour requirements to construct the assemblies drawn and defined in these Construction Graphic Standards are shown in the accompanying tables. The man-hours per unit for installing each component of an assembly was determined by first creating a typical

installing crew. This crew consisted of tradesmen, laborers, helpers, foremen, or any combination of these used for a typical installation.

The man-hours for the installation per unit of a component is determined by dividing the number of man-hours per day in the crew by the number of units installed in one day by that crew.

To use the table, multiply the number of units of each component of your assembly by the man-hours per unit. The result is the man-hours required for each component. The total of these component man-hours is a good estimate of the total man-hours required to construct your assembly.

Description
The tasks, or components, of the assembly shown in the drawing.

m/hr
The number of man-hours per unit to install each component.

Unit
The unit of measure for each component.

Man-hours

Description	m/hr	Unit
Forms in Place		
Columns Square	.136	sfca
Round Fiber Tube	.221	lf
Round Steel	.256	lf
Capitols	2.667	Ea.
Flat Plate	.086	sf
Edge Forms	.091	sfca
Reinforcing Columns		
#3 to #7	21.333	ton
#8 to #14	13.913	ton
Spirals	14.545	ton
Butt Splice, Clamp Sleeve and Wedge	.373	Ea.
Mechanical Full Tension Splice with Filler Metal	.903	Ea.
Elevated Slab	11.034	ton
Hoisting Reinforcing	.609	ton
Place Concrete		
Pumped	.492	cu yd
With Crane and Bucket	.582	cu yd
Steel Trowel Finish	.015	sf
Concrete in Place Including		
Forms, Reinforcing and Finish	6.467	cu yd

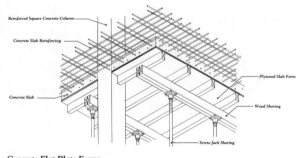

Concrete Flat Plate Forms

ABBREVIATIONS (Used in This Book)

A, Amp	ampere	Eq.	equation
ABS	Acrylonitrile Butadiene Styrene	erg	erg
ac	alternating current	esu	electrostatic units
alt	altitude		
a.m.	ante meridiem	F	fahrenheit
approx	approximate	f.c.c.	face-centered cubic
avg	average	Fig.	figure
AWG	American wire gauge	ft (also, 5′)	foot
		ft kips (also, ′k)	foot kips
b.c.c.	body-centered cubic	ft lb	foot-pound
BF	board feet	ft/min	foot per minute
Btu	British thermal units	ft/s	foot per second
Btu/hr			
		G	gauss
C	centigrade, hundred	g	gram
calc	calculated	ga.	gauge
C/C	center to center	gal.	gallon
CCF	hundred cubic feet	gal./hr, GPH	gallons per hour
cd	candela		
cd/sf	candela per square foot	H	henry
CF	hundred feet	HID	high intensity discharge
cf	cubic foot	horiz.	horizontal
c.g.	center of gravity	hp	horsepower
cgs	centimeter-gram-second	hr	hour
C.I.	cast iron	HVAC	heating, ventilating, and air conditioning
C.I.P.	concrete in place	Hz	hertz (cycle per second)
CLF	hundred linear feet		
cm	centimeter	ID, I.D.	inside diameter
cm³	cubic centimeter	in. (also, 5″)	inch
cm²	square centimeter	in./min	inches per minute
cos	cosine	IPS	iron pipe size
cot	cotangent		
CPM	Critical Path Method	J	joule
cps	cycles per second		
CRT	cathode-ray tube	k	kilo
csc	cosecant	kG	kilogauss
CSF	hundred square feet	kg	kilogram
CTS	copper tube size	kgf	kilogram force
cu	cubic	kHz	kilohertz
cu in.	cubic inch	$k\,\Omega$	kilohm
cu yd	cubic yard	kips	kips
cu yd/h	cubic yard per hour	kJ	kilojoule
cw	continuous wave	klf	kips per linear foot
		km	kilometer
day	day (s)	ksf	kips per square foot
dB	decibel	ksi	kips per square inch
dc	direct current	kV	kilovolt
degree	degree (temperature)	KVA	kilovolt amperes
diam	diameter	kW	kilowatt
DWV	drain waste vent	kWh	kilowatt-hour
dyn	dyne		
		lat	latitude
E	east	lb	pound
Ea.	each	lbf/sq in.	pound-force per square inch
EDP	electronic data processing	lb/lf	pounds per linear foot

lf	linear foot	sfca	square foot contact area
lm	lumen	sin	sine
lm/sf	lumen per square foot	sq	square
log	logrithm	sq in.	square inch
		sq yd	square yard
M Ω	megohm	STP	standard temperature and pressure
M	thousand		
mA	milliampere	tan	tangent
max.	maximum	ton	ton
MBF	thousand board feet	T&G	tongue & groove
M Btu/hr	thousand Btu per hour		
MCFM	thousand cubic feet per minute	V	volt
meter	meter	V/A	volt/ampere
MF	thousand feet	vlf	vertical linear foot
mg	milligram		
mho	mho	W	watt, west
m/hr	manhour(s)	WF	wide flange beam
MHz	megahertz	wt%	weight percent
mile	mile		
min	minute	yd	yard
min.	minimum	yr	year
mks	meter-kilogram-second		
ml	milliliter	@	at
mph	miles per hour	°C	degrees Celsius
MPT	male-pipe thread	°F	degrees Fahrenheit
ms	millisecond	..	degree (plane angle)
MSF	thousand square feet	>	greater than
MV	megavolt	<	less than
MW	megawatt	...′	minute (plane angle)
mW	milliwatt	Ω	ohm
MYD	thousand yards	...″	second (plane angle)
N	north		
nA	nanoampere		
nm	nanometer		
No.	number		
NRS	non-rising stem		
ns	nanosecond		
nW	nanowatt		
OC	on center		
OD, O.D.	outside diameter		
Opng.	opening		
oz	ounce		
PVC	polyvinyl chloride		
psf	pounds per square foot		
psi	pounds per square inch		
rms	root mean square		
rpm	revolutions per minute		
S	south		
s	second		
sf	square foot		

DIVISION 1

FOUNDATIONS

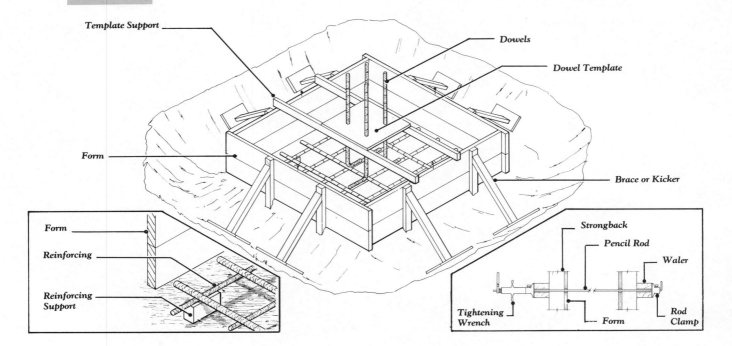

Spread footings are used to convert a concentrated pier, grade beam, or column load into an allowable area load on the supporting subgrade. Where suitable soil conditions exist, spread footings are the most widely used type of footing, because excavation, backfill, and construction materials are minimized.

Spread footings may be square, rectangular, triangular, or any common polygon and may be used to support single or multiple concentrated loads. Spread footings should be placed below maximum frost penetration on suitable soils.

Spread footings may be formed with planks, or prefabricated forms. They may be externally braced using stakes and kickers, or tied across utilizing rods or wire with suitable hardware.

If the soil permits, spread footings may be excavated to neat lines to eliminate the need for formwork. For cost effectiveness, reinforcing steel may be tied or assembled in an adjacent area and placed in the forms by hand for smaller footings, and by crane for larger footings with heavier reinforcing. If the reinforcing mat is preassembled, it must be placed before the dowel or anchor bolt template is in position. Reinforcing mats are usually supported above the subgrade by concrete bricks or by a manufactured support system.

Concrete may be pumped, placed by direct chute, by buggies, or with a crane and bucket, with direct chute being the most economical. Special care should be afforded to the location of templates for anchor bolts and dowels, and the elevation of anchor bolts should be carefully checked.

Man-hours

Description	m/hr	Unit
Formwork	.105	sfca
Reinforcing		
#4 to #7	15.239	ton
#8 to #14	8.889	ton
Placing Concrete under 1 cu yd		
Direct Chute	.873	cu yd
Pumped	1.280	cu yd
Crane and Bucket	1.422	cu yd
Over 5 cu yd		
Direct Chute	.436	cu yd
Pumped	.610	cu yd
Crane and Bucket	.640	cu yd
Anchor Bolt or Dowel Templates	1.000	Ea.

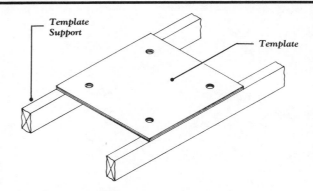

Anchor Bolt or Dowel Template

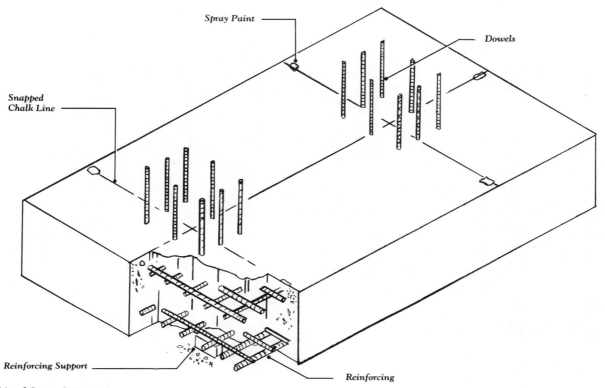

Combined Spread Footing

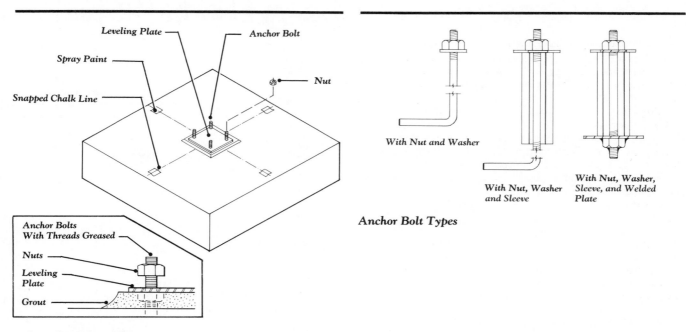

Bolt and Leveling Plate

Anchor Bolt Types

With Nut and Washer

With Nut, Washer and Sleeve

With Nut, Washer, Sleeve, and Welded Plate

1 FOUNDATIONS
STRIP FOOTINGS

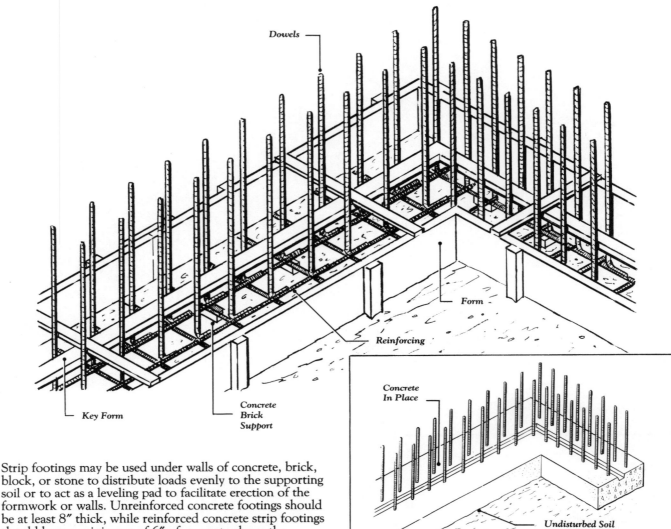

Dowels

Form

Reinforcing

Key Form

Concrete Brick Support

Concrete In Place

Undisturbed Soil

Strip footings may be used under walls of concrete, brick, block, or stone to distribute loads evenly to the supporting soil or to act as a leveling pad to facilitate erection of the formwork or walls. Unreinforced concrete footings should be at least 8″ thick, while reinforced concrete strip footings should have a minimum of 6″ of concrete above the reinforcing.

Strip footings normally are not narrower then twice the wall thickness, nor less in height than the wall thickness. The bottom of the footings should be placed on undisturbed soil and are usually located 12″ below average frost penetration.

Side forms may be made of plywood, planks, or prefabricated material. If soil conditions permit, the ditch can be excavated with the sides of the trench "neat," eliminating the use of forms. This can produce considerable savings. Keyway forms, if required, may be constructed of wood (tapered to facilitate removal), preformed metal, or plastic. Reinforcing, if required, should be supported at a constant height above the subgrade by utilizing concrete bricks or by employing a manufactured support system. Depending on location and accessibility, concrete may be placed by direct chute, by pumping, or by crane and bucket, with direct chute being the most economical.

Man-hours

Description	m/hr	Unit
Formwork	.066	sfca
Formwork Keyway	.015	lf
Reinforcing		
#4 to #7	15.238	ton
#8 to #14	8.889	ton
Dowels		
#4	.128	Ea.
#6	.152	Ea.
Place Concrete,		
Direct Chute	.400	cu yd
Pumped	.640	cu yd
Crane and Bucket	.711	cu yd
Concrete in Place		
36″ x 12″ Reinforced	1.630	cu yd

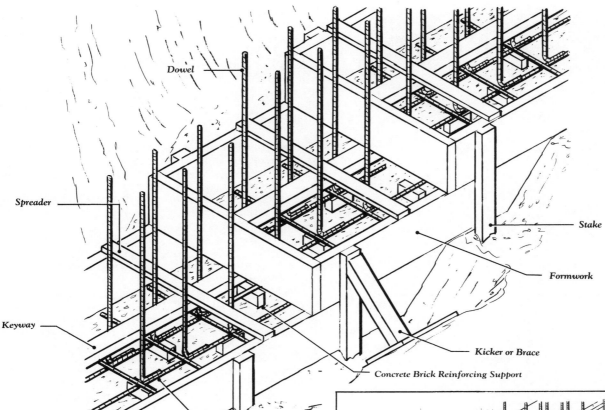

Dowel

Spreader

Keyway

Reinforcing

Stake

Formwork

Kicker or Brace

Concrete Brick Reinforcing Support

Sloping terrain or differences in subsurface elevations of the foundation wall require steps in strip footings to change depth. Sloped footings would create undesirable horizontal thrust on the support material. The excavation must mainly be done by hand to guard against erosion and to maintain specified lengths of the steps. Steps should measure at least 2′ horizontally, and each vertical step should measure no greater than three-fourths the horizontal distance between steps. Vertical risers should be a minimum of 6″ thick and match the footing's width. Step footing horizontal lengths are usually matched to standard widths of prefabricated wall forms whenever possible to save special wall form fabrication. This results in careful layout and excavation and may necessitate additional concrete in the risers to maintain the horizontal dimensions. Step footings are time consuming and costly to install when compared with strip footings.

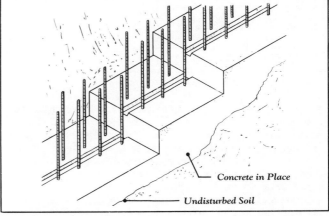

Concrete in Place

Undisturbed Soil

Man-hours

Description	m/hr	Unit
Formwork (Sides and Steps)	.085	sfca
Formwork (Keyway)	.015	lf
Reinforcing #4 to #7	15.238	ton
Reinforcing (Dowels, 2′ Long)	.128	Ea.
Place Concrete,		
Direct Chute	.400	cu yd
Pumped	.640	cu yd
Crane and Bucket	.711	cu yd
Concrete in Place	3.900	cu yd

1 FOUNDATIONS
FOUNDATION WALLS

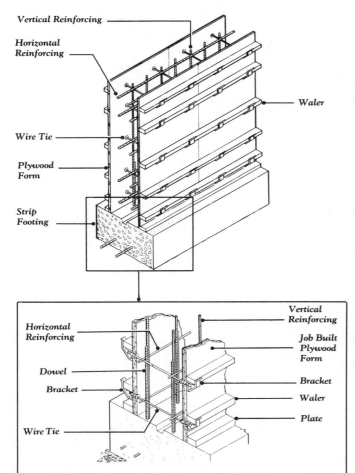

Figure labels (upper):
Vertical Reinforcing
Horizontal Reinforcing
Wire Tie
Plywood Form
Strip Footing
Waler

Figure labels (lower):
Horizontal Reinforcing
Dowel
Bracket
Wire Tie
Vertical Reinforcing
Job Built Plywood Form
Bracket
Waler
Plate

The sequence of erection for reinforced walls generally proceeds as follows:

Step 1: Form one side with ties extended
Step 2: Place reinforcing
Step 3: Double up or place the opposite side
Step 4: Align and straighten the entire forming system

Walls may be unreinforced, but the forces exerted by shrinkage and temperature require longitudinal reinforcing, while vertical and lateral loads make vertical reinforcing necessary.

The introduction of piers or buttresses interrupts the sequence of wall forming. These forms are costly and time-consuming to install. Brick shelves, corbels, openings, and bulkheads are other wall-forming operations that require special consideration, and should be priced accordingly in any cost estimate.

Due to considerable horizontal thrust from backfill material, basement walls are not usually designed as retaining walls. The structure at the top of the wall must be in place before backfilling, unless special consideration is given to bracing.

Concrete may be placed by direct chute, pumped, or placed with a crane and bucket. Some specifications may require use of an elephant trunk or placing ports to limit the height of drop of the concrete in high walls in order to prevent segregation. Vertical placing rates may be limited by form design, but care must be taken to avoid the formation of "cold joints." Internal vibration, or hitting the forms on the exposed face side, are two methods used to prevent honeycombing and to ensure an even finish. Tie holes are usually filled and patched, and the concrete is then rubbed and finished on the exposed faces.

Foundation walls are generally constructed of cast-in-place concrete or concrete block. The wall height depends on depth of frost penetration, basement configuration, or the architectural requirements of the structure. Minimum wall thickness is determined by code requirements, unsupported height or length, structural load considerations, or the thickness of the wall to be supported. Expansion or control joints at approximately 30′ intervals in segments of long, exposed walls help to relieve internal stresses and control crack lines. Walls may be formed by using job-built plyform with stud framing, modular prefabricated plywood, steel-framed plywood, or prefabricated aluminum or steel forms.

The forms for long or large concrete walls may be ganged or unitized and moved as a unit with a crane to increase productivity and lower costs on repetitive pours. Additional hardware is required to maintain the form as a unit and to allow stripping and moving as a unit.

The lateral pressure of the uncured concrete and the vibration used to consolidate the concrete are resisted by wire ties or rods that extend through both sides of the form and connect to a system of clips and walers.

Man-hours

Description	m/hr	Unit
Forms		
Job-Built Plywood		
1 Use/Month	.130	sfca
4 Uses/Month	.095	sfca
Modular Prefabricated Plywood		
1 Use/Month	.053	sfca
4 Uses/Month	.049	sfca
Steel Framed Plywood		
1 Use/Month	.080	sfca
4 Uses/Month	.072	sfca
Box Out Openings to 10 sf	2.000	Ea.
Brick Shelf	.200	sfca
Bulkhead	.181	lf
Corbel to 12″ Wide	.320	lf
Pilasters	.178	sfca
Waterstop Dumbbell	.055	lf
Reinforcing		
#3 to #7	10.667	ton
#8 to #14	8.000	ton
Place Concrete 12″ Walls		
Direct Chute	.480	cu yd
Pumped	.674	cu yd
With Crane and Bucket	.711	cu yd
Finish, Break Ties and Patch Voids	.015	sf
Burlap Rub with Grout	.018	sf
Concrete in Place 12″ Thick	5.926	cu yd

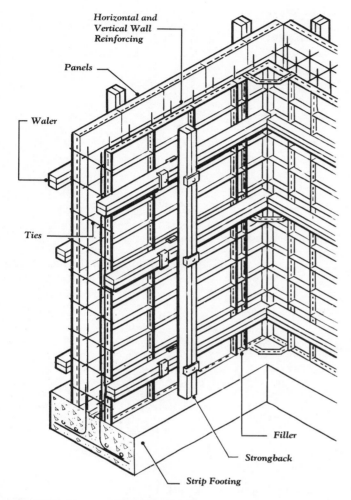

Horizontal and Vertical Wall Reinforcing

Panels

Waler

Ties

Filler

Strongback

Strip Footing

Steel-Ply Concrete Forming System

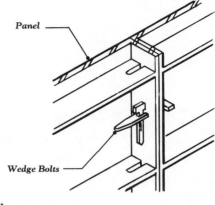

Panel

Wedge Bolts

Panel Attachment

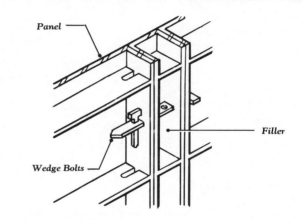

Panel

Filler

Wedge Bolts

Panel Filler Attachment

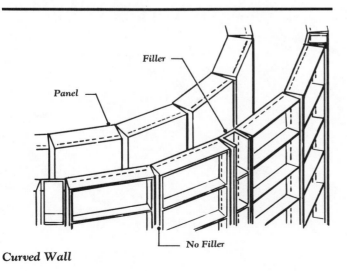

Filler

Panel

No Filler

Curved Wall

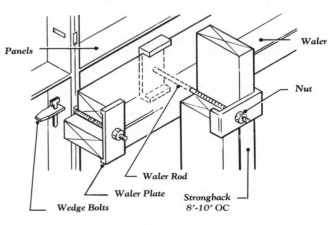

Panels

Waler

Nut

Wedge Bolts

Waler Plate

Waler Rod

Strongback 8'-10' OC

Strongback and Waler Attachment

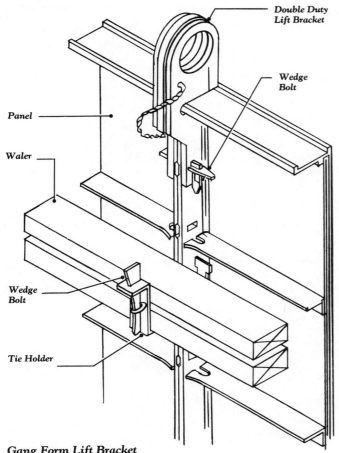

Gang Form Lift Bracket

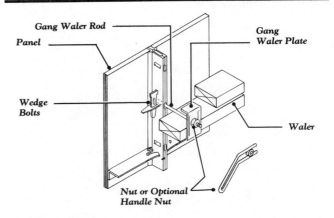

Alternate Waler Attachment

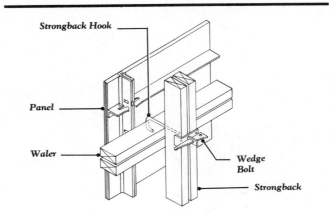

Strongback Attachment

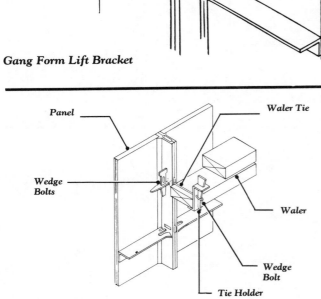

Waler Attachment

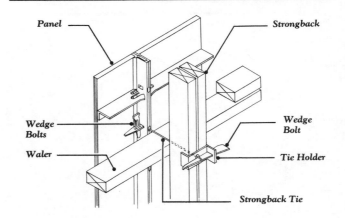

Alternate Strongback Attachment

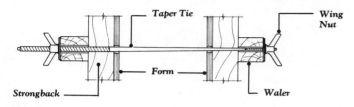

Taper Tie

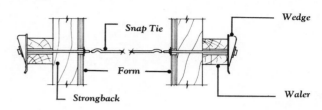

Snap Tie

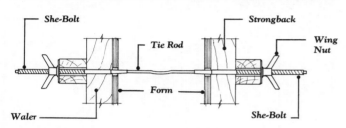

She-Bolts

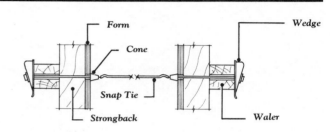

Snap Tie With Cones

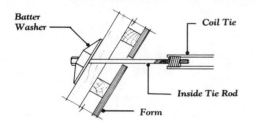

Battered Washer

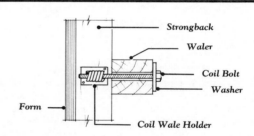

Wale Holder

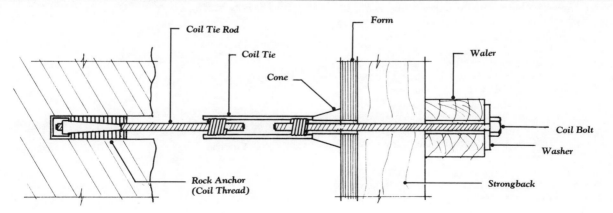

Coil Type System

1 FOUNDATIONS
GRADE BEAMS

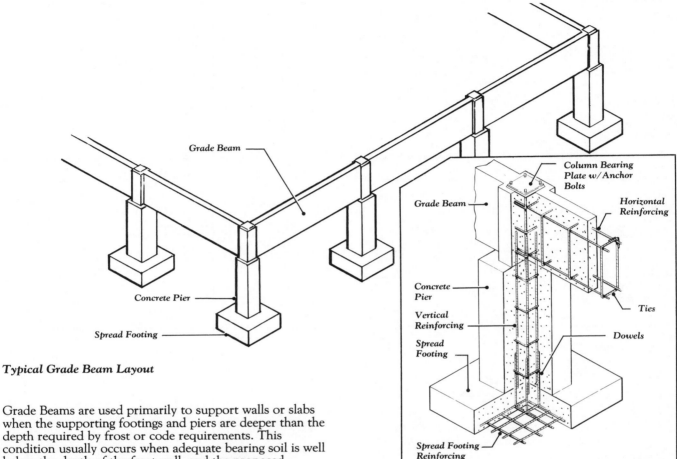

Grade Beam

Concrete Pier

Spread Footing

Typical Grade Beam Layout

Column Bearing Plate w/Anchor Bolts

Grade Beam

Horizontal Reinforcing

Concrete Pier

Vertical Reinforcing

Spread Footing

Ties

Dowels

Spread Footing Reinforcing

Grade Beams are used primarily to support walls or slabs when the supporting footings and piers are deeper than the depth required by frost or code requirements. This condition usually occurs when adequate bearing soil is well below the depth of the frost wall, and the proposed structure has no basement below grade. Grade beams may be designed as single-span beams that span from pier to pier and support wall or floor loads. The piers carry the load from the grade beams to the supporting footing. Grade beams do not depend on the soil for direct support.

Forming is usually accomplished with plywood (job-fabricated, modular prefabricated, or steel-framed) utilizing ties and walers. Grade beams differ from foundation walls because the concrete on the bottom is generally placed against the earth in lieu of strip footings.

Reinforcing for grade beams is usually designed with ties and is sometimes pre-assembled and placed by crane or some other moveable lifting device. Concrete may be placed by direct chute, pumped, or with a crane and bucket.

Heavy column loads and light wall loads tend to make grade beams an economical consideration, when compared to other foundation systems.

Man-hours

Description	m/hr	Unit
Forms	.083	sfca
Brick or Slab Shelf	.200	sfca
Bulkhead	.181	lf
Reinforcing		
#3 to #7	20.000	ton
#8 to #18	11.852	ton
Place Concrete		
Direct Chute	.320	cu yd
Pumped	.492	cu yd
With Crane and Bucket	.533	cu yd
Finish, Break Ties and Patch Voids	.015	sf

1 FOUNDATIONS
WATERPROOFING AND UNDERDRAIN

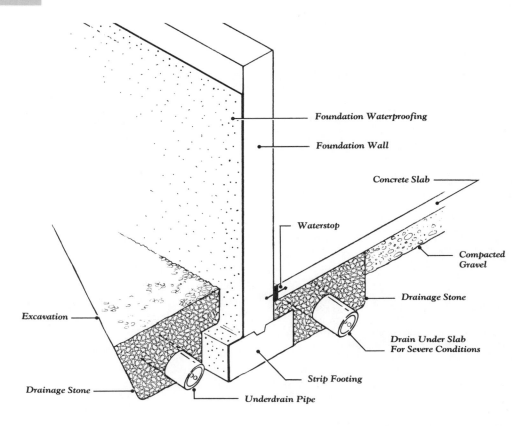

Foundation Waterproofing

Foundation Wall

Concrete Slab

Waterstop

Compacted Gravel

Excavation

Drainage Stone

Drain Under Slab For Severe Conditions

Strip Footing

Drainage Stone

Underdrain Pipe

To be effective, foundation waterproofing requires a protective coating or covering on the exterior of the wall, a collecting underdrain, and porous backfill material to direct water to the underdrain. The protective coating may consist of bituminous coating applied by brush or spray, troweled on, asphalt protective board and mastic, or a membrane. An application of protective iron coating may also be used for waterproofing but requires an additional chipping of the concrete surface before installation.

The underdrain conduit should either be porous or contain openings to ensure collection of water. The conduit should be surrounded by coarse rock or gravel and laid with sufficient slope to direct the water to the collection area. The underdrain materials may be asbestos cement, bituminous fiber, corrugated metal (asphalt-coated), corrugated polyethylene corrugated tubing, porous concrete, vitrified clay, or PVC.

The porous backfill should extend up to a minimum depth below the final grade as required by site conditions. Where gravel or stone are uneconomical or unavailable, concrete block installed dry with spaced joints provides both protection for the waterproofing and a channel for excess water.

Man-hours

Description	m/hr	Unit
Asphalt Coating, Troweled On		
1/16" Thick	.016	sf
1/8" Thick	.020	sf
1/2" Thick	.023	sf
Bituminous Coating		
Brushed On		
1 Coat	.012	sf
2 Coats	.016	sf
Sprayed On		
1 Coat	.010	sf
2 Coats	.016	sf
Protective Board, Asphalt Coated, in Mastic	.018	sf
Metallic Coating Iron Compound		
5/8" Thick	.229	sf
1" Thick	.250	sf
Underdrain	.060	lf
Perforated,		
4" Diameter	.060	lf
6" Diameter	.076	lf
8" Diameter	.083	lf
Bituminous		
4" Diameter	.032	lf
6" Diameter	.035	lf
Porous Concrete		
4" Diameter	.072	lf
6" Diameter	.076	lf
8" Diameter	.090	lf
Drainage Stone 3/4" Diameter	.092	cu yd

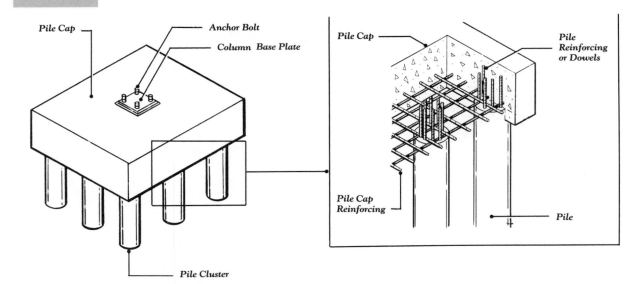

Pile Cap — Anchor Bolt — Column Base Plate — Pile Cluster

Pile Cap — Pile Reinforcing or Dowels — Pile Cap Reinforcing — Pile

Piles are column-like shafts which receive superstructure loads, overturning forces, or uplift forces from isolated column or pier foundations (pile caps), foundation walls, grade beams, or foundation mats. The piles then transfer these loads through poor soil strata to deeper soil of adequate support strength and acceptable settlement.

Several pile types have been developed to suit different ground conditions and are designed to support loads by end bearing or friction. End-bearing piles have shafts that pass through inadequate soil and bear on bedrock or penetrate some distance into a dense, suitable soil. Friction piles have shafts that may be entirely embedded in cohesive soil. They develop required capacity mainly by adhesion or "skin-friction" between soil and shaft area.

Piles pass through soil by one of two ways:

1. Displacement piles force soil out of the way. This may cause compaction, ground heaving, remolding of sensitive soils, damage to adjacent structures, or hard driving.
2. Non-displacement piles have either a hole bored and the pile cast or placed in the hole, or open-ended pipe (casing) driven and the soil core removed. They tend to eliminate heaving or lateral pressure and damage to adjacent structures or piles. Steel HP piles are considered small displacement piles.

Placement of piles (attitude) is most often vertical; however, they are sometimes battered (placed at a small angle from vertical) for lateral loads.

Piles are seldom installed singly, but rather in clusters and capped with a reinforced concrete footing (pile cap) or slab that evenly distributes loads to the individual piles.

Typical Boring Log

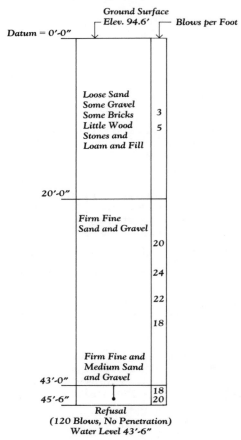

Datum = 0'-0"

Ground Surface
Elev. 94.6'   Blows per Foot

Loose Sand
Some Gravel
Some Bricks
Little Wood
Stones and
Loam and Fill   3   5

20'-0"

Firm Fine
Sand and Gravel   20   24   22   18

Firm Fine and
Medium Sand
and Gravel

43'-0"   18
45'-6"   20

Refusal
(120 Blows, No Penetration)
Water Level 43'-6"

Soil boring log used to determine the need for and design of pile foundations.

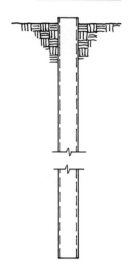

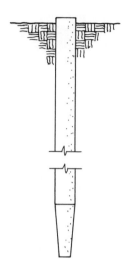

Cast-in-Place Concrete Pile

Precast Concrete Pile

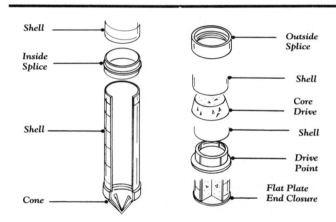

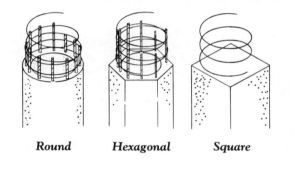

Round *Hexagonal* *Square*

Cast-in-place concrete piles are usually formed with a steel shell, but piles up to 25′ in length are sometimes augered or bored with a drilling rig and filled with concrete directly against the earth. If a steel shell is used to form the concrete pile it is usually driven with a mandrel (a filler that maintains the shell configuration). When the shell and mandrel have reached specified capacity or location, the mandrel is withdrawn from the shell, the shell inspected and then filled with concrete.

Cast-in-place piles are permanent, can be treated for sea water, and are easily altered in length.

Precast concrete piles are cast by a specialty contractor to a specified length and trucked to the job site. They may be reinforced or posttensioned. Precast piles are durable and can be treated for sea water, but they are difficult and expensive to extend and require heavy equipment for handling and driving.

Man-hours

Description	m/hr	Unit
Precast, Prestressed 14″ Square	.107	vlf
Mobilization		
10,000 lf Job	.019	vlf
25,000 lf Job	.008	vlf

Man-hours

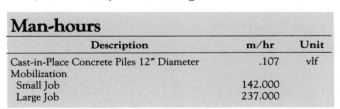

Description	m/hr	Unit
Cast-in-Place Concrete Piles 12″ Diameter	.107	vlf
Mobilization		
Small Job	142.000	
Large Job	237.000	

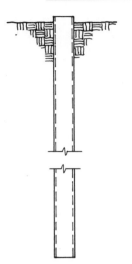

Steel Pipe Pile

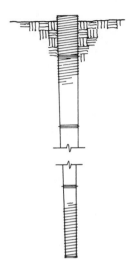

Step-Tapered Pile

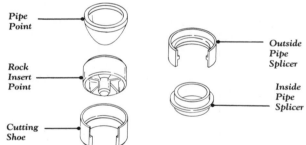

Pipe piles have high structural strength, are flexurally strong, and can be easily spliced. They are normally concrete filled and sometimes classified as "composite." They may be driven with a standard hammer rather than with a mandrel and can withstand rough handling and driving conditions.

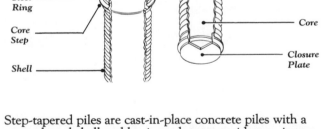

Step-tapered piles are cast-in-place concrete piles with a tapered steel shell and horizontal steps or ridges to increase friction. They are usually driven with a mandrel to prevent damage to the shell during the driving procedure. When the pile reaches the required capacity, the mandrel is withdrawn and the shell is filled with concrete. Step-tapered piles are easily spliced and inspected and may be treated for sea water.

Man-hours

Description	m/hr	Unit
Pipe Piles 44 lb per lf		
No Concrete	.135	vlf
Concrete Filled	.154	vlf
Splices	2.111	Ea.
Standard Points	1.975	Ea.
Heavy Duty Points	3.960	Ea.
Mobilization		
Small Job	142.000	
Large Job	237.000	

Man-hours

Description	m/hr	Unit
Cast-in-Place Step-Tapered Pile 12" Diameter	.107	vlf
Mobilization		
Small Job	142.000	
Large Job	237.000	

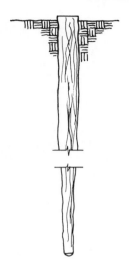

Treated Wood Pile

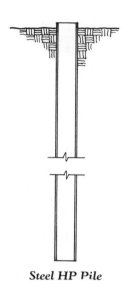

Steel HP Pile

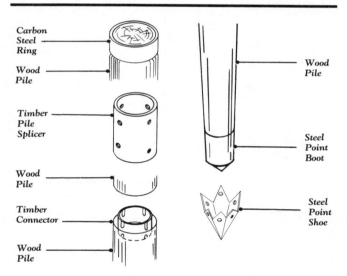

Carbon Steel Ring
Wood Pile
Timber Pile Splicer
Wood Pile
Timber Connector
Wood Pile

Wood Pile
Steel Point Boot
Steel Point Shoe

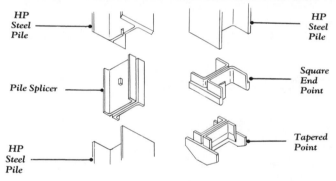

HP Steel Pile
Pile Splicer
HP Steel Pile

HP Steel Pile
Square End Point
Tapered Point

Pressure creosoted wood piles are relatively inexpensive and easy to handle. Although they cut easily, they are normally used as a single length, because extensions are hard to achieve and expensive. They are not usually driven through hard strata or boulders, as they are prone to driving damage.

HP piles are seldom used as friction piles, but are economical for long lengths driven to refusal in rock or hardpan. They are essentially wide-flange shapes with parallel flange surfaces and equal web and flange thicknesses. They are rolled expressly for bearing pile use. They are easily spliced and can withstand rough handling and driving conditions.

Man-hours

Description	m/hr	Unit
Wood Piles Treated 12 lb Creosote per cf		
12" Butts 8" Points up to 30' Long	.102	vlf
Boot for Pile Tip	.300	Ea.
Point for Pile Tip	.450	Ea.
Mobilization for 10,000 lf Job	.019	

Man-hours

Description	m/hr	Unit
H Sections 12" x 12", 53 lb per lf	.108	vlf
Splice or Standard Points	2.000	Ea.
Heavy Duty Points	2.286	Ea.
Mobilization		
Small Job	142.000	
Large Job	237.000	

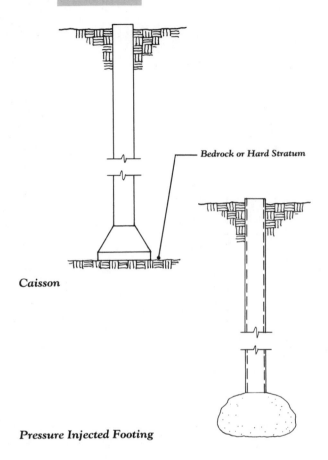

Bedrock or Hard Stratum

Caisson

Pressure Injected Footing

Caissons are drilled cylindrical foundation shafts which function primarily as compression members. Like columns, they transfer building loads through inadequate soils to bedrock or hard stratum. They may be either reinforced or unreinforced, and either straight or belled out at the bearing level.

Shaft diameters range in size from 20" to 84" with the most usual sizes beginning at 34". If inspection of the bottom is required, the minimum diameter practical is 30". If handwork is required (in addition to mechanical belling, etc.) the minimum diameter is 32". A frequently used shaft diameter is 36" with a 5' or 6' bell diameter. The maximum practical bell diameter is three times the shaft diameter.

Plain concrete is commonly used and poured directly against the excavated face of soil. Permanent casings add to the cost and should be avoided for economic reasons. Wet or loose strata is undesirable and the associated installation sometimes involves a mudding operation with a bentonite clay slurry to keep the walls of excavation stable.

In the case of heavy loads, reinforcement is sometimes used. Reinforcing also is required if uplift, bending moment, or lateral loads exist. A small amount of reinforcement is desirable at the top portion of each caisson, even if the above conditions are not present. There

are three basic types of caisson and these are briefly described in the following paragraphs.

Belled caissons are generally recommended to provide reduced bearing pressure on soil. These are not for shallow depths or poor soils. Good soils for belling include most clays, hardpan, soft shale, and decomposed rock. Soils not recommended include sand, gravel, silt, and igneous rock. Water in the bearing strata is undesirable.

Straight shafted caissons have no bell, but the entire length is enlarged to permit safe bearing pressures. They are most economical for light loads on high bearing capacity soils.

Socketed (or keyed) caissons are used for extremely heavy loads. They involve sinking the shaft into rock for combined friction and bearing support action. Reinforcement of the shaft is usually necessary, and wide flange cores are frequently used.

The advantages of using caissons over piles include no soil heaving, displacement vibration, or noise during placement. Also the bearing strata can be visually inspected and tested.

Man-hours

Description	m/hr	Unit
Caissons Open Style Machine Drilled to 50' Deep in Stable Ground, No Casings or Ground Water		
36" Diameter	.384	vlf
8' Bell Diameter Add	20.000	Ea.
Open Style Machine Drilled to 50' Deep in Wet Ground Pulled Casing and Pumping		
36" Diameter	.933	vlf
8' Bell Diameter Add	23.333	Ea.
Open Style Machine Drilled to 50' Deep in Soft Rock and Medium Hard Shale		
36" Diameter	5.867	vlf
8' Bell Diameter Add	67.692	Ea.
Mobilization	65.000	
Bottom Inspection	6.667	Ea.

Pressure-injected footings or bulb end piles are placed in the same way as cast-in-place concrete piles but differ in that a plug of uncured concrete is driven through the casing to form a bulb in the surrounding earth at the base of the pile. The bulb forms a footing with the pile acting as a column or pier. Piles of 25' or less may have the casing withdrawn, but those of 25' or more are usually cased with metal shells.

Man-hours

Description	m/hr	Unit
Pressure Injected Footings 12" Diameter Shaft	.400	vlf
Pressure Injected Footing 12" to 18"		
Diameter to 40'	.278	vlf
Pile Cutoff Concrete Pile with Thin Shell	.211	Ea.
Mobilization		
Small Job	142.000	
Large Job	237.000	

DIVISION 2

SUBSTRUCTURES

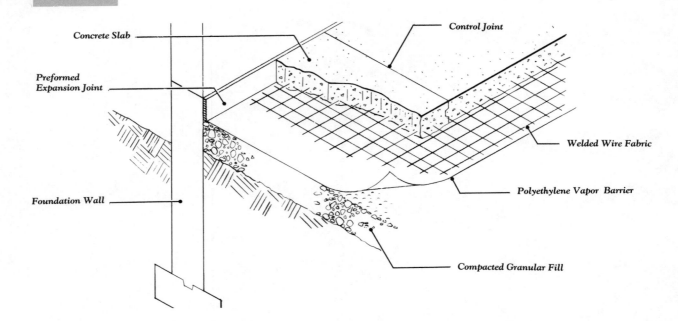

Concrete Slab

Preformed Expansion Joint

Foundation Wall

Control Joint

Welded Wire Fabric

Polyethylene Vapor Barrier

Compacted Granular Fill

Slab-on-grade should be placed on compacted, granular fill such as gravel or crushed stone, which in turn is covered by a polyethylene vapor barrier. The slab thickness is based on use, density of the supporting subsoil, or, in limited cases, hydrostatic pressure. Slabs may be unreinforced or reinforced. Reinforcing is usually provided by welded wire fabric, primarily to provide assurance that any crack that does occur will be tightly closed.

Recommended minimum concrete strength is 3500 psi. Concrete floors are usually placed in strips which extend across the building at column lines or at 20' to 30' widths.

Construction joints may be keyed or straight and may contain smooth or deformed reinforcing which, in turn, can be wrapped or greased on one side to allow horizontal movement and to control cracking.

Control joints hopefully limit cracking to designated lines and may be established by saw-cutting the partially cured concrete slab to a specified depth or by applying a preformed metal strip to create a crack line. Many specifications require a boxed out section around columns, with that area to be concreted after the slab has been placed and cured.

Expansion joints are generally used against confining walls, foundations, etc., and are commonly made from a preformed expansion material. The finish of the slabs is usually dictated by their use and varies between a screed finish (associated with two-course floors) to a steel trowel-treated finish (common to exposed concrete floors). Dropped areas or deeper slabs may be used under concentrated loads, such as masonry walls. Formed depressions to receive other floor materials, such as mud-set ceramic tile or terrazzo, may also be necessary, but add an additional cost to the slab-on-grade system.

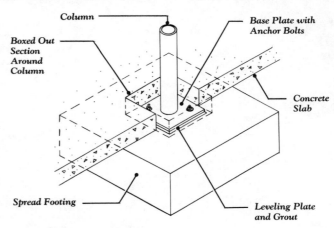

Column

Boxed Out Section Around Column

Base Plate with Anchor Bolts

Concrete Slab

Spread Footing

Leveling Plate and Grout

Control Joint Around Column

Man-hours

Description	m/hr	Unit
Slab-on-Grade		
Fine Grade	.010	sq yd
Gravel Under Floor Slab 6″ Deep Compacted	.005	sf
Polyethylene Vapor Barrier	.216	sq
Reinforcing WWF 6 x 6 (W1.4/W1.4)	.457	CSF
Place and Vibrate Concrete 4″ Thick Direct Chute	.436	cu yd
Expansion Joint Premolded Bituminous Fiber		
1/2″ x 6″	.021	lf
Edge Forms in Place to 6″ High 4 Uses on Grade	.053	lf
Curing w/Sprayed Membrane Curing Compound	.168	CSF
Finishing Floor		
Monolithic Screed Finish	.009	sf
Steel Trowel Finish	.015	sf

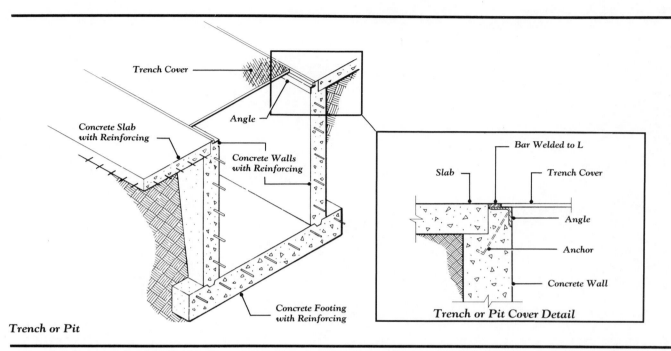

Trench Cover

Concrete Slab
with Reinforcing

Angle

Concrete Walls
with Reinforcing

Concrete Footing
with Reinforcing

Trench or Pit

Bar Welded to L

Slab

Trench Cover

Angle

Anchor

Concrete Wall

Trench or Pit Cover Detail

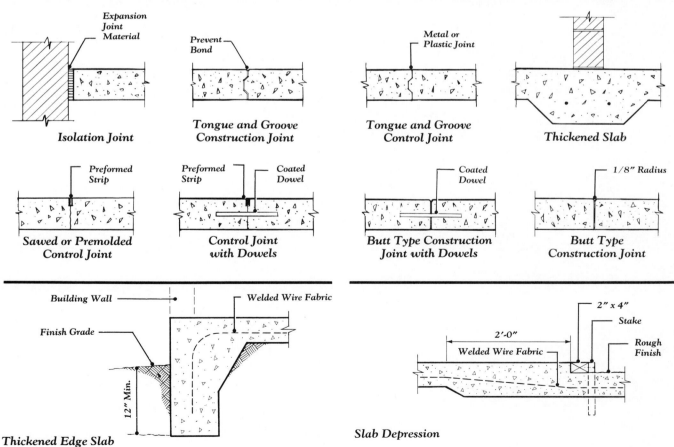

Expansion
Joint
Material

Isolation Joint

Prevent
Bond

**Tongue and Groove
Construction Joint**

Metal or
Plastic Joint

**Tongue and Groove
Control Joint**

Thickened Slab

Preformed
Strip

**Sawed or Premolded
Control Joint**

Preformed
Strip

Coated
Dowel

**Control Joint
with Dowels**

Coated
Dowel

**Butt Type Construction
Joint with Dowels**

1/8" Radius

**Butt Type
Construction Joint**

Building Wall

Welded Wire Fabric

Finish Grade

12" Min.

Thickened Edge Slab

2" x 4"

Stake

2'-0"

Welded Wire Fabric

Rough
Finish

Slab Depression

DIVISION 3
SUPERSTRUCTURE

Forming for suspended concrete slabs varies with job conditions, type of slab, degree of repetition, and weight. Form materials may consist of wood, plywood, steel, aluminum, or a combination. Shoring may be accomplished with wood posts, adjustable jacks, scaffolding, adjustable aluminum or steel beams, or with flying truss systems. Edge forms may be prefabricated and reused. Good practice dictates a crown or rise in the center of exposed beams and slabs to offset the appearance of deflection or sagging.

The placing sequence for slab reinforcement usually follows this order: the bottom steel, electrical conduit or horizontal pipes, the top steel. Reinforcing varies with the type of slab, but most systems require top and bottom steel supported by slab or beam bolsters and high chairs, which may be individual or continuous.

Beam and column reinforcing may be prefabricated and placed in the forms either by hand or, preferably, by crane.

Depending on location, concrete may be placed by direct chute, buggies and runways, pumping, or by crane and bucket. Screed materials may be steel or aluminum pipe or wood, with adjustable screed holders. Wet screeds may also be used, and consist of an area or strip of wet concrete troweled to the desired elevation and used as a guide for the remaining concrete to be placed.

Slab finishes vary from monolithic screed or darby finish (for two-course floors) to steel-troweled finish (for finish floor). Steel-troweled finish on concrete floors may limit the area of floor that may be placed in a day. Curing may be accomplished by using wet burlap, waterproof curing paper, or by a sprayed membrane curing compound.

Many multiple slab concrete buildings require reshoring after form removal to resist loads imposed by the formwork and concrete of higher floors. Superimposed design loads, strength of the supporting slab, and the dead load of the concrete and supporting formwork should all be carefully analyzed to determine reshore requirements.

Bend, Place and Tie Reinforcing

Placing and tieing by rodmen for footings and slabs runs from 9 hours per ton for heavy bars to 15 hours per ton for light bars. For beams, columns and walls, production runs from 8 hours per ton for heavy bars to 20 hours per ton for light bars. Overall average for typical reinforced concrete buildings is about 14 hours per ton. These production figures include placing of accessories and usual inserts. Equipment handling is necessary for the larger size bars so that installation costs for the very heavy bars will not decrease proportionately.

For tie wire, figure 5 lbs, 16 gauge black annealed wire for each ton of bars. Use of commercial accessories improves placing accuracy and reduced labor time.

Materials for One Cubic Yard of Concrete

This is an approximate method of figuring quantities of cement, sand and gravel for a field mix with waste allowance included. With gravel as coarse aggregate for barrels of cement required, divide 10 by total mix; that is, for 1:2:4 mix, 10 divided by 7 = 1-3/7 barrels.

For tons of sand multiply barrels of cement by parts of sand and then by 0.2; that is, for the 1:2:4 mix, as above, 1-3/7 x 2 x .2 = .57 tons.

Tons of gravel are in the same ratio to tons of sand as parts in mix, or 4/2 x .57 = 1.14 tons.

If course aggregate is crushed stone, use 10-1/2 instead of 10 as given for gravel.

1 bag cement = 94#
4 bags = 1 barrel
1 cu yd sand or gravel = 2700#
1 ton = 20 cf
1 cu yd crushed stone = 2575#
1 ton = 21 cf

Average carload of cement is 692 bags; sand or gravel is 56 tons.

Do not stack stored cement over 10 bags high.

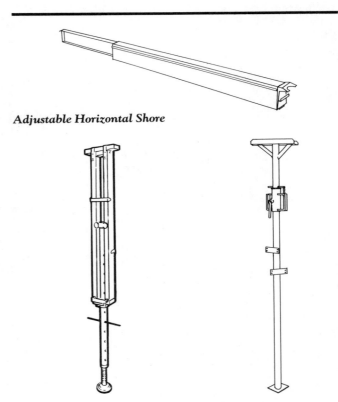

Adjustable Horizontal Shore

Composite Metal and Wood Shore   *Steel Shore*

Proportionate Quantities

The tables below show both quantities per square foot of floor area as well as form and reinforcing quantities per cubic yard. Unusual structural requirements would increase the ratios below. High strength reinforcing would reduce the steel weights. Figures are for 3000 psi concrete and 60,000 psi reinforcing unless specified otherwise.

Type of Construction	Live Load	Span	Per sf of Floor Area				Per cu yd of Concrete		
			Concrete	Forms	Reinf.	Pans	Forms	Reinf.	Pans
Flat Plate	50 psf	15 ft	.46 cf	1.06 sf	1.71 lb		62 sf	101 lb	
		20	.63	1.02	2.4		44	104	
		25	.79	1.02	3.03		35	104	
	100	15	.46	1.04	2.14		61	126	
		20	.71	1.02	2.72		39	104	
		25	.83	1.01	3.47		33	113	
Flat Plate (waffle construction) 20″ domes	50	20	.43	1.0	2.1	.84 sf	63	135	53 sf
		25	.52	1.0	2.9	.89	52	150	46
		30	.64	1.0	3.7	.87	42	155	37
	100	20	.51	1.0	2.3	.84	53	125	45
		25	.64	1.0	3.2	.83	42	135	35
		30	.76	1.0	4.4	.81	36	160	29
Waffle Construction 30″ domes	50	25	.69	1.06	1.83	.68	42	72	40
		30	.74	1.06	2.39	.69	39	87	39
		35	.86	1.05	2.71	.69	33	85	39
		40	.78	1.0	4.8	.68	35	165	40
Flat Slab (two way with drop panels)	50	20	.62	1.03	2.34		45	102	
		25	.77	1.03	2.99		36	105	
		30	.95	1.03	4.09		29	116	
	100	20	.64	1.03	2.83		43	119	
		25	.79	1.03	3.88		35	133	
		30	.96	1.03	4.66		29	131	
	200	20	.73	1.03	3.03		38	112	
		25	.86	1.03	4.23		32	133	
		30	1.06	1.03	5.3		26	135	
One Way Joists 20″ pans	50	15	.36	1.04	1.4	.93	78	105	70
		20	.42	1.05	1.8	.94	67	120	60
		25	.47	1.05	2.6	.94	60	150	54
	100	15	.38	1.07	1.9	.93	77	140	66
		20	.44	1.08	2.4	.94	67	150	58
		25	.52	1.07	3.5	.94	55	185	49
One Way Joists 8″ x 16″ filler blocks	50	15	.34	1.06	1.8	.81 Ea.	84	145	64 Ea.
		20	.40	1.08	2.2	.82	73	145	55
		25	.46	1.07	3.2	.83	63	190	49
	100	15	.39	1.07	1.9	.81	74	130	56
		20	.46	1.09	2.8	.82	64	160	48
		25	.53	1.10	3.6	.83	56	190	42
One Way Beam and Slab	50	15	.42	1.30	1.73		84	111	
		20	.51	1.28	2.61		68	138	
		25	.64	1.25	2.78		53	117	
	100	15	.42	1.30	1.9		84	122	
		20	.54	1.35	2.69		68	154	
		25	.69	1.37	3.93		54	154	
	200	15	.44	1.31	2.24		80	137	
		20	.58	1.40	3.30		65	163	
		25	.69	1.42	4.89		53	183	
Two Way Beam and Slab	100	15	.47	1.20	2.26		69	130	
		20	.63	1.29	3.06		55	131	
		25	.83	1.33	3.79		43	123	
	200	15	.49	1.25	2.70		41	149	
		20	.66	1.32	4.04		54	165	
		25	.88	1.32	6.08		41	187	

Proportionate Quantities (cont.)

4000 psi Concrete and 60,000 psi Reinforcing — Form and Reinforcing Quantities per cu yd					
Item	Size	Forms		Reinforcing	Minimum	Maximum
	10″ x 10″	130 sfca		#5 to #11	220 lbs	875 lbs
	12″ x 12″	108		#6 to #14	200	955
	14″ x 14″	92		#7 to #14	190	900
	16″ x 16″	81		#6 to #14	187	1082
	18″ x 18″	72		#6 to #14	170	906
	20″ x 20″	65		#7 to #18	150	1080
Columns (square tied)	22″ x 22″	59		#8 to #18	153	902
	24″ x 24″	54		#8 to #18	164	884
	26″ x 26″	50		#9 to #18	169	994
	28″ x 28″	46		#9 & #18	147	864
	30″ x 30″	43		#10 to #18	146	983
	32″ x 32″	40		#10 to #18	175	866
	34″ x 34″	38		#10 to #18	157	772
	36″ x 36″	36		#10 to #18	175	852
	38″ x 38″	34		#10 to #18	158	765
	40″ x 40″	32		#10 to #18	143	692

Item	Size		Forms	Spirals	Reinforcing	Minimum	Maximum
	12″ Diameter		34.5 lf	190 lb	#4 to #11	165 lb	1505 lb
			34.5	190	#14 & #18	—	1100
	14″		25	170	#4 to #11	150	970
			25	170	#14 & #18	800	1000
	16″		19	160	#4 to #11	160	950
			19	160	#14 & #18	605	1080
	18″		15	150	#4 to #11	160	915
			15	150	#14 & #18	480	1075
	20″		12	130	#4 to #11	155	865
			12	130	#14 & #18	385	1020
	22″		10	125	#4 to #11	165	775
			10	125	#14 & #18	320	995
Columns (spirally reinforced)	24″		9	120	#4 to #11	195	800
			9	120	#14 & #18	290	1150
	26″		7.3	100	#4 to #11	200	729
			7.3	100	#14 & #18	235	1035
	28″		6.3	95	#4 to #11	175	700
			6.3	95	#14 & #18	200	1075
	30″		5.5	90	#4 to #11	180	670
			5.5	90	#14 & #18	175	1015
	32″		4.8	85	#4 to #11	185	615
			4.8	85	#14 & #18	155	955
	34″		4.3	80	#4 to #11	180	600
			4.3	80	#14 & #18	170	855
	36″		3.8	75	#4 to #11	165	570
			3.8	75	#14 & #18	155	865
	40″		3.0	70	#4 to #11	165	500
			3.0	70	#14 & #18	145	765

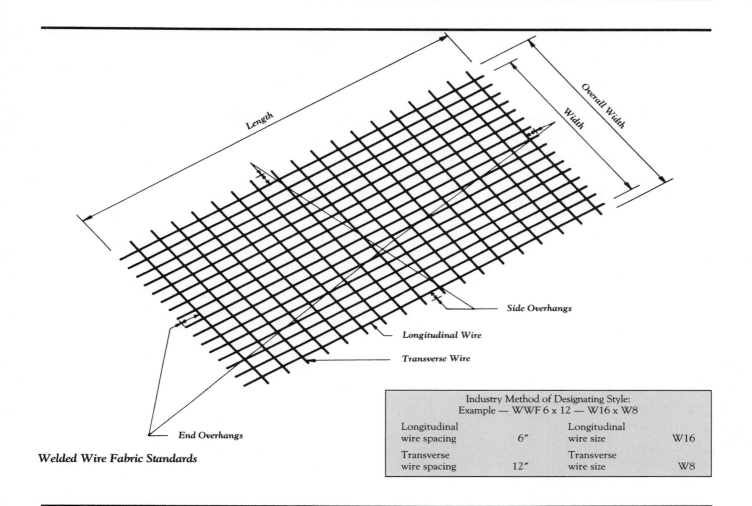

Welded Wire Fabric Standards

Industry Method of Designating Style: Example — WWF 6 x 12 — W16 x W8			
Longitudinal wire spacing	6″	Longitudinal wire size	W16
Transverse wire spacing	12″	Transverse wire size	W8

ASTM Standard Reinforcing Bars			
Bar Size Designation	Area* Square Inches	Weight Pounds Per Foot	Diameter* Inches
#3	.11	.376	.375
#4	.20	.668	.500
#5	.31	1.043	.625
#6	.44	1.502	.750
#7	.60	2.044	.875
#8	.79	2.670	1.000
#9	1.00	3.400	1.128
#10	1.27	4.303	1.270
#11	1.56	5.313	1.410
#14	2.25	7.650	1.693
#18	4.00	13.600	2.257

Current ASTM Specifications cover bar sizes #14 and #18 in A615 Grade 60 and in A706 only.
*Nominal dimensions.

Type of Steel: **S** for Billet meeting supplementary requirement (S1) of (A 615)

N for Billet (A 615)

R for Rail (A 616) meeting bend test requirements of ASTM A 617, Grade 60 [per ACI 318-83]

I for Rail (A 616)

A for Axle (A 617)

W for Low-Alloy (A 706)

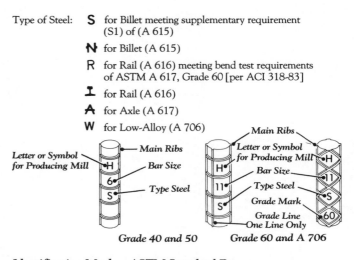

Identification Marks - ASTM Standard Bars

Symbol	Bar Support Illustration	Bar Support Illustration Plastic Capped or Dipped	Type of Support	Sizes
SB		CAPPED	Slab Bolster	3/4, 1, 1-1/2, and 2″ heights in 5′ and 10′ lengths
SBU*			Slab Bolster Upper	Same as SB
BB		CAPPED	Beam Bolster	1, 1-1/2, 2, over 2″ to 5″ heights in increments of 1/4″ in lengths of 5′
BBU*			Beam Bolster Upper	Same as BB
BC		DIPPED	Individual Bar Chair	3/4, 1, 1-1/2, and 1-3/4″ heights
JC		DIPPED DIPPED	Joist Chair	4, 5, and 6″ widths and 3/4, 1, and 1-1/2″ heights
HC		CAPPED	Individual High Chair	2 to 15″ heights in increments of 1/4″
HCM*			High Chair for Metal Deck	2 to 15″ heights in increments of 1/4″
CHC		CAPPED	Continuous High Chair	Same as HC in 5′ and 10′ lengths
CHCU*			Continuous High Chair Upper	Same as CHC
CHCM*			Continuous High Chair for Metal Deck	Up to 5″ heights in increments of 1/4″
JCU**		DIPPED	Joist Chair Upper	14″ span. Heights - 1″ through +3-1/2″ vary in 1/4″ increments

* Usually available in Class 3 only, except on special order.
** Usually Available in Class 3 only, with upturned or end bearing legs.

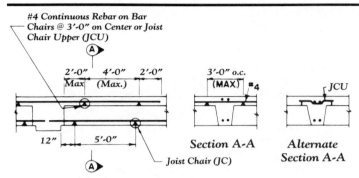

#4 Continuous Rebar on Bar Chairs @ 3'-0" on Center or Joist Chair Upper (JCU)

2'-0" Max 4'-0" (Max.) 2'-0"

12" 5'-0"

Joist Chair (JC)

Section A-A

3'-0" o.c. (MAX.) #4

Alternate Section A-A

JCU

JCU in Place

JCU - Joist Chair uppper - is available on special order only.

One-Way Joist Construction

Joist Bar Supports

Bar supports are generally not provided for temperature welded wire fabric or bars in concrete joist slabs. It is recommended that temperature bars be tied, and spaced with #3 bars centered on alternate rows of forms, i.e., about 4'-2" to 6'-0" centers at right angles to temperature bars.

Top bars are normally supported either by bars on individual chairs or by "joist chairs upper" (JCU) shown.* For two-way joist construction (waffle slabs), the bar supports in the ribs in one direction can usually be made the same as for one-way concrete joist construction. Bar supports can be omitted in ribs at right angles as these bars are supported on the bottom bars running in the first direction, except top bars in the middle strips.

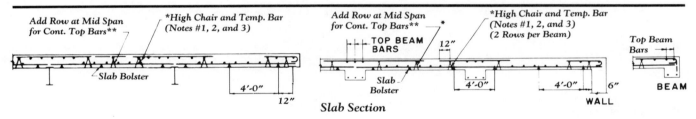

Add Row at Mid Span for Cont. Top Bars**

*High Chair and Temp. Bar (Notes #1, 2, and 3)

Slab Bolster

4'-0"

12"

Add Row at Mid Span for Cont. Top Bars**

TOP BEAM BARS

12"

*High Chair and Temp. Bar (Notes #1, 2, and 3) (2 Rows per Beam)

Slab Bolster

4'-0" 4'-0" 6"

WALL

Top Beam Bars

BEAM

Slab Section

One-Way Solid Slab

NOTES:
1. A line of properly lapped support bars can replace an equal amount of temperature steel. Temperature bars to be used for support bars—

    Use Class B or Class C tension lap splices, depending on the percentage of bars spliced at any point.
2. For #5 temperature bars use high chairs, HC, @ 4'-0" center to center.

    For #4 temperature bars use high chairs, HC, @ 3'-0" center to center.

3. Do not use #3 temp. bar for support bar, substitute one #4 bar (properly lapped) with chairs, BC, @ 3'-0" center to center.

*Continuous high chairs, CHC, may be used in lieu of support bar and HC.
**Exceptions: Not required if adjacent rows are spaced 4'-0" or less apart. With #3 continuous top bars provide rows of support @ 2'-0" center to center.

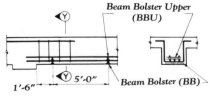

Beam Bolster Upper (BBU)

1'-6" 5'-0" Beam Bolster (BB)

Section Y-Y

Beams and Girders

Beam Bar Supports

Beam bolsters transversely spaced at a maximum of 5'-0" on centers, and, for bars in two layers, "beam bolsters upper" at the same spacing, are current field placing practices. Beam bolsters for use longitudinally are supplied only upon special arrangements between contractor and supplier, if approved by the engineer or architect.

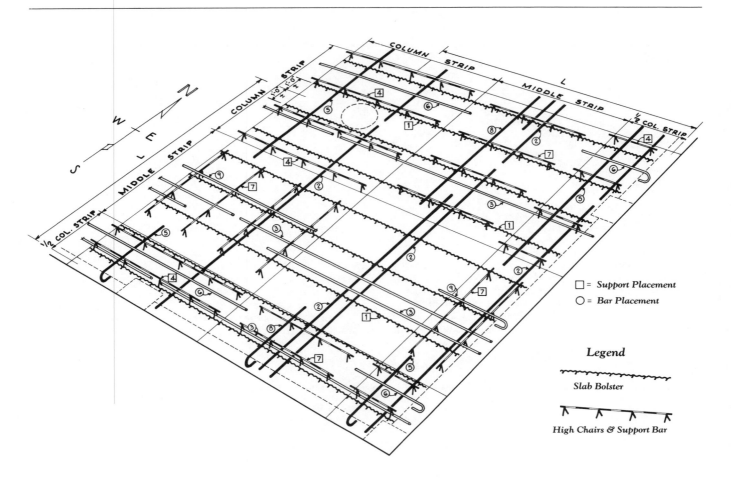

= Support Placement

○ = Bar Placement

Legend

Slab Bolster

High Chairs & Support Bar

Sequence of Placing Bar Supports and Bars in Two-Way Flat Plate

1. Place continuous lines of slab bolsters in E-W direction at 4'-0" maximum on center between columns. Begin spacing 1'-0" from center line of columns.

2. Set N-S bottom bars in column and middle strips.

3. Set E-W bottom bars in column and middle strips.

4. Place 3 or more rows of #4 support bars (length .5L) at 4'-0" maximuum on center on high chairs at 3'-0" maximum on center in E-W direction at each column head.

5. Set N-S top bars in column strips.

6. Set E-W top bars in column strips.

7. Place 3 or more rows of #4 support bars (length approximately .4L) at 4'-0" maximum on center on high chairs at 3'-0" maximum on center between columns lengthwise in N-S and E-W column strips. Place 2 rows at all slab edges.

8. Set N-S middle strip top bars between ends of E-W column strip top bars.

9. Set E-W middle strip top bars.

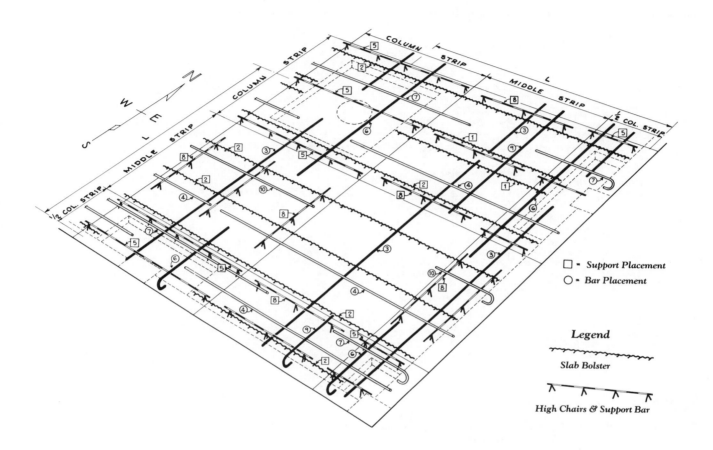

☐ = *Support Placement*

○ = *Bar Placement*

Legend

Slab Bolster

High Chairs & Support Bar

Sequence of Placing Bar Supports and Bars in Two-Way Flat Slab

1. Place a single line of slab bolsters in E-W direction on each side adjacent to column center line between drop panels.

2. Place continuous lines of slab bolsters in E-W direction of 4'-0" maximum on center between drop panels. Begin spacing 3" outside drop panels. Add one E-W slab bolster at slab edges between drop panels.

3. Set N-S bottom bars, column and middle strips.

4. Set E-W bottom bars, column and middle strips.

5. Place 3 rows of #4 support bars (length .50L) on high chairs at 3'-0" maximum on center in E-W direction at each column head. Tie middle support bar to column verticals.

6. Set N-S column strip top bars.

7. Set E-W column strip top bars.

8. Place 3 or more rows of #4 support bars (length .32L) at 4'-0" maximum on center in N-S and E-W column strips, parallel to the strips. Place 2 rows at all slab edges.

9. Set N-S middle strip top bars between ends of E-W column strip top bars.

10. Set E-W middle strip top bars.

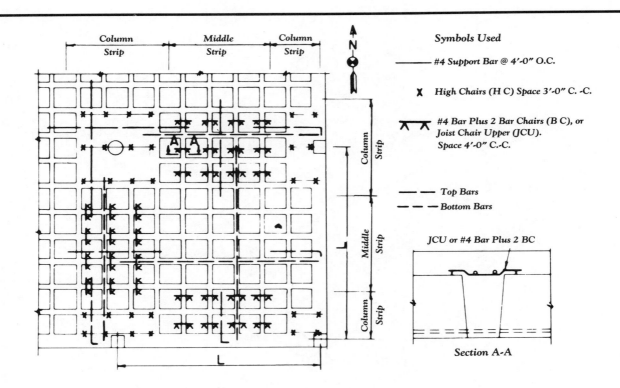

Symbols Used

——————— #4 Support Bar @ 4'-0" O.C.

X   High Chairs (H C) Space 3'-0" C.-C.

#4 Bar Plus 2 Bar Chairs (B C), or
Joist Chair Upper (JCU).
Space 4'-0" C.-C.

– – –   Top Bars

— – —   Bottom Bars

JCU or #4 Bar Plus 2 BC

Section A-A

Sequence of Placing Bar Supports and Bars in Two-Way Waffle Flat Plate

1. Place standard joist chairs in joist rib bottom @ 5'-0" center to center in N-S column strip and N-S middle strip (full length).
2. Set N-S column and middle strip bottom bars.
3. Set E-W column and middle strip bottom bars.
4. At column heads, place 3 (or more) rows of #4 support bars (length full width of column head) on high chairs @ 3'-0" center to center in E-W direction.
5. Set N-S column strip top bars.
6. Place #4 bar plus two BC (or JCU) @ 4'-0" center to center in N-S and E-W middle strip. If top bars are spaced over the entire middle strip, use #4 support bars at 4'-0" center to center with BC @ 3'-0" center to center.
7. Set N-S and E-W middle strip top bars.
8. Set E-W column strip top bars; tie column strip N-S bars to support bars; tie column strip E-W bars to N-S bars.

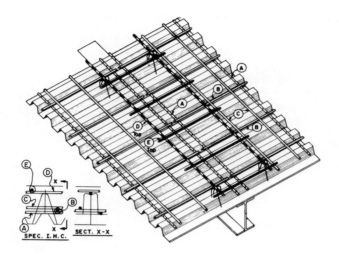

One-Way Slabs on Corrugated Steel Forms

Bar Supports for Special Conditions

A. Slabs on corrugated steel forms placing sequence

1. Place corrugated steel form and fasten to supporting members.

2. Set #3 support bars (A) @ 5'-0" on steel form.

3. Set main bars (B) (positive reinforcing) over valleys. Tie to support bars. NOTE: Main bar spacing should be a multiple of steel form pitch.

4. Set temperature bars (C). Tie to main bars.

5. Place special individual high chairs @ 3'-0" on center.*

Note: for continuous top bars place extra row of high chairs at midspan.

6. Place #4 support bars (D) on chairs. (A line of properly lapped support bars can replace an equal amount of temperature steel.)

7. Set top bars (E) (negative reinforcing). Tie to support bars.

*Special continuous high chairs may be used in lieu of support bars and high chairs.

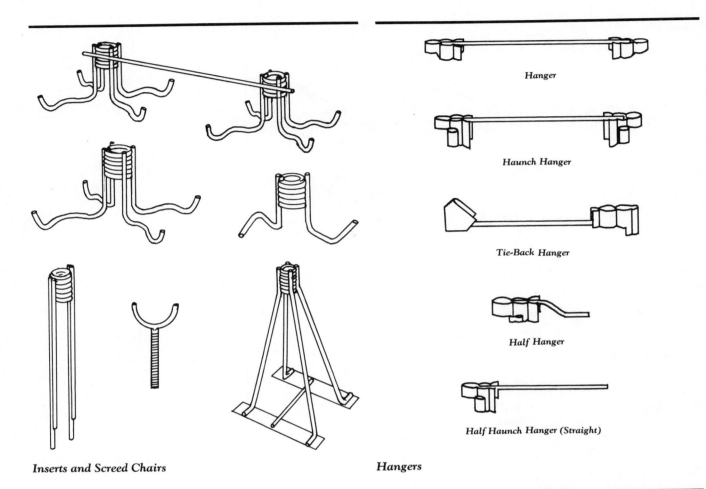

Hanger

Haunch Hanger

Tie-Back Hanger

Half Hanger

Half Haunch Hanger (Straight)

Inserts and Screed Chairs

Hangers

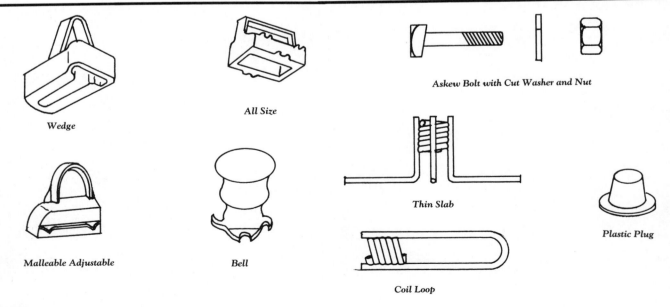

Wedge

All Size

Askew Bolt with Cut Washer and Nut

Malleable Adjustable

Bell

Thin Slab

Plastic Plug

Coil Loop

Inserts

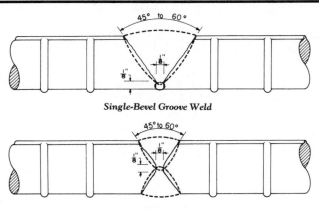

Single-Bevel Groove Weld

Double-Bevel Groove Weld

Direct Butt Splices - Horizontal

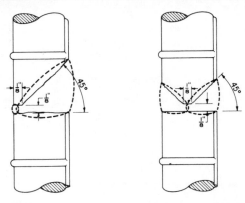

Single-Vee Groove Weld   Double-Vee Groove Weld

Direct Butt Splices - Vertical

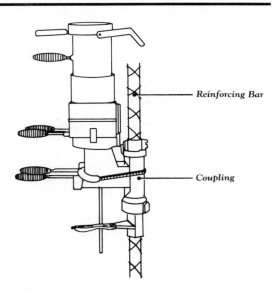

Reinforcing Bar

Coupling

Mechanical Butt Splicer

End Bearing Wedge Lock Clamp

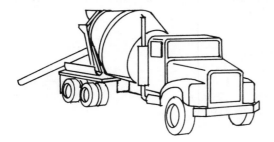

Concrete Truck

Concrete Buggy

Concrete Bucket

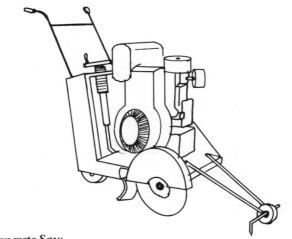

Concrete Saw

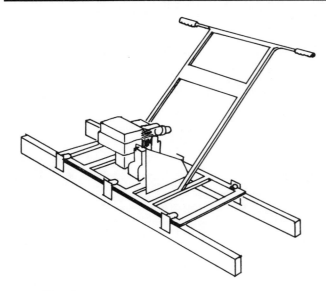

Power Screed

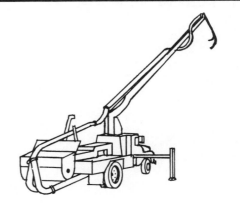

Concrete Pump

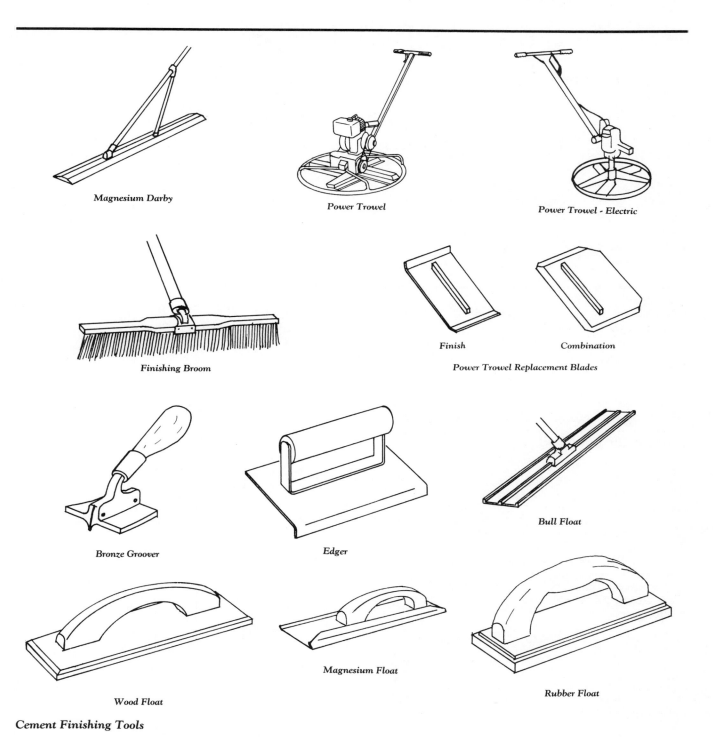

Magnesium Darby

Power Trowel

Power Trowel - Electric

Finishing Broom

Finish

Combination

Power Trowel Replacement Blades

Bronze Groover

Edger

Bull Float

Wood Float

Magnesium Float

Rubber Float

Cement Finishing Tools

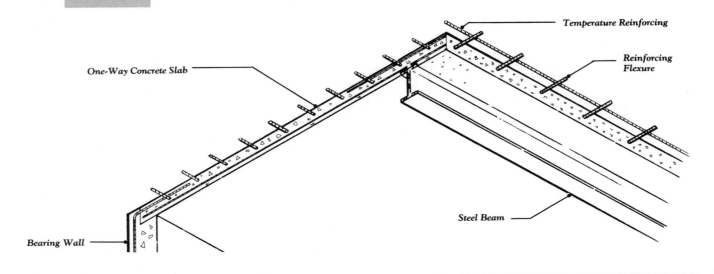

One-Way Concrete Slab

Temperature Reinforcing

Reinforcing Flexure

Steel Beam

Bearing Wall

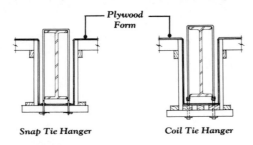

Plywood Form

Snap Tie Hanger Coil Tie Hanger

Floor Slab Hung from Steel Beam

One-way slabs are solid poured-in-place concrete slabs of uniform depth. They may be single or multiple span and are usually supported by bearing walls or beams. Formwork (or centering) for slabs may be constructed of plywood and wood framing, steel deck, adjustable beams and plywood, or with patented forming systems.

For flexure, one-way slabs are reinforced in the direction of the span. For temperature and shrinkage reinforcement they are reinforced perpendicular to the span. Multiple-span slabs usually require top and bottom flexure reinforcement.

Man-hours

Description	m/hr	Unit
Forms in Place Elevated Slabs		
Column Forms	.134	sfca
Floor Slab Hung from Steel Beams	.085	sfca
Flat Slab, Shored	.086	sfca
Edge Forms	.091	sfca
Elevated Slabs Reinforcing in Place	11.034	ton
Columns	13.913	ton
Hoisting Reinforcing	.609	ton
Placing Concrete Pumped	.492	cu yd
With Crane and Bucket	.582	cu yd
Steel Trowel Finish	.015	sf
Concrete in Place 8″ Slab	6.480	cu yd

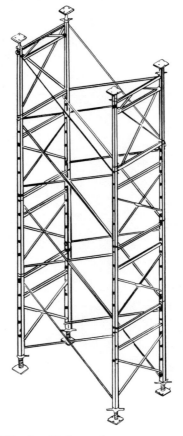

Tubular Steel Shoring/Falsework to Support Elevated Slab Forms

3 SUPERSTRUCTURE
ONE-WAY CONCRETE BEAMS AND SLABS

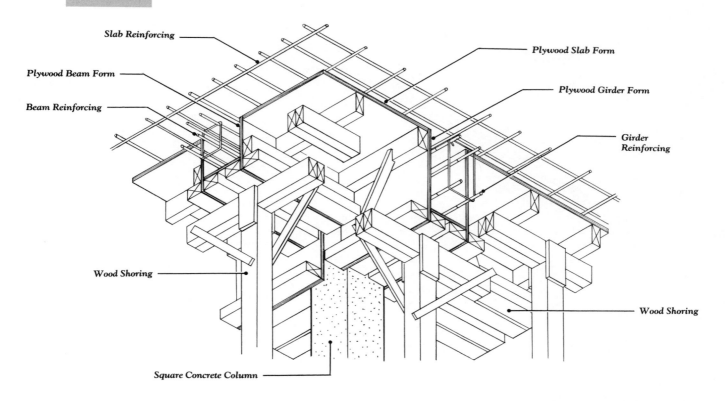

- Slab Reinforcing
- Plywood Beam Form
- Beam Reinforcing
- Plywood Slab Form
- Plywood Girder Form
- Girder Reinforcing
- Wood Shoring
- Wood Shoring
- Square Concrete Column

One-way concrete girders, beams, and slabs are solid, monolithically cast systems. They are an effective support system for heavy or concentrated loads and are most cost effective at spans under 20'. The floor system is relatively deep and the formwork is complex and costly. Form materials usually consist of plywood and wood framing, with shores or scaffolding as the support system. Slabs are reinforced in the direction of the span for flexure and perpendicular to the span for temperature and shrinkage.

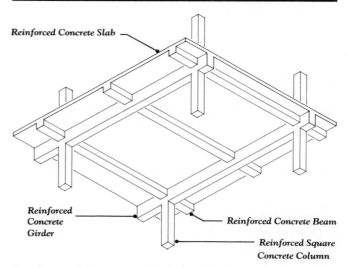

- Reinforced Concrete Slab
- Reinforced Concrete Girder
- Reinforced Concrete Beam
- Reinforced Square Concrete Column

One-Way Concrete Beam and Slab System

Man-hours

Description	m/hr	Unit
Forms-in-Place		
Square Column	.136	sfca
Beam and Girder	.120	sfca
Slab	.086	sfca
Reinforcing Columns		
#3 to #7	21.333	ton
#8 to #14	13.913	ton
Beams and Girders		
#3 to #7	20.000	ton
#8 to #14	11.852	ton
Elevated Slabs	11.034	ton
Hoisting	.609	ton
Placing Concrete Pumped	.492	cu yd
With Crane and Bucket	.582	cu yd
Steel Trowel Finish	.015	sf
Concrete in Place Including		
Forms, Reinforcing and Finish	9.880	cu yd

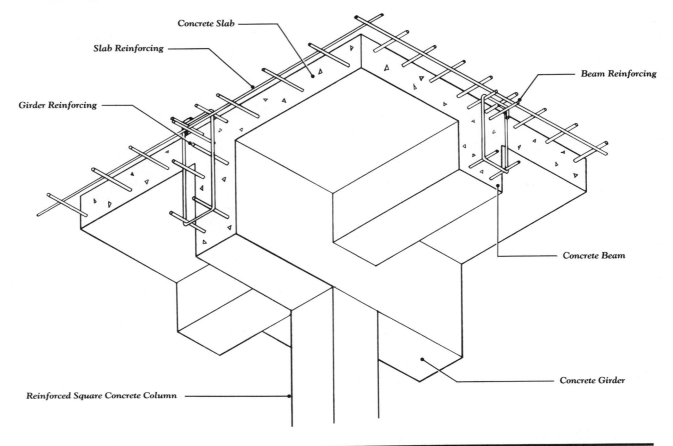

3 SUPERSTRUCTURE
TWO-WAY CONCRETE BEAMS AND SLABS

Concrete Slab

Slab Reinforcing

Beam Reinforcing

Girder Reinforcing

Concrete Beam

Concrete Girder

Reinforced Square Concrete Column

Two-way concrete beams and slabs are solid slabs cast monolithically with support beams on columns. They are an effective support system for heavy or concentrated loads. They are most cost effective in square bays with spans under 30′. Slabs are reinforced for flexure in both directions. Beams in square bays assume equal loading for uniform superimposed loads.

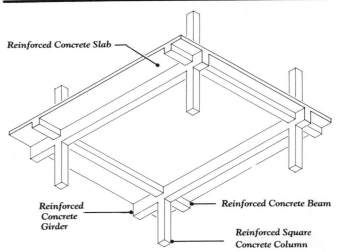

Reinforced Concrete Slab

Reinforced Concrete Girder

Reinforced Concrete Beam

Reinforced Square Concrete Column

Two-Way Concrete Beam and Slab

Man-hours

Description	m/hr	Unit
Forms in Place		
Column Square	.136	sfca
Beams	.120	sfca
Slab	.086	sfca
Reinforcing Columns		
#3 to #7	21.333	ton
#8 to #14	13.913	ton
Beams		
#3 to #7	20.000	ton
#8 to #14	11.852	ton
Elevated Slabs	11.034	ton
Hoisting Reinforcing	.609	ton
Placing Concrete Pumped	.492	cu yd
With Crane and Bucket	.582	cu yd
Steel Trowel Finish	.015	sf
Concrete in Place Including		
Forms, Reinforcing and Finish	8.630	cu yd

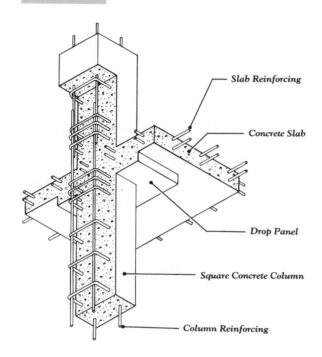

Slab Reinforcing

Concrete Slab

Drop Panel

Square Concrete Column

Column Reinforcing

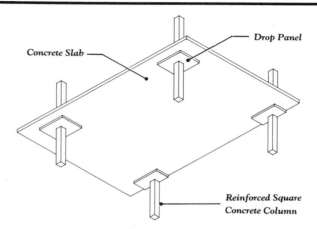

Concrete Slab

Drop Panel

Reinforced Square
Concrete Column

Concrete Flat Slab with Drop Panel System

Concrete flat slabs are solid slabs cast monolithically with a drop panel. The drop panel is a thick area of concrete surrounding the column where it meets the slab. Flat slabs will sustain heavy loads with long spans, use less concrete and reinforcing, and usually require smaller columns. Flat slabs are most cost efficient for square bays. Flat slabs with no spandrels can be formed with flying forms and are reinforced for flexure in both directions.

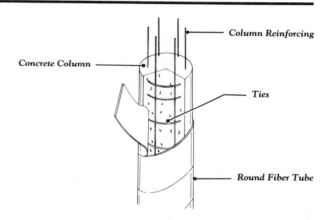

Concrete Column

Column Reinforcing

Ties

Round Fiber Tube

Column - Round Tied

Man-hours

Description	m/hr	Unit
Forms in Place		
Columns Square	.136	sfca
Round Fiber Tube	.221	lf
Round Steel	.256	lf
Capitols	2.667	Ea.
Flat Slab with Drops	.088	sf
Edge Forms	.091	sfca
Reinforcing Columns		
#3 to #7	21.333	ton
#8 to #14	13.913	ton
Spirals	14.545	ton
Butt Splice Clamp Sleeve and Wedge	.373	Ea.
Mechanical Full Tension Splice with Filler Metal	.903	Ea.
Elevated Slab	11.034	ton
Hoisting Reinforcing	.609	ton
Place Concrete		
Pumped	.492	cu yd
With Crane and Bucket	.582	cu yd
Steel Trowel Finish	.015	sf
Concrete in Place Including		
Forms, Reinforcing and Finish	5.403	cu yd

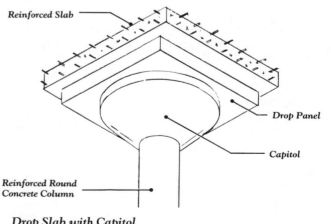

Reinforced Slab

Drop Panel

Capitol

Reinforced Round
Concrete Column

Drop Slab with Capitol

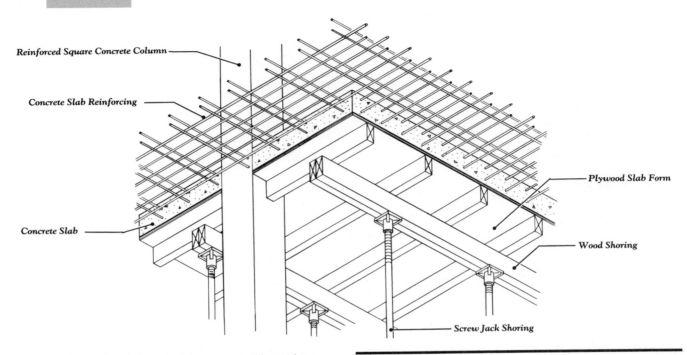

Reinforced Square Concrete Column

Concrete Slab Reinforcing

Concrete Slab

Plywood Slab Form

Wood Shoring

Screw Jack Shoring

Flat-plate slabs are solid, uniform two-way slabs without drops or interior beams. They are economically efficient for moderate uniform loads and spans and are of minimum floor depth, so they save on building height. Flexibility in column and opening location and the adaptability to flying-form construction make flat slabs a popular choice for multistory buildings with repetitive floors. Flat-plate slabs are reinforced for flexure in both directions.

Column size is determined by slab thickness unless shear heads or special methods to distribute loads at the top of columns are provided. Flat-plates require larger columns than flat slabs.

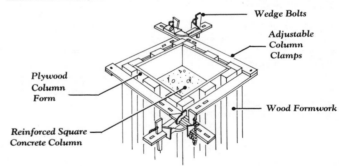

Wedge Bolts

Adjustable Column Clamps

Plywood Column Form

Reinforced Square Concrete Column

Wood Formwork

Square Column Form

Man-hours

Description	m/hr	Unit
Forms in Place		
Columns Square	.136	sfca
Round Fiber Tube	.221	lf
Round Steel	.256	lf
Capitols	2.667	Ea.
Flat Plate	.086	sf
Edge Forms	.091	sfca
Reinforcing Columns		
#3 to #7	21.333	ton
#8 to #14	13.913	ton
Spirals	14.545	ton
Butt Splice, Clamp Sleeve and Wedge	.373	Ea.
Mechanical Full Tension Splice with Filler Metal	.903	Ea.
Elevated Slab	11.034	ton
Hoisting Reinforcing	.609	ton
Place Concrete		
Pumped	.492	cu yd
With Crane and Bucket	.582	cu yd
Steel Trowel Finish	.015	sf
Concrete in Place Including		
Forms, Reinforcing and Finish	6.467	cu yd

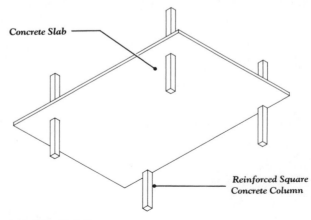

Concrete Slab

Reinforced Square Concrete Column

Concrete Flat Slab System

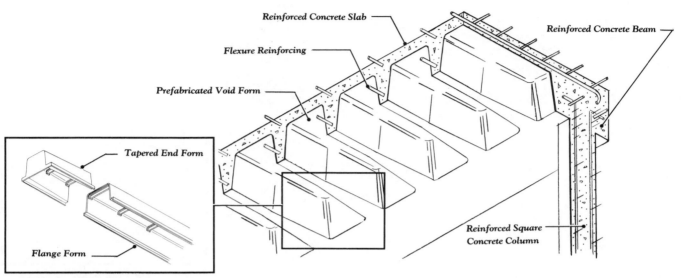

Prefabricated Void Forms

One-way concrete joist construction consists of a monolithic combination of regularly spaced concrete joists and a thin slab. The voids between joists are created by prefabricated void forms shaped like pans. The forms may be metal, fiberglass, or paper. The joists are arranged spanning in one direction between parallel supports. The joists may be supported on integrally placed concrete beams of equal or greater depth, or they may be supported on bearing walls.

Standard void forms are 20″ and 30″ wide and are available in 6″, 8″, 10″, 12″, 14″, 16″, or 20″ depths. Span lengths are varied by using telescoping forms. The joist bottoms may be formed by installing an overall plywood forming system similar to the form used for flat-plate construction, or by using centering material to form the bottom of each joist. For spans over 20′, the floor system requires distribution ribs at least 4″ wide placed perpendicular to the span.

Joists are reinforced for flexure in the direction of the span. For ease in placing reinforcing bars with the required cover, it is desirable to use a minimum joist width of 5″. Temperature reinforcing in the slab may be accomplished with bars or welded-wire fabric.

Maximum economy is realized in multiple joist slab construction by selecting pan forms of sufficient depth to maintain a uniform dimension between beams and joists and consistant layouts throughout the structure.

In comparison with one-way slabs, one-way joist construction is lightweight and permits long spans and heavy uniform superimposed loading. Special architectural ceiling effects may be achieved by integrating the joist and rib patterns with acoustical material and lighting fixtures.

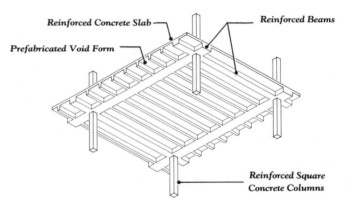

One-Way Concrete Joist Slab System

Man-hours

Description	m/hr	Unit
Forms in Place		
Joists and Pans	.096	sf
Beam Bottoms	.166	sfca
Beam Sides	.108	sfca
Edge Forms	.091	sfca
Reinforcing		
Joists and Ribs		
#3 to #11	20.000	ton
Slabs, Bars		
#3 to #7	11.034	ton
WWF 6 x 6 w4/w4	.593	csf
Placing Concrete		
Pumped	.582	cu yd
With Crane and Bucket	.674	cu yd
Steel Trowel Finish	.015	sf
Concrete in Place Incl.		
Forms, Reinforcing and Finish	8.056	cu yd

3 SUPERSTRUCTURE
CONCRETE WAFFLE SLAB

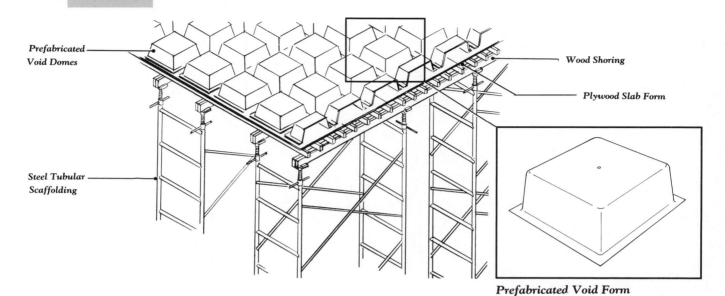

Prefabricated Void Domes

Steel Tubular Scaffolding

Wood Shoring

Plywood Slab Form

Prefabricated Void Form

Two-way waffle slab construction consists of evenly spaced joists at right angles to each other, integrally placed with a thin slab. The joists are formed by using a standard square dome which is usually prefabricated from steel or fiberglass. Domes are omitted around the columns to form solid heads. Standard forms to create voids between ribs are 19″ or 30″ square and are usually available in 6″, 8″, 10″, and 12″ depths for 19″ domes and 8″, 10″, 12″, 14″, 16″, and 20″ depths for 30″ domes. Nineteen-inch square domes create a 24″ module, and 30″ domes create a 36″ module. For maximum economy, the same size dome form should be maintained throughout the structure, and the solid heads should be in the same plain as the domes.

Joists are reinforced for flexure in both directions, creating a two-way slab with voids. Bay sizes should be relatively square in configuration.

Waffle slab construction is considerably lighter than solid flat-slab or flat-plate construction and allows longer spans and heavier superimposed loads. The geometrical shape formed is often architecturally desirable for exposed ceilings.

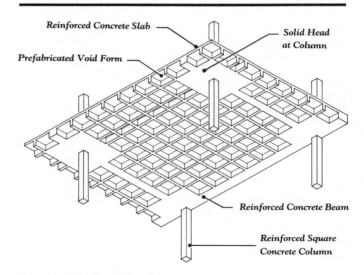

Reinforced Concrete Slab

Prefabricated Void Form

Solid Head at Column

Reinforced Concrete Beam

Reinforced Square Concrete Column

Concrete Waffle Slab System

Man-hours

Description	m/hr	Unit
Forms in Place		
Joists and 19″ Domes	.097	sf
Joists and 30″ Domes	.102	sf
Edge Forms	.091	sfca
Reinforcing		
Joists	20.000	ton
Slabs 6 x 6 w4/w4	.593	CSF
Placing Concrete		
Pumped	.582	cu yd
With Crane and Bucket	.674	cu yd
Steel Trowel Finish	.015	sf
Concrete in Place Including		
Forms, Reinforcing and Finish	5.950	cu yd

46

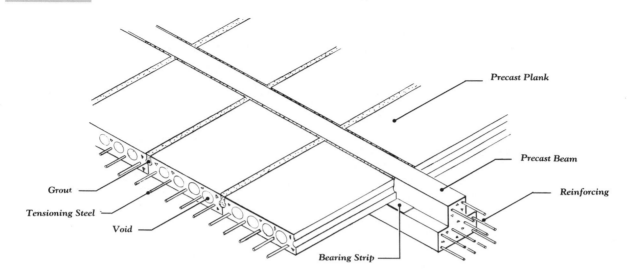

Precast concrete planks are plant-produced, tensioned structural members that are trucked to the site and erected. They may be supported on steel beams, concrete beams, concrete walls, or block walls. The planks are generally produced on long beds and cut to specified lengths. Standard widths are 2'-0", 3'-4", 4'-0", 5'-0", and 8'-0". Cross sections of the plank vary with patented manufacturers' systems. The shape of voids (used to lighten the weight of the plank) varies according to the methods used by the different manufacturers.

The plank may be erected with ends bearing on masonite or plastic strips to allow sliding, and the joints between the planks are grouted with a cement-sand mixture. Where planks are to be used as exposed structures, or where spans change direction, they should be leveled by using screw jacks or other suitable devices before grouting. Planks are used untopped and grouted or with structural concrete topping, usually 2" thick and reinforced with welded wire fabric for both floor and roof structures.

Precast, prestressed planks are quickly erected and produce a large workable floor area in a minimum amount of time. In relation to span, they are lightweight when compared to poured-in-place systems. Plank bottoms are often painted and used as the exposed ceiling for the area below. Voids in the plank are often used to carry utilities.

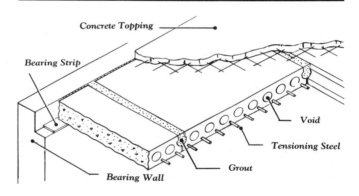

Precast Plank

Man-hours

Description	m/hr	Unit
Erect and Grout Prestressed Slabs		
6" Deep	1.029	Ea.
8" Deep	1.108	Ea.
10" Deep	1.200	Ea.
12" Deep	1.309	Ea.
Average		
8" Deep	.010	sf

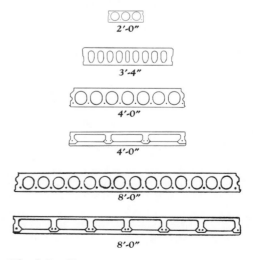

Precast Plank Profiles

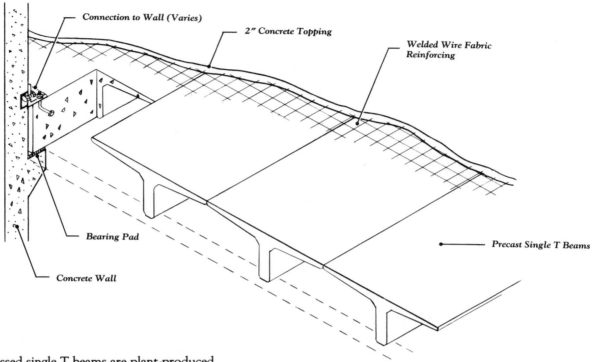

3 SUPERSTRUCTURE
PRECAST SINGLE T BEAMS

Connection to Wall (Varies)

2″ Concrete Topping

Welded Wire Fabric Reinforcing

Bearing Pad

Concrete Wall

Precast Single T Beams

Precast, prestressed single T beams are plant-produced structural members that are transported to the site and erected. Long span T beams require heavy erection equipment. T's are available in both normal and lightweight concrete in 8′ and 10′ widths for spans of over 100′ (roofs). They are usually supported on bearing walls, beams, or columns. T's used for floors are normally topped with 2″ of concrete reinforced with welded-wire fabric. Single T's may take more time to erect than double T's because they are usually welded to adjacent members for stability before being released from the crane. Long lengths require special delivery equipment, escorts in front and behind the carrier, and special travel permits.

Some specialized site conditions require the use of two cranes to place the T's in position. The capacity and operating radius of the lifting equipment should be carefully evaluated when erecting precast T's.

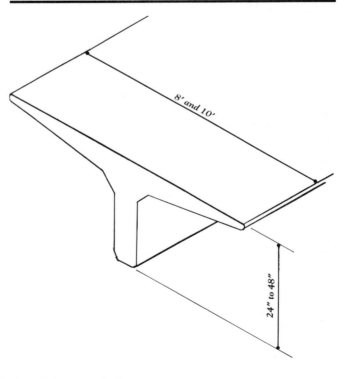

Single T Size Variations

8′ and 10′

24″ to 48″

Man-hours

Description	m/hr	Unit
Single T Erection		
Span		
40′	7.200	Ea.
80′	9.000	Ea.
100′	12.000	Ea.
120′	14.400	Ea.
Welded Wire Fabric 6 x 6-10/10	.457	CSF
Place Concrete		
Pumped	.582	cu yd
With Crane and Bucket	.674	cu yd
Broom Finish	.012	sf
Steel Trowel Finish	.015	sf

3 SUPERSTRUCTURE
PRECAST COLUMNS, BEAMS, AND DOUBLE T'S

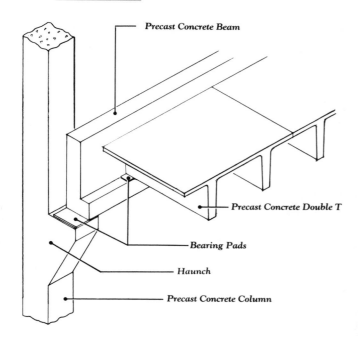

Precast Concrete Beam

Precast Concrete Double T

Bearing Pads

Haunch

Precast Concrete Column

Man-hours

Description	m/hr	Unit
Columns		
12' High	1.895	Ea.
24' High	2.400	Ea.
Beams		
20' Span		
12" x 20"	2.250	Ea.
18" x 36"	3.000	Ea.
24" x 44"	3.273	Ea.
30' Span		
12" x 36"	3.000	Ea.
18" x 44"	3.600	Ea.
24" x 52"	4.500	Ea.
40' Span		
12" x 52"	3.600	Ea.
18" x 52"	4.500	Ea.
24" x 52"	6.000	Ea.
Double T's		
45' Span		
12" Deep x 8' Wide	3.273	Ea.
18" Deep x 8' Wide	3.600	Ea.
50' Span, 24" Deep x 8' Wide	4.500	Ea.
60' Span, 32" Deep x 10' Wide	5.143	Ea.
Welded Wire Fabric 6 x 6-10/10	.457	CSF
Place Concrete		
Pumped	.582	cu yd
With Crane and Bucket	.674	cu yd
Broom Finish	.012	sf
Steel Trowel Finish	.015	sf

Precast concrete columns, beams, and double T's are plant-produced, reinforced or prestressed members that are transported to the site and erected. Columns may be one or several stories high and may be concentrically or eccentrically loaded by using haunches. Haunches are reinforced with concrete-encased structural steel or with galvanized reinforcing. Precast members may be connected to foundations with cast-in-place base plates and secured with anchor bolts or by a socket in the foundation (which is grouted after erection). Columns, in many cases, require temporary shores accomplished by the use of adjustable rigid braces or guys until the remaining structure is in place and connected.

Beams may be prestressed or reinforced rectangular T-shaped or L-shaped members, depending on the design or building configuration. They may be welded, bolted, or pin-connected at the bearing points.

Double T's are commonly produced in 8' and 10' widths and are available in spans in excess of 80' in normal or lightweight concrete. They may be used untopped for roof construction or with a structural concrete topping for floors and roofs. They may also be used for building fascia or exterior walls. Double T's are generally connected with weld plates, spaced along the flange, to provide continuity with adjacent members. Stems commonly bear on neoprene pads to ensure equal bearing.

Care should be exercised when erecting precast members in cold weather to ensure that the openings at connections are free of collected water which may cause freeze cracking. Special attention should be paid to rigid connections at haunches. Lifting loops should be checked on heavy lifts to ascertain that they are located at the center of gravity. This precaution will ensure level lifting without the use of special devices.

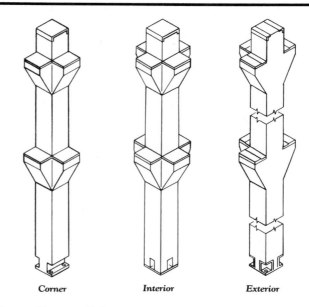

Corner   Interior   Exterior

Precast Concrete Columns

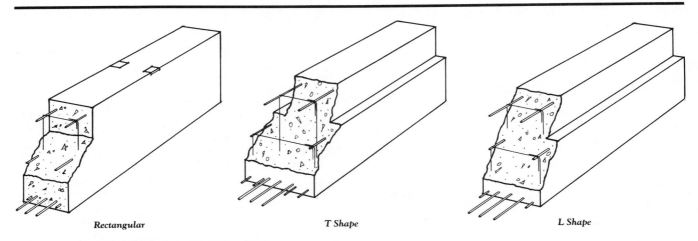

Rectangular *T Shape* *L Shape*

Precast Concrete Prestressed or Reinforced Beams

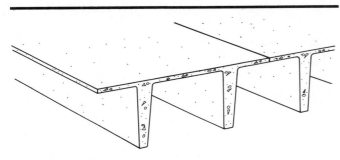

Double T with No Topping

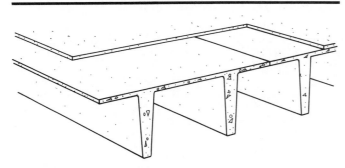

Double T with 2″ Topping

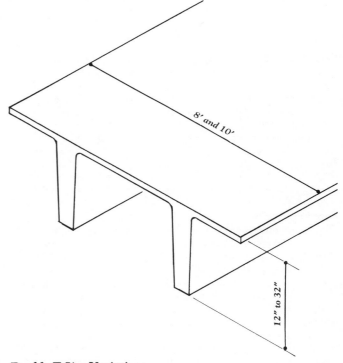

8′ and 10′

12″ to 32″

Double T Size Variations

3 SUPERSTRUCTURE
POSTTENSIONED SLABS AND BEAMS

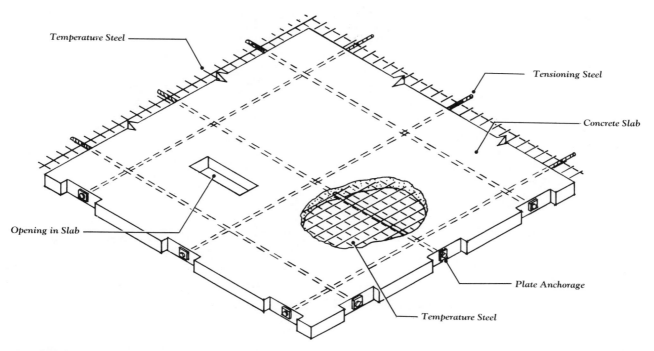

Temperature Steel

Tensioning Steel

Concrete Slab

Opening in Slab

Plate Anchorage

Temperature Steel

Posttensioned Slab

Posttensioning of concrete slabs or beams may be accomplished with most concrete structural systems. Longer spans, heavier loading and reduced member sizes are possible when high strength concrete is used to fully utilize the tensioning steel.

Steel for tensioning is commonly high-strength bars or strands. Bars are best suited for vertical prestressing because they provide their own support. Bars vary from 3/4" to 1-3/8" diameter and may be supplied with couplings for connections and wedged to allow tensioning. Tendons are available in various sizes in long lengths. Commonly used diameters are 0.5" and 0.6".

The bars or strands are normally tensioned using pressure jacking equipment after the concrete has reached approximately three quarters of its ultimate strength. The cableways may be grouted after tensioning to provide bond between the steel and concrete, or bond may be prevented by coating the tendons with corrosion preventative grease and wrapping with waterproof paper or plastic. Grouting cableways is usually accomplished using a pressure grout pump with threaded fittings that attach to a plate at the end

of the cableway. A hole with the capacity of being plugged at the opposite end of the cableway ensures completion of the grouting procedure.

Simple span beams usually require one end stressing. Continuous beams with draped stands require two end stressing. Long slabs are usually placed from the center outward and stressed in increments.

In addition to the tensioning steel, post-tensioned members normally require conventional reinforcing for temperature.

Flat slab or flat plate construction with a span to depth ratio between 36' to 44' usually requires 1 lb of post-tensioning steel and 1/2 lb of reinforcing steel per square foot for 24' to 28' bays. Pan and joist construction with a span to depth ratio of 28' to 30' usually requires 0.8 lb of posttensioning steel and 1 lb of reinforcing steel per square foot for normal spans.

Posttension suppliers will usually provide engineering services at no cost and supply jacks and equipment required to accomplish the posttensioning procedure at minimal rental.

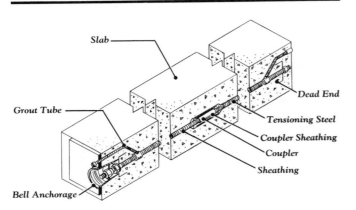

Posttensioning Assembly

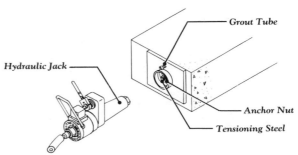

Hydraulic Tensioning Jack

Man-hours

Description	m/hr	Unit
Prestressing Steel Post-Tensioned in Field		
Grouted Strand 100 kip		
50' Span	.053	lb
100' Span	.038	lb
200' Span	.024	lb
300 kip		
50' Span	.024	lb
100' Span	.020	lb
200' Span	.018	lb
Ungrouted Strand 100 kip		
50' Span	.025	lb
100' Span or 200' Span	.021	lb
200 kip		
50' Span	.022	lb
100' Span or 200' Span	.019	lb
Grouted Bars 42 kip		
50' Span	.025	lb
75' Span	.020	lb
143 kip		
50' Span	.020	lb
75' Span	.015	lb
Ungrouted Bars 42 kip		
50' Span	.023	lb
75' Span	.018	lb
143 kip		
50' Span	.019	lb
75' Span	.015	lb
Ungrouted Single Strand, 100' Slab		
25 kip	.027	lb
35 kip	.022	lb

Man-hours per Tendon and Labor Costs per Pound of Prestress Steel						
Length	100' Beam		75' Beam		100' Slab	
Type Steel	Strand		Bars		Strand	
Diameter	0.5"		3/4"	1-3/8"	0.5"	.06"
Number	12	24	1	1	1	1
Force in Kips	298	595	42	143	25	35
Prep and Placing Cables	6.6	10.0	0.7	2.3	0.8	0.8
Stressing Cables	2.4	3.0	0.6	1.3	0.4	0.4
Grouting, if Required	3.0	3.5	0.5	1.0		
Total Man-hours	12.0	16.5	1.8	4.6	1.2	1.2
Prestressing Steel Wts.	640#	1280#	115#	380#	53#	74#

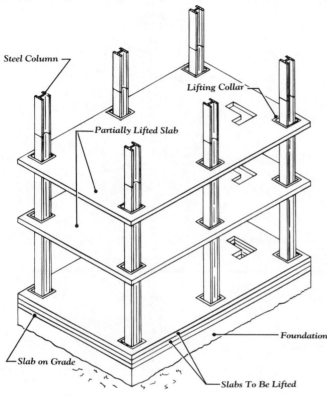

Lift-Slab Assembly

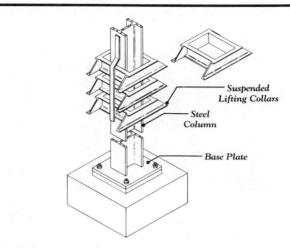

Method of Suspending
Lifting Collars

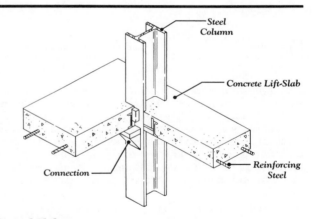

Typical Slab to
Column Connection

Lift-slab construction is accomplished by stacking the floor and roof slabs at ground level and then lifting them to their final elevation with jacks attached to the pre-erected steel columns. The system eliminates flat-plate formwork and allows reinforcing as well as mechanical and electrical roughing in the slab to be placed virtually at ground level.

The sequence of operations usually proceeds as follows: erecting the column; placing the slab on grade (used as a casting bed); casting the suspended slabs and roof; lifting the slabs to their final position in the specified sequence; and adding column extensions (if required). Separation of the slabs is ensured by application of a separating material or bond breaker applied between slabs.

Large-area slabs may be raised in sections and connected with a pour strip 3′ to 6′ wide, placed after the lifting and connection operations has been completed. Lifting of the slabs is usually carried out by a specialty contractor, who also normally supplies the lifting jacks and the required collars. Lifting collars are cast in the slabs at each column to provide connections between the jacks and slab and to function as weldments for the final connection, slab to column.

Reinforcing for lift slabs may be fabricated from conventional bars. Slabs may also be posttensioned and stressed prior to lifting.

Man-hours

Description	m/hr	Unit
Edge Forms 7″ to 12″ 4 uses	.074	sfca
Box Out for Slab Openings	.120	lf
Reinforcing #3 to #7	13.913	ton
Post Tensioning Strand		
100 kip	.021	lb
200 kip	.019	lb
Placing Concrete		
Direct Chute	.291	cu yd
Pumped	.388	cu yd
Steel Trowel Finish	.015	sf

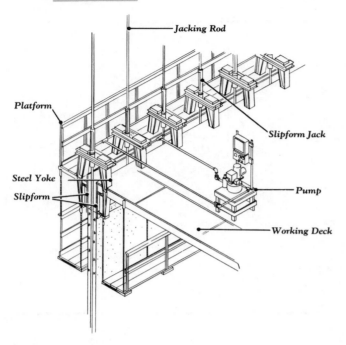

Slipforms

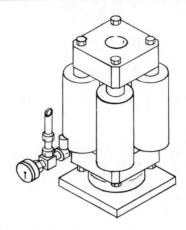

Four-Cylinder Hydraulic Jack

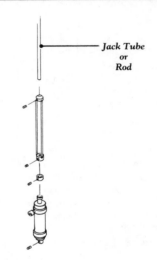

Hydraulic Jack, Jacking Rod

The slipform method of forming may be used for circular silo and multi-celled storage structures over 30' high, as well as building core shear walls over eight stories high. Such shear walls usually enclose elevator shafts, stairwells, mechanical spaces, and toilet rooms. Reuse of the form on duplicate structures will lower the cost per use. Slipform systems can be used to cast chimneys, towers, piers, dams, underground shafts, or other structures capable of being extruded.

Slipforms are usually 4' high and are raised semi-continuously by jacks climbing on rods embedded in the concrete. The jacks, powered by hydraulic, pneumatic, or electric motors, are available in 3, 6, and 22 ton capacities. Interior work decks and exterior scaffolds must be provided for the placing of inserts, embedded items, and reinforcing steel. Scaffolds below the form may be required for finishers. The interior work decks are often used as roof slab forms on silos and bins.

Forms will travel up the structure at a rate of 6" to 20" per hour for silos, 6" to 30" per hour for buildings, and 6" to 48" per hour for shaft work. Reinforcing bars and stress strands are usually hoisted by crane or gin pole, and the concrete by crane, winch, or pump. The slipform system is operated on a continuous 24-hour day when a monolithic structure is desired. For least cost, however, the system is operated only during normal working hours.

Man-hours

Description	m/hr	Unit
Slipforms		
Silos Average	.047	sfca
Buildings Average	.057	sfca
Reinforcing		
#3 to #7	21.333	ton
#8 to #14	13.913	ton
Posttensioning Grounted Strand		
100 kips	.053	lb
300 kips	.024	lb
Placing Concrete		
Minimum	.500	cu yd
Maximum	1.250	cu yd

3 SUPERSTRUCTURE
STRUCTURAL STEEL: GENERAL

Structural steel framing is versatile and may be designed in unlimited combinations of columns, girders, beams, and miscellaneous shapes. It is usually shop-fabricated to conform with shop drawings, with individual pieces numbered to agree with an erection drawing. Shop operations may consist of shearing or cutting to length, punching holes, coping for connection clearances, attaching connection materials, web stiffners, etc., and painting. Steel purchased from the mill is usually shipped, cut to length, as ordered from a cutting list prepared with the shop drawings. Most shapes are available in lengths of 60' to 75'. Shapes purchased from a warehouse are standard length, and often produce waste when cut to length, which increases material costs.

The erection procedure normally includes: shakeout or unloading (spreading or sequencing the pieces); lifting and connecting; plumbing and guying (accomplished with wire cables and turnbuckles); and bolting up, riveting, or welding the connections.

Hoisting, lifting, or raising is accomplished by using truck-mounted cable or hydraulic cranes, track cable cranes, stiff leg derricks, guy derricks, climbing cranes, or sometimes, chainfalls or forklifts in existing buildings.

Coordination of the erection sequence and the location and required reach of erection equipment are important considerations in the structural steel erection procedure.

Many codes and/or safety regulations, both federal and state, specify procedures and regulations that are costly and which interrupt the erection sequence. Planking, decking, or safety net requirements, mandated for every other floor in high-rise buildings, along with perimeter rails, add considerable expense to the structural steel erection.

Man-hours

Description	m/hr	Unit	m/hr	Unit
Beams Wide Flange				
W6 x 9	.949	Ea.	.093	lf
W10 x 22	1.037	Ea.	.085	lf
W12 x 26	1.037	Ea.	.064	lf
W14 x 34	1.333	Ea.	.069	lf
W16 x 31	1.333	Ea.	.062	lf
W18 x 50	2.162	Ea.	.088	lf
W21 x 62	2.222	Ea.	.077	lf
W24 x 76	2.353	Ea.	.072	lf
W27 x 94	2.581	Ea.	.067	lf
W30 x 108	2.857	Ea.	.067	lf
W33 x 130	3.200	Ea.	.071	lf
W36 x 300	3.810	Ea.	.077	lf

Man-hours

Description	m/hr	Unit
Light Framing		
Angles 4" and Larger	.011	lb
Less than 4"	.018	lb
Channels 8" and Larger	.009	lb
Less than 8"	.016	lb
Cross Bracing Angles	.009	lb
Rods	.034	lb
Hanging Lintels	.028	lb
High-Strength Bolts in Place		
3/4" Bolts	.097	Ea.
7/8" Bolts	.100	Ea.

Man-hours

Description	m/hr	Unit
Columns		
Steel Concrete Filled		
3-1/2" Diameter	.933	Ea.
6-5/8" Diameter	1.120	Ea.
Steel Pipe		
3" Diameter	.933	Ea.
8" Diameter	1.120	Ea.
12" Diameter	1.244	Ea.
Structural Tubing		
4" x 4"	.966	Ea.
8" x 8"	1.120	Ea.
12" x 8"	1.167	Ea.
Wide Flange 2 Tier		
W8 x 31	.052	lf
W8 x 67	.057	lf
W10 x 45	.054	lf
W10 x 112	.058	lf
W12 x 50	.054	lf
W12 x 190	.061	lf
W14 x 74	.057	lf
W14 x 176	.061	lf

Man-hours

Description	m/hr	Unit	m/hr	Unit
Apartments, Nursing Homes, Etc.				
1-2 Stories	4.211	Piece	7.767	ton
3-6 Stories	4.444	Piece	7.921	ton
7-15 Stories	4.923	Piece	9.014	ton
Over 15 Stories	5.333	Piece	9.209	ton
Offices Hospitals, Etc.				
1-2 Stories	4.211	Piece	7.767	ton
3-6 Stories	4.741	Piece	8.889	ton
7-15 Stories	4.923	Piece	9.014	ton
Over 15 Stories	5.120	Piece	9.209	ton
Industrial Buildings				
1 Story	3.478	Piece	6.202	ton

Hot rolled structural steel is available in the following shapes.

Shape & Designation	Name & Characteristics	Shape & Designation	Name & Characteristics
W	Wide Flange Parallel flange surfaces	M C	Miscellaneous Channel Infrequently rolled by some producers
S	American Standard Beam (I Beam) Sloped inner flange	L	Angle Equal or unequal legs, constant thickness
M	Miscellaneous Beams Cannot be classified as W, HP or S infrequently rolled by some producers	T	Structural Tee Cut from W, M or S on center of web
C	American Standard Channel Sloped inner flange	H P	Bearing Pile Parallel flanges and equal flange and web thickness

Common drawing designations follow.

Wide Flange
W 18 x 35 ← Weight in Pounds Per Foot
└ Nominal Depth in Inches (Actual 17-3/4")

American Standard Beam
S   12 x 31.8
↑ └ Weight in Pounds Per Foot
└ Depth in Inches

Miscellaneous Beam
M   8 x 6.5
└ Weight in Pounds Per Foot
└ Depth in Inches

American Standard Channel
C   8 x 11.5
└ Weight in Pounds Per Foot
└ Depth in Inches

Miscellaneous Channel
MC 8 x 22.8
└ Weight in Pounds Per Foot
└ Depth in Inches

Angle
┌ Length of One Leg in Inches
L 6 x 3-1/2 x 3/8 ← Thickness of Each Leg in Inches
└ Length of Other Leg in Inches

Tee Cut From W16 x 100
WT 8 x 50
└ Weight in Pounds Per Foot
└ Nominal Depth in Inches (Actual 8-1/2")

Tee Cut From S   12 x 35
ST 6 x 17.5
↑ └ Weight in Pounds Per Foot
└ Depth in Inches

Tee Cut From M   10 x 9
MT 5 x 4.5
↑ └ Weight in Pounds Per Foot
└ Depth in Inches

Bearing Pile
HP   12 x 84
↑ └ Weight in Pounds Per Foot
└ Nominal Depth in Inches (Actual 12-1/4")

Hot rolled structural shapes are generally available in the following ASTM specifications.

Steel Type	ASTM Designation	Minimum Yield Stress KSI	Characteristics
Carbon	A36	36	
	A529	42	
High-Strength Low Alloy	A441	50	Structural Manganese Vanadium Steel
	A572	42	Columbium-Vanadium Steel
		50	
		60	
		65	
Corrosion Resistant High Strength Low Alloy	A242	50	Corrosion Resistant
	A588	50	Corrosion Resistant to 4" Thick

W Shapes Dimensions				
		Web	**Flange**	
Designation	**Depth (in.)**	**Thickness (in.)**	**Width (in.)**	**Thickness (in.)**
W 36x300	36-3/4	15/16	16-5/8	1-11/16
x280	36-1/2	7/8	16-5/8	1-9/16
x260	36-1/4	13/16	16-1/2	1-7/16
x245	36-1/8	13/16	16-1/2	1-3/8
x230	35-7/8	3/4	16-1/2	1-1/4
W 36x210	36-3/4	13/16	12-1/8	1-3/8
x194	36-1/2	3/4	12-1/8	1-1/4
x182	36-3/8	3/4	12-1/8	1-3/8
x170	36-1/8	11/16	12	1-1/8
x160	36	5/8	12	1
x150	35-7/8	5/8	12	15/16
x135	35-1/2	5/8	12	13/16
W 33x241	34-1/8	13/16	15-7/8	1-3/8
x221	33-7/8	3/4	15-3/4	1-1/4
x201	33-5/8	11/16	15-3/4	1-1/8
W 33x152	33-1/2	5/8	11-5/8	1-1/16
x141	33-1/4	5/8	11-1/2	15/16
x130	33-1/8	9/16	11-1/2	7/8
x118	32-7/8	9/16	11-1/2	3/4
W 30x211	31	3/4	15-1/8	1-5/16
x191	30-5/8	11/16	15	1-3/16
x173	30-1/2	5/8	15	1-1/16
W 30x132	30-1/4	5/8	10-1/2	1
x124	30-1/8	9/16	10-1/2	15/16
x116	30	9/16	10-1/2	7/8
x108	29-7/8	9/16	10-1/2	3/4
x 99	29-5/8	1/2	10-1/2	11/16
W 27x178	27-3/4	3/4	14-1/8	1-3/16
x161	27-5/8	11/16	14	1-1/16
x146	27-3/8	5/8	14	1
W 27x114	27-1/4	9/16	10-1/8	15/16
x102	27-1/8	1/2	10	13/16
x 94	26-7/8	1/2	10	3/4
x 84	26-3/4	7/16	10	5/8
W 24x162	25	11/16	13	1-1/4
x146	24-3/4	5/8	12-7/8	1-1/16
x131	24-1/2	5/8	12-7/8	15/16
x117	24-1/4	9/16	12-3/4	7/8
x104	24	1/2	12-3/4	3/4
W 24x 94	24-1/4	1/2	9-1/8	7/8
x 84	24-1/8	1/2	9	3/4
x 76	23-7/8	7/16	9	11/16
x 68	23-3/4	7/16	9	9/16
W 24x 62	23-3/4	7/16	7	9/16
x 55	23-5/8	3/8	7	1/2
W 21x147	22	3/4	12-1/2	1-1/8
x132	21-7/8	5/8	12-1/2	1-1/16
x122	21-5/8	5/8	12-3/8	15/16
x111	21-1/2	9/16	12-3/8	7/8
x101	21-3/8	1/2	12-1/4	13/16

W Shapes Dimensions				
		Web	**Flange**	
Designation	**Depth (in.)**	**Thickness (in.)**	**Width (in.)**	**Thickness (in.)**
W 21x 93	21-5/8	9/16	8-3/8	15/16
x 83	21-3/8	1/2	8-3/8	13/16
x 73	21-1/4	7/16	8-1/4	3/4
x 68	21-1/8	7/16	8-1/4	11/16
x 62	21	3/8	8-1/4	5/8
W 21x 57	21	3/8	6-1/2	5/8
x 50	20-7/8	3/8	6-1/2	9/16
x 44	20-5/8	3/8	6-1/2	7/16
W 18x119	19	5/8	11-1/4	1-1/16
x106	18-3/4	9/16	11-1/4	15/16
x 97	18-5/8	9/16	11-1/8	7/8
x 86	18-3/8	1/2	11-1/8	3/4
x 76	18-1/4	7/16	11	11/16
W 18x 71	18-1/2	1/2	7-5/8	13/16
x 65	18-3/8	7/16	7-5/8	3/4
x 60	18-1/4	7/16	7-1/2	11/16
x 55	18-1/8	3/8	7-1/2	5/8
x 50	18	3/8	7-1/2	9/16
W 18x 46	18	3/8	6	5/8
x 40	17-7/8	5/16	6	1/2
x 35	17-3/4	5/16	6	7/16
W 16x100	17	9/16	10-3/8	1
x 89	16-3/4	1/2	10-3/8	7/8
x 77	16-1/2	7/16	10-1/4	3/4
x 67	16-3/8	3/8	10-1/4	11/16
W 16x 57	16-3/8	7/16	7-1/8	11/16
x 50	16-1/4	3/8	7-1/8	5/8
x 45	16-1/8	3/8	7	9/16
x 40	16	5/16	7	1/2
x 36	15-7/8	5/16	7	7/16
W 16x 31	15-7/8	1/4	5-1/2	7/16
x 26	15-3/4	1/4	5-1/2	3/8
W 14x730	22-3/8	3-1/16	17-7/8	4-15/16
x665	21-5/8	2-13/16	17-5/8	4-1/2
x605	20-7/8	2-5/8	17-3/8	4-3/16
x550	20-1/4	2-3/8	17-1/4	3-13/16
x500	19-5/8	2-3/16	17	3-1/2
x455	19	2	16-7/8	3-3/16
W 14x426	18-5/8	1-7/8	16-3/4	3-1/16
x398	18-1/4	1-3/4	16-5/8	2-7/8
x370	17-7/8	1-5/8	16-1/2	2-11/16
x342	17-1/2	1-9/16	16-3/8	2-1/2
x311	17-1/8	1-7/16	16-1/4	2-1/4
x283	16-3/4	1-5/16	16-1/8	2-1/16
x257	16-3/8	1-3/16	16	1-7/8
x233	16	1-1/16	15-7/8	1-3/4
x211	15-3/4	1	15-3/4	1-9/16
x193	15-1/2	7/8	15-3/4	1-7/16
x176	15-1/4	13/16	15-5/8	1-5/16
x159	15	3/4	15-5/8	1-3/16
x145	14-3/4	11/16	15-1/2	1-1/16

W Shapes Dimensions		Web	Flange	
Designation	Depth (in.)	Thickness (in.)	Width (in.)	Thickness (in.)
W 14x132	14-5/8	5/8	14-3/4	1
x120	14-1/2	9/16	14-5/8	15/16
x109	14-3/8	1/2	14-5/8	7/8
x 99	14-1/8	1/2	14-5/8	3/4
x 90	14	7/16	14-1/2	11/16
W 14x 82	14-1/4	1/2	10-1/8	7/8
x 74	14-1/8	7/16	10-1/8	13/16
x 68	14	7/16	10	3/4
x 61	13-7/8	3/8	10	5/8
W 14x 53	13-7/8	3/8	8	11/16
x 48	13-3/4	5/16	8	5/8
x 43	13-5/8	5/16	8	1/2
W 14x 38	14-1/8	5/16	6-3/4	1/2
x 34	14	5/16	6-3/4	7/16
x 30	13-7/8	1/4	6-3/4	3/8
W 14x 26	13-7/8	1/4	5	7/16
x 22	13-3/4	1/4	5	5/16
W 12x336	16-7/8	1-3/4	13-3/8	2-15/16
x305	16-3/8	1-5/8	13-1/4	2-11/16
x279	15-7/8	1-1/2	13-1/8	2-1/2
x252	15-3/8	1-3/8	13	2-1/4
x230	15	1-5/16	12-7/8	2-1/16
x210	14-3/4	1-3/16	12-3/4	1-7/8
x190	14-3/8	1-1/16	12-5/8	1-3/4
x170	14	15/16	12-5/8	1-9/16
x152	13-3/4	7/8	12-1/2	1-3/8
x136	13-3/8	13/16	12-3/8	1-1/4
x120	13-1/8	11/16	12-3/8	1-1/8
x106	12-7/8	5/8	12-1/4	1
x 96	12-3/4	9/16	12-1/8	7/8
x 87	12-1/2	1/2	12-1/8	13/16
x 79	12-3/8	1/2	12-1/8	3/8
x 72	12-1/4	7/16	12	11/16
x 65	12-1/8	3/8	12	5/8
W 12x 58	12-1/4	3/8	10	5/8
x 53	12	3/8	10	9/16
W 12x 50	12-1/4	3/8	8-1/8	5/8
x 45	12	5/16	8	9/16
x 40	12	5/16	8	1/2
W 12x 35	12-1/2	5/16	6-1/2	1/2
x 30	12-3/8	1/4	6-1/2	7/16
x 26	12-1/4	1/4	6-1/2	3/8
W 12x 22	12-1/4	1/4	4	7/16
x 19	12-1/8	1/4	4	3/8
x 16	12	1/4	4	1/4
x 14	11-7/8	3/16	4	1/4
W 10x112	11-3/8	3/4	10-3/8	1-1/4
x100	11-1/8	11/16	10-3/8	1-1/8
x 88	10-7/8	5/8	10-1/4	1
x 77	10-5/8	1/2	10-1/4	7/8
x 68	10-3/8	1/2	10-1/8	3/4
x 60	10-1/4	7/16	10-1/8	11/16
x 54	10-1/8	3/8	10	5/8
x 49	10	5/16	10	9/16

W Shapes Dimensions		Web	Flange	
Designation	Depth (in.)	Thickness (in.)	Width (in.)	Thickness (in.)
W 10x 45	10-1/8	3/8	8	5/8
x 39	9-7/8	5/16	8	1/2
x 33	9-3/4	5/16	8	7/16
W 10x 30	10-1/2	5/16	5-3/4	1/2
x 26	10-3/8	1/4	5-3/4	7/16
x 22	10-1/8	1/4	5-3/4	3/8
W 10x 19	10-1/4	1/4	4	3/8
x 17	10-1/8	1/4	4	5/16
x 15	10	1/4	4	1/4
x 12	9-7/8	3/16	4	3/16
W 8x 67	9	9/16	8-1/4	15/16
x 58	8-3/4	1/2	8-1/4	13/16
x 48	8-1/2	3/8	8-1/8	11/16
x 40	8-1/4	3/8	8-1/8	9/16
x 35	8-1/8	5/16	8	1/2
x 31	8	5/16	8	7/16
W 8x 28	8	5/16	6-1/2	7/16
x 24	7-7/8	1/4	6-1/2	3/8
W 8x 21	8-1/4	1/4	5-1/4	3/8
x 18	8-1/8	1/4	5-1/4	5/16
W 8x 15	8-1/8	1/4	4	5/16
x 13	8	1/4	4	1/4
x 10	7-7/8	3/16	4	3/16
W 6x 25	6-3/8	5/16	6-1/8	7/16
x 20	6-1/4	1/4	6	3/8
x 15	6	1/4	6	1/4
W 6x 16	6-1/4	1/4	4	3/8
x 12	6	1/4	4	1/4
x 9	5-7/8	3/16	4	3/16
W 5x 19	5-1/8	1/4	5	7/16
x 16	5	1/4	5	3/8
W 4x 13	4-1/8	1/4	4	3/8

M Shapes Dimensions		Web	Flange	
Designation	Depth (in.)	Thickness (in.)	Width (in.)	Thickness (in.)
M 14x 18	14	3/16	4	1/4
M 12x 11.8	12	3/16	3-1/8	1/4
M 10x 9	10	3/16	2-3/4	3/16
M 8x 6.5	8	1/8	2-1/4	3/16
M 6x 20	6	1/4	6	3/8
x 4.4	6	1/8	1-7/8	3/16
M 5x 18.9	5	5/16	5	7/16
M 4x 13	4	1/4	4	3/8

		S Shapes Dimensions			
		Web		Flange	
Designation	Depth (in.)	Thickness (in.)	Width (in.)	Thickness (in.)	
S 24x121	24-1/2	13/16	8	1-1/16	
x106	24-1/2	5/8	7-7/8	1-1/16	
x100	24	3/4	7-1/4	7/8	
x 90	24	5/8	7-1/8	7/8	
x 80	24	1/2	7	7/8	
S 20x 96	20-1/4	13/16	7-1/4	15/16	
x 86	20-1/4	11/16	7	15/16	
x 75	20	5/8	6-3/8	13/16	
x 66	20	1/2	6-1/4	13/16	
S 18x 70	18	11/16	6-1/4	11/16	
x 54.7	18	7/16	6	11/16	
S 15x 50	15	9/16	5-5/8	5/8	
x 42.9	15	7/16	5-1/2	5/8	
S 12x 50	12	11/16	5-1/2	11/16	
x 40.8	12	7/16	5-1/4	11/16	
x 35	12	7/16	5-1/8	9/16	
x 31.8	12	3/8	5	9/16	
S 10x 35	10	5/8	5	1/2	
x 25.4	10	5/16	4-5/8	1/2	
S 8x 23	8	7/16	4-1/8	7/16	
x 18.4	8	1/4	4	7/16	
S 7x 20	7	7/16	3-7/8	3/8	
x 15.3	7	1/4	3-5/8	3/8	
S 6x 17.25	6	7/16	3-5/8	3/8	
x 12.5	6	1/4	3-3/8	3/8	
S 5x 14.75	5	1/2	3-1/4	5/16	
x 10	5	3/16	3	5/16	
S 4x 9.5	4	5/16	2-3/4	5/16	
x 7.7	4	3/16	2-5/8	5/16	
S 3x 7.5	3	3/8	2-1/2	1/4	
x 5.7	3	3/16	2-3/8	1/4	

		H P Shapes Dimensions			
		Web		Flange	
Designation	Depth (in.)	Thickness (in.)	Width (in.)	Thickness (in.)	
HP 14x117	14-1/4	13/16	14-7/8	13/16	
x102	14	11/16	14-3/4	11/16	
x 89	13-7/8	5/8	14-3/4	5/8	
x 73	13-5/8	1/2	14-5/8	1/2	
HP 13x100	13-1/8	3/4	13-1/4	3/4	
x 87	13	11/16	13-1/8	11/16	
x 73	12-3/4	9/16	13	9/16	
x 60	12-1/2	7/16	12-7/8	7/16	
HP 12x 84	12-1/4	11/16	12-1/4	11/16	
x 74	12-1/8	5/8	12-1/4	5/8	
x 63	12	1/2	12-1/8	1/2	
x 53	11-3/4	7/16	12	7/16	
HP 10x 57	10	9/16	10-1/4	9/16	
x 42	9-3/4	7/16	10-1/8	7/16	
HP 8x 36	8	7/16	8-1/8	7/16	

Channels
American Standard
Dimensions

Designation	Depth (in.)	Web Thickness (in.)	Flange Width (in.)	Flange Thickness (in.)
C 15x 50	15	11/16	3-3/4	5/8
x 40	15	1/2	3-1/2	5/8
x 33.9	15	3/8	3-3/8	5/8
C 12x 30	12	1/2	3-1/8	1/2
x 25	12	3/8	3	1/2
x 20.7	12	5/16	3	1/2
C 10x 30	10	11/16	3	7/16
x 25	10	1/2	2-7/8	7/16
x 20	10	3/8	2-3/4	7/16
x 15.3	10	1/4	2-5/8	7/16
C 9x 20	9	7/16	2-5/8	7/16
x 15	9	5/16	2-1/2	7/16
x 13.4	9	1/4	2-3/8	7/16
C 8x 18.75	8	1/2	2-1/2	3/8
x 13.75	8	5/16	2-3/8	3/8
x 11.5	8	1/4	2-1/4	3/8
C 7x 14.75	7	7/16	2-1/4	3/8
x 12.25	7	5/16	2-1/4	3/8
x 9.8	7	3/16	2-1/8	3/8
C 6x 13	6	7/16	2-1/8	5/16
x 10.5	6	5/16	2	5/16
x 8.2	6	3/16	1-7/8	5/16
C 5x 9	5	5/16	1-7/8	5/16
x 6.7	5	3/16	1-3/4	5/16
C 4x 7.25	4	5/16	1-3/4	5/16
x 5.4	4	3/16	1-5/8	5/16
C 3x 6	3	3/8	1-5/8	1/4
x 5	3	1/4	1-1/2	1/4
x 4.1	3	3/16	1-3/8	1/4

Channels
American Standard
Dimensions

Designation	Depth (in.)	Web Thickness (in.)	Flange Width (in.)	Flange Thickness (in.)
MC18x 58	18	11/16	4-1/4	5/8
x 51.9	18	5/8	4-1/8	5/8
x 45.8	18	1/2	4	5/8
x 42.7	18	7/16	4	5/8
MC13x 50	13	13/16	4-3/8	5/8
x 40	13	9/16	4-1/8	5/8
x 35	13	7/16	4-1/8	5/8
x 31.8	13	3/8	4	5/8
MC12x 50	12	13/16	4-1/8	11/16
x 45	12	11/16	4	11/16
x 40	12	9/16	3-7/8	11/16
x 35	12	7/16	3-3/4	11/16
x 37	12	5/8	3-5/8	5/8
x 32.9	12	1/2	3-1/2	5/8
x 30.9	12	7/16	3-1/2	5/8
x 10.6	12	3/16	1-1/2	5/16
MC10x 41.1	10	13/16	4-3/8	9/16
x 33.6	10	9/16	4-1/8	9/16
x 28.5	10	7/16	4	9/16
x 28.3	10	1/2	3-1/2	9/16
x 25.3	10	7/16	3-1/2	1/2
x 24.9	10	3/8	3-3/8	9/16
x 21.9	10	5/16	3-1/2	1/2
x 8.4	10	3/16	1-1/2	1/4
x 6.5	10	1/8	1-1/8	3/16
MC 9x 25.4	9	7/15	3-1/2	9/16
x 23.9	9	3/8	3-1/2	9/16
MC 8x 22.8	8	7/16	3-1/2	1/2
x 21.4	8	3/8	3-1/2	1/2
x 20	8	3/8	3	1/2
x 18.7	8	3/8	3	1/2
x 8.5	8	3/16	1-7/8	5/16
MC 7x 22.7	7	1/2	3-5/8	1/2
x 19.1	7	3/8	3-1/2	1/2
x 17.6	7	3/8	3	1/2
MC 6x 18	6	3/8	3-1/2	1/2
x 16.3	6	3/8	3	1/2
x 15.3	6	5/16	3-1/2	3/8
x 15.1	6	5/16	3	1/2
x 12	6	5/16	2-1/2	3/8

Angles
Equal Legs and Unequal Legs
Properties for Designing

Size and Thickness (in.)	Weight per Foot (lb)	Size and Thickness (in.)	Weight per Foot (lb)
L 9 x4 x 5/8	26.3	L 4 x4 x 3/4	18.5
9/16	23.8	5/8	15.7
1/2	21.3	1/2	12.8
L 8 x8 x1-1/8	56.9	7/16	11.3
1	51.0	3/8	9.8
7/8	45.0	5/16	8.2
3/4	38.9	1/4	6.6
5/8	32.7	L 4 x3-1/2x 5/8	14.7
9/16	29.6	1/2	11.9
1/2	26.4	7/16	10.6
L 8 x6 x1	44.2	3/8	9.1
7/8	39.1	5/16	7.7
3/4	33.8	1/4	6.2
5/8	28.5	L 3 x2-1/2x 1/2	8.5
9/16	25.7	7/16	7.6
1/2	23.0	3/8	6.6
7/16	20.2	5/16	5.6
L 8 x4 x1	37.4	1/4	4.5
3/4	28.7	3/16	3.39
9/16	21.9	L 3 x2 x 1/2	7.7
1/2	19.6	7/16	6.8
L 7 x4 x 3/4	26.2	3/8	5.9
5/8	22.1	5/16	5.0
1/2	17.9	1/4	4.1
3/8	13.6	3/16	3.07
L 5 x3-1/2x 3/4	19.8	L 2-1/2x2-1/2x 1/2	7.7
5/8	16.8	3/8	5.9
1/2	13.6	5/16	5.0
7/16	12.0	1/4	4.1
3/8	10.4	3/16	3.07
5/16	8.7	L 2-1/2x2 x 3/8	5.3
1/4	7.0	5/16	4.5
L 5x x3 x 5/8	15.7	1/4	3.62
1/2	12.8	3/16	2.75
7/16	11.3	L 2 x2 x 3/8	4.7
3/8	9.8	5/16	3.92
5/16	8.2	1/4	3.19
1/4	6.6	3/16	2.44
		1/8	1.65

Structural Tees Cut from W Shapes Dimensions				
Designation	Depth (in.)	Stem Thickness (in.)	Flange Width (in.)	Flange Thickness (in.)
WT 18 x150	18-3/8	15/16	16-5/8	1-11/16
x140	18-1/4	7/8	16-5/8	1-9/16
x130	18-1/8	13/16	16-1/2	1-7/16
x122.5	18	13-16	16-1/2	1-3/8
x115	18	3/4	16-1/2	1-1/4
x105	18-3/8	13/16	12-1/8	1-3/8
x 97	18-1/4	3/4	12-1/8	1-1/4
x 91	18-1/8	3/4	12-1/8	1-3/16
x 85	18-1/8	11/16	12	1-1/8
x 80	18	5/8	12	1
x 75	17-7/8	5/8	12	15/16
x 67.5	17-3/4	5/8	12	13/16
WT 16.5x120.5	17-1/8	13/16	15-7/8	1-3/8
x110.5	17	3/4	15-3/4	1-1/4
x100.5	16-7/8	11/16	15-3/4	1-1/8
x 76	16-3/4	5/8	11-5/8	1-1/16
x 70.5	16-5/8	5/8	11-1/2	15/16
x 65	16-1/2	9/16	11-1/2	7/8
x 59	16-3/8	9/16	11-1/2	3/4
WT 15 x105.5	15-1/2	3/4	15-1/8	1-5/16
x 95.5	15-3/8	11/16	15	1-3/16
x 86.5	15-1/4	5/8	15	1-1/16
x 66	15-1/8	5/8	10-1/2	1
x 62	15-1/8	9/16	10-1/2	15/16
x 58	15	9/16	10-1/2	7/8
x 54	14-7/8	9/16	10-1/2	3/4
x 49.5	14-7/8	1/2	10-1/2	11/16
WT 13.5x 89	13-7/8	3/4	14-1/8	1-3/16
x 80.5	13-3/4	11/16	14	1-1/16
x 73	13-3/4	5/8	14	1
x 57	13-5/8	9/16	10-1/8	15/16
x 51	13-1/2	1/2	10	13/16
x 47	13-1/2	1/2	10	3/4
x 42	13-3/8	7/16	10	5/8
WT 12 x 81	12-1/2	11/16	13	1-1/4
x 73	12-3/8	5/8	12-7/8	1-1/16
x 65.5	12-1/4	5/8	12-7/8	15/16
x 58.5	12-1/8	9/16	12-3/4	7/8
x 52	12	1/2	12-3/4	3/4
x 47	12-1/8	1/2	9-1/8	7/8
x 42	12	1/2	9	3/4
x 38	12	7/16	9	11/16
x 34	11-7/8	7/16	9	9/16
x 31	11-7/8	7/16	7	9/16
x 27.5	11-3/4	3/8	7	1/2

Structural Tees Cut from W Shapes Dimensions				
Designation	Depth (in.)	Stem Thickness (in.)	Flange Width (in.)	Flange Thickness (in.)
WT 10.5x 73.5	11	3/4	12-1/2	1-1/8
x 66	10-7/8	5/8	12-1/2	1-1/16
x 61	10-7/8	5/8	12-3/8	15/16
x 55.5	10-3/4	9/16	12-3/8	7/8
x 50.5	10-5/8	1/2	12-1/4	13/16
x 46.5	10-3/4	9/16	8-3/8	15/16
x 41.5	10-3/4	1/2	8-3/8	13/16
x 36.5	10-5/8	7/16	8-1/4	3/4
x 34	10-5/8	7/16	8-1/4	11/16
x 31	10-1/2	3/8	8-1/4	5/8
x 28.5	10-1/2	3/8	6-1/2	5/8
x 25	10-3/8	3/8	6-1/2	9/16
x 22	10-3/8	3/8	6-1/2	7/16
WT 9x 59.5	9-1/2	5/8	11-1/4	1-1/16
x 53	9-3/8	9/16	11-1/4	15/16
x 48.5	9-1/4	9/16	11-1/8	7/8
x 43	9-1/4	1/2	11-1/8	3/4
x 38	9-1/8	7/16	11	11/16
x 35.5	9-1/4	1/2	7-5/8	13/16
x 32.5	9-1/8	7/16	7-5/8	3/4
x 30	9-1/8	7/16	7-1/2	11/16
x 27.5	9	3/8	7-1/2	5/8
x 25	9	3/8	7-1/2	9/16
x 23	9	3/8	6	5/8
x 20	9	5/16	6	1/2
x 17.5	8-7/8	5/16	6	7/16
WT 8x 50	8-1/2	9/16	10-3/8	1
x 44.5	8-3/8	1/2	10-3/8	7/8
x 38.5	8-1/4	7/16	10-1/4	3/4
x 33.5	8-1/8	3/8	10-1/4	11/16
x 28.5	8-1/4	7/16	7-1/8	11/16
x 25	8-1/8	3/8	7-1/8	5/8
x 22.5	8-1/8	3/8	7	9/16
x 20	8	5/16	7	1/2
x 18	7-7/8	5/16	7	7/16
x 15.5	8	1/4	5-1/2	7/16
x 13	7-7/8	1/4	5-1/2	3/8

Structural Tees Cut from W Shapes Dimensions				
Designation	Depth (in.)	Stem Thickness (in.)	Flange Width (in.)	Flange Thickness (in.)
WT 7 x365	11-1/4	3-1/16	17-7/8	4-15/16
x332.5	10-7/8	2-13/16	17-5/8	4-1/2
x302.5	10-1/2	2-5/8	17-3/8	4-3/16
x275	10-1/8	2-3/8	17-1/4	3-13/16
x250	9-3/4	2-3/16	17	3-1/2
x227.5	9-1/2	2	16-7/8	3-3/16
x213	9-3/8	1-7/8	16-3/4	3-1/16
x199	9-1/8	1-3/4	16-5/8	2-7/8
x185	9	1-5/8	16-1/2	2-11/16
x171	8-3/4	1-9/16	16-3/8	2-1/2
x155.5	8-1/2	1-7/16	16-1/4	2-1/4
x141.5	8-3/8	1-5/16	16-1/8	2-1/16
x128.5	8-1/4	1-3/16	16	1-7/8
x116.5	8	1-1/16	15-7/8	1-3/4
x105.5	7-7/8	1	15-3/4	1-9/16
x 96.5	7-3/4	7/8	15-3/4	1-7/16
x 88	7-5/8	13/16	15-5/8	1-5/16
x 79.5	7-1/2	3/4	15-5/8	1-3/16
x 72.5	7-3/8	11/16	15-1/2	1-1/16
x 66	7-3/8	5/8	14-3/4	1
x 60	7-1/4	9/16	14-5/8	15/16
x 54.5	7-1/8	1/2	14-5/8	7/8
x 49.5	7-1/8	1/2	14-5/8	3/4
x 45	7	7/16	14-1/2	11/16
x 41	7-1/8	1/2	10-1/8	7/8
x 37	7-1/8	7/16	10-1/8	13/16
x 34	7	7/16	10	3/4
x 30.5	7	3/8	10	5/8
x 26.5	7	3/8	8	11/16
x 24	6-7/8	5/16	8	5/8
x 21.5	6-7/8	5/16	8	1/2
x 19	7	5/16	6-3/4	1/2
x 17	7	5/16	6-3/4	7/16
x 15	6-7/8	1/4	6-3/4	3/8
x 13	7	1/4	5	7/16
x 11	6-7/8	1/4	5	5/16

Structural Tees Cut from W Shapes Dimensions				
Designation	Depth (in.)	Stem Thickness (in.)	Flange Width (in.)	Flange Thickness (in.)
WT 6 x168	8-3/8	1-3/4	13-3/8	2-15/16
x152.5	8-1/8	1-5/8	13-1/4	2-11/16
x139.5	7-7/8	1-1/2	13-1/8	2-1/2
x126	7-3/4	1-3/8	13	2-1/4
x115	7-1/2	1-5/16	12-7/8	2-1/16
x105	7-3/8	1-3/16	12-3/4	1-7/8
x 95	7-1/4	1-1/16	12-5/8	1-3/4
x 85	7	15/16	12-5/8	1-9/16
x 76	6-7/8	7/8	12-1/2	1-3/8
x 68	6-3/4	13/16	12-3/8	1-1/4
x 60	6-1/2	11/16	12-3/8	1-1/8
x 53	6-1/2	5/8	12-1/4	1
x 48	6-3/8	9/16	12-1/8	7/8
x 43.5	6-1/4	1/2	12-1/8	13/16
x 39.5	6-1/4	1/2	12-1/8	3/4
x 36	6-1/8	7/16	12	11/16
x 32.5	6	3/8	12	5/8
x 29	6-1/8	3/8	10	5/8
x 26.5	6	3/8	10	9/16
x 25	6-1/8	3/8	8-1/8	5/8
x 22.5	6	5/16	8	9/16
x 20	6	5/16	8	1/2
x 17.5	6-1/4	5/16	6-1/2	1/2
x 15	6-1/8	1/4	6-1/2	7/16
x 13	6-1/8	1/4	6-1/2	3/8
x 11	6-1/8	1/4	4	7/16
x 9.5	6-1/8	1/4	4	3/8
x 8	6	1/4	4	1/4
x 7	6	3/16	4	1/4

Structural Tees Cut from W Shapes Dimensions

Designation	Depth (in.)	Stem Thickness (in.)	Flange Width (in.)	Flange Thickness (in.)
WT 5 x 56	5-5/8	3/4	10-3/8	1-1/4
x 50	5-1/2	11/16	10-3/8	1-1/8
x 44	5-3/8	5/8	10-1/4	1
x 38.5	5-1/4	1/2	10-1/4	7/8
x 34	5-1/4	1/2	10-1/8	3/4
x 30	5-1/8	7/16	10-1/8	11/16
x 27	5	3/8	10	5/8
x 24.5	5	5/16	10	9/16
x 22.5	5	3/8	8	5/8
x 19.5	5	5/16	8	1/2
x 16.5	4-7/8	5/16	8	7/16
x 15	5-1/4	5/16	5-3/4	1/2
x 13	5-1/8	1/4	5-3/4	7/16
x 11	5-1/8	1/4	5-3/4	3/8
x 9.5	5-1/8	1/4	4	3/8
x 8.5	5	1/4	4	5/16
x 7.5	5	1/4	4	1/4
x 6	4-7/8	3/16	4	3/16
WT 4 x 33.5	4-1/2	9/16	8-1/4	15/16
x 29	4-3/8	1/2	8-1/4	13/16
x 24	4-1/4	3/8	8-1/8	11/16
x 20	4-1/8	3/8	8-1/8	9/16
x 17.5	4	5/16	8	1/2
x 15.5	4	5/16	8	7/16
x 14	4	5/16	6-1/2	7/16
x 12	4	1/4	6-1/2	3/8
x 10.5	4-1/8	1/4	5-1/4	3/8
x 9	4-1/8	1/4	5-1/4	5/16
x 7.5	4	1/4	4	5/16
x 6.5	4	1/4	4	1/4
x 5	4	3/16	4	3/16
WT 3 x 12.5	3-1/4	5/16	6-1/8	7/16
x 10	3-1/8	1/4	6	3/8
x 7.5	3	1/4	6	1/4
x 8	3-1/8	1/4	4	3/8
x 6	3	1/4	4	1/4
x 4.5	3	3/16	4	3/16
WT 2.5x 9.5	2-5/8	1/4	5	7/16
x 8	2-1/2	1/4	5	3/8
WT 2 x 6.5	2-1/8	1/4	4	3/8

Structural Tees Cut from M Shapes Dimensions

Designation	Depth (in.)	Stem Thickness (in.)	Flange Width (in.)	Flange Thickness (in.)
MT 7 x 9	7	3/16	4	1/4
MT 6 x 5.9	6	3/16	3-1/8	1/4
MT 5 x 4.5	5	3/16	2-3/4	3/16
MT 4 x 3.25	4	1/8	2-1/4	3/16
MT 3 x10	3	1/4	6	3/8
x 2.2	3	1/8	1-7/8	3/16
MT 2.5x 9.45	2-1/2	5/16	5	7/16
MT 2 x 6.5	2	1/4	4	3/8

Structural Tees Cut from S Shapes Dimensions

Designation	Depth (in.)	Stem Thickness (in.)	Flange Width (in.)	Flange Thickness (in.)
ST 12 x60.5	12-1/4	13/16	8	1-1/16
x53	12-1/4	5/8	7-7/8	1-1/16
x50	12	3/4	7-1/4	7/8
x45	12	5/8	7-1/8	7/8
x40	12	1/2	7	7/8
ST 10 x48	10-1/8	13/16	7-1/4	15/16
x43	10-1/8	11/16	7	15/16
x37.5	10	5/8	6-3/8	13/16
x33	10	1/2	6-1/4	13/16
ST 9 x35	9	11/16	6-1/4	11/16
x27.35	9	7/16	6	11/16
ST 7.5x25	7-1/2	9/16	5-5/8	5/8
x 21.45	7-1/2	7/16	5-1/2	5/8
ST 6 x25	6	11/16	5-1/2	11/16
x20.4	6	7/16	5-1/4	11/16
x17.5	6	7/16	5-1/8	9/16
x15.9	6	3/8	5	9/16
ST 5 x17.5	5	5/8	5	1/2
x12.7	5	5/16	4-5/8	1/2
ST 4 x11.5	4	7/16	4-1/8	7/16
x 9.2	4	1/4	4	7/16
ST 3.5x10	3-1/2	7/16	3-7/8	3/8
x 7.65	3-1/2	1/4	3-5/8	3/8
ST 3 x 8.625	3	7/16	3-5/8	3/8
x 6.25	3	1/4	3-3/8	3/8
ST 2.5x 7.375	2-1/2	1/2	3-1/4	5/16
x 5	2-1/2	3/16	3	5/16
ST 2 x 4.75	2	5/16	2-3/4	5/16
x 3.85	2	3/16	2-5/8	5/16
ST 1.5x 3.75	1-1/2	3/8	2-1/2	1/4
x 2.85	1-1/2	3/16	2-3/8	1/4

Surface Areas and Box Areas
W Shapes
Square feet per foot of length

Designation	Case A	Case B	Case C	Case D
W 36x300	9.99	11.40	7.51	8.90
x280	9.95	11.30	7.47	8.85
x260	9.90	11.30	7.42	8.80
x245	9.87	11.20	7.39	8.77
x230	9.84	11.20	7.36	8.73
x210	8.91	9.93	7.13	8.15
x194	8.88	9.89	7.09	8.10
x182	8.85	9.85	7.06	8.07
x170	8.82	9.82	7.03	8.03
x160	8.79	9.79	7.00	8.00
x150	8.76	9.76	6.97	7.97
x135	8.71	9.70	6.92	7.92
W 33x241	9.42	10.70	7.02	8.34
x221	9.38	10.70	6.97	8.29
x201	9.33	10.60	6.93	8.24
x152	8.27	9.23	6.55	7.51
x141	8.23	9.19	6.51	7.47
x130	8.20	9.15	6.47	7.43
x118	8.15	9.11	6.43	7.39
W 30x211	8.71	9.97	6.42	7.67
x191	8.66	9.92	6.37	7.62
x173	8.62	9.87	6.32	7.57
x132	7.49	8.37	5.93	6.81
x124	7.47	8.34	5.90	6.78
x116	7.44	8.31	5.88	6.75
x108	7.41	8.28	5.84	6.72
x 99	7.37	8.25	5.81	6.68
W 27x178	7.95	9.12	5.81	6.98
x161	7.91	9.08	5.77	6.94
x146	7.87	9.03	5.73	6.89
x114	6.88	7.72	5.39	6.23
x102	6.85	7.68	5.35	6.18
x 94	6.82	7.65	5.32	6.15
x 84	6.78	7.61	5.28	6.11
W 24x162	7.22	8.30	5.25	6.33
x146	7.17	8.24	5.20	6.27
x131	7.12	8.19	5.15	6.22
x117	7.08	8.15	5.11	6.18
x104	7.04	8.11	5.07	6.14
x 94	6.16	6.92	4.81	5.56
x 84	6.12	6.87	4.77	5.52
x 76	6.09	6.84	4.74	5.49
x 68	6.06	6.80	4.70	5.45
x 62	5.57	6.16	4.54	5.13
x 55	5.54	6.13	4.51	5.10

Surface Areas and Box Areas
W Shapes
Square feet per foot of length

Designation	Case A	Case B	Case C	Case D
W 21x147	6.61	7.66	4.72	5.76
x132	6.57	7.61	4.68	5.71
x122	6.54	7.57	4.65	5.68
x111	6.51	7.54	4.61	5.64
x101	6.48	7.50	4.58	5.61
x 93	5.54	6.24	4.31	5.01
x 83	5.50	6.20	4.27	4.96
x 73	5.47	6.16	4.23	4.92
x 68	5.45	6.14	4.21	4.90
x 62	5.42	6.11	4.19	4.87
x 57	5.01	5.56	4.06	4.60
x 50	4.97	5.51	4.02	4.56
x 44	4.94	5.48	3.99	4.53
W 18x119	5.81	6.75	4.10	5.04
x106	5.77	6.70	4.06	4.99
x 97	5.74	6.67	4.03	4.96
x 86	5.70	6.62	3.99	4.91
x 76	5.67	6.59	3.95	4.87
x 71	4.85	5.48	3.71	4.35
x 65	4.82	5.46	3.69	4.32
x 60	4.80	5.43	3.67	4.30
x 55	4.78	5.41	3.65	4.27
x 50	4.76	5.38	3.62	4.25
x 46	4.41	4.91	3.52	4.02
x 40	4.38	4.88	3.48	3.99
x 35	4.34	4.84	3.45	3.95
W 16x100	5.28	6.15	3.70	4.57
x 89	5.24	6.10	3.66	4.52
x 77	5.19	6.05	3.61	4.47
x 67	5.16	6.01	3.57	4.43
x 57	4.39	4.98	3.33	3.93
x 50	4.36	4.95	3.30	3.89
x 45	4.33	4.92	3.27	3.86
x 40	4.31	4.89	3.25	3.83
x 36	4.28	4.87	3.23	3.81
x 31	3.92	4.39	3.11	3.57
x 26	3.89	4.35	3.07	3.53

Case A: Shape perimeter, minus one flange surface.
Case B: Shape perimeter.
Case C: Box perimeter, equal to one flange surface plus twice the depth.
Case D: Box perimeter, equal to two flange surfaces plus twice the depth.

Surface Areas and Box Areas W Shapes Square feet per foot of length				
Designation	Case A	Case B	Case C	Case D
W 14x730	7.61	9.10	5.23	6.72
x665	7.46	8.93	5.08	6.55
x605	7.32	8.77	4.94	6.39
x550	7.19	8.62	4.81	6.24
x500	7.07	8.49	4.68	6.10
x455	6.96	8.36	4.57	5.98
x426	6.89	8.28	4.50	5.89
x398	6.81	8.20	4.43	5.81
x370	6.74	8.12	4.36	5.73
x342	6.67	8.03	4.29	5.65
x311	6.59	7.94	4.21	5.56
x283	6.52	7.86	4.13	5.48
x257	6.45	7.78	4.06	5.40
x233	6.38	7.71	4.00	5.32
x211	6.32	7.64	3.94	5.25
x193	6.27	7.58	3.89	5.20
x176	6.22	7.53	3.84	5.15
x159	6.18	7.47	3.79	5.09
x145	6.14	7.43	3.76	5.05
x132	5.93	7.16	3.67	4.90
x120	5.90	7.12	3.64	4.86
x109	5.86	7.08	3.60	4.82
x 99	5.83	7.05	3.57	4.79
x 90	5.81	7.02	3.55	4.76
x 82	4.75	5.59	3.23	4.07
x 74	4.72	5.56	3.20	4.04
x 68	4.69	5.53	3.18	4.01
x 61	4.67	5.50	3.15	3.98
x 53	4.19	4.86	2.99	3.66
x 48	4.16	4.83	2.97	3.64
x 43	4.14	4.80	2.94	3.61
x 38	3.93	4.50	2.91	3.48
x 34	3.91	4.47	2.89	3.45
x 30	3.89	4.45	2.87	3.43
x 26	3.47	3.89	2.74	3.16
x 22	3.44	3.86	2.71	3.12

Surface Areas and Box Areas W Shapes Square feet per foot of length				
Designation	Case A	Case B	Case C	Case D
W 12x336	5.77	6.88	3.92	5.03
x305	5.67	6.77	3.82	4.93
x279	5.59	6.68	3.74	4.83
x252	5.50	6.58	3.65	4.74
x230	5.43	6.51	3.58	4.66
x210	5.37	6.43	3.52	4.58
x190	5.30	6.36	3.45	4.51
x170	5.23	6.28	3.39	4.43
x152	5.17	6.21	3.33	4.37
x136	5.12	6.15	3.27	4.30
x120	5.06	6.09	3.21	4.24
x106	5.02	6.03	3.17	4.19
x 96	4.98	5.99	3.13	4.15
x 87	4.95	5.96	3.10	4.11
x 79	4.92	5.93	3.07	4.08
x 72	4.89	5.90	3.05	4.05
x 65	4.87	5.87	3.02	4.02
x 58	4.39	5.22	2.87	3.70
x 53	4.37	5.20	2.84	3.68
x 50	3.90	4.58	2.71	3.38
x 45	3.88	4.55	2.68	3.35
x 40	3.86	4.52	2.66	3.32
x 35	3.63	4.18	2.63	3.18
x 30	3.60	4.14	2.60	3.14
x 26	3.58	4.12	2.58	3.12
x 22	2.97	3.31	2.39	2.72
x 19	2.95	3.28	2.36	2.69
x 16	2.92	3.25	2.33	2.66
x 14	2.90	3.23	2.32	2.65

Case A: Shape perimeter, minus one flange surface.
Case B: Shape perimeter.
Case C: Box perimeter, equal to one flange surface plus twice the depth.
Case D: Box perimeter, equal to two flange surfaces plus twice the depth.

Surface Areas and Box Areas
W Shapes
Square feet per foot of length

Designation	Case A	Case B	Case C	Case D
W 10x112	4.30	5.17	2.76	3.63
x100	4.25	5.11	2.71	3.57
x 88	4.20	5.06	2.66	3.52
x 77	4.15	5.00	2.62	3.47
x 68	4.12	4.96	2.58	3.42
x 60	4.08	4.92	2.54	3.38
x 54	4.06	4.89	2.52	3.35
x 49	4.04	4.87	2.50	3.33
x 45	3.56	4.23	2.35	3.02
x 39	3.53	4.19	2.32	2.98
x 33	3.49	4.16	2.29	2.95
x 30	3.10	3.59	2.23	2.71
x 26	3.08	3.56	2.20	2.68
x 22	3.05	3.53	2.17	2.65
x 19	2.63	2.96	2.04	2.38
x 17	2.60	2.94	2.02	2.35
x 15	2.58	2.92	2.00	2.33
x 12	2.56	2.89	1.98	2.31
W 8x 67	3.42	4.11	2.19	2.88
x 58	3.37	4.06	2.14	2.83
x 48	3.32	4.00	2.09	2.77
x 40	3.28	3.95	2.05	2.72
x 35	3.25	3.92	2.02	2.69
x 31	3.23	3.89	2.00	2.67
x 28	2.87	3.42	1.89	2.43
x 24	2.85	3.39	1.86	2.40
x 21	2.61	3.05	1.82	2.26
x 18	2.59	3.03	1.79	2.23
x 15	2.27	2.61	1.69	2.02
x 13	2.25	2.58	1.67	2.00
x 10	2.23	2.56	1.64	1.97
W 6x 25	2.49	3.00	1.57	2.08
x 20	2.46	2.96	1.54	2.04
x 15	2.42	2.92	1.50	2.00
x 16	1.98	2.31	1.38	1.72
x 12	1.93	2.26	1.34	1.67
x 9	1.90	2.23	1.31	1.64
W 5x 19	2.04	2.45	1.28	1.70
x 16	2.01	2.43	1.25	1.67
W 4x 13	1.63	1.96	1.03	1.37

Case A: Shape perimeter, minus one flange surface.
Case B: Shape perimeter.
Case C: Box perimeter, equal to one flange surface plus twice the depth.
Case D: Box perimeter, equal to two flange surfaces plus twice the depth.

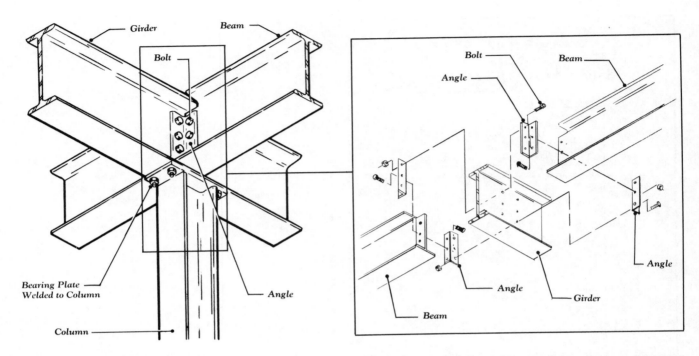

Roof Girder, Beam, Column Assembly

Steel Pipe, Tubing, Bars and Plates

Steel pipe is generally available in standard, extra strong, and double extra strong weights with a minimum yield stress value of 36 ksi.

Square and rectangular tubing is generally available in various sizes and wall thicknesses of cold-formed steel with a minimum yield stress value of 46 ksi.

Bars and plates are available in the ASTM specifications shown for hot-rolled shapes and minimum yields. Bars 6″ or less in width are generally classified .203″ and over in thickness. Bars 6″ to 8″ wide are classified .230″ and over in thickness.

Plates 8″ to 48″ in width are normally defined as over .203″ thick and plates over 48″ wide are defined as over .18″ thick.

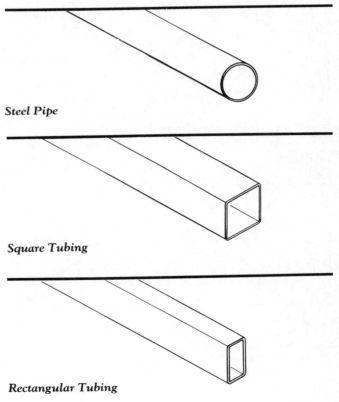

Steel Pipe

Square Tubing

Rectangular Tubing

Pipe Dimensions

Nominal Diameter (in.)	Outside Diameter (in.)	Inside Diameter (in.)	Wall Thickness (in.)
Standard Weight			
1/2	.840	.622	.109
3/4	1.050	.824	.113
1	1.315	1.049	.133
1-1/4	1.660	1.380	.140
1-1/2	1.900	1.610	.145
2	2.375	2.067	.154
2-1/2	2.875	2.469	.203
3	3.500	3.068	.216
3-1/2	4.000	3.548	.226
4	4.500	4.026	.237
5	5.563	5.047	.258
6	6.625	6.065	.280
8	8.625	7.981	.322
10	10.750	10.020	.365
12	12.750	12.000	.375
Extra Strong			
1/2	.840	.546	.147
3/4	1.050	.742	.154
1	1.315	.957	.179
1-1/4	1.660	1.278	.191
1-1/2	1.900	1.500	.200
2	2.375	1.939	.218
2-1/2	2.875	2.323	.276
3	3.500	2.900	.300
3-1/2	4.000	3.364	.318
4	4.500	3.826	.337
5	5.563	4.813	.375
6	6.625	5.761	.432
8	8.625	7.625	.500
10	10.750	9.750	.500
12	12.750	11.750	.500
Double-Extra Strong			
2	2.375	1.503	.436
2-1/2	2.875	1.771	.552
3	3.500	2.300	.600
4	4.500	3.152	.674
5	5.563	4.063	.750
6	6.625	4.897	.864
8	8.625	6.875	.875

Structural Tubing Rectangular Dimensions		
Nominal Size	Wall Thickness	Weight per Foot
(in.)	(in.)	(lb)
20 x 12	1/2	103.30
	3/8	78.52
	5/16	65.87
20 x 8	1/2	89.68
	3/8	68.31
	5/16	57.36
20 x 4	1/2	76.07
	3/8	58.10
	5/16	48.86
18 x 6	1/2	76.07
	3/8	58.10
	5/16	48.86
16 x 12	1/2	89.68
	3/8	68.31
	5/16	57.36
16 x 8	1/2	76.07
	3/8	58.10
	5/16	48.86
16 x 4	1/2	62.46
	3/8	47.90
	5/16	40.35
14 x 10	1/2	76.07
	3/8	58.10
	5/16	48.86
14 x 6	1/2	62.46
	3/8	47.90
	5/16	40.35
	1/4	32.63
14 x 4	1/2	55.66
	3/8	42.79
	5/16	36.10
	1/4	29.23
12 x 8	5/8	76.33
	1/2	62.46
	3/8	47.90
	5/16	40.35
	1/4	32.63
12 x 6	1/2	55.66
	3/8	42.79
	5/16	36.10
	1/4	29.23
	3/16	22.18
12 x 4	1/2	48.85
	3/8	37.69
	5/16	31.84
	1/4	25.82
	3/16	19.63
12 x 2	1/4	22.42
	3/16	17.08

Structural Tubing Rectangular Dimensions		
Nominal Size	Wall Thickness	Weight per Foot
(in.)	(in.)	(lb)
10 x 6	1/2	48.85
	3/8	37.69
	5/16	31.84
	1/4	25.82
	3/16	19.63
10 x 4	1/2	42.05
	3/8	32.58
	5/16	27.59
	1/4	22.42
	3/16	17.08
10 x 2	3/8	27.48
	5/16	23.34
	1/4	19.02
	3/16	14.53
8 x 6	1/2	42.05
	3/8	32.58
	5/16	27.59
	1/4	22.42
	3/16	17.08
8 x 4	1/2	35.24
	3/8	27.48
	5/16	23.34
	1/4	19.02
	3/16	14.53
8 x 3	3/8	24.93
	5/16	21.21
	1/4	17.32
	3/16	13.25
8 x 2	3/8	22.37
	5/16	19.08
	1/4	15.62
	3/16	11.97
7 x 5	1/2	35.24
	3/8	27.48
	5/16	23.34
	1/4	19.02
	3/16	14.53
7 x 4	3/8	24.93
	5/16	21.21
	1/4	17.32
	3/16	13.25
7 x 3	3/8	22.37
	5/16	19.08
	1/4	15.62
	3/16	11.97
6 x 4	1/2	28.43
	3/8	22.37
	5/16	19.08
	1/4	15.62
	3/16	11.97

Structural Tubing Rectangular Dimensions

Nominal Size (in.)	Wall Thickness (in.)	Weight per Foot (lb)
6 x 3	3/8	19.82
	5/16	16.96
	1/4	13.91
	3/16	10.70
6 x 2	3/8	17.27
	5/16	14.83
	1/4	12.21
	3/16	9.42
5 x 4	3/8	19.82
	5/16	16.96
	1/4	13.91
	3/16	10.70
5 x 3	1/2	21.63
	3/8	17.27
	5/16	14.83
	1/4	12.21
	3/16	9.42
5 x 2	5/16	12.70
	1/4	10.51
	3/16	8.15
4 x 3	5/16	12.70
	1/4	10.51
	3/16	8.15
4 x 2	5/16	10.58
	1/4	8.81
	3/16	6.87
3 x 2	1/4	7.11
	3/16	5.59

Structural Tubing Square Dimensions

Nominal Size (in.)	Wall Thickness (in.)	Weight per Foot (lb)
16 x 16	1/2	103.30
	3/8	78.52
	5/16	65.87
14 x 14	1/2	89.68
	3/8	68.31
	5/16	57.36
12 x 12	1/2	76.07
	3/8	58.10
	5/16	48.86
	1/4	39.43
10 x 10	5/8	76.33
	1/2	62.46
	3/8	47.90
	5/16	40.35
	1/4	32.63
8 x 8	5/8	59.32
	1/2	48.85
	3/8	37.69
	5/16	31.84
	1/4	25.82
	3/16	19.63
7 x 7	1/2	42.05
	3/8	32.58
	5/16	27.59
	1/4	22.42
	3/16	17.08
6 x 6	1/2	35.24
	3/8	27.48
	5/16	23.34
	1/4	19.02
	3/16	14.53
5 x 5	1/2	28.43
	3/8	22.37
	5/16	19.08
	1/4	15.62
	3/16	11.97
4 x 4	1/2	21.63
	3/8	17.27
	5/16	14.83
	1/4	12.21
	3/16	9.42
3.5 x 3.5	5/16	12.70
	1/4	10.51
	3/16	8.15
3 x 3	5/16	10.58
	1/4	8.81
	3/16	6.87
2.5 x 2.5	1/4	7.11
	3/16	5.59
2 x 2	1/4	5.41
	3/16	4.32

Weight of Rectangular Sections
Pounds per linear foot

Width In.	Thickness, Inches													
	3/16	1/4	5/16	3/8	7/16	1/2	9/16	5/8	11/16	3/4	13/16	7/8	15/16	1
1/4	0.16	0.21	0.27	0.32	0.37	0.43	0.48	0.53	0.58	0.64	0.69	0.74	0.80	0.85
1/2	0.32	0.43	0.53	0.64	0.74	0.85	0.96	1.06	1.17	1.28	1.38	1.49	1.60	1.70
3/4	0.48	0.64	0.80	0.96	1.12	1.28	1.44	1.60	1.75	1.91	2.07	2.23	2.39	2.55
1	0.64	0.85	1.06	1.28	1.49	1.70	1.91	2.13	2.34	2.55	2.76	2.98	3.19	3.40
1-1/4	0.80	1.06	1.33	1.60	1.86	2.13	2.39	2.66	2.92	3.19	3.46	3.72	3.99	4.25
1-1/2	0.96	1.28	1.60	1.91	2.23	2.55	2.87	3.19	3.51	3.83	4.15	4.47	4.79	5.10
1-3/4	1.12	1.49	1.86	2.23	2.61	2.98	3.35	3.72	4.09	4.47	4.84	5.21	5.58	5.95
2	1.28	1.70	2.13	2.55	2.98	3.40	3.83	4.25	4.68	5.10	5.53	5.95	6.38	6.81
2-1/4	1.44	1.91	2.39	2.87	3.35	3.83	4.31	4.79	5.26	5.74	6.22	6.70	7.18	7.66
2-1/2	1.60	2.13	2.66	3.19	3.72	4.25	4.79	5.32	5.85	6.38	6.91	7.44	7.98	8.51
2-3/4	1.75	2.34	2.92	3.51	4.09	4.68	5.26	5.85	6.43	7.02	7.60	8.19	8.77	9.36
3	1.91	2.55	3.19	3.83	4.47	5.10	5.74	6.38	7.02	7.66	8.29	8.93	9.57	10.2
3-1/4	2.07	2.76	3.46	4.15	4.84	5.53	6.22	6.91	7.60	8.29	8.99	9.68	10.4	11.1
3-1/2	2.23	2.98	3.72	4.47	5.21	5.95	6.70	7.44	8.19	8.93	9.68	10.4	11.2	11.9
3-3/4	2.39	3.19	3.99	4.79	5.58	6.38	7.18	7.98	8.77	9.57	10.4	11.2	12.0	12.8
4	2.55	3.40	4.25	5.10	5.95	6.81	7.66	8.51	9.36	10.2	11.1	11.9	12.8	13.6
4-1/4	2.71	3.62	4.52	5.42	6.33	7.23	8.13	9.04	9.94	10.8	11.8	12.7	13.6	14.5
4-1/2	2.87	3.83	4.79	5.74	6.70	7.66	8.61	9.57	10.5	11.5	12.4	13.4	14.4	15.3
4-3/4	3.03	4.04	5.05	6.06	7.07	8.08	9.09	10.1	11.1	12.1	13.1	14.1	15.2	16.2
5	3.19	4.25	5.32	6.38	7.44	8.51	9.57	10.6	11.7	12.8	13.8	14.9	16.0	17.0
5-1/4	3.35	4.47	5.58	6.70	7.82	8.93	10.0	11.2	12.3	13.4	14.5	15.6	16.7	17.9
5-1/2	3.51	4.68	5.85	7.02	8.19	9.36	10.5	11.7	12.9	14.0	15.2	16.4	17.5	18.7
5-3/4	3.67	4.89	6.11	7.34	8.56	9.78	11.0	12.2	13.5	14.7	15.9	17.1	18.3	19.6
6	3.83	5.10	6.38	7.66	8.93	10.2	11.5	12.8	14.0	15.3	16.6	17.9	19.1	20.4
6-1/4	3.99	5.32	6.65	7.98	9.30	10.6	12.0	13.3	14.6	16.0	17.3	18.6	19.9	21.3
6-1/2	4.15	5.53	6.91	8.29	9.68	11.1	12.4	13.8	15.2	16.6	18.0	19.4	20.7	22.1
6-3/4	4.31	5.74	7.18	8.61	10.0	11.5	12.9	14.4	15.8	17.2	18.7	20.1	21.5	23.0
7	4.47	5.95	7.44	8.93	10.4	11.9	13.4	14.9	16.4	17.9	19.4	20.8	22.3	23.8
7-1/4	4.63	6.17	7.71	9.25	10.8	12.3	13.9	15.4	17.0	18.5	20.0	21.6	23.1	24.7
7-1/2	4.79	6.38	7.98	9.57	11.2	12.8	14.4	16.0	17.5	19.1	20.7	22.3	23.9	25.5
7-3/4	4.94	6.59	8.24	9.89	11.5	13.2	14.8	16.5	18.1	19.8	21.4	23.1	24.7	26.4
8	5.10	6.81	8.51	10.2	11.9	13.6	15.3	17.0	18.7	20.4	22.1	23.8	25.5	27.2
8-1/2	5.42	7.23	9.04	10.8	12.7	14.5	16.3	18.1	19.9	21.7	23.5	25.3	27.1	28.9
9	5.74	7.66	9.57	11.5	13.4	15.3	17.2	19.1	21.1	23.0	24.9	26.8	28.7	30.6
9-1/2	6.06	8.08	10.1	12.1	14.1	16.2	18.2	20.2	22.2	24.2	26.3	28.3	30.3	32.3
10	6.38	8.51	10.6	12.8	14.9	17.0	19.1	21.3	23.4	25.5	27.6	29.8	31.9	34.0
10-1/2	6.70	8.93	11.2	13.4	15.6	17.9	20.1	22.3	24.6	26.8	29.0	31.3	33.5	35.7
11	7.02	9.36	11.7	14.0	16.4	18.7	21.1	23.4	25.7	28.1	30.4	32.8	35.1	37.4
11-1/2	7.34	9.78	12.2	14.7	17.1	19.6	22.0	24.5	26.9	29.3	31.8	34.2	36.7	39.1
12	7.66	10.2	12.8	15.3	17.9	20.4	23.0	25.5	28.1	30.6	33.2	35.7	38.3	40.8

Square and Round Bars Weight		
Size Inches	Weight Lb per Foot	
	■	●
0		
1/16	0.013	0.010
1/8	0.053	0.042
3/16	0.120	0.094
1/4	0.213	0.167
5/16	0.332	0.261
3/8	0.479	0.376
7/16	0.651	0.512
1/2	0.851	0.668
9/16	1.077	0.846
5/8	1.329	1.044
11/16	1.608	1.263
3/4	1.914	1.503
13/16	2.246	1.764
7/8	2.605	2.046
15/16	2.991	2.349
1	3.403	2.673
1/16	3.841	3.017
1/8	4.307	3.382
3/16	4.798	3.769
1/4	5.317	4.176
5/16	5.862	4.604
3/8	6.433	5.053
7/16	7.032	5.523
1/2	7.656	6.013
9/16	8.308	6.525
5/8	8.985	7.057
11/16	9.690	7.610
3/4	10.421	8.185
13/16	11.179	8.780
7/8	11.963	9.396
15/16	12.774	10.032
2	13.611	10.690
1/16	14.475	11.369
1/8	15.366	12.068
3/16	16.283	12.788
1/4	17.227	13.530
5/16	18.197	14.292
3/8	19.194	15.075
7/16	20.217	15.879
1/2	21.267	16.703
9/16	22.344	17.549
5/8	23.447	18.415
11/16	24.577	19.303
3/4	25.734	20.211
13/16	26.917	21.140
7/8	28.126	22.090
15/16	29.362	23.061

Square and Round Bars Weight		
Size Inches	Weight Lb per Foot	
	■	●
3	30.63	24.05
1/16	31.91	25.07
1/8	33.23	26.10
3/16	34.57	27.15
1/4	35.94	28.23
5/16	37.34	29.32
3/8	38.76	30.44
7/16	40.21	31.58
1/2	41.68	32.74
9/16	43.19	33.92
5/8	44.71	35.12
11/16	46.27	36.34
3/4	47.85	37.58
13/16	49.46	38.85
7/8	51.09	40.13
15/16	52.76	41.43
4	54.44	42.76
1/16	56.16	44.11
1/8	57.90	45.47
3/16	59.67	46.86
1/4	61.46	48.27
5/16	63.28	49.70
3/8	65.13	51.15
7/16	67.01	52.63
1/2	68.91	54.12
9/16	70.83	55.63
5/8	72.79	57.17
11/16	74.77	58.72
3/4	76.78	60.30
13/16	78.81	61.90
7/8	80.87	63.51
15/16	82.96	65.15
5	85.07	66.81
1/16	87.21	68.49
1/8	89.38	70.20
3/16	91.57	71.92
1/4	93.79	73.66
5/16	96.04	75.43
3/8	98.31	77.21
7/16	100.61	79.02
1/2	102.93	80.84
9/16	105.29	82.69
5/8	107.67	84.56
11/16	110.07	86.45
3/4	112.50	88.36
13/16	114.96	90.29
7/8	117.45	92.24
15/16	119.96	94.22

Square and Round Bars Weight

Size Inches	Weight Lb per Foot ■	Weight Lb per Foot ●
6	122.50	96.21
1/16	125.07	98.23
1/8	127.66	100.26
3/16	130.28	102.32
1/4	132.92	104.40
5/16	135.59	106.49
3/8	138.29	108.61
7/16	141.02	110.75
1/2	143.77	112.91
9/16	146.55	115.10
5/8	149.35	117.30
11/16	152.18	119.52
3/4	155.04	121.77
13/16	157.92	124.03
7/8	160.83	126.32
15/16	163.77	128.63
7	166.74	130.95
1/16	169.73	133.30
1/8	172.74	135.67
3/16	175.79	138.06
1/4	178.86	140.48
5/16	181.96	142.91
3/8	185.08	145.36
7/16	188.23	147.84
1/2	191.41	150.33
9/16	194.61	152.85
5/8	197.84	155.38
11/16	201.10	157.94
3/4	204.38	160.52
13/16	207.69	163.12
7/8	211.03	165.74
15/16	214.39	168.38
8	217.78	171.04
1/16	221.19	173.73
1/8	224.64	176.43
3/16	228.11	179.15
1/4	231.60	181.90
5/16	235.12	184.67
3/8	238.67	187.45
7/16	242.25	190.26
1/2	245.85	193.09
9/16	249.48	195.94
5/8	253.13	198.81
11/16	256.82	201.70
3/4	260.53	204.62
13/16	264.26	207.55
7/8	268.02	210.50
15/16	271.81	213.48

Square and Round Bars Weight

Size Inches	Weight Lb per Foot ■	Weight Lb per Foot ●
9	275.63	216.48
1/16	279.47	219.49
1/8	283.33	222.53
3/16	287.23	225.59
1/4	291.15	228.67
5/16	295.10	231.77
3/8	299.07	234.89
7/16	303.07	238.03
1/2	307.10	241.20
9/16	311.15	244.38
5/8	315.24	247.59
11/16	319.34	250.81
3/4	323.48	254.06
13/16	327.64	257.33
7/8	331.82	260.61
15/16	336.04	263.92
10	340.28	267.25
1/16	344.54	270.60
1/8	348.84	273.98
3/16	353.16	277.37
1/4	357.50	280.78
5/16	361.88	284.22
3/8	366.28	287.67
7/16	370.70	291.15
1/2	375.16	294.65
9/16	379.64	298.17
5/8	384.14	301.70
11/16	388.67	305.26
3/4	393.23	308.84
13/16	397.82	312.45
7/8	402.43	316.07
15/16	407.07	319.71
11	411.74	323.38
1/16	416.43	327.06
1/8	421.15	330.77
3/16	425.89	334.49
1/4	430.66	338.24
5/16	435.46	342.01
3/8	440.29	345.80
7/16	445.14	349.61
1/2	450.02	353.44
9/16	454.92	357.30
5/8	459.85	361.17
11/16	464.81	365.06
3/4	469.80	368.98
13/16	474.81	372.91
7/8	479.84	376.87
15/16	484.91	380.85
12	490.00	384.85

Wire and Sheet Metal Gauges In Comparison to Fractions and Decimals of an Inch

Fractions of an Inch	Gauge Number	U.S. Standard Gauge for Uncoated Hot and Cold Rolled Sheets — Decimals of an Inch	Galvanized Sheet Gauge for Hot Dipped Zinc Coated Sheets — Decimals of an Inch
	28	.0149	.0187
1/64″		.015625	
	27	.0164	.0202
	26	.0179	.0217
	25	.0209	.0247
	24	.0239	.0276
	23	.0269	.0306
	22	.0299	.0336
1/32″		.03125	
	21	.0329	.0366
	20	.0359	.0396
	19	.0418	.0456
3/64″		.046875	
	18	.0478	.0516
	17	.0538	.0575
	16	.0598	.0635
1/16″		.0625	
	15	.0673	.0710
	14	.0747	.0785
5/64″		.078125	
	13	.0897	.0934
3/32″		.09375	
	12	.1046	.1084
7/64″		.109375	
	11	.1196	.1233
1/8″		.125	
	10	.1345	.1382
9/64″		.140625	
	9	.1495	.1532
5/32″		.15625	
	8	.1644	.1681
11/64″		.171875	
	7	.1793	
3/16″		.1875	
	6	.1943	
13/64″		.203125	
	5	.2092	
7/32″		.21875	
	4	.2242	
15/64″		.234375	
	3	.2391	
1/4″		.250	

Connections are commonly provided by bolting, riveting, welding, or by a combination of shop-welded and field-bolted.

Bolts may be common (ASTM A307) or high strength (ASTM A325 or A490). High-strength bolts may be specified with friction-type connections or with bearing-type connections with threads included or excluded from the shear plane.

Connection details vary with the type and number of connectors. Some commonly used types are as follows:

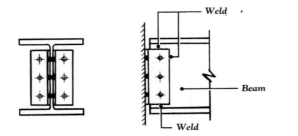

Framed Beam Connection - Bolted

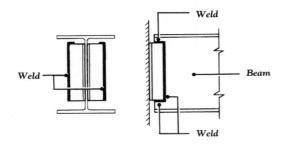

Framed Beam Connection - Welded

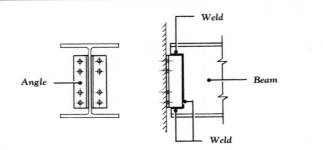

Framed Beam Connection - Combination

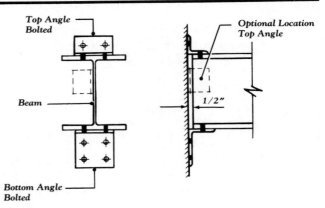

Seated Beam Connection - Bolted

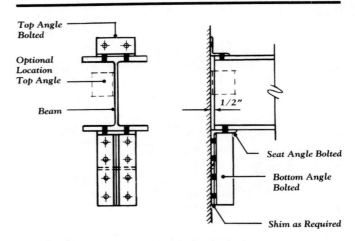

Stiffened Seated Beam Connection - Bolted

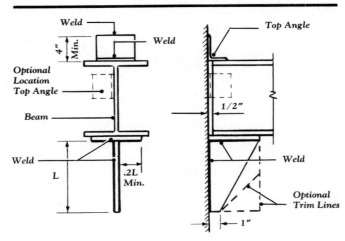

Stiffened Seated Beam Connection - Welded

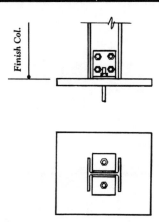

Column Base Plate

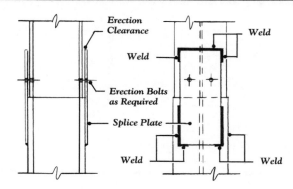

Column Splice - Welded

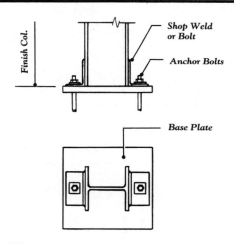

Shop Weld
or Bolt

Anchor Bolts

Base Plate

Column Base Plate

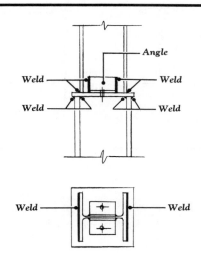

Angle

Weld — Weld

Weld — Weld

Weld — Weld

Column Butt Splice - Welded

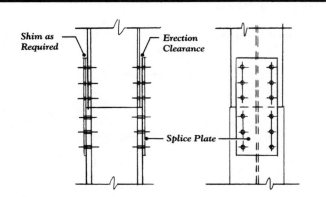

Shim as
Required

Erection
Clearance

Splice Plate

Column Splice - Bolted

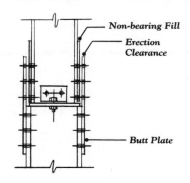

Non-bearing Fill

Erection
Clearance

Butt Plate

Column Butt Splice - Bolted

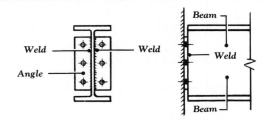

End Plate Shear Connection - Bolted

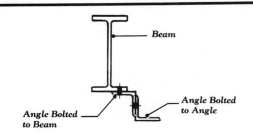

Zee Connection

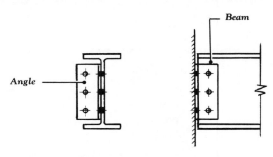

One Sided Connection - Bolted

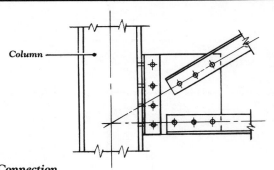

Truss Connection

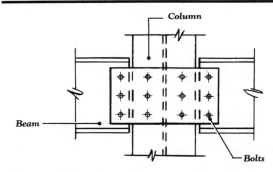

Symmetrical Beam to Column Connection

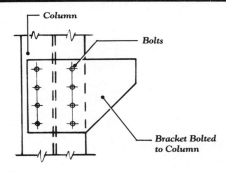

Bracket Plate Connection

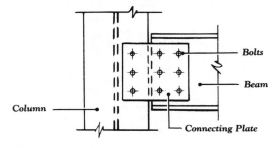

Beam to Column Connection

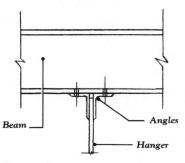

Hanger Type Connection

78

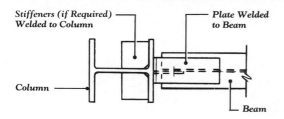

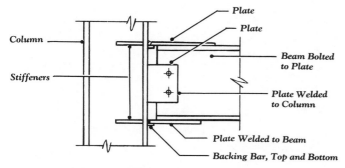

Moment Connection - Welded

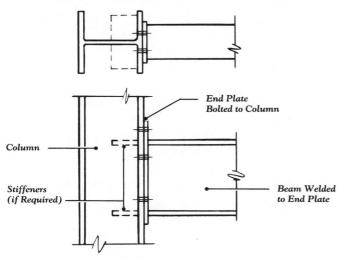

Moment Connection - End Plate

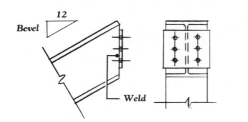

Sloped Beam Connection

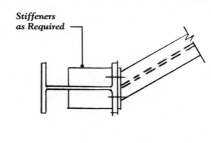

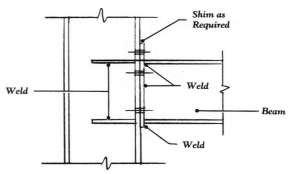

Skewed Beam Connection

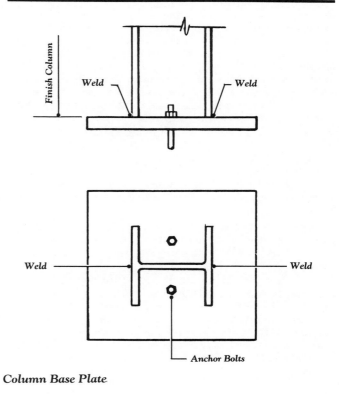

Column Base Plate

3 SUPERSTRUCTURE
COMPOSITE WF BEAMS, DECK, AND SLAB

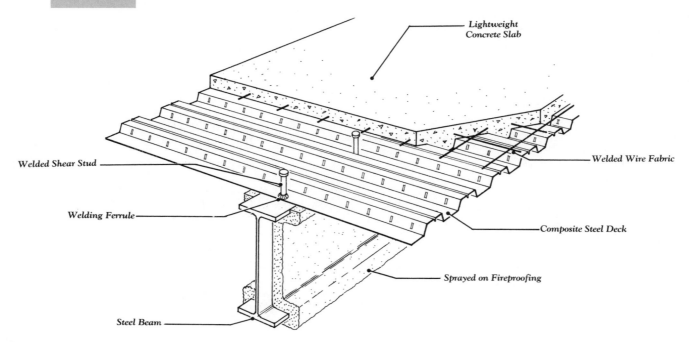

Composite Beam, Deck, and Slab

Composite construction, as applied to floor or roof systems, consists of steel beams and girders with shear connectors (metal studs) welded to the top flange of the beam and encased by a concrete slab. This system causes the concrete slab and the steel beam or girder to act as a unit. Because the connectors are sized and spaced to resist horizontal shear between the steel beam and the concrete slab, the effective depth of the system is increased. The concrete slab may also be of composite design by utilizing a composite steel deck for formwork.

In some designs the steel beams require temporary shoring until the concrete has attained a specified compressive strength. In addition, some composite beams require partial cover plates (a section of steel plate welded to the bottom flange of the beam) to increase the effective tension area of the beam and to balance the effective compressive area of the concrete slab.

Because a composite beam has greater stiffness than a non-composite beam of equal size, it is an effective system to use for heavy loading, relatively long span, and wide beam spacing.

The concrete slab depth usually ranges from 4″ to 5-1/2″ and may be made from regular or lightweight concrete. The use of 4″ lightweight concrete usually conforms to required fire-resistance codes for concrete decks, without using sprayed on fireproofing on the metal deck form. Shear connectors are generally welded studs of 3/4″ or 7/8″ applied to the beams after erection, for safety reasons, with a stud welder.

Man-hours

Description	m/hr	Unit
WF Beams Average	4.000	ton
Shear Studs	.016	Ea.
Non-cellular Composite Deck 2″ Deep		
22 Gauge	.008	sf
18 and 20 Gauge	.009	sf
16 Gauge	.010	sf
3″ Deep 22 Gauge	.010	sf
18 and 20 Gauge	.011	sf
16 Gauge	.012	sf

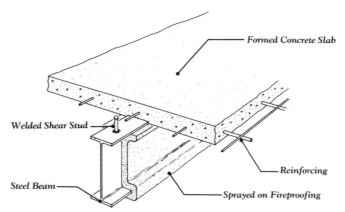

Formed in Place Concrete Slab

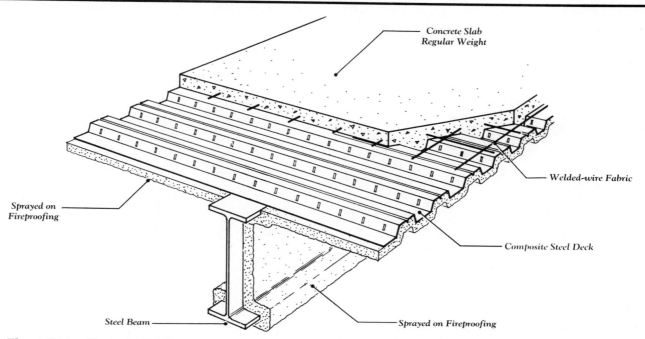

Concrete Slab
Regular Weight

Welded-wire Fabric

Sprayed on
Fireproofing

Composite Steel Deck

Steel Beam

Sprayed on Fireproofing

Wide Flange Beam, Composite Deck, and Slab

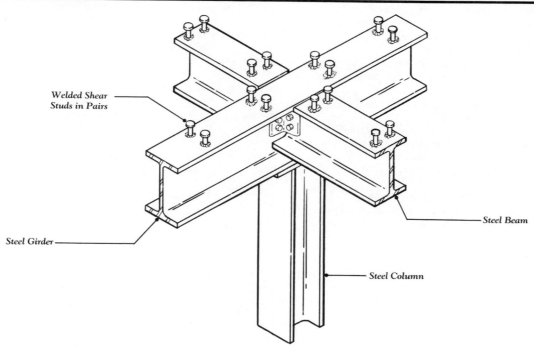

Welded Shear
Studs in Pairs

Steel Beam

Steel Girder

Steel Column

Paired Shear Studs on Beam and Girder

3 SUPERSTRUCTURE
STEEL FLOOR DECK

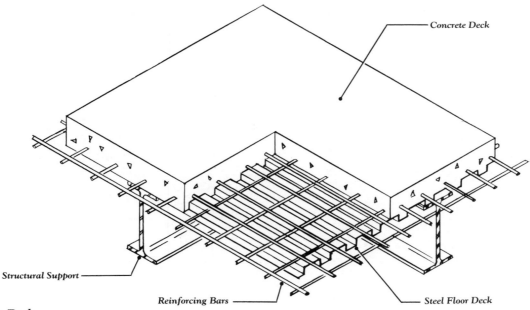

Concrete Deck

Structural Support

Reinforcing Bars

Steel Floor Deck

Steel Floor Deck

Steel floor deck may be composite or non-composite. In section it may be cellular or non-cellular. When suitably fastened, the steel deck acts as a working platform for the various trades, provides decking as required by OSHA code, and provides the form for the concrete deck.

Composite-steel floor deck is cold-formed steel that acts as the permanent form and as the positive bending reinforcement for the concrete slab. It is normally available in 14, 16, 18, 20, and 22 gauge and is galvanized so as to last the life of the structure.

Composite decks vary in depth from 1-1/2" to 3" and have cover widths of 12", 24", 30", and 36". They are available from some manufacturers in a form that allows blending of cellular and non-cellular decks to provide raceways in the floor. Composite cellular deck is available in different gauge steel for the formed section and the flat cover plate.

The deck should be erected in accordance with an approved manufacturer's drawing and should span three or more supports, where practical. Deck should be attached to all supporting members, including bearing walls, with a minimum 5/8" puddle weld at a maximum spacing of 12" and at side laps by welds or screws at a maximum spacing of 3". The minimum compressive strength of the concrete used with composite decks should be no less than 3,000 psi. Admixtures containing chloride salts should not be used. Wire mesh or two-way reinforcing bars should be installed near the top of the slab for temperature and crack control.

Slab form or form deck is formed steel centering for concrete slabs. When suitably fastened to the supporting members, the deck provides a working platform for the various trades and provides the cover required by OSHA code.

Form deck is normally available galvanized, uncoated, or painted with one coat of primer in 22, 24, 26, and 28 gauge steel. Some available depths include 9/16", 19/32", 1-5/16", 1-1/2", and 2". Coverage widths include 27" and 30".

Form deck should be welded to the supporting steel immediately after alignment using welding washers for all decks less than 22 gauge in thickness. Deck ends should be lapped a minimum of 2″ over a support. Welds should be pattened in accord with the following drawings. For spans greater than 5′, side laps should be fastened by welding, crimping, or mechanical fasteners.

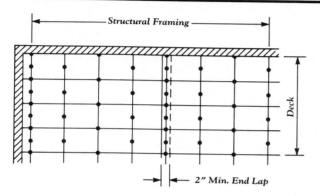

Pattern A, for Deck Spans up to 4′-6″

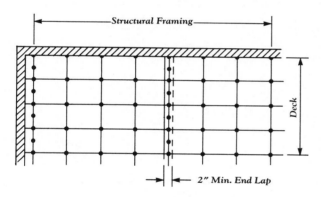

Pattern B, for Deck Spans up to 4′-6″

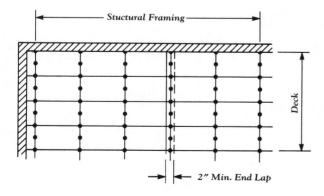

Pattern C, for Deck Spans from 4′-6″ to 8′-0″

Non Composite Form Deck Welding Patterns

To prevent concrete leakage at side laps, it is suggested that concrete be placed in the opposite direction to which the sheets are erected, so the concrete will flow away from the lap. Slabs are normally reinforced with welded-wire fabric that may be draped over beam bolsters, or placed over the supports, for slabs of 3″ or thicker.

Inverted, wide-rib roof deck may also be used for non-composite floor forming. The deck should be welded to the structural supports using 5/8″-diameter puddle welds, spaced no more than 12″ on center. Side laps should be fastened at mid span by welding or with screws. The deck acts only as a form. Slabs must be reinforced for flexure and temperature, as in a conventionally formed one-way slab.

Many manufacturers can supply accessories that may be used with the deck, including metal edge forms for various slab thicknesses, hangar tabs, piercing hangar tabs, column closures, and rubber end closures that seal the flutes of the deck.

Man-hours

Description	m/hr	Unit
Metal Decking		
Open Type, Galvanized 1-1/2″ Deep		
22 Gauge	.007	sf
20 Gauge or 18 Gauge	.008	sf
3″ Deep		
22 Gauge or 20 Gauge	.009	sf
18 Gauge	.010	sf
16 Gauge	.011	sf
4-1/2″ Deep Long Span		
20 Gauge	.012	sf
18 Gauge	.013	sf
16 Gauge	.014	sf
6″ Deep Long Span		
18 Gauge	.016	sf
16 Gauge or 14 Gauge	.017	sf
7-1/2″ Deep Long Span		
18 Gauge	.019	sf
16 Gauge	.020	sf
Non-cellular Composite Deck Galvanized 2″ Deep		
22 Gauge	.008	sf
20 Gauge or 18 Gauge	.009	sf
16 Gauge	.010	sf
3″ Deep		
22 Gauge	.010	sf
20 Gauge or 18 Gauge	.011	sf
16 Gauge	.012	sf
Steel Slab Form Uncoated or Galvanized	.006	sf
Sheet Metal Edge Closure Form 12″ Wide with 2 Bends	.022	lf
Welded Wire Fabric 6 x 6-W1.4/W1.4	.457	CSF

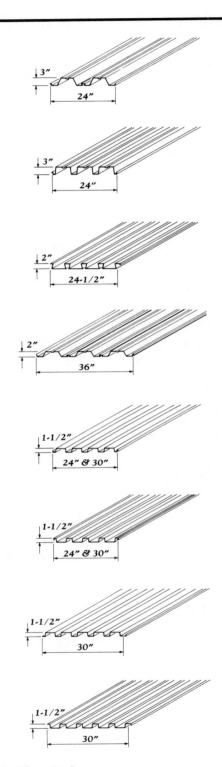

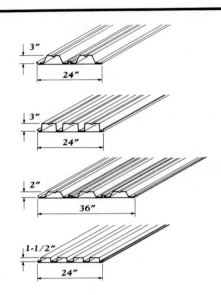

Cellular Deck

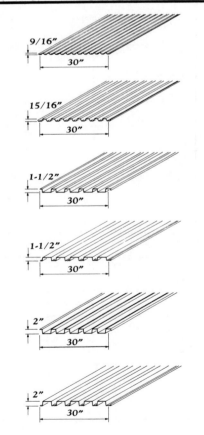

Composite Floor Deck

Form Deck

Open-web joists are parallel chord members suitable for the support of floors and roof decks. These joists are normally manufactured in various classifications: H and K series, in spans of 8' to 60' and depths of 8" to 30"; LH series, in spans of 25' to 96' and depths of 18" to 48"; and DLH series in spans of 89' to 144' and depths of 52" to 72". H and LH series joists are suitable for the support of floors and roof decks; DLH series are suitable for the support of roof decks only. Joists are manufactured of high-strength steels in accordance with specifications adopted by the Steel Joist Institute. Joist designations are as follows:

18 H 7 are 18" deep, H series, 7 chord size
(approximate weight 10.4 lbs per ft)
40 LH 11 are 40" deep, LH series, 11 chord size
(approximate weight 27 lbs per ft)
60 DLH 13 are 60" deep, DLH series, 13 chord size
(approximate weight 36 lbs per ft)

Weights of joists are usually not shown on drawings, but are available in manufacturers' catalogs. This may be important to the estimator, since joists are generally priced by the pound, or ton.

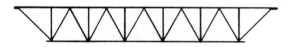

Parallel Chords, Underslung

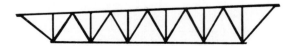

Top Chord Single Pitched, Underslung

Top Chord Double Pitched, Underslung

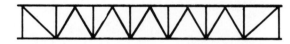

Parallel Chords, Square Ends

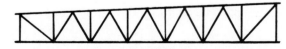

Top Chord Single Pitched, Square Ends

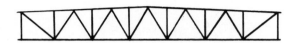

Top Chord Double Pitched, Square Ends

Standard Joist Details

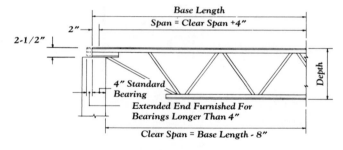

H & K Series

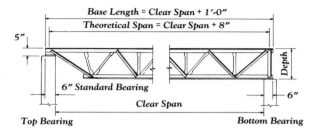

LH and DLH Series

Standard Joist Details

Longspan joists can be furnished with either underslung or square ends with parallel chords, or with single- or double-pitched top chords to provide for roof drainage. The standard pitch of a pitched joist is 1/8" per foot.

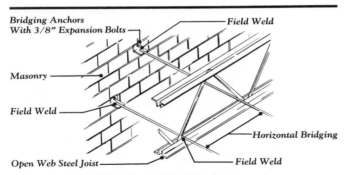

Short Span Joist Horizontal Bridging

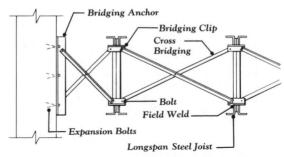

Bolted Cross Bridging - LH and DLH Series (Alternate)

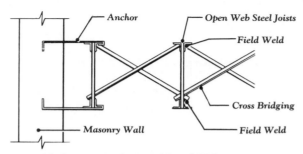

Welded Cross Bridging - H and K Series

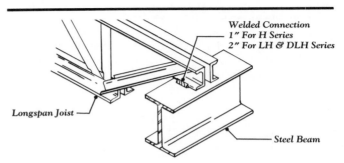

Welded Connection to Steel

Joists require bridging to brace against lateral movement during the construction period and to maintain their position. The bridging for H and K series may be constructed of horizontal members welded or mechanically fastened to the top and bottom chords. Diagonal bridging is usually fabricated from steel angles welded or bolted to the top and bottom chords.

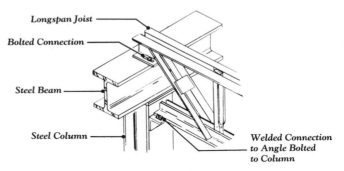

Bolted Connection to Steel

Anchorage to Masonry (Alternates) - H and K Series

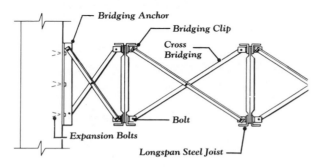

Bolted Cross Bridging — LH and DLH Series

Bridging for LH and DLH series joists consist of cross bracing, welded or bolted to the joists. Cross bridging members are usually angles.

Joists are anchored to steel by welding in all but the following case. The Occupational Safety and Health Administration Standard (OSHA), Paragraph 1910.12 refers to Paragraph 1518.751 of "Construction Standards," which states:

"In steel framing, where bar joists are utilized, and columns are not framed in at least two directions with structural steel members, bar joists shall be field-bolted at columns to provide lateral stability during construction."

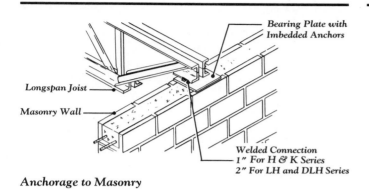

Bearing Plate with
Imbedded Anchors

Longspan Joist

Masonry Wall

Welded Connection
1" For H & K Series
2" For LH and DLH Series

Anchorage to Masonry

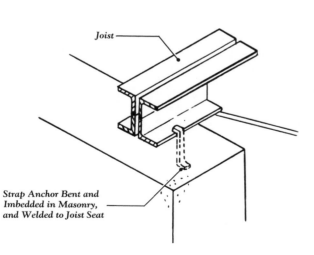

Joist

Bar Anchor Imbedded
in Masonry

Joist

Strap Anchor Bent and
Imbedded in Masonry,
and Welded to Joist Seat

Anchorage to Masonry (Alternates) - H and K Series

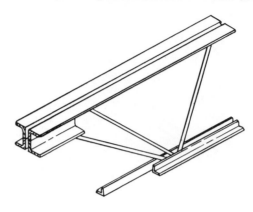

Ceiling Extension

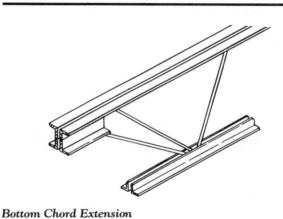

Bottom Chord Extension

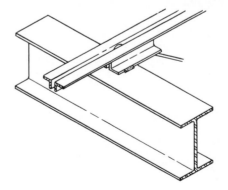

Top Chord Extension

Joists anchored to masonry may be welded to an imbedded anchored plate or by the use of masonry anchors for H and K series joists.

Joists may be ordered with ceiling extensions (extensions of the bottom chord from the first web member to the joist end) to allow attachment of ceiling furring.

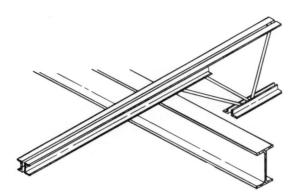

Extended Ends

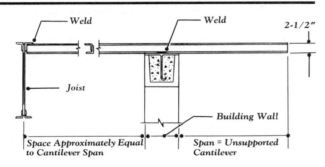

Loose Outriggers

Joist floor systems include the following types: concrete slabs on steel-slabform, installed on the joists by welding or with self-taping screws; precast plank sytems of lightweight concrete or gypsum, usually in conjunction with concrete fill; and plywood installed on the joist with screws or power-activated fasteners.

Roof systems include steel deck, bulb T's, formboards and lightweight fill, or plywood deck.

Some advantages of open-web joists include lightweight all-weather construction, long spans for interior layout flexibility, fast erection, and open construction for the mechanical and electrical components used in the building.

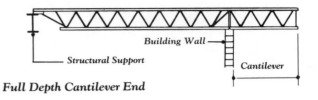

Full Depth Cantilever End

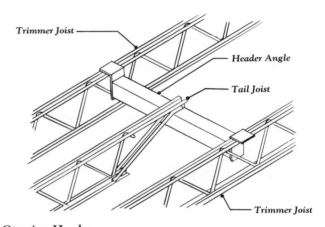

Opening Header

Man-hours

Description	m/hr	Unit
Open Web Joists H and K Series Horizontal		
Bridging Span to 30'	6.667	ton
Span 30' to 50'	6.353	ton
LH Series Bolted Cross Bridging		
Span to 96'	6.154	ton
DLH Series Bolted Cross Bridging		
Span to 144'	6.154	ton
Joist Girders	6.154	ton
Trusses Factory Fabricated WT Chords	7.273	ton

3 SUPERSTRUCTURE
STEEL ROOF DECK

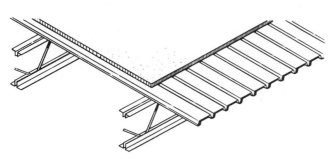

Roof Deck System with Insulation

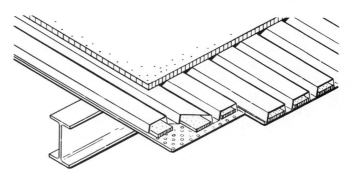

Acoustic Deck System

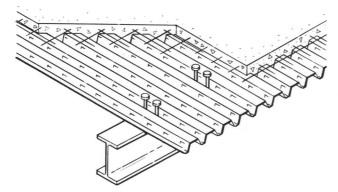

Composite Beam, Deck and Slab

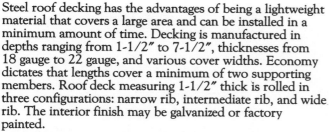

Cellular Deck System

Steel roof decking has the advantages of being a lightweight material that covers a large area and can be installed in a minimum amount of time. Decking is manufactured in depths ranging from 1-1/2″ to 7-1/2″, thicknesses from 18 gauge to 22 gauge, and various cover widths. Economy dictates that lengths cover a minimum of two supporting members. Roof deck measuring 1-1/2″ thick is rolled in three configurations: narrow rib, intermediate rib, and wide rib. The interior finish may be galvanized or factory painted.

Steel deck stored on site should be blocked off the ground with one end higher than the other to provide drainage. They should be covered with ventilated waterproof material for protection from the elements. The deck may be welded to supporting members, including bearing walls, or fastened with screws. Welds are made from the topside of the deck with puddle welds at least 1/2″ diameter and fillet welds at least 1″ long. Screws should be a minimum size No. 12. Spacing of welds or screws are as follows for all widths of deck: all side laps plus a sufficient number of interior ribs to limit the spacing between adjacent points of attachment to 18″. For spans greater than 5′, side laps shall be fastened between supports, center-to-center, at a maximum spacing of 3′.

Deck sheets should be placed in accordance with an approved erection-layout drawing supplied by the deck manufacturer and in accordance with the deck manufacturers standards. Roofs having a slope of 1/4″ in

12″ or more should be erected from the low side up to produce a shingle effect. The ends of the sheet should lap a minimum of 2″ and be located over a support.

The deck erector normally cuts openings in the roof deck which are shown on the erection drawings and are less than 16 square feet in area, as well as skew cuts. Openings for stacks, conduits, vents, etc. should be cut (and reinforced if necessary) by the trades requiring the openings.

Man-hours

Description	m/hr	Unit
Roof Deck Open Type		
1-1/2″ Deep		
22 Gauge	.007	sf
18 and 20 Gauge	.008	sf
3″ Deep		
22 and 20 Gauge	.009	sf
18 Gauge	.010	sf
16 Gauge	.011	sf
4-1/2″ Deep		
20 Gauge	.012	sf
6″ Deep		
18 Gauge	.016	sf
7-1/2″ Deep		
18 Gauge	.019	sf
Roof Deck Cellular Units Galv.		
3″ Deep 20-20 Gauge	.023	sf
4-1/2″ Deep 20-18 Gauge	.029	sf

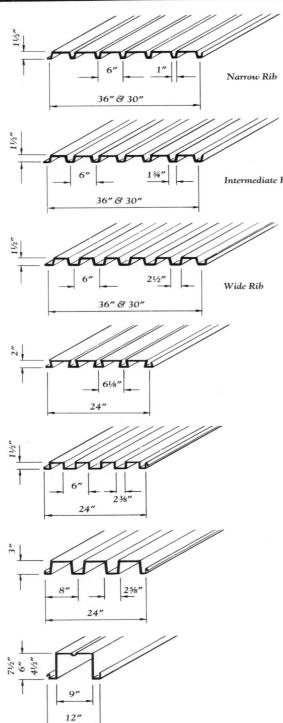

Narrow Rib

1½"   6"   1"   36" & 30"

Intermediate Rib

1½"   6"   1¾"   36" & 30"

Wide Rib

1½"   6"   2½"   36" & 30"

2"   6⅛"   24"

1½"   6"   2⅜"   24"

3"   8"   2⅝"   24"

7½" 6" 4½"   9"   12"

Roof Deck

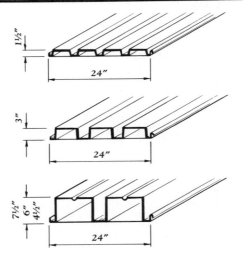

1½"   24"

3"   24"

7½" 6" 4½"   24"

Acoustic Deck

1½"   24"

3"   24"

7½" 6" 4½"   24"

Cellular Deck

3 SUPERSTRUCTURE
CONCRETE FLOORS ON SLAB FORM, STEEL JOISTS, COLUMNS, BEAMS, OR BEARING WALLS

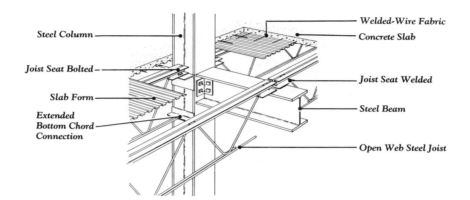

Steel Floor Systems

Concrete floors on slab form or centering, joists, or beams and columns (or columns and bearing walls) is a widely used structural system because of its light weight, fast erection time, and flexibility in bay sizes. The joists are normally welded to the steel beams or girders and are attached to bearing walls with masonry anchors, or welded to preset embedded plates. When joists are attached to columns and form a part of the framing system, OSHA requires a bolted connection for both the top and bottom chords. Concrete slabs are normally placed over steel forms attached to the joists by welding or by self-tapping screws. In multistoried buildings the deck is normally installed immediately following joist attachment to provide a working platform and to satisfy safety requirements.

Slabs are normally reinforced with welded-wire fabric. They may be also placed over a centering of lightweight tongue and groove planks, channel slabs, or hollow slabs. They are attached by welding where the planks are metal edged, or by clips when no welding attachments are included.

Plywood installed directly on the joists creates a very economical sub-flooring system and working deck. New developments in fastening systems make connection of the plywood to the steel joists fast and economical.

The system may be fire rated by using a rated suspended ceiling of acoustical tile or gypsum board, or a plaster ceiling on metal lath.

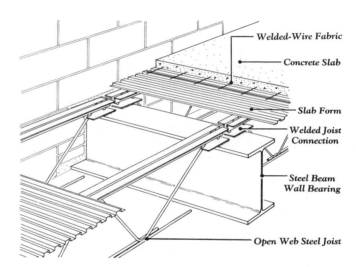

Deck and Joists on Steel Beams

Man-hours

Description	m/hr	Unit
Steel Joists H and K Series		
Horizontal Bridging to 30' Span	6.667	ton
30' to 50' Span (Includes One Row of Bolted Cross Bridging for Spans Over 40' Where Required)	6.353	ton
LH Series Bolted Cross Bridging Spans to 96'	6.154	ton
DLH Series Bolted Cross Bridging Spans to 144' Shipped in 2 Pieces	6.154	ton
Slab Form Steel		
28 Gauge 9/16" Deep to 24 Gauge 1-5/16" Deep	.006	sf
Welded-Wire Fabric Rolls 6 x 6-10/10	.457	CSF
2-1/2" Thick Concrete Floor Fill Including Finish	.021	sf

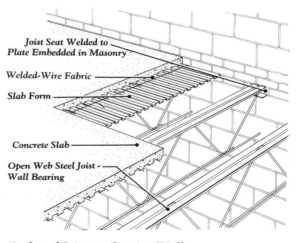

Deck and Joists on Bearing Walls

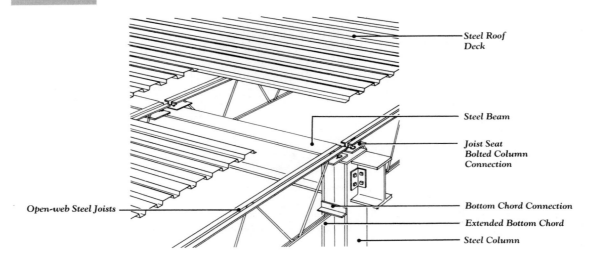

Steel Roof Deck

Steel Beam

Joist Seat Bolted Column Connection

Bottom Chord Connection

Extended Bottom Chord

Steel Column

Open-web Steel Joists

Steel roof deck on open-web joists is a widely used roof-support system due to its light weight, fast erection time, and span flexibility. The deck is normally attached to the joists by welding or by self-drilling, self-tapping screws. For economy, joist spacing should be close to the maximum allowable span for the deck used. Spans may be limited by construction and maintenance loads or by insurance regulations. When the deck bearing terminates at a bearing wall, the deck specifications require a positive attachment at the wall, such as a continuous embedded plate for welding or other suitable attachment devices. Steel deck is typically covered by rigid insulation boards, but may also be insulated with vermiculite, perlite, or cellular types of roof fill which are then covered by the roofing material.

Forming roof decks over open-web joists can also be accomplished using bulb T's, formboards and gypsum fill, structural wood fiber planks, or plywood. The roof deck system may be fire-rated with the use of a rated, suspended acoustical-ceiling system, gypsum board or plaster.

Man-hours

Description	m/hr	Unit
Columns Steel Concrete Filled		
4" Diameter	.072	lf
5" Diameter	.055	lf
6-5/8" Diameter	.047	lf
Steel Pipe		
6" Diameter	6.000	ton
12" Diameter	2.000	ton
Structural Tubing		
6" x 6"	6.000	ton
10" x 10"	2.000	ton
Wide Flange		
W8 x 31	3.355	ton
W10 x 45	2.412	ton
W12 x 50	2.171	ton
W14 x 74	1.538	ton
Beams WF Average	4.000	ton
Steel Joists H and K Series Horizontal Bridging		
To 30' Span	6.667	ton
30' to 50' Span	6.353	ton
(Includes One Row of Bolted Cross Bridging for Spans Over 40' Where Required)		
LH Series Bolted Cross Bridging		
Spans to 96'	6.154	ton
DLH Series Bolted Cross Bridging		
Spans to 144' Shipped in 2 Pieces	6.154	ton
Metal Decking Open Type		
1-1/2" Deep		
22 Gauge	.007	sf
18 and 20 Gauge	.008	sf
3" Deep		
20 and 22 Gauge	.009	sf
18 Gauge	.010	sf
16 Gauge	.011	sf
4-1/2" Deep		
20 Gauge	.012	sf
18 Gauge	.013	sf
16 Gauge	.014	sf
7-1/2 Deep		
18 Gauge	.019	sf
16 Gauge	.020	sf

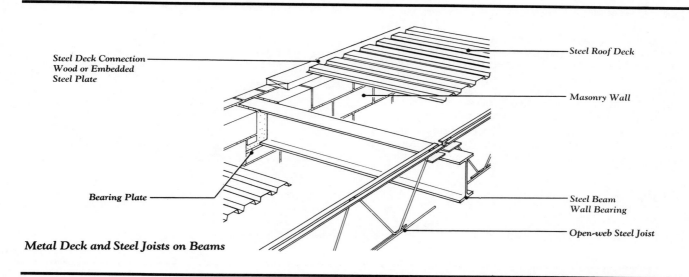

Steel Deck Connection
Wood or Embedded
Steel Plate

Steel Roof Deck

Masonry Wall

Bearing Plate

Steel Beam
Wall Bearing

Open-web Steel Joist

Metal Deck and Steel Joists on Beams

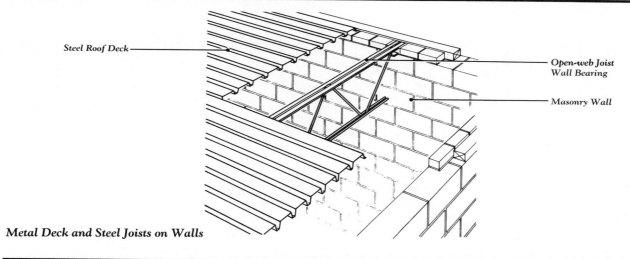

Steel Roof Deck

Open-web Joist
Wall Bearing

Masonry Wall

Metal Deck and Steel Joists on Walls

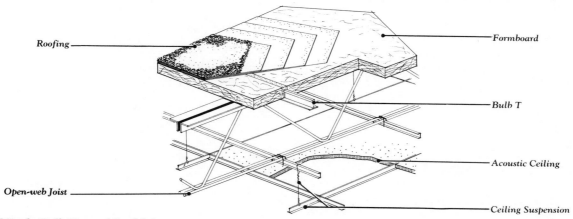

Roofing

Formboard

Bulb T

Acoustic Ceiling

Open-web Joist

Ceiling Suspension

Formboard Deck, Bulb T's, and Steel Joists

3 SUPERSTRUCTURE
JOIST GIRDERS

Joist girders are framing members that span between columns and support floor or roof joists. They are often used in place of steel beams or girders. Joist girders are steel trusses that look like longspan, open-web joists. They are designed as single-span members and support equally spaced open-web joists at the panel points. Spans vary in length from 20′ to 100′ and in depth from 20″ to 96″. The standard designation and a sketch follow. Approximate weights per foot are shown in manufacturers' catalogs.

Joist girders allow larger bay sizes, fewer columns, and more usable floor space. They are easily erected, form rigid connections at columns, and provide openings for passage of ducts and piping.

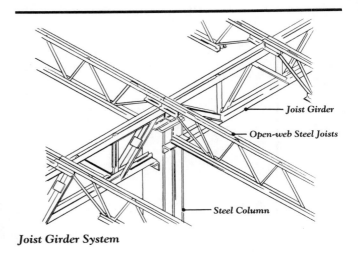

Joist Girder System

Man-hours

Description	m/hr	Unit
Joist Girders Average	6.154	Ton

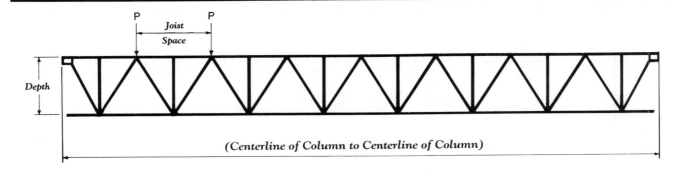

(Centerline of Column to Centerline of Column)

Standard Designation

48G	8N	8.8K
Depth in Inches	Number of Joist Spaces	Kip Load on Each Panel Point (One Kip = 1000 lbs)

3 SUPERSTRUCTURE
STEEL ROOF DECK ON STEEL JOISTS, JOIST GIRDERS, AND COLUMNS

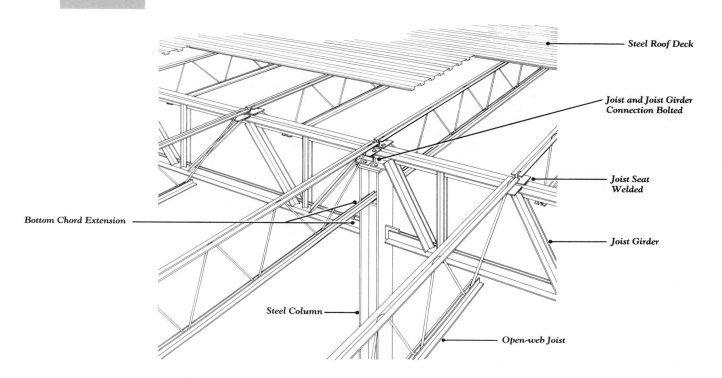

Steel Roof Deck

Joist and Joist Girder Connection Bolted

Joist Seat Welded

Joist Girder

Bottom Chord Extension

Steel Column

Open-web Joist

Steel roof deck on open-web joists, joist girders, and columns allows for long spans in both directions with openings for pipes, conduit, and ductwork through all of the roof-support members. The system is economical and lightweight, and can be quickly erected in most types of weather conditions.

*Note: The following regulations were developed by OSHA: "As soon as joists are erected, all bridging shall be completely installed and the joists permanently fastened into place before the application of any loads except the weight of the erectors. Many joists exhibit some degree of lateral instability under the weight of an erector until bridging is installed. Therefore, where three or more rows of bridging are required caution shall be exercised by the erectors until all bridging is completely and properly installed.

Where five rows of bridging are required in spans of over 40 feet, each joist shall be adequately braced laterally before the next joist is erected and before any loads are applied. Hoisting cables shall not be released until support has been provided by the center row of diagonal bridging and the bridging line has been anchored to prevent lateral movement, and where joists are bottom bearing, their ends have been restrained laterally."

Man-hours

Description	m/hr	Unit
Columns Steel Structural Tubing		
6" x 6"	6.000	ton
10" x 10"	2.000	ton
Wide Flange		
W8 x 31	3.355	ton
W10 x 45	2.412	ton
W12 x 50	2.171	ton
W14 x 74	1.538	ton
Joist Girders Average	6.154	ton
Steel Joists H and K Series Horizontal Bridging		
To 30' Span	6.667	ton
30' to 50' Span	6.353	ton
LH Series Bolted Cross Bridging Spans to 96'	6.154	ton
DLH Series Bolted Cross Bridging Spans to 144'	6.154	ton
Metal Decking Open Type		
1-1/2" Deep		
22 Gauge	.007	sf
18 and 20 Gauge	.008	sf
3" Deep		
20 and 22 Gauge	.009	sf
18 Gauge	.010	sf
16 Gauge	.011	sf

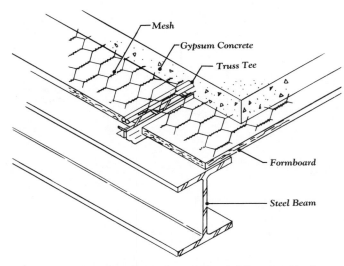

Truss Tees, Formboard, Mesh, and Poured Gypsum Deck

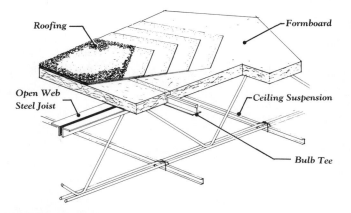

Bulb Tees and 3″ Thick Deck

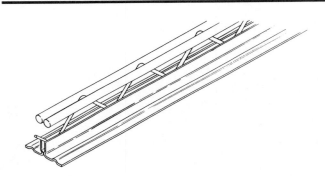

Truss Tee

Subpurlins are high-strength rolled bulb Tees. Truss Tees are fabricated open-web Tees that are welded to the supporting structure. They are used in conjunction with formboards, mesh, and gypsum concrete to provide a monolithic, lightweight roof deck system. Formboards are manufactured from asbestos cement, fiberglass polystyrene, or either mineral or wood fibers bonded with Portland cement.

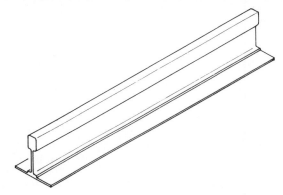

Bulb Tee

Man-hours

Description	m/hr	Unit
Bulb Tee Subpurlins and 1″ Thick Formboard	.009	sf
Bulb Tee Subpurlins and 3″ Thick Deck	.024	sf
Poured Gypsum		
2″ Thick	.008	sf
3″ Thick	.010	sf

3 SUPERSTRUCTURE
PRE-ENGINEERED STEEL BUILDINGS

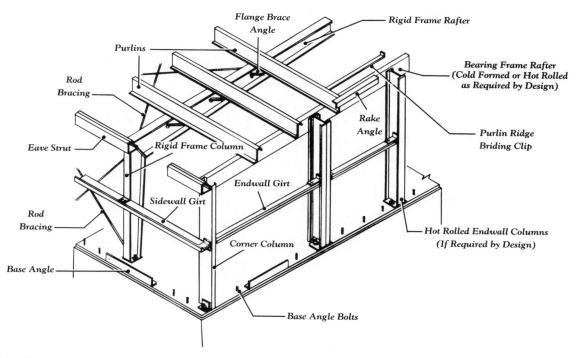

Pre-Engineered Building

Pre-engineered steel buildings are manufactured by many companies and are normally erected by franchised dealers. They are manufactured of pre-engineered components, which allow flexibility in the choice of configuration for one- or two-story buildings. Some systems are available with provisions for cranes, balconies, and mezzanines. There are four basic types: rigid frame, truss type, post and beam, and sloped beam. Roof pitches vary, but the most popular type is a low pitch of 1″ in 12″.

Eave heights are available in increments from 10′ to 24′. Rigid-frame, clear-span buildings are manufactured in widths of 30′ to 130′, and tapered-beam, clear-span buildings in widths of 30′ to 80′. Post and beam building widths, with one post at center, measure from 80′ to 120′; with two posts, from 120′ to 180′, and with three posts, from 160′ to 240′. Bay sizes are usually 20′ to 24′, but may be extended to 30′.

Roofs and sidewalls are normally covered with 26-gauge colored steel siding with various insulation options. Some manufacturers offer precast concrete and masonry siding options. Other options include eave overhangs, entrance canopies, end-wall overhangs, doors and windows, gutters and leaders, skylights, and roof vents.

Pre-engineered buildings are relatively low in construction costs. They are used extensively for industrial, commercial, institutional, and recreational facilities.

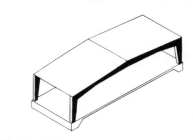

Low Profile Rigid Frame

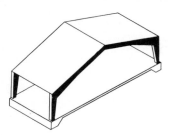

High Profile Rigid Frame

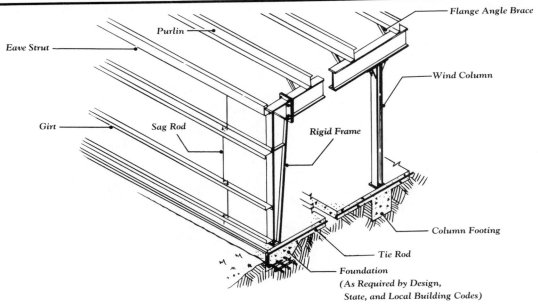

Post and Beam Construction

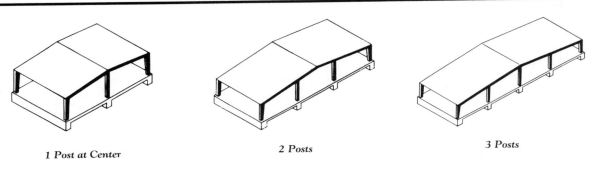

1 Post at Center   2 Posts   3 Posts

Post and Beam Types

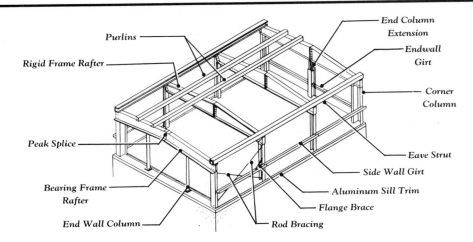

Sloped Beam Roof Construction

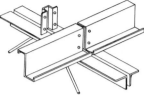

Purlin to Beam Connection

Purlin Spacing

Lateral Bracing

Endwall Beam to Column Connection

Endwall Girt Connection

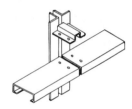

Girt to Column Connection

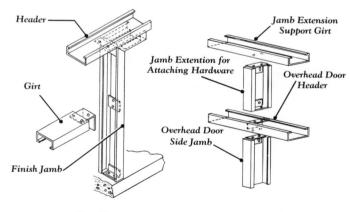

Header
Girt
Finish Jamb
Pass Door

Jamb Extension Support Girt
Jamb Extention for Attaching Hardware
Overhead Door Header
Overhead Door Side Jamb
Overhead Door

Typical Connections

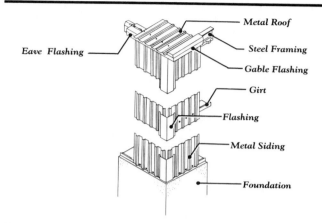

Eave Flashing
Metal Roof
Steel Framing
Gable Flashing
Girt
Flashing
Metal Siding
Foundation

Covering and Flashing at Corner

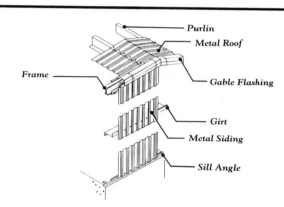

Frame
Purlin
Metal Roof
Gable Flashing
Girt
Metal Siding
Sill Angle

Covering and Flashing at Endwall

Man-hours

Description	m/hr	Unit
Pre-Engineered Building Shell		
Above Foundation Average	.044	sf floor
Eave Overhang 4′ Wide with Soffit	.224	lf
End Wall Overhang 4′ Wide with Soffit	.112	lf
Door 3′ x 7′	3.200	opng.
Door Framing Only 10′ x 10′	2.667	opng.
Sash		
3′ x 3′	1.714	opng.
4′ x 3′	1.846	opng.
6′ x 4′	2.000	opng.
Gutter	.050	lf
Skylight, Fiberglass Panel to 30 sf	2.400	Ea.
Roof Vents, Circular 20″ Diameter	2.667	Ea.
Continuous 12″ Wide 10′ Long	4.000	Ea.

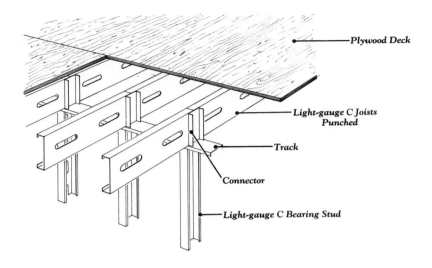

Plywood Deck

Light-gauge C Joists
Punched

Track

Connector

Light-gauge C Bearing Stud

Light-Gauge Steel

Light-gauge steel construction is a building system utilizing galvanized steel studs for bearing walls and galvanized C joists (single or double) for support of the floor system. Decking may consist of plywood, steel slab form and concrete, steel deck and concrete, or precast concrete. Studs are also used as a backup system for preformed fascia of various materials and allow for easy attachment and bracing to resist wind or earthquake forces.

Fastening is accomplished by using self-drilling screws or by welding. Plywood decks are fastened using screws or spiral shank nails. Adhesives may also be used in conjunction with screws or nails. When welds are used with galvanized members, weld areas should be touched up with a suitable paint.

Some cost advantages of light-gauge steel include lightweight construction, rapid all-weather installation, uniform depths, and good control of material waste.

Man-hours

Description	m/hr	Unit
3-5/8" Stud 18 Gauge	.017	sf
6" Stud 18 Gauge	.018	sf
C or Double Joists	.027	lf
3/4" Plywood	.013	sf
Slab Form	.006	sf

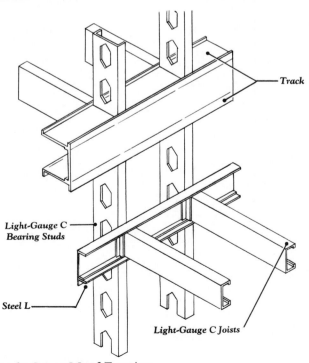

Track

Light-Gauge C
Bearing Studs

Steel L

Light-Gauge C Joists

Light-Gauge Metal Framing

Concrete Deck

Track

Slabform

Light-Gauge C Joist

Light-Gauge L Clip

Light-Gauge C Bearing Stud

Light-Gauge

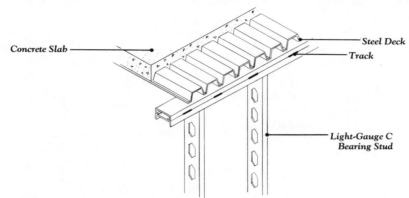

Concrete Slab

Steel Deck

Track

Light-Gauge C Bearing Stud

Light-Gauge Metal Framing - Walls

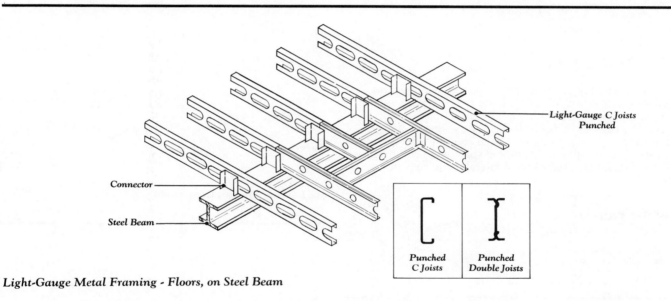

Light-Gauge C Joists Punched

Connector

Steel Beam

Punched C Joists	Punched Double Joists

Light-Gauge Metal Framing - Floors, on Steel Beam

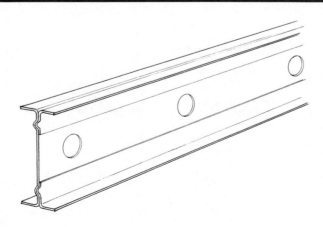

Light-Gauge Double Steel Joists

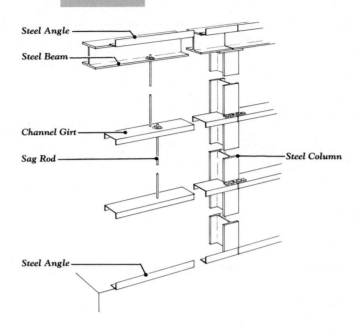

Girts used to support aluminum, steel, or composition siding are usually channels with the stronger axis oriented horizontally to resist wind load. Bending or sagging is generally resisted by sag rods threaded at each end for connections and adjustment. When column spacing is greater than the allowable channel span, wind columns, or vertical members used to resist lateral loads, may be introduced.

Man-hours

Description	m/hr	Unit
Channel Framing Including Sag Rods	.933	Ea.
Wind Columns	.054	Ea.

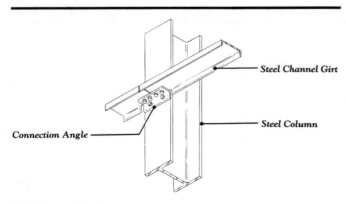

Girt Connection

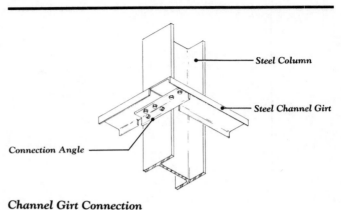

Channel Girt Connection

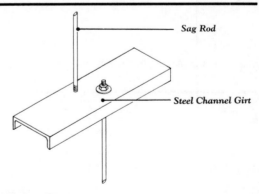

Steel Siding Zee Girt Connection

Steel Siding Girt Support

SUPERSTRUCTURE
FIREPROOFING STRUCTURAL STEEL

Fire resistant encasement is required for structural members in some steel-framed buildings in order to comply with building code requirements. No construction material, however, can totally resist damage or impairment from a fire that burns at high intensity for a long duration. Fire resistance ratings of the materials used for protection are therefore expressed in hours, with the rating depending on the applied thickness of the material.

Ceiling protection for floor systems may provide the required fire resistance by the use of rated suspension systems and acoustical tile or steel pans with mineral wool, mineral fibers on metal or gypsum lath, gypsum board on furring, or plaster on gypsum or metal lath.

Beams, columns and miscellaneous shapes may be fire-rated by enclosing the members with masonry, furring and gypsum board or metal lath and plaster. Encasement with concrete, masonry, or poured gypsum is also an option. Possible sprayed applications include cementitious mixtures, mineral fibers, mastic, and intumescent coatings.

Intumescent coatings are mixtures of resins, binders, pigments, ceramics and refractory fillers. They are spray applied and harden to a textured finish similar to paint. At elevated temperatures a reaction occurs causing the coating to expand and form an insulating blanket that protects the steel against heat penetration.

Man-hours

Description	m/hr	Unit
Fireproofing - 10″ Column Encasements		
Perlite Plaster	.273	vlf
1″ Perlite on 3/8″ Gypsum Lath	.345	vlf
Sprayed Fiber	.131	vlf
Concrete 1-1/2″ Thick	.716	vlf
Gypsum Board 1/2″ Fire Resistant,		
1 Layer	.364	vlf
2 Layer	.428	vlf
3 Layer	.530	vlf
Fireproofing - 16″ x 7″ Beam Encasements		
Perlite Plaster on Metal Lath	.453	lf
Gypsum Plaster on Metal Lath	.408	lf
Sprayed Fiber	.079	lf
Concrete 1-1/2″ Thick	.554	lf
Gypsum Board 5/8″ Fire Resistant	.488	lf

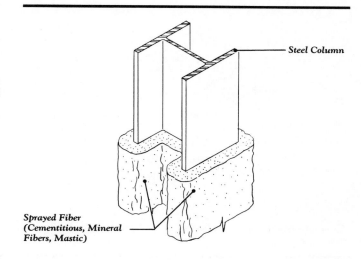

Sprayed Fiber on Columns

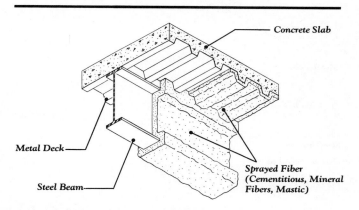

Sprayed Fiber on Beams and Girders

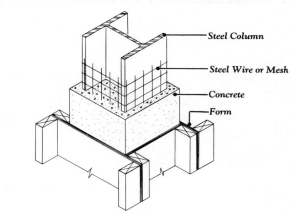

Concrete Encasement on Columns

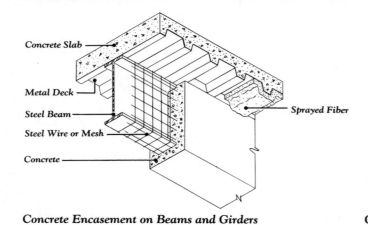

Concrete Slab
Metal Deck
Steel Beam
Steel Wire or Mesh
Concrete
Sprayed Fiber

Concrete Encasement on Beams and Girders

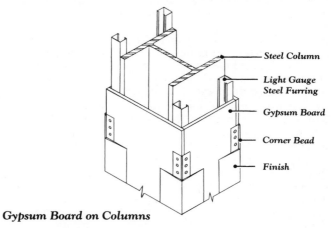

Steel Column
Light Gauge Steel Furring
Gypsum Board
Corner Bead
Finish

Gypsum Board on Columns

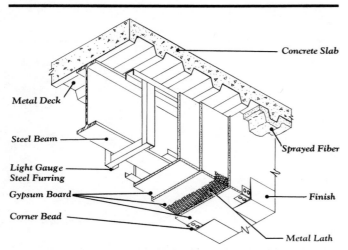

Concrete Slab
Metal Deck
Steel Beam
Light Gauge Steel Furring
Gypsum Board
Corner Bead
Sprayed Fiber
Finish
Metal Lath

Gypsum Board on Beams and Girders

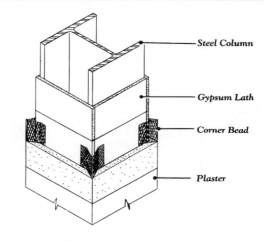

Steel Column
Gypsum Lath
Corner Bead
Plaster

Plaster on Gypsum Lath - Columns

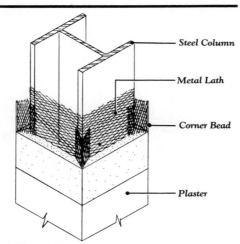

Steel Column
Metal Lath
Corner Bead
Plaster

Plaster on Metal Lath - Columns

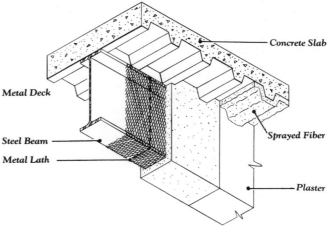

Concrete Slab
Metal Deck
Steel Beam
Metal Lath
Sprayed Fiber
Plaster

Plaster on Metal Lath - Beams and Girders

104

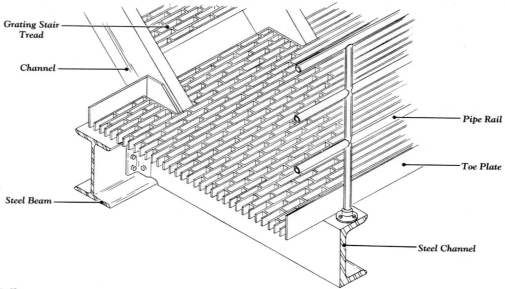

Grating Stair Tread

Channel

Steel Beam

Pipe Rail

Toe Plate

Steel Channel

Platforms and Walkways

Platforms, walkways, and catwalks may be fabricated on site. They may also be prefabricated in sections, and then shipped and erected. Some commonly used supporting members are angles, channels, beams, or wide-flange sections, and pipe and tube columns constructed of steel or aluminum. Materials used for cover plates may be aluminum or steel checkered plate, or aluminum, fiberglass, or steel grating. Open areas are usually surrounded by pipe rails with toe plates, as required by OSHA regulations.

Platforms and walkways are used throughout manufacturing plants for machinery access, boiler and power-plant walkways, and piping and valve access. Access to the platforms is usually provided by steel ladders, stairs, or ship's ladders. Ship's ladders are factory or shop-assembled with rails attached. They are used mainly for platform access.

Man-hours

Description	m/hr	Unit
Light Framing, Steel		
L's 4″ and Larger	.011	lb
Less than 4″	.018	lb
C's 8″ and Larger	.009	lb
Less than 8″	.016	lb
Aluminum Shapes	.053	lb
Grating, Aluminum		
1-1/4″ Deep	.032	sf
2-1/4″ Deep	.037	sf
Fiberglass	.064	sf
Steel	.070	sf
Steel Pipe Railing w/Toe Plate	.160	lf

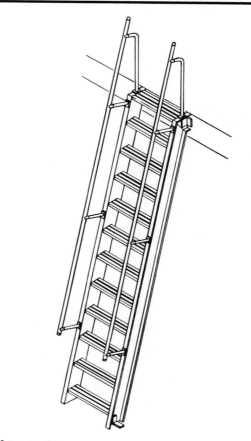

Ship's Ladder

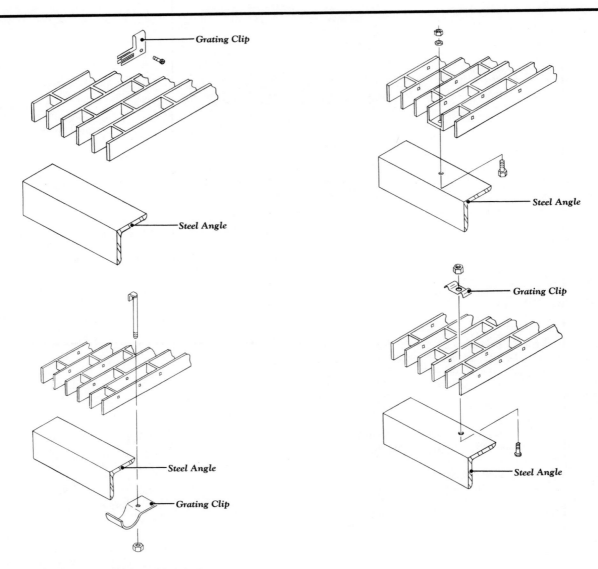

Platforms and Walkways Anchoring Methods

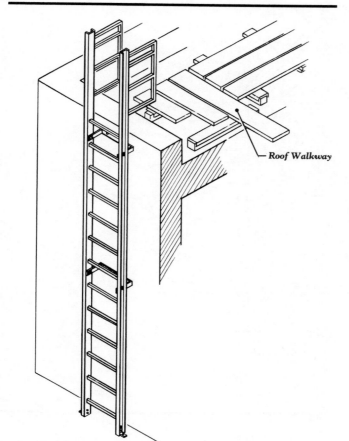

— Roof Walkway

Fixed Access Ladder

Grating Stair Tread

— Abrasive Nosing

Platforms and Walkways
Stair Treads

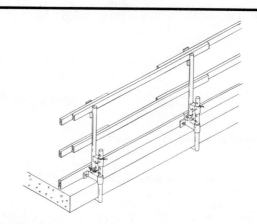

Guard Rail Stanchions
Platforms and Walkways

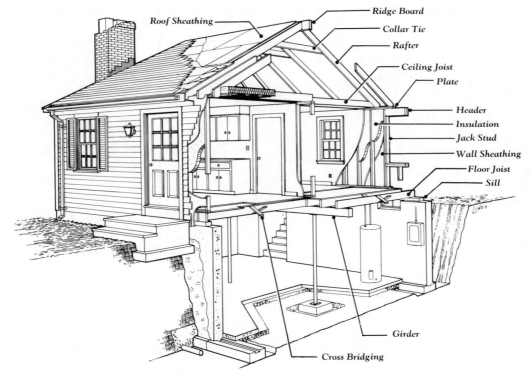

Roof Sheathing — Ridge Board — Collar Tie — Rafter — Ceiling Joist — Plate — Header — Insulation — Jack Stud — Wall Sheathing — Floor Joist — Sill — Girder — Cross Bridging

Wood Framing System - Residential

Wood, unlike most processed building materials, is an organic material that can be used in its natural state. Factors that influence its strength are density, natural defects (knots, grain, etc.), and moisture content. Wood can be easily shaped or cut to size on the site or prefabricated in the shop. Structurally, wood may be used for joists, posts, columns, beams, and trusses. It may also be laminated into columns, beams, rigid frames, arches, vaults, and folded-plate configurations. Various framing systems may be used with fiberboard, particle board, plywood, or wood decking (plain or laminated) to form composite panels. Wood may be exposed and finished for interior or exterior use. It may be pressure-treated to provide resistance to moisture, rot, or exposure.

Structural lumber is stress-graded for bending, tension parallel to the grain, horizontal shear, compression perpendicular to the grain, compression parallel to the grain, and modulus of elasticity.

Structural lumber is usually dressed on four sides and, therefore, reduced in its normal size by approximately 1/2″ in each direction. For example, the actual dimension of a 6″ x 10″ beam dressed is 5-1/2″ x 9-1/2″.

Wood structures are fastened by nails, pins, dowels, screws, bolts, and adhesives, or by fabricated metal connectors, tailored to perform specific connection functions.

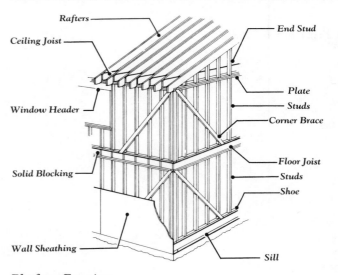

Rafters — Ceiling Joist — Window Header — Solid Blocking — Wall Sheathing — End Stud — Plate — Studs — Corner Brace — Floor Joist — Studs — Shoe — Sill

Platform Framing

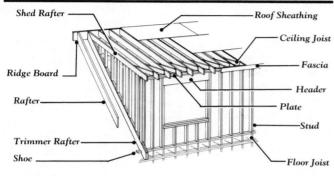

Shed Dormer Framing

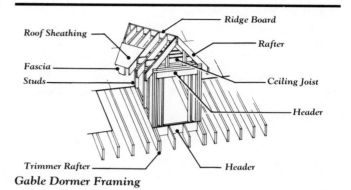

Gable Dormer Framing

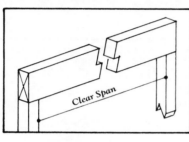

Single Beam

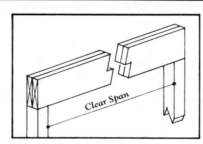

Double Beam

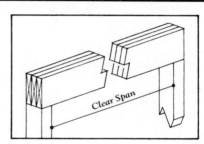

Triple Beam

Wood Beams and Columns

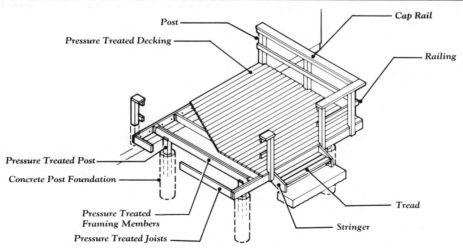

Wood Deck Construction

3 SUPERSTRUCTURE
WOOD JOISTS AND PLYWOOD DECK

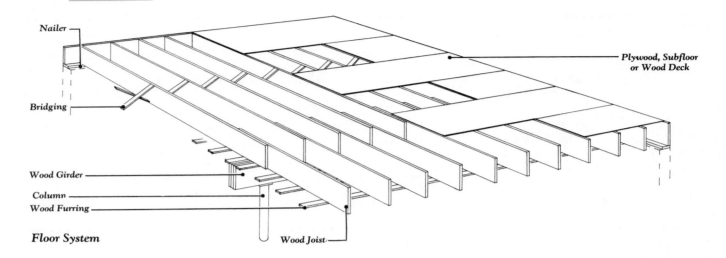

Nailer

Bridging

Wood Girder

Column

Wood Furring

Plywood, Subfloor or Wood Deck

Wood Joist

Floor System

Wood joists may be used with all types of bearing wall or support systems. They may also be used in conjunction with various deck materials to provide economical floor and roof systems with moderate spans and loadings. The spacing of the wood joists may be varied to suit deck span or loading requirements.

Man-hours

Description	m/hr	Unit
Joist Framing		
2" x 6"	.013	lf
2" x 8"	.015	lf
2" x 10"	.018	lf
2" x 12"	.018	lf
2" x 14"	.021	lf
3" x 8"	.017	lf
3" x 12"	.027	lf
4" x 8"	.026	lf
4" x 12"	.036	lf
Bridging Wood or Steel, Joists		
16" On Center	.062	pr.
24" On Center	.057	pr.
Sub Floor Plywood CDX		
1/2" Thick	.011	sf
5/8" Thick	.012	sf
3/4" Thick	.013	sf
Boards Diagonal		
1" x 8"	.019	sf
1" x 10"	.018	sf
Wood Fiber T & G		
2' x 8' Planks		
1" Thick	.016	sf
1-3/8" Thick	.018	sf

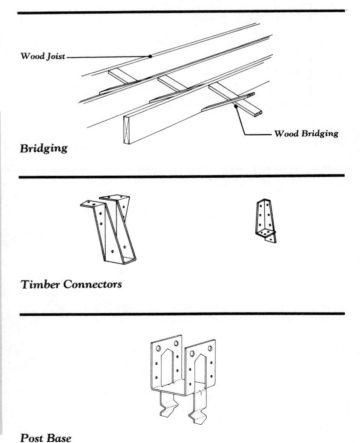

Wood Joist

Wood Bridging

Bridging

Timber Connectors

Post Base

3 SUPERSTRUCTURE
WOOD COLUMNS, GIRDERS, BEAMS AND WOOD DECK

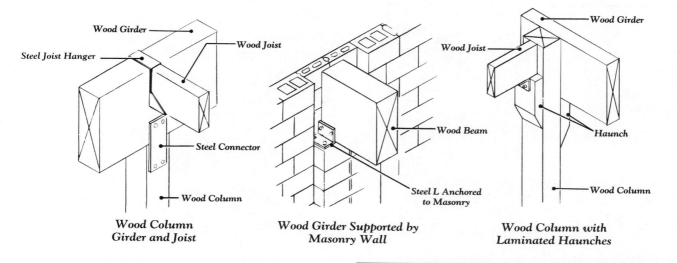

Wood Column Girder and Joist

Wood Girder Supported by Masonry Wall

Wood Column with Laminated Haunches

Wood columns, girders, and beams may be used with wood deck, or joists and plywood deck, to provide a floor or roof-framing system. The system is sometimes called "post and beam construction." This system is primarily used in situations where the floor or roof loading is relatively moderate, and where clear spans are not excessive.

Man-hours

Description	m/hr	Unit
Columns		
6″ x 6″	.074	lf
8″ x 8″	.122	lf
10″ x 10″	.178	lf
12″ x 12″	.240	lf
Beams and Girders		
6″ x 10″	.073	lf
8″ x 16″	.142	lf
12″ x 12″	.240	lf
10″ x 16″	.213	lf
Wood Deck		
3″ Nominal	.050	sf
4″ Nominal	.064	sf
Laminated Wood Deck		
3″ Nominal	.038	sf
4″ Nominal	.049	sf

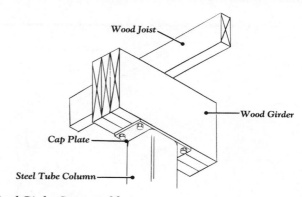

Wood Girder Supported by Square Tube Column

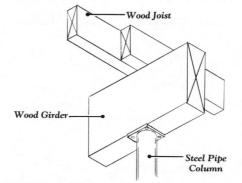

Wood Girder Supported by Pipe Column

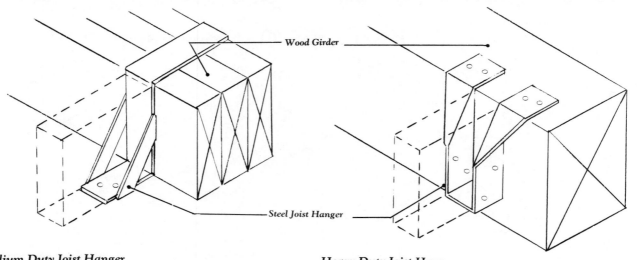

Wood Girder

Steel Joist Hanger

Medium Duty Joist Hanger

Heavy Duty Joist Hanger

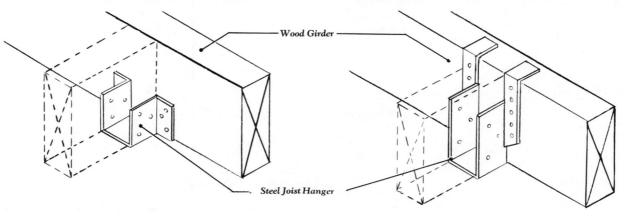

Wood Girder

Steel Joist Hanger

Light Duty Joist Hanger on Timber Joist

Medium Duty Joist Hanger

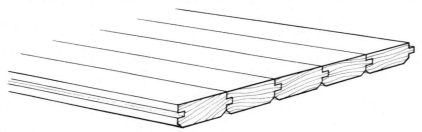

Tongue & Groove Solid Wood Decking

3 SUPERSTRUCTURE
LAMINATED COLUMNS, GIRDERS, BEAMS, AND WOOD DECK

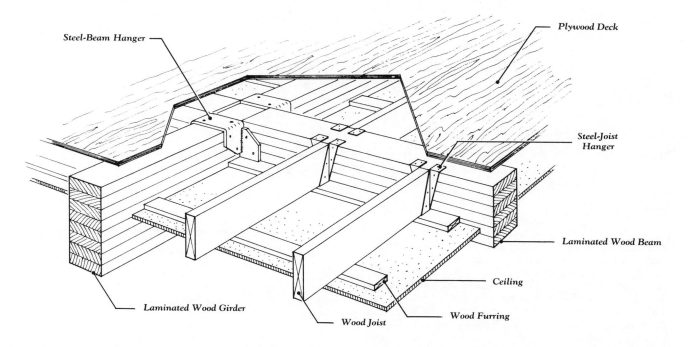

Laminated wood members and decking (or plywood) may be used to frame floors and roofs with varied spans and loadings. Both the framing members and decking may be left unfinished or be supplied factory stained. The connectors are usually fabricator-designed and supplied.

Because of their light weight, these wood members may be erected without the assistance of expensive lifting equipment, thus decreasing the installed cost of the floor or roof system.

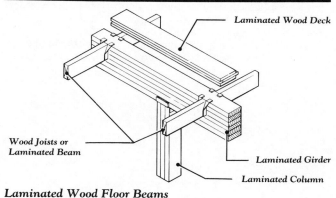

Laminated Wood Floor Beams

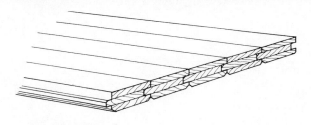

Laminated Wood Deck

Man-hours

Description	m/hr	Unit
Laminated Framing Roof Beams		
20' Span		
8' On Center	.016	sf floor
16' On Center	.013	sf floor
40' Span		
8' On Center	.013	sf floor
16' On Center	.010	sf floor
60' Span		
8' On Center	.017	sf floor
16' On Center	.013	sf floor
Columns	.020	MBF
Beams	.011	MBF
Wood Deck		
3" Nominal	.050	sf
4" Nominal	.064	sf
Laminated Wood Deck		
3" Nominal	.038	sf
4" Nominal	.049	sf

113

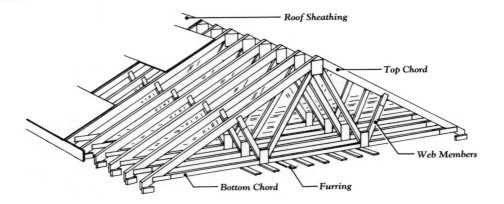

Truss Roof

Wood roof trusses are factory fabricated of dimension lumber for high or low pitched roofs, hipped roofs, mansard roofs, and flat roofs. They are available from manufacturers in many different configurations. Members are connected with wood or metal gussetts and are glued and/or power nailed in a jig to ensure uniformity.

The clear span characteristics of trusses provides flexibility for interior planning and partition layout. A considerable cost savings results from the fact that ceilings or decking can be applied directly to the truss chords. Truss spacing is dictated by maximum allowable span of the sheathing material. Normal span wood trusses can be erected by hand due to their light weight, but are also efficiently erected utilizing a boom truck or crane and a small erection crew.

Split Ring Connector

4" Shear Plate Connector

2-5/8" Shear Plate Connector

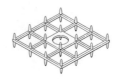

Flat Spiked Grid Connector
Connectors, Used in Trusses Fabricated from Timbers

Man-hours

Description	m/hr	Unit
Wood Roof Trusses		
4/12 Pitch		
20' Span	.645	Ea.
28' Span	.755	Ea.
36' Span	.870	Ea.
8/12 Pitch		
20' Span	.702	Ea.
28' Span	.816	Ea.
36' Span	.976	Ea.
Fink or King Post 30' to 60' Span	.013	sf floor
Plywood Roof Sheathing		
1/2" Thick	.011	sf
5/8" Thick	.012	sf
3/4" Thick	.013	sf
Stressed Skin, Plywood Roof Panels 4' x 8'		
4-1/4" Deep	.019	sf roof
6-1/2" Deep	.023	sf roof
Wood Roof Decks		
3" Thick Nominal	.050	sf
4" Thick Nominal	.064	sf
Laminated Wood Roof Deck		
3" Thick Nominal	.038	sf
4" Thick Nominal	.049	sf

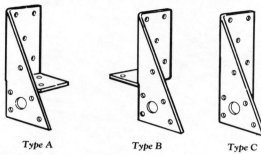

Type A   Type B   Type C

Framing Anchors Used in Truss Connections

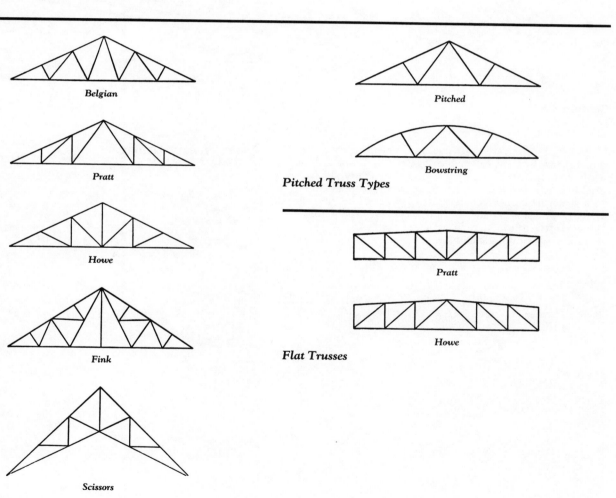

Belgian

Pratt

Howe

Fink

Scissors

Pitched Truss Types

Pitched

Bowstring

Pitched Truss Types

Pratt

Howe

Flat Trusses

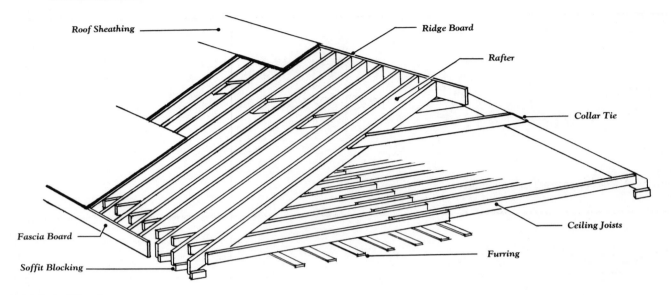

Roof Sheathing — Ridge Board

Rafter

Collar Tie

Fascia Board

Soffit Blocking

Ceiling Joists

Furring

Gable End Roof

Wood roof rafters are fabricated in the field from dimensional lumber, for high or low pitched roofs, hipped roofs, mansard roofs, and flat roofs. They are typically used in conjunction with plywood decking to provide a roof structure compatible with many types of bearing wall systems. The spacing and slope of wood rafter systems may be varied to suit the loading span and aesthetic requirements of the installation.

Man-hours

Description	m/hr	Unit
Rafters, Common, to 4 in 12 Pitch, 2" x 6"	.016	lf
2" x 8"	.017	lf
2" x 10"	.025	lf
2" x 12"	.028	lf
On Steep Roofs, 2" x 6"	.020	lf
2" x 8"	.021	lf
2" x 10"	.032	lf
2" x 12"	.035	lf
On Dormers or Complex Roofs, 2" x 6"	.027	lf
2" x 8"	.030	lf
2" x 10"	.038	lf
2" x 12"	.041	lf
Hip and Valley, to 4 in 12 Pitch, 2" x 6"	.021	lf
2" x 8"	.022	lf
2" x 10"	.028	lf
2" x 12"	.030	lf
Hip and Valley, on Steep Roofs, 2" x 6"	.027	lf
2" x 8"	.029	lf
2" x 10"	.036	lf
2" x 12"	.039	lf
Hip and Valley, on Dormers/Complex Roofs,		
2" x 6"	.031	lf
2" x 8"	.034	lf
2" x 10"	.042	lf
2" x 12"	.045	lf

Man-hours (cont.)

Description	m/hr	Unit
Hip and Valley Jacks to 4 in 12 Pitch, 2" x 6"	.027	lf
2" x 8"	.033	lf
2" x 10"	.036	lf
2" x 12"	.043	lf
Hip and Valley Jacks on Steep Roofs, 2" x 6"	.034	lf
2" x 8"	.042	lf
2" x 10"	.046	lf
2" x 12"	.054	lf
Hip and Valley Jacks on Dormers/Complex Roofs, 2" x 6"	.039	lf
2" x 8"	.048	lf
2" x 10"	.052	lf
2" x 12"	.063	lf
Collar Ties, 1" x 4"	.020	lf
Ridge Board, 1" x 6"	.027	lf
1" x 8"	.029	lf
1" x 10"	.032	lf
2" x 6"	.032	lf
2" x 8"	.036	lf
2" x 10"	.040	lf
Sub-Fascia, 2" x 8"	.071	lf
2" x 10"	.089	lf

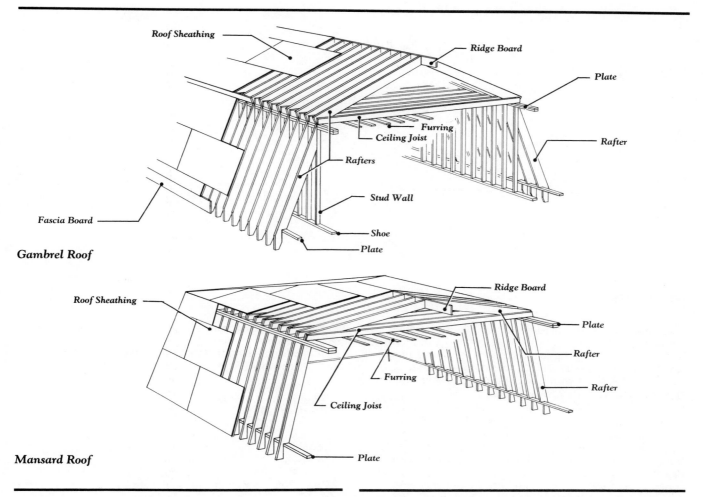

Gambrel Roof

Roof Sheathing
Ridge Board
Plate
Furring
Ceiling Joist
Rafter
Rafters
Stud Wall
Fascia Board
Shoe
Plate

Mansard Roof

Roof Sheathing
Ridge Board
Plate
Rafter
Furring
Rafter
Ceiling Joist
Plate

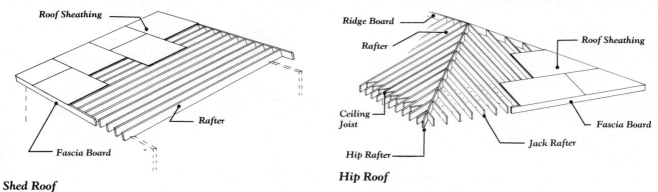

Shed Roof

Roof Sheathing
Rafter
Fascia Board

Hip Roof

Ridge Board
Rafter
Roof Sheathing
Ceiling Joist
Fascia Board
Hip Rafter
Jack Rafter

3 SUPERSTRUCTURE
STAIRS

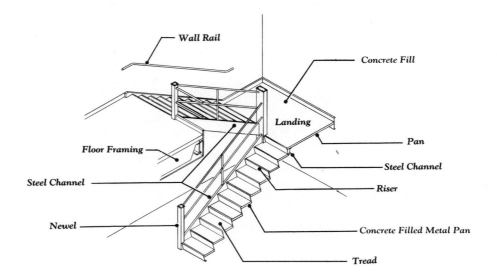

Wall Rail

Concrete Fill

Landing

Floor Framing

Pan

Steel Channel

Steel Channel

Riser

Newel

Concrete Filled Metal Pan

Tread

Steel Pan Stair

Stairs may be prefabricated or built in place from aluminum, concrete, cast iron, steel, or wood. In many instances stairs are designed to conform to the structural framing system: concrete stairs with a concrete-framed building; steel stairs with a steel-framed building; and wood stairs with a wood-framed building.

Steel stair assemblies are usually prefabricated of channel stringers with treads of light gauge metal called "pans", cast iron, checkered plates, or gratings. The treads may be combined with non-slip nosings of various materials. The treads may also be treated with applied non-slip treatments. Pan stairs require concrete fill in the pans to provide the contact area for the treads. Landings are designed and fabricated of various structural shapes and usually covered with the same materials used in the stair treads.

Concrete stair systems may be formed and cast-in-place, or prefabricated and erected. Concrete stair systems allow design flexibility in the creation of aesthetic shapes and orientations. Precast treads of concrete or terrazzo in combination with cast-in-place stringers allow cantilevered open stairways.

Wood stairs may be prefabricated box or basement types, but their application varies with the available space and the aesthetic effect desired. Unlimited shapes, orientations, and railing arrangements are available in various species of wood both as prefabricated kits or built in place.

Prefabricated spiral stairs are available in aluminum, cast iron, steel, and wood in both stock and custom designs with many tread designs and materials. Railings are normally prebent or prefabricated to follow the stair curvature but may be customized to suit the application.

Pipe and ornamental stair rails may be stair supported or wall attached, in countless designs, materials and configurations. Stair rails are available prefabricated or may be erected from components, available from manufacturers, that are designed to satisfy any condition. Rail design is subject to "OSHA" regulations.

Man-hours

Description	m/hr	Unit
Concrete		
Stairs C.I.P.	.600	lf tread
Landings C.I.P.	.253	sf
Steel Custom Stair 3'-6" Wide	.914	riser
Steel Pan		
Stair Shop Fabricated 3'-6" Wide	.376	riser
Landing Shop Fabricated	.125	sf
Concrete Fill Pans	.110	sf
Spiral Stair		
Aluminum 5'-0" Diameter	.711	riser
Cast Iron 4'-0" Diameter	.711	riser
Steel Industrial 6'-0" Diameter	.800	riser
Included Ladder (Ships' Stair) 3'-0" Wide	.320	vlf
Wood Box Stair Prefabricated 3'-6" Wide		
4' High	4.000	flight
8' High	5.333	flight
Open Stair Prefabricated 8" High	5.333	flight
Curved Stair 3'-3" Wide		
Open 1 Side 10' High	22.857	flight
Open 2 Sides 10' High	32.000	flight
Railings 1-1/2" Pipe		
Aluminum	.164	lf
Steel	.160	lf
Wall Mounted	.125	lf
Rails Ornamental		
Bronze or Stainless	.611	lf
Aluminum	.767	lf
Wrought Iron	.834	lf
Composite Metal Wood or Glass	1.467	lf

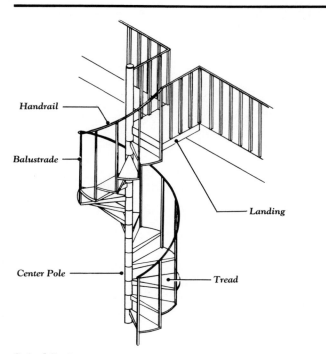

Spiral Stair

- Handrail
- Balustrade
- Center Pole
- Landing
- Tread

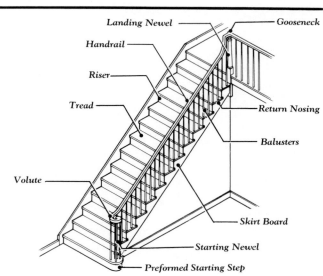

Wood Stair

- Landing Newel
- Gooseneck
- Handrail
- Riser
- Tread
- Return Nosing
- Balusters
- Volute
- Skirt Board
- Starting Newel
- Preformed Starting Step

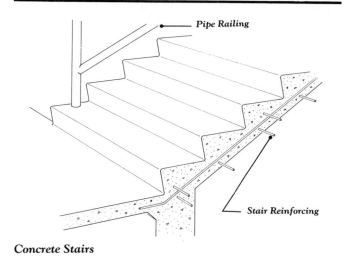

Concrete Stairs

- Pipe Railing
- Stair Reinforcing

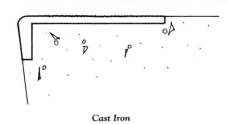

Cast Iron

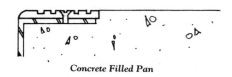

Concrete Filled Pan

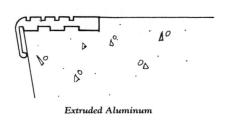

Extruded Aluminum

Stair Tread Nosings

DIVISION 4

EXTERIOR CLOSURE

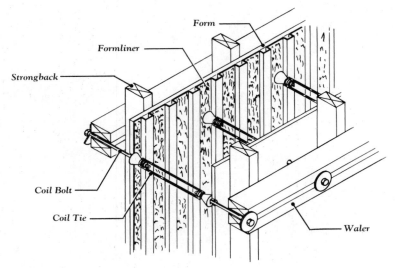

Textured Cast-in-Place Concrete Wall

Exposed cast-in-place concrete walls may function as the structural components of a building, as well as form the building's skin. Formwork may be job-built plyform or modular steel framed plywood, plywood, or aluminum.

Formwork can be used to create different textures, patterns, reveals, and edges on concrete surfaces. Wood, metal, and sometimes fiberglass, have been traditionally used for forming systems and formliners. Special plastic and elastomeric materials that can be formed in many patterns and textures are also available for liners. The cost of liners, as well as the time to install them, must be added to the cost and installation time for the back-up forms that hold them in place. Adhesives and/or mechanical fasteners may be used to attach the liners to the backing. Tapes and sealers are sometimes necessary at formliner joints to prevent leakage.

A wide range of special patterns, costly to produce in wood or metal, are made possible by the plastic and elastomeric liners. The finishes are very consistent, and the liners quite durable. Typical materials include fiberglass, rigid PVC (polyvinyl-chloride), ABS (acrylonitrile-butadiene-styrene), and rubber-like, highly flexible elastomerics, such as

urethane. The number of possible reuses is determined by the material chosen and the care exercised in cleaning and handling the liner, with up to 200 reuses possible. Form release agents are usually employed to facilitate quick cleaning for reuse.

Many additional forming accessories are available to create special reveals, beveled edges, rounded corners, and similar features. Form tie-holes can be hidden in rustication grooves and patched, or they can be deliberately highlighted with wood or plastic cone ties. Other accessories, such as dovetail anchor slots, keyways, waterstops, flashing reglets, and lock-strip glazing channels, as well as control, construction, and isolation joints, must be coordinated with all forms and liners. Even though formliners may be used, abrasion or other finishing or cleaning methods may still be employed. Surface sealers may also be specified as an additional step in the finishing process.

Exposed concrete walls may be reinforced to resist temperature variations or, if used as structural elements, to resist vertical or horizontal loads. The reinforcing steel may be galvanized or epoxy coated to combat rust staining of

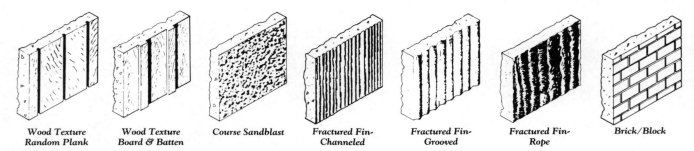

| Wood Texture Random Plank | Wood Texture Board & Batten | Course Sandblast | Fractured Fin-Channeled | Fractured Fin-Grooved | Fractured Fin-Rope | Brick/Block |

Various Texture Patterns

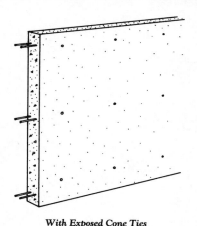

With Exposed Cone Ties

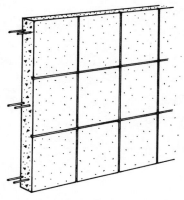

With Rustication Grooves

Reinforced Cast-in-Place Concrete Walls

the exposed concrete. In some instances the walls may be posttensioned to control shrinkage cracks.

In addition to the various sculptured effects of special forms and formliners, a variety of abrasive methods may be used to finish the surfaces of exposed concrete walls. These methods create different textures, expose the underlying aggregate, and generally offer the additional benefit of removing or concealing surface defects. Retarding agents applied before placement and special sealers sprayed on after the finishing process provide additional options to enhance the surface appearance.

Many abrasive methods of concrete wall finishing exist—some relatively simple and others complex. All of them are labor intensive and some require special tools or machinery. Simple hand rubbing, either with a special synthetic stone or with burlap and cement paste, is the mildest abrasive finishing method. Power tools may be used to grind the surface to expose and flatten the large aggregate. Bush-hammering, either by hand or with air-powered tools, is another method which exposes the aggregate and creates a rough-textured surface finish. Sandblasting also produces a mildly textured surface while exposing the aggregate, but air pollution control methods

must be implemented in populated areas when this process is used. The effect of rough, highly textured ribs can be acquired by carefully breaking back the ridges in walls cast with serrations.

Special retarding agents, applied to the formwork prior to the placing of the concrete, are also used to carry out the surface-finishing process. These retarding agents weaken the final surface and facilitate its removal after the formwork has been stripped. If enough retarder is used, the surface may simply be washed down with pressurized water to expose the aggregate without implementing abrasive methods. Acid etching is also a suitable option for removing the superficial layer of sand and cement to expose the aggregate.

Special sealers may be sprayed onto the surface of concrete walls to minimize discoloration, especially in polluted environments. This method of finishing is particularly effective on walls where the aggregate is light in color or where the final surface is rough. In most cases, these sealers require reapplication every few years. If short-term curing compounds are used, they must be compatible with the long-term sealers.

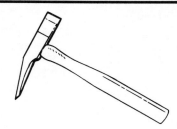

Masonry Hammer

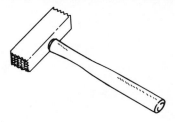

Bush Hammer

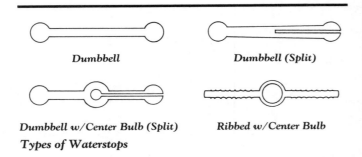

Dumbbell   **Dumbbell (Split)**

Dumbbell w/Center Bulb (Split)   **Ribbed w/Center Bulb**

Types of Waterstops

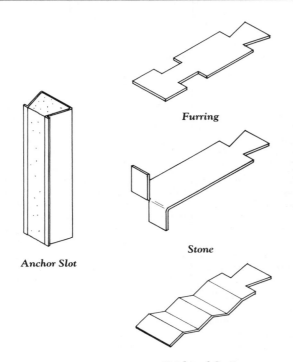

Anchor Slot

Furring

Stone

Brick and Cavity

Dovetail Anchors

Man-hours

Description	m/hr	Unit
Forms		
Job-Built Plywood		
1 Use/Month	.130	sfca
4 Uses/Month	.095	sfca
Modular Prefabricated Plywood		
1 Use/Month	.053	sfca
4 Uses/Month	.049	sfca
Steel Framed Plywood		
1 Use/Month	.080	sfca
4 Uses/Month	.072	sfca
Box Out Openings to 10 sf	2.000	Ea.
Bulkhead	.181	lf
Corbel to 12″ wide	.320	lf
Pilasters	.178	sfca
Waterstop Dumbbell	.055	lf
Reinforcing		
#3 to #7	10.667	ton
#8 to #14	8.000	ton
Place Concrete 12″ Walls		
Direct Chute	.480	cu yd
Pumped	.674	cu yd
With Crane and Bucket	.711	cu yd
Concrete in Place 12″ Thick	5.926	cu yd
Liners for Forms and Accessories		
ABS Plastic		
Aged Wood 4″ Wide or Fractured Rope Rib		
or Rustic Brick Pattern		
1 Use	.032	sfca
4 Uses	.011	sfca
Ribbed Look 1/2″ and 3/4″ Deep or		
Striated Random 3/8″ x 3/8″ Deep		
1 Use	.027	sfca
4 Uses	.010	sfca
Rustication Strips ABS Plastic 2 Piece Snap-on		
1″ Deep x 1-3/8″ Wide		
1 Use	.120	lf
2 Uses	.080	lf
4 Uses	.060	lf
Wood Beveled Edge		
3/4″ Deep 1 Use	.080	lf
1″ Deep 1 Use	.107	lf
Dovetail Anchor Slot	.020	lf
Flashing Reglet	.053	lf
Keyway	.016	lf
Waterstop		
4″ Wide	.052	lf
6″ Wide	.055	lf
9″ Wide	.059	lf
Finishing Walls Burlap Rub with Grout	.018	sf
Carborundum Rub Dry	.030	sf
Wet Rub	.046	sf
Bush Hammer Green Concrete	.047	sf
Cured Concrete	.073	sf
Sandblast Light Penetration	.022	sf
Heavy Penetration	.064	sf

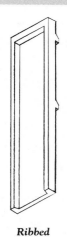

Ribbed

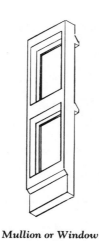

Mullion or Window

Flat

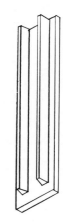

Double Tee

Panel Types

Factory precast wall panels require a minimum of on-site work and facilitate quick, all-weather erection. The curing and quality control of finished panels are more easily accomplished in the plant and, consequently, a more uniform product results. Also, because the in-plant work force is more knowledgeable, and skilled in the casting process, sophisticated finishes and shapes which are difficult to achieve in the field become more feasible. A wide variety of finishes is available on all types of precast panels, including: textured, plain, sandblasted, bush hammered, exposed aggregate, stone-faced, and brick-faced. Reveals, returns, or sculptured faces in any shippable configuration are also available.

Although precast wall panels may be manufactured to function in loadbearing situations, non-loadbearing types attached to a frame structure are more commonly used. Welded or bolted matching plates or clips are used to connect non-leadbearing panels to the building frame. The panels are usually joined with tongue-and-groove or butt joints, which are then caulked or grouted depending on structural requirements.

Precast wall panels are manufactured in a variety of specialized concrete weights and materials, including: normal weight, lightweight polymer-based, glass fiber-reinforced (GFRC), or a combination of these concrete types. Polymer-based concrete panels are sometimes reinforced with woven fiberglass cloth and then used to form uninsulated facing panels measuring 8' in width and 20' in height, but only 3/4" in thickness. Thin polymer or GFRC panels are usually manufactured by spraying. Panels are usually supported on a prewelded frame of light gauge metal framing.

Insulation for precast wall panels may be installed by sandwiching the insulating material inside the panel during casting or by applying it to the interior face of the panel at the factory or site. Rigid, foamed, batt, and sprayed-on systems may be used for panel insulation.

Erection of Precast Panels Daily Production Range		
Panel Size Ft	Low Rise	High Rise
	Daily Production Range Pieces	Daily Production Range Pieces
4' x 4'	8 - 20	7 - 18
4' x 8'	8 - 19	7 - 16
4' x 12'	8 - 18	7 - 16
4' x 16'	8 - 16	7 - 14
4' x 20'	8 - 16	7 - 14
8' x 8'	7 - 18	7 - 14
8' x 12'	7 - 16	6 - 12
8' x 16'	7 - 12	5 - 10
8' x 20'	6 - 10	5 - 9
Col. Covers	7 - 20	6 - 16
Soffits	3 - 10	2 - 10

Man-hours

Description	m/hr	Unit
Precast Double T Wall Members up to 55' High	.020	sf
Rigid Insulation Board	.010	sf
Sprayed Insulation	.012	sf
Fiberglass Insulation, Batts	.005	sf
Furring, Metal, 16" On Center	.038	sf
Polyethylene Vapor Barrier	.216	CSF
Foam Adhesive Caulking	.055	lf
Precast Wall Panels		
Uninsulated 4" Thick Smooth Gray Low Rise		
4' x 8'	.225	sf
8' x 8'	.125	sf
8' x 16'	.070	sf
High Rise		
4' x 8'	.250	sf
8' x 8'	.141	sf
8' x 16'	.094	sf
Rigid Insulation Board	.010	sf
Sprayed Insulation	.012	sf
Caulking, Bulk, in Place		
1/4" Wide Bead	.055	lf
3/4" Wide Bead	.062	lf

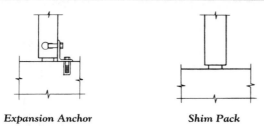

Expansion Anchor **Shim Pack**

Direct Bearing Connections

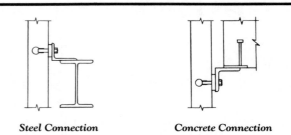

Steel Connection **Concrete Connection**

Bolted and Welded Tie-back Connections

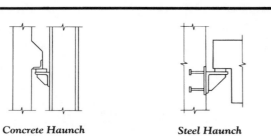

Concrete Haunch **Steel Haunch**

Haunch Bearing Connections

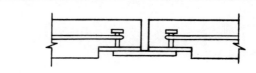

Welded Alignment Connection

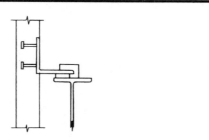

Angle Seat Bearing Connection

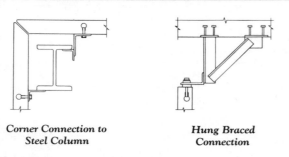

Corner Connection to **Hung Braced**
Steel Column **Connection**

Welded Angle Tie-back Connections

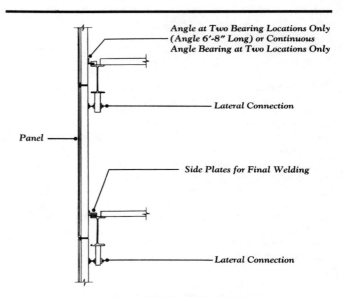

Angle at Two Bearing Locations Only
(Angle 6'-8" Long) or Continuous
Angle Bearing at Two Locations Only

Lateral Connection

Panel

Side Plates for Final Welding

Lateral Connection

Solid Wall or Window Wall Panel

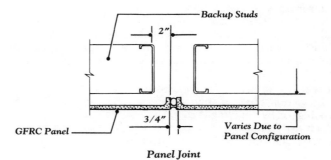

Backup Studs

2"

GFRC Panel

3/4"

Varies Due to
Panel Configuration

Panel Joint

Glass Fiber Reinforced Concrete (GFRC) Panels

4 EXTERIOR CLOSURE
TILT-UP CONCRETE PANELS

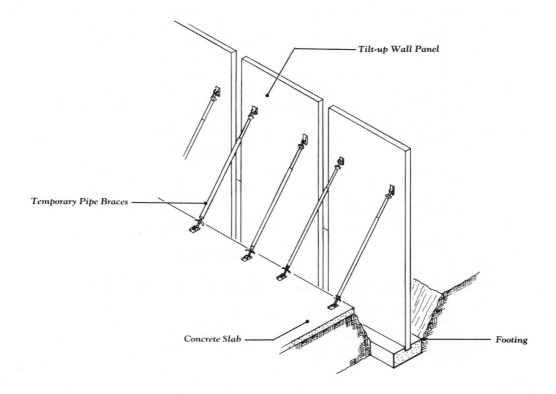

Tilt-up Wall Panel

Temporary Pipe Braces

Concrete Slab

Footing

Tilt-up wall panels, because they are cast at the construction site, provide the advantages of avoiding shipping size restrictions and creating the opportunity for relatively quick, ground-level casting operations. Another advantage is that the size and mass of the panels are restricted only by the capacity of the lifting equipment and the limits of tolerable stress in concrete. They are cast (usually 4" to 8" in thickness) on the building slab itself or on adjacent casting beds and then tilted and lifted into position by a crane. Depending on the radius capacity and/or reach of the lifting equipment and the amount of available space at the site, the panels may be cast in single units or in layers or stacks. Liquid bond breaker or plastic sheeting should be used to prevent bonding of the panel to the building slab or casting bed.

During the casting process, lifting inserts or loops are placed at designed locations in the panel to provide connections to the crane lifting assembly. Lifting plates, cast in the edge of the panel, may also be used for connections. Large panels may require several double insets and rolling hitches on a lifting beam to accomplish the lifting procedure safely. Panels with large openings may require the temporary installation of wood or metal strongbacks to prevent cracking when the panel is being tilted and raised into position. These stiffeners are removed after the panel has been positioned and secured.

When sufficient strength is achieved in the cured concrete, the panel is ready to be tilted into position. The crane

assembly is attached to the inserts or edge lifting plates; the top of the panel is lifted; and the unit is raised to its final position. Temporary bracing is then installed with adjustable braces which have been anchored to the building slab prior to the tilting operation. Pre-erected wide flange, steel, or concrete columns support the panel after placement. The panel is then bolted or welded in its permanent position between the supporting columns.

Man-hours

Description	m/hr	Unit
Tilt-up Wall Panel		
5-1/2" Thick	.062	sf
7-1/2" Thick	.073	sf
Concrete Surface Treatments, Bond Breaker	.170	CSF
Forms in Place, Edge Forms, to 6" High, 4 Uses	.053	lf
Reinforcing in Place, #3 to #7	13.910	ton
Lightgauge Framing Angles Less than 4"	.018	lb
Lifting Inserts	.133	Ea.
Place and Vibrate Concrete, 4" Thick, Direct Chute	.436	cu yd
Finish Floor Monolithic Steel Trowel Finish	.015	sf
Cure with Sprayed Membrane Curing Compound	.168	CSF

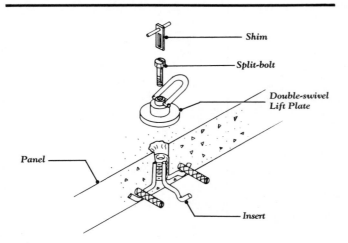

Cast-in-Place Lifting Insert

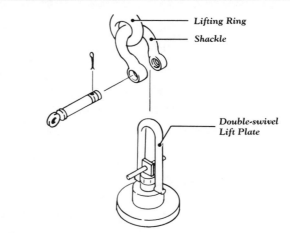

Lifting Plate Connection

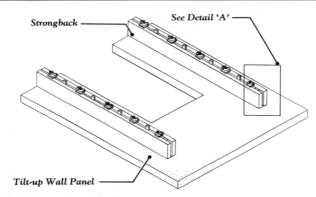

Temporary Strongback

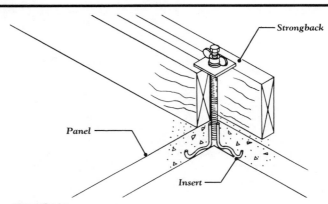

Detail 'A'

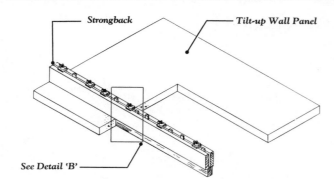

Large Opening Temporary Strongback

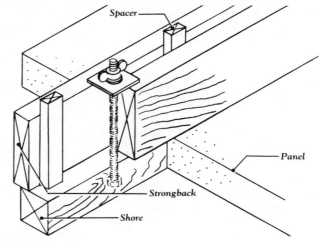

Detail 'B'

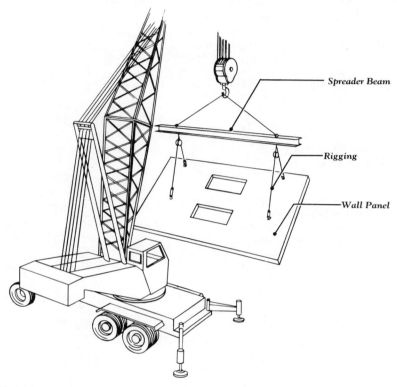

Spreader Beam

Rigging

Wall Panel

Lifting Procedure with Crane Assembly

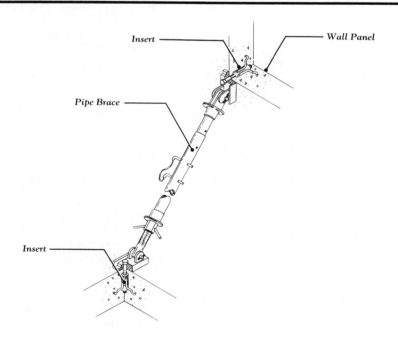

Insert

Wall Panel

Pipe Brace

Insert

Temporary Brace Assembly

4 EXTERIOR CLOSURE
CONCRETE BLOCK WALLS

Because of their strength, versatility, and economy, concrete blocks are among the most frequently used materials for constructing masonry walls and partitions. They may be used for many different types of bearing and nonbearing wall structures, including foundation walls, exterior and interior bearing walls, infill panels, interior partitions, and fire walls. They may also function effectively as backup walls for composite and cavity-design wall structures with brick or other veneer facings. If the concrete block wall is to serve as the finished wall, many architectural finishes are available, including ground (exposed aggregate), split, scored, and split-ribbed surfaces. Slump block, which features a distinctive bulging face, and blocks with geometric faces or embossed patterns provide other options for enhancing the appearance of exposed block-wall surfaces.

Concrete blocks are manufactured in two types, solid and hollow, and in various strength ratings. If the cross-sectional area, exclusive of voids, is 75% or greater than the gross area of the block, then it is classified as "solid block". If the area is below the 75% figure, then the block is classified as "hollow block". The strength of concrete block is determined by the compressive strength of the type of concrete used in its manufacture, or by the equivalent compressive strength, which is based on the gross area of the block, including voids.

There are several special aggregates that can be used to manufacturer lightweight blocks. These blocks can be identified by the weight of the concrete mixture used in their manufacture. Regular weight block is made from 125 lb per cubic foot concrete (PCF) and lightweight block from 105 to 85 PCF concrete. Costs for lightweight concrete block as compared to regular weight range from 20% more for 105 PCF block to 50% more for 85 PCF block.

Care should be exercised when installing concrete block walls to prevent cracking caused by block shrinkage, temperature expansion and contraction, excessive stress in a particular area, or excessive moisture. The cracking can be controlled by selecting and installing blocks of the proper moisture content for the locality and by employing a sufficient amount of horizontal joint reinforcing. Cracking can also be controlled with vertical control joints, which are typically placed at intervals ranging from 20' to 40'. The spacing intervals of these control joints depends on the wall height and the amount of joint reinforcing used. Corners and openings generally require control joints as well. Mortar joints throughout the wall should be made weather tight (to restrict moisture invasion) by being tooled and compressed into a concave shape.

Joint reinforcing and individual ties serve as important components of the various types of concrete block walls. Two types of joint reinforcing are available: the truss type and the ladder type. Because the truss type provides better load distribution, it is normally used in bearing walls. The ladder type is usually installed in light-duty walls that serve nonbearing functions. Both types of joint reinforcing may also be used to tie together the inner and outer wythes of composite or cavity-design walls. Corrugated strips, as well as Z-type, rectangular, and adjustable wall ties, may also be

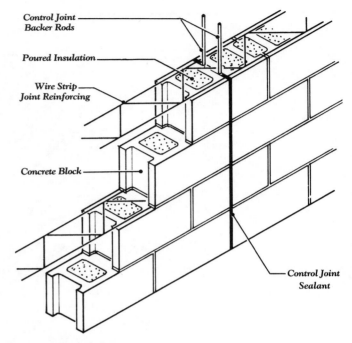

Control Joint
Backer Rods

Poured Insulation

Wire Strip
Joint Reinforcing

Concrete Block

Control Joint
Sealant

Concrete Block Wall

used for this purpose. Generally, one metal wall tie should be installed for each 4-1/2 square feet of veneer. Although both types of joint reinforcing may be used as ties, individual ties should not be used as joint reinforcing to control cracking.

Structural reinforcement is also commonly required in concrete block walls, especially in those that are loadbearing. Deformed steel bars may be used as vertical reinforcement when grouted into the block voids and as horizontal reinforcement when installed above openings and in bond beams. Horizontal and vertical bars may be grouted into the void normally used as the collar joint in a composite wall. Lintels should be installed to carry the weight of the wall above openings. Steel angles, built-up steel members, bond beams filled with steel bars and grout, and precast shapes may function as lintels.

Various methods may be employed for insulating concrete block walls. Rigid foam inserts or loose perlite can be used to fill the voids in the blocks of single wythe structures. Composite and single wythe walls can also be insulated by installing the insulating material between the furring strips for the interior wall facing. For cavity walls, rigid-board insulation may be attached within the cavity to the surface of the inner wythe.

Different cleaning and finishing techniques may be used to enhance the appearance of block walls and to protect them from the elements. Small amounts of dried mortar and soiling can be cleaned from the block surface with a stiff brush. Exterior and interior block walls can be painted or coated with a variety of opaque and clear sealers.

Man-hours

Description	m/hr	Unit
Foundation Walls, Trowel Cut Joints, Parged		
1/2" Thick, 1 Side, 8" x 16" Face		
Hollow		
8" Thick	.093	sf
12" Thick	.122	
Solid		
8" Thick	.096	sf
12" Thick	.126	sf
Back up Walls, Tooled Joint 1 Side,		
8" x 16" Face		
4" Thick	.091	sf
8" Thick	.100	sf
Partition Walls, Tooled Joint 2 Sides		
8" x 16" Face		
Hollow		
4" Thick	.093	sf
8" Thick	.107	sf
12" Thick	.141	sf
Solid		
4" Thick	.096	sf
8" Thick	.111	sf
12" Thick	.148	sf
Stud Block Walls, Tooled Joints 2 Sides		
8" x 16" Face		
6" Thick and 2", Plain	.098	sf
Embossed	.103	sf
10" Thick and 2", Plain	.108	sf
Embossed	.114	sf
6" Thick and 2" Each Side, Plain	.114	sf
Acoustical Slotted Block Walls		
Tooled 2 Sides		
4" Thick	.127	sf
8" Thick	.151	sf
Glazed Block Walls, Tooled Joint 2 Sides		
8" x 16", Glazed 1 Face		
4" Thick	.116	sf
8" Thick	.129	sf
12" Thick	.171	sf
8" x 16", Glazed 2 Faces		
4" Thick	.129	sf
8" Thick	.148	sf
8" x 16", Corner		
4" Thick	.140	Ea.
Structural Facing Tile, Tooled 2 Sides		
5" x 12", Glazed 1 Face		
4" Thick	.182	sf
8" Thick	.222	sf
5" x 12", Glazed 2 Faces		
4" Thick	.205	sf
8" Thick	.246	sf
8" x 16", Glazed 1 Face		
4" Thick	.116	sf
8" Thick	.129	sf
8" x 16", Glazed 2 Faces		
4" Thick	.123	sf
8" Thick	.137	sf
Exterior Walls, Tooled Joint 2 Sides, Insulated		
8" x 16" Face, Regular Weight		
8" Thick	.110	sf
12" Thick	.145	sf
Lightweight		
8" Thick	.104	sf
12" Thick	.137	sf
Architectural Block Walls, Tooled Joint 2 Sides		

Man-hours (cont.)

Description	m/hr	Unit
8" x 16" Face		
4" Thick	.116	sf
8" Thick	.138	sf
12" Thick	.181	sf
Interlocking Block Walls, Fully Grouted		
Vertical Reinforcing		
8" Thick	.131	sf
12" Thick	.145	sf
16" Thick	.173	sf
Bond Beam, Grouted, 2 Horizontal Rebars		
8" x 16" Face, Regular Weight		
8" Thick	.133	lf
12" Thick	.192	lf
Lightweight		
8" Thick	.131	lf
12" Thick	.188	lf
Lintels, Grouted, 2 Horizontal Rebars		
8" x 16" Face, 8" Thick	.119	lf
16" x 16" Face, 8" Thick	.131	lf
Control Joint 4" Wall	.013	lf
8" Wall	.020	lf
Grouting Bond Beams and Lintels		
8" Deep Pumped 8" Thick	.018	lf
12" Thick	.025	lf
Concrete Block Cores Solid		
4" Thick By Hand	.035	sf
8" Thick Pumped	.038	sf
Cavity Walls 2" Space Pumped	.016	sf
6" Space	.034	sf
Joint Reinforcing		
Wire Strips Regular Truss to 6" Wide	.267	CLF
12" Wide	.400	CLF
Cavity Wall with Drip Section to 6" Wide	.267	CLF
12" Wide	.400	CLF
Lintels Steel Angles Minimum	.008	lb
Maximum	.016	lb
Wall Ties	.762	C
Coping For 12" Wall Stock Units, Aluminum	.200	lf
Precast Concrete	.188	lf
Structural Reinforcing, Placed Horizontal,		
#3 and #4 Bars	.018	lb
#5 and #6 Bars	.010	lb
Placed Vertical, #3 and #4 Bars	.023	lb
#5 and #6 Bars	.012	lb

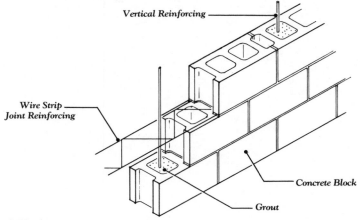

Grouted and Reinforced Block

Vertical Reinforcing

Wire Strip Joint Reinforcing

Concrete Block

Grout

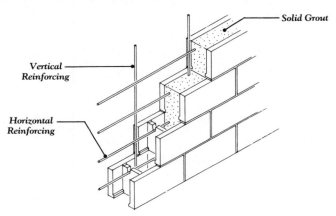

Interlocking Concrete Block

Solid Grout

Vertical Reinforcing

Horizontal Reinforcing

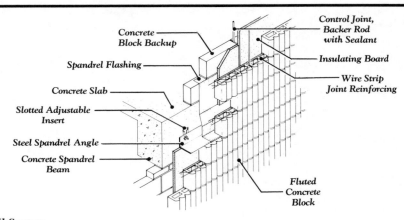

Block Face Cavity Wall System

Concrete Block Backup

Spandrel Flashing

Concrete Slab

Slotted Adjustable Insert

Steel Spandrel Angle

Concrete Spandrel Beam

Control Joint, Backer Rod with Sealant

Insulating Board

Wire Strip Joint Reinforcing

Fluted Concrete Block

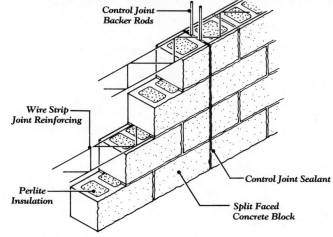

Ground Face Block Wall
Control Joint Backer Rods
Wire Strip Joint Reinforcing
Control Joint Sealant
Ground Face Concrete Block

Ground Face Block Wall

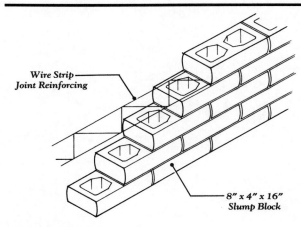

Control Joint Backer Rods
Wire Strip Joint Reinforcing
Perlite Insulation
Control Joint Sealant
Split Faced Concrete Block

Split Faced Block Wall

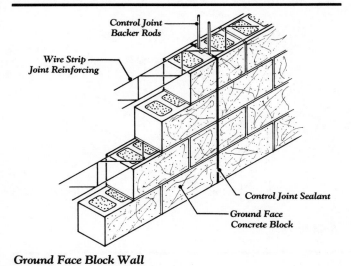

Wire Strip Joint Reinforcing
Perlite Insulation
Split Ribbed Concrete Block

Split Ribbed Block Wall

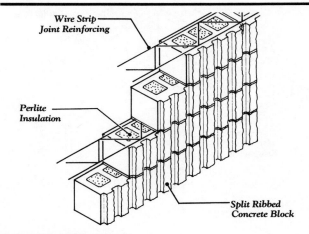

Wire Strip Joint Reinforcing
8" x 4" x 16" Slump Block

Slump Block Wall

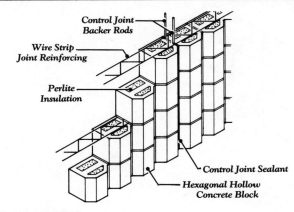

Control Joint Backer Rods
Wire Strip Joint Reinforcing
Perlite Insulation
Control Joint Sealant
Hexagonal Hollow Concrete Block

Hexagonal Block

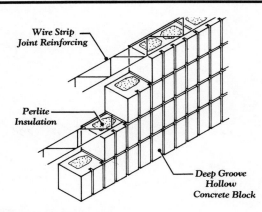

Wire Strip Joint Reinforcing
Perlite Insulation
Deep Groove Hollow Concrete Block

Deep Groove Hollow Block

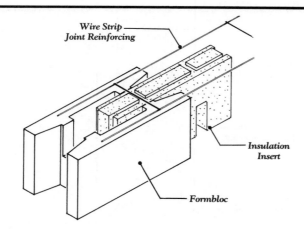

Formbloc

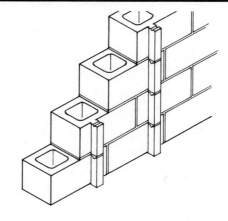

Stud Block

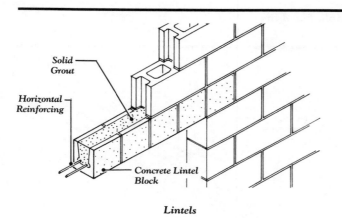

Lintels

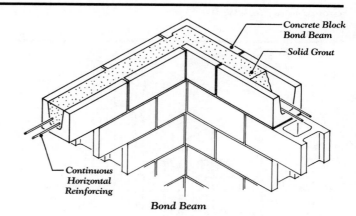

Bond Beam

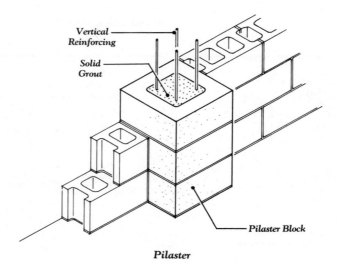

Pilaster

Concrete Block

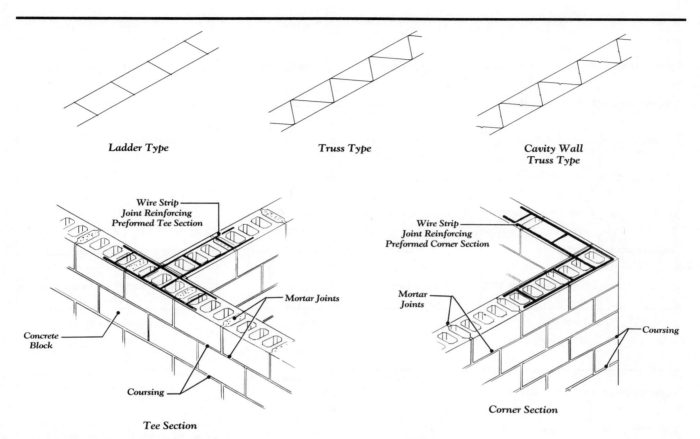

Ladder Type

Truss Type

**Cavity Wall
Truss Type**

Wire Strip
Joint Reinforcing
Preformed Tee Section

Mortar Joints

Concrete
Block

Coursing

Tee Section

Wire Strip
Joint Reinforcing
Preformed Corner Section

Mortar
Joints

Coursing

Corner Section

Wire Strip Joint Reinforcing Types

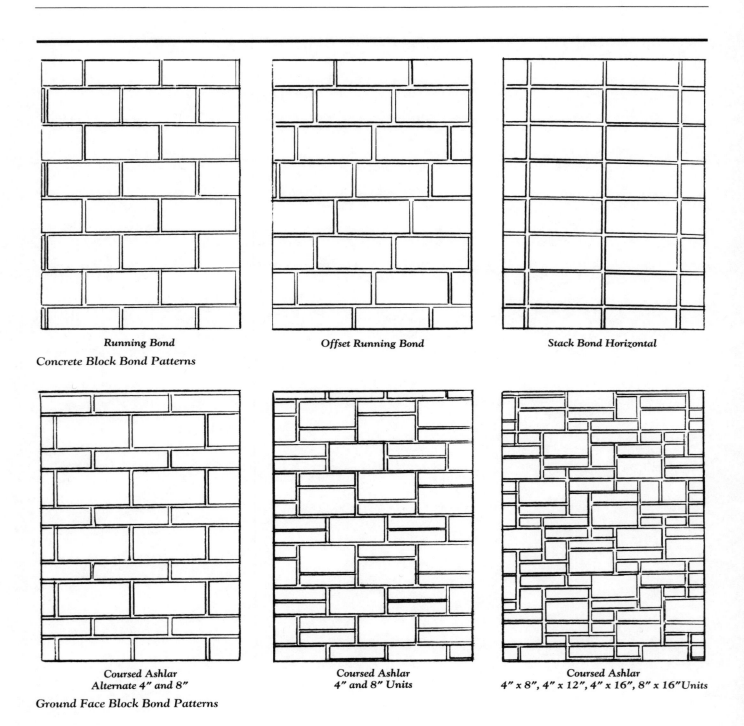

Running Bond

Offset Running Bond

Stack Bond Horizontal

Concrete Block Bond Patterns

**Coursed Ashlar
Alternate 4″ and 8″**

**Coursed Ashlar
4″ and 8″ Units**

**Coursed Ashlar
4″ x 8″, 4″ x 12″, 4″ x 16″, 8″ x 16″ Units**

Ground Face Block Bond Patterns

4 EXTERIOR CLOSURE
BRICK WALLS

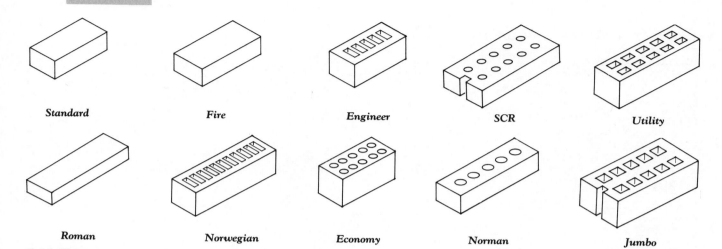

Standard   Fire   Engineer   SCR   Utility

Roman   Norwegian   Economy   Norman   Jumbo

Brick Shapes

Brick is among the most popular of wall materials because it is durable, economical to maintain, readily available in most areas, and varied in size and color. Also, brick can be installed rapidly with local labor and with a limited number of specialized tools and equipment. Brick can be used alone in single wythe walls or as the facing material in composite or cavity walls. Although the insulating value of a single wythe brick wall is a low 1.6R rating, it has the relatively high fire rating of one hour.

Brick walls may function as bearing structures if the proper guidelines are followed for the type of brick, the type of mortar, and the reinforcement methods and materials. For example, brick units with a compressive strength between 2,000 and 14,000 psi, when installed with a mortar of commensurate compressive strength, produce a wall with a strength between 500 and 3,000 psi. Horizontal joint reinforcement can be placed within the wall to control shrinkage cracks and to tie the face wythe to the backup wythe in composite or cavity bearing wall systems. To provide resistance to lateral and flexural loads, bar reinforcement may be grouted vertically into the brick cores or vertically and horizontally into the collar joint between two wythes of bearing wall.

Because brick walls are not waterproof, provisions must be made to limit the amount of water penetration through the exterior face. Flashing should be installed at the junction of walls and floors, as well as over and under openings. In cavity walls, the outside face of the backup wall or the insulation between the wythes should be waterproofed. Weep holes should be located above the flashing at brick shelves, relieving angles, and lintels to provide a means of escape for moisture that has penetrated the wall.

One of the greatest advantages of brick as a wall material is the availability of different sizes and variations of length-to-width-to-height dimensions. The most common size brick measures nominal 4″ wide x 8″ long, with heights of 2-2/3″ (standard), 3-1/5″ (engineer), 4″ (economy), 5-1/3″ (double), and 8″ (square or panel). The next most commonly used size measures 4″ in width by 12″ in length, with heights of 2″ (Roman), 2-2/3″ (Norman), 3-1/5″

(Norwegian), 4″ (utility), 5-1/3″ (triple), and 12″ (square or panel). The heights of courses of brick may also vary slightly with the thickness requirements of mortar joints. For example, a 1/2″ mortar joint requires a brick that is 1/16″ thinner than does a 3/8″ mortar joint to maintain the same modular coursing in the wall. Because the two apparent dimensions in any brick wall structure are its length and height, these two measurements are critical in determining the number of units of brick material to be used. Therefore, bricks with larger length and height dimensions are more economically installed because fewer of them have to be laid per square foot of wall area.

In addition to the advantage of the variety of sizes and dimensions, bricks may also be installed in many different positions and patterns within the wall to enhance its structure and appearance. A brick may be laid in one of six basic positions: stretcher, header, soldier, sailor, rowlock, and shiner. A brick unit is called a "stretcher" when its length is horizontal and aligned with the plane of the wall. This method of positioning bricks is the most commonly employed format. A brick unit is called a "header" when its length runs perpendicular to, and its width is aligned with, the plane of the wall. Before the development of brick ties, headers were used to attach the brick face to the backup wall. "Soldier" brick describes the positioning of the unit with its length placed vertically in the plane of the wall. This method of positioning is often used to accent the fascia above windows or to create a horizontal belt around the building at floor levels.

The various combinations of stretcher and header patterns within brick walls determine its bond. Running bond describes courses of stretchers that are laid without any headers and with alternate courses staggered in alignment. If the stretchers line up vertically, the format is called "stack bond." "Common bond" describes the combination of five courses of stretchers and one course of headers. "Flemish bond" is composed of stretchers and headers laid alternately in the same course; "English bond" consists of alternating courses of stretchers and headers.

Brick and Mortar Quantities

Running Bond							For Other Bonds Standard Size Add to SF Quantities in Table to Left		
Number of Brick per SF of Wall - Single Wythe with 1/2" Joints				CF of Mortar per M Bricks, Waste Included					
Type Brick	Nominal Size (Incl. Mortar) L H W	Modular Coursing	Number of Brick per SF	1 wythe	2 wythe	Bond Type	Description	Factor	
Standard	8 x 2-2/3 x 4	3C=8"	6.75	12.9	16.5	Common	full header every fifth course	+20%	
Economy	8 x 4 x 4	1C=4"	4.50	14.6	19.6		full header every sixth course	+16.7%	
Engineer	8 x 3-1/5 x 4	5C=16"	5.63	13.6	17.6	English	full header every second course	+50%	
Fire	9 x 2-1/2 x 4-1/2	2C=5"	6.40	550# fireclay	—	Flemish	alternate headers every course	+33.3%	
Jumbo	12 x 4 x 6 or 8	1C=4"	3.00	34.0	41.4		every sixth course	+5.6%	
Norman	12 x 2-2/3 x 4	3C=8"	4.50	17.8	22.8	Header = W x H exposed		+100%	
Norwegian	12 x 3-1/5 x 4	5C=16"	3.75	18.5	24.4	Rowlock = H x W exposed		+100%	
Roman	12 x 2 x 4	2C=4"	6.00	17.0	20.7	Rowlock stretcher = L x W exposed		+33.3%	
SCR	12 x 2-2/3 x 6	3C=8"	4.50	26.7	31.7	Soldier = H x L exposed		—	
Utility	12 x 4 x 4	1C=4"	3.00	19.4	26.8	Sailor = W x L exposed		–33.3%	

Man-hours

Description	m/hr	Unit
Brick Wall,		
Veneer		
4" Thick		
Running Bond		
Standard Brick (6.75/sf)	.182	sf
Engineer Brick (5.63/sf)	.154	sf
Economy Brick (4.50/sf)	.129	sf
Roman Brick (6.00/sf)	.160	sf
Norman Brick (4.50/sf)	1.25	sf
Norwegian Brick (3.75/sf)	.107	sf
Utility Brick (3.00/sf)	.089	sf
Common Bond, Standard Brick (7.88/sf)	.216	sf
Flemish Bond, Standard Brick (9.00/sf)	.267	sf
English Bond, Standard Brick (10.13/sf)	.286	sf
Stack Bond, Standard Brick (6.75/sf)	.200	sf
6" Thick		
Running Bond		
S.C.R. Brick (4.50/sf)	.129	sf
Jumbo Brick (3.00/sf)	.092	sf
Backup		
4" Thick		
Running Bond		
Standard Brick (6.75/sf)	.167	sf
Solid, Unreinforced		
8" Thick Running Bond (13.50/sf)	.296	sf
12" Thick, Running Bond (20.25/sf)	.421	sf
Solid, Rod Reinforced		
8" Thick, Running Bond	.308	sf
12" Thick, Running Bond	.444	sf
Cavity		
4" Thick		
4" Backup	.242	sf
6" Backup	.276	sf
Brick Chimney		
16" x 16", Standard Brick w/8" x 8" Flue	.889	vlf
16" x 16", Standard Brick w/8" x 12" Flue	1.000	vlf
20" x 20", Standard Brick w/12" x 12" Flue	1.140	vlf
Brick Column		
8" x 8", Standard Brick 9.0 vlf	.286	vlf
12" x 12", Standard Brick 20.3 vlf	.640	vlf

Man-hours (cont.)

Description	m/hr	Unit
20" x 20", Standard Brick 56.3 vlf	1.780	vlf
Brick Coping		
Precast, 10" Wide, or Limestone, 4" Wide	.178	lf
Precast 14" Wide, or Limestone, 6" Wide	.200	lf
Brick Fireplace		
30" x 24" Opening, Plain Brickwork	40.000	Ea.
Firebox Only, Fire Brick (110/Ea.)	8.000	Ea.
Brick Prefabricated Wall Panels, 4" Thick		
Minimum	.093	sf
Maximum	.144	sf
Brick Steps	53.330	M
Window Sill		
Brick on Edge	.200	lf
Precast, 6" Wide	.229	lf
Needle Brick and Shore, Solid Brick		
8" Thick	6.450	Ea.
12" Thick	8.160	Ea.
Repoint Brick		
Hard Mortar		
Running Bond	.100	sf
English Bond	.123	sf
Soft Mortar		
Running Bond	.080	sf
English Bond	.098	sf
Toothing Brick		
Hard Mortar	.267	vlf
Soft Mortar	.200	vlf
Sandblast Brick		
Wet System		
Minimum	.024	sf
Maximum	.057	sf
Dry System		
Minimum	.013	sf
Maximum	.027	sf
Sawing Brick, Per Inch of Depth	.027	lf
Steam Clean, Face Brick	.033	sf
Wash Brick, Smooth	.014	sf

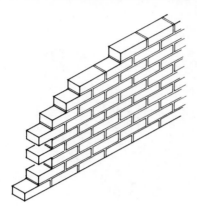

Standard Running Bond

Full Header Every 6th Course

English, Full Header Every 2nd Course

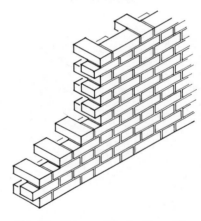

Flemish, Alternate Header Every Course

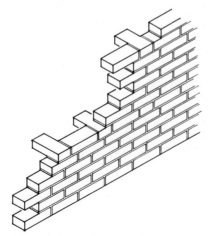

Flemish Alternate Header Every 6th Course

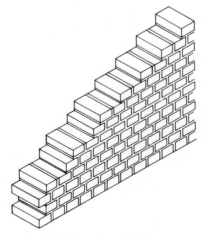

Brick Veneer, Full Header Throughout

Brick Bond Patterns

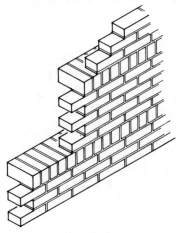

Rowlock Course

Shiner Course or Rowlock Stretcher

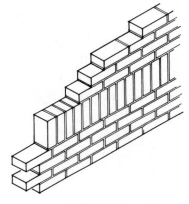

Soldier Course

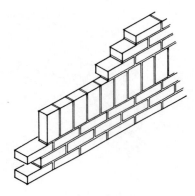

Sailor Course

Brick Bond Patterns

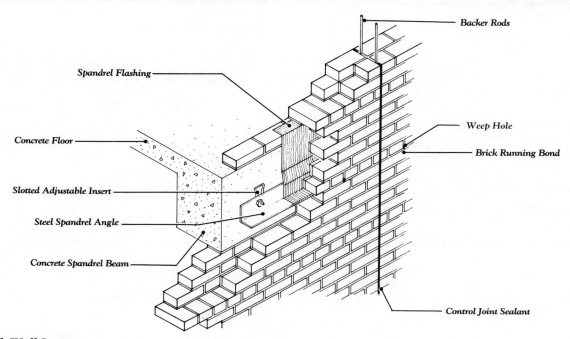

Backer Rods

Spandrel Flashing

Concrete Floor

Slotted Adjustable Insert

Steel Spandrel Angle

Concrete Spandrel Beam

Weep Hole

Brick Running Bond

Control Joint Sealant

Solid Brick Wall System

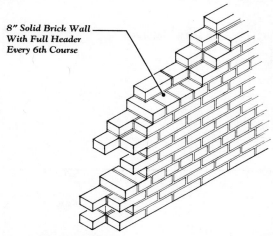

8" Solid Brick Wall
With Full Header
Every 6th Course

8" Solid Brick Wall

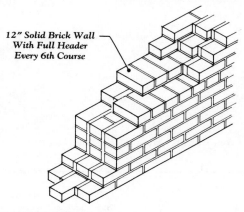

12" Solid Brick Wall
With Full Header
Every 6th Course

12" Solid Brick Wall

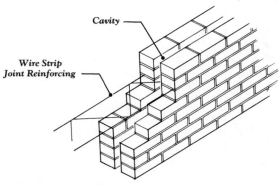

Brick Cavity Wall

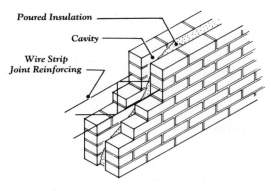

Insulated Cavity Wall

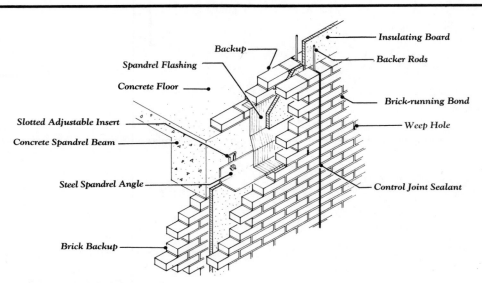

Brick Face Cavity Wall System with Brick Backup

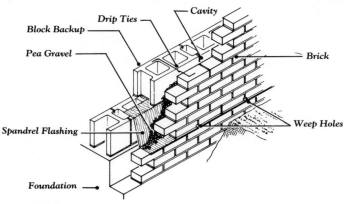

Brick Face Cavity Wall System with Block Backup

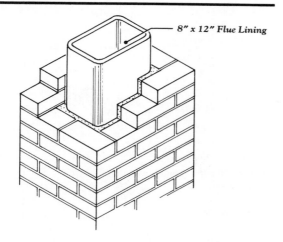

Brick Chimney

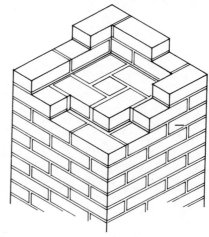

Brick Column

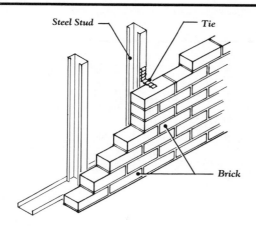

Interior Brick Veneer/Metal Stud Backup

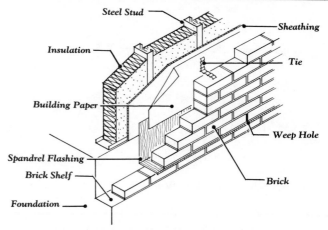

Brick Veneer/Metal Stud Backup Wall System

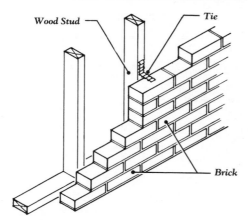

Interior Brick Veneer/Wood Stud Backup

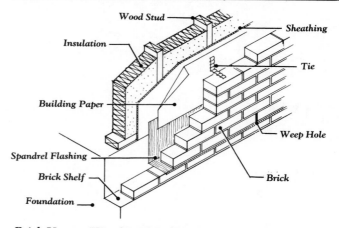

Brick Veneer/Wood Stud Backup System

143

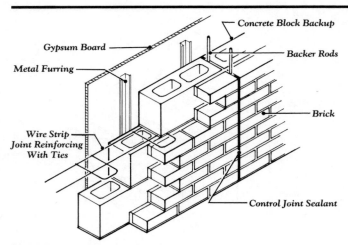

Brick Face Composite Wall

Gypsum Board
Metal Furring
Wire Strip Joint Reinforcing With Ties
Concrete Block Backup
Backer Rods
Brick
Control Joint Sealant

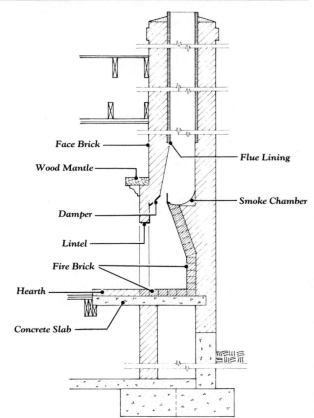

Brick Face Composite Wall System

Poured Insulation
Concrete Block Backup
Wire Strip Joint Reinforcing With Ties
Backer Rods
Concrete Floor
Spandrel Flashing
Slotted Adjustable Insert
Concrete Spandrel Beam
Steel Spandrel Angle
Weep Hole
Control Joint Sealant
Brick-running Bond

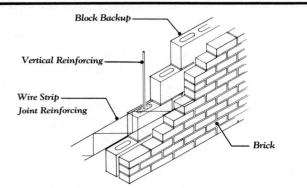

Brick and Reinforced Block Wall

Block Backup
Vertical Reinforcing
Wire Strip Joint Reinforcing
Brick

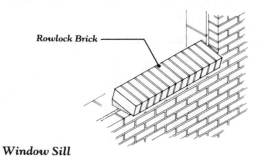

Window Sill

Rowlock Brick

Brick Fireplace

Face Brick
Wood Mantle
Damper
Lintel
Fire Brick
Hearth
Concrete Slab
Flue Lining
Smoke Chamber

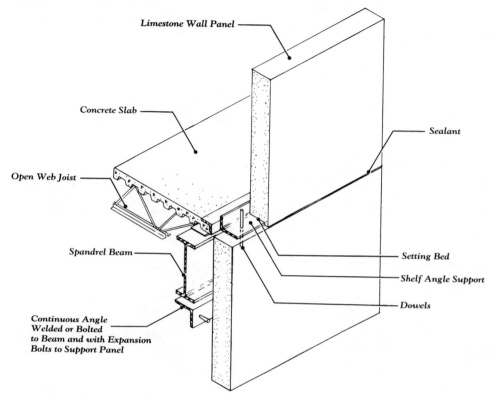

Stone Panel Connection at Floor Slab

Labels: Limestone Wall Panel, Concrete Slab, Open Web Joist, Spandrel Beam, Continuous Angle Welded or Bolted to Beam and with Expansion Bolts to Support Panel, Sealant, Setting Bed, Shelf Angle Support, Dowels

Because of its durability, strength, and unique appearance, stone provides a wide range of structural and decorative applications as a building material. It can be installed in small units, referred to as "building stone," which can be assembled in many different formats, with or without mortar, to create walls and veneers of all sorts. Stone can also be employed as a material in larger units, such as stone panels, which are installed with elaborate anchor and framing systems and used as decorative wall facings in high-rise office and other commercial buildings. The cost of stone building material varies considerably with location and depends on the available supply and the distance that it must be transported. Other significant cost factors include the extent of quarrying and subsequent processing required to extract and produce the finished stone product.

Building stone may be used for many different types of structures, in random sizes and shapes, or in pre-cut sizes and shapes. Small, irregular building stone that has been quarried in random sizes is called "rubble" or "fieldstone." This material, which is sold by the ton, is commonly installed with mortar to create small retaining or freestanding walls. Flat, random pieces of rubble can also be assembled without mortar and will derive strength from the interlocking of adjacent pieces within the structure. Fieldstone may be split by hand on site to provide a flat exterior surface for a patterned wall or fireplace.

Small stone units may also be quarry split and processed to meet aesthetic needs. For example, decorative building stone can be purchased by the ton in 4″ thick slabs, available in lengths ranging from 6″ to 14″ and in heights ranging from 2″ to 16″. These pieces are commonly installed with mortar to create veneer walls of varying patterns, such as ledge stone, spider web, uncoursed rectangular, and squared. Ashlar stone, also priced by the ton, is the name given to building stones that have been sawn on the edges so as to produce a rectangular face. This shape makes ashlar stone another veneer material because the pieces can be arranged in either a regular or random-coursed pattern within the face of a wall. Stone veneer can be tied to the backup wall with galvanized ties or 8″ stone headers in a method similar to that used in brick veneer walls. The coverage of stone veneer ranges from 35 to 50 square feet per ton for 4″ wide veneer, with correspondingly reduced coverages per ton for veneers of 6″ and 8″ in width.

Large stone facing panels can be installed as decorative features in many types of commercial buldings. These panels, which are usually priced by the square foot, are available in widths of up to 5′ and in thicknesses of approximately 1″ to 5″. Panel faces may be clear or patterned with split, sawn, or sand-rubbed surface finish. The edges of the panel are saw cut and the back is planed.

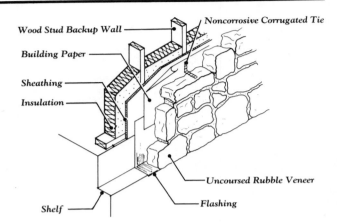

Exterior Stone Veneer Wall

Wood Stud Backup Wall
Building Paper
Sheathing
Insulation
Noncorrosive Corrugated Tie
Uncoursed Rubble Veneer
Flashing
Shelf

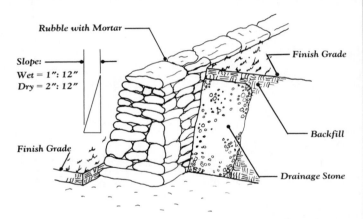

Rubble Retaining Wall

Rubble with Mortar
Slope:
Wet = 1": 12"
Dry = 2": 12"
Finish Grade
Finish Grade
Backfill
Drainage Stone

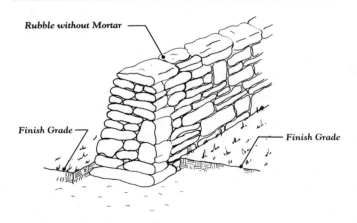

Rubble Freestanding Wall

Rubble without Mortar
Finish Grade
Finish Grade

The method selected for anchoring and supporting these panels is determined by the size of the panel, the conditions of its placement, and its location. In some installations the weight of the panel can be supported on steel angles that are attached at the bottom to the structural frame of the building. Horizontal movement at the top is prevented with an anchor rod, strap, or grip stay which is also attached to the structural frame. Small pieces and other sections of panel that cannot be attached to the structural frame are supported by doweling them into adjacent secured panels. Thin veneer panels that tend to crack or fracture are backed with durable stone strip liners that reinforce the panel and serve as the means of attachment to the structure. An alternative to the conventional methods of securing stone facing panels directly to the structure is to connect them during manufacture to a strut subframing system, preassembled steel frame, or precast concrete panel.

Man-hours

Description	m/hr	Unit
Stone Walls, Random Field Stone		
To 18" Thick	.533	cf
Over 18" Thick	.508	cf
Including Footing, 3' Below Grade, Exposed		
Face, Building Stone		
To 6' High		
Dry Set	.457	sf
Mortar Set	.400	sf
6' to 10' High		
Dry Set	.356	sf
Mortar Set	.320	sf
Stone Veneer Walls, Building Stone, Ashlar or		
Rubble, Mortar Set		
Minimum	.320	sf
Maximum	.333	sf
Stone Facing Panels		
Granite		
3/4" to 2-1/2" Thick	.308	sf
2-1/2" to 4" Thick	.364	sf
Limestone		
2" Thick,		
3' x 5' Panel	.308	sf
3" Thick		
4' x 9' Panel	.178	sf
5" Thick		
5' x 14' Panel	.145	sf
Marble		
3/4" Thick	.308	sf
1-1/4" Thick	.320	sf
2" Thick	.333	sf
Sandstone		
2-1/2" Thick		
2' x 4' Panel	.308	sf
4" Thick		
3'-6" x 8' Panel	.400	sf
Slate		
1" Thick		
9 sf	.500	sf
4' x 6' Panel	.222	sf
Stone Anchors, L, T, Split Bend		
Twisted or Plate/Rod	.762	C

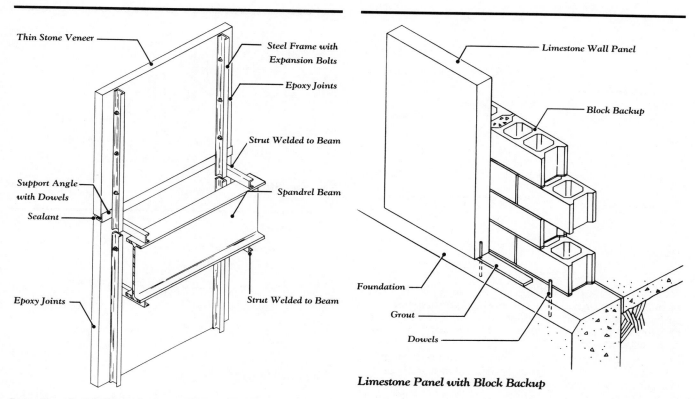

Thin Stone Veneer

Steel Frame with Expansion Bolts

Epoxy Joints

Strut Welded to Beam

Support Angle with Dowels

Spandrel Beam

Sealant

Epoxy Joints

Strut Welded to Beam

Stone Panel with Plant Assembled Steel Frame

Limestone Wall Panel

Block Backup

Foundation

Grout

Dowels

Limestone Panel with Block Backup

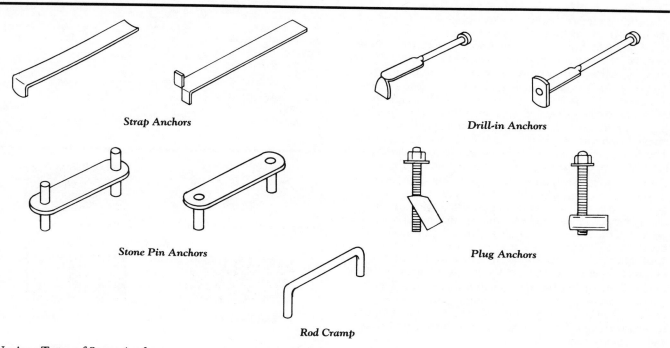

Strap Anchors

Drill-in Anchors

Stone Pin Anchors

Plug Anchors

Rod Cramp

Various Types of Stone Anchors

147

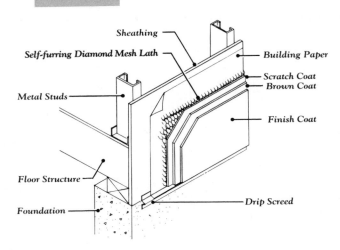

Stucco

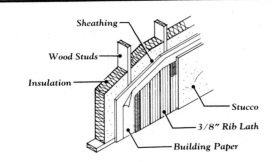

Stucco on Rib Lath

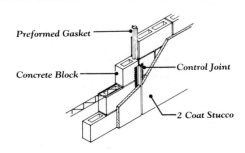

Stucco on Masonry

Stucco

Stucco is a facing material applied like plaster to the exterior of buildings to provide a decorative and weather-resistant surface. Stucco is usually a mixture of sand, cement, lime, and water, but it may be composed of patented mixes with manufactured additives.

Stucco mixes differ from plaster mixes because portland cement and lime are the basic ingredients in lieu of gypsum. Portland cement plasters are more difficult to trowel and finish, but are durable, relatively unaffected by water, and withstand repeated cycles of wetting, drying, freezing, and thawing.

The two common methods of applying stucco are the open-frame method and the direct-application system. In the open-frame method, the stucco is applied to galvanized metal lath attached directly to wood or steel studs. Three coats of stucco are normally required: (1) a scratch or first coat, (2) a brown or leveling coat, and (3) a finish coat.

The frame should be sufficiently braced to prevent movement. The flashings and drips should be carefully placed to prevent water penetration. Waterproof building paper applied over the studs or self-furring lath with integral waterproof paper are used for moisture protection. The system may also be waterproofed by back plastering or applying a coat of stucco on the interior side of the wall.

Directly applied stucco may be placed over sheathed wood or steel stud systems, or applied on concrete or masonry walls. One coat of stucco, with a bonding agent, or two coats of stucco may be used if the backing material possesses the surface characteristics required for adequate bonding. If the surface is not suitable for direct stucco application the surface should first be covered with building paper and then furred with self-furring metal lath.

Exterior stucco finishes are usually highly textured and coarse in appearance, to conceal staining and shrinkage cracks. The finish stucco coat may be tinted, painted, or treated with colorless coating materials that help prevent water penetration.

Man-hours

Description	m/hr	Unit
Stucco 3 Coats, 1″ Thick, Float Finish	.923	sq yd
Stucco with Bonding Agent 1 Coat	.167	sq yd
Lath		
Nailed to Wood		
1.8 lb	.133	sq yd
3.6 lb	.145	sq yd
Wired to Steel		
1.8 lb	.151	sq yd
3.6 lb	.160	sq yd

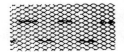

Self-furring Diamond Mesh Lath

Small Diamond Mesh Lath

Flat Rib Lath

3/8″ Rib Lath

Types of Metal Lath

4 EXTERIOR CLOSURE
GLASS BLOCK

Glass block, which was a popular building material in the 1930's, has recently undergone a revival of usage in the construction industry, particularly for the construction of exterior walls. It has the dual advantage of admitting light but preventing heat transmission. The maximum freestanding wall area allowed using glass block construction is 144 square feet, but larger areas are feasible using intermediate bracing.

The more decorative types of glass block are finding acceptance as interior walls as well. They can provide privacy yet allow limited visual accessibility, and are therefore frequently used in places such as reception areas. Patterns may be introduced into the glass during the forming process. Ceramic design can be fused to the surface for a decorative finish, but this will increase the cost of the block.

Glass block can be placed on a raised base, plate, or sill, provided that the surfaces to be mortared are primed with asphalt emulsion. Wall recesses and channel tract that receive the glass block should be lined with expansion strips prior to oakum filler and caulking. Horizontal joint reinforcing is specified for flexural as well as shrinkage control and is laid in the joints along with the mortar. End blocks are anchored to the adjacent construction with metal anchors, if no other provisions for attachment exist. If intermediate support is required, vertical I-shaped stiffeners can either be installed in the plane of the wall or adjacent to it, but the stiffeners should be tied to the wall with wire anchors. The top of the wall is supported between angles or in a channel track similar to the jambs.

Glass block is manufactured in sizes from 6″ by 6″ to 12″ by 12″, with thickness from 3″ to 4″. The block may be hollow or fused brick; the latter allows vision through the block. Inserts can be manufactured into the block to reduce solar transmission to the building interior.

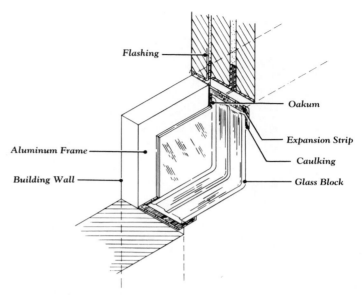

Glass Block Head Section

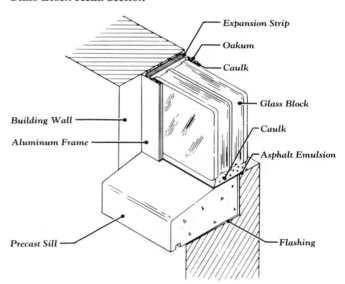

Glass Block Sill Section

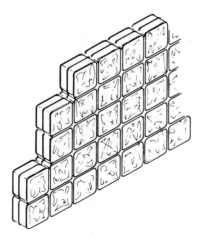

Glass Blocks

Man-hours

Description	m/hr	Unit
Glass Block, Including Scaffolding		
Under 1000 sf		
6″ x 6″	.347	sf
8″ x 8″	.250	sf
12″ x 12″	.228	sf
Over 5000 sf		
6″ x 6″	.275	sf
8″ x 8″	.186	sf
12″ x 12″	.166	sf
Aluminum Columns	.045	lb

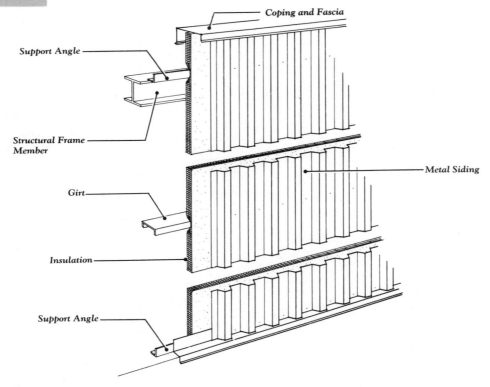

Coping and Fascia

Support Angle

Structural Frame Member

Girt

Insulation

Support Angle

Metal Siding

Metal Siding System

Formed metal panels are used for mansard roofing and building fascia, as well as for siding. The panels are made from sheet metal, aluminum, or steel and can span distances from 3′ to 12′. The ability to transfer wind loads across these spans depends on the gauge of the sheet metal and the profile that is formed with the sheet during rolling. Where structural strength is not required, the corrugation can be regarded as merely an architectural feature.

Sheet metal can be finished with a variety of coatings for corrosion protection. Steel panels may be galvanized, aluminized, or stainless. Both aluminum and steel panels can have a baked enamel or porcelain finish.

Canopy fascia and mansard panels can be fastened or clipped to extruded or channel-shaped metal girts. They are placed horizontally at the required span distance and are supported on channel truss frames. Panels may also be fastened to solid backing (sheathing mounted on stud framing) directly or with metal clips. Where a backup wall exists, metal panels may simply be mounted on channel or stud furring. The top of the panels are supplied with matching coping or gravel stops, and the panel bottom with nose or sill trim. Conversion trim pieces are available for

slope transitions. Inside and outside corners are finished with post or cap assemblies. Smooth or ventilated ceiling panels are used for soffits.

For a more custom installation, panel ribs and flat sheets can be supplied as separate components to be installed to a stud and sheathing backup system within the constraints of the field conditions. The exposed open ends of ribs can be closed with "plugs", should coping not be desirable. Ribs may also be field crimped at slope transitions. A very dramatic vertical rib effect can be obtained by mounting a standard 4″ wide panel on horizontal carriers that will support the ribs in a variety of angles to the plane of the exterior wall, giving a "louvered" appearance.

A total siding system involves the metal panels fastened to a structural girt support, as mentioned above, with the addition of insulation and backup (liner) panels. In some metal siding systems, the liner panel and insulation are installed to the girt first. The face panel then interlocks with grooves in the liner panel. The thickness of the face and liner panels range from 24 to 20 gauge, and the span can reach 8′.

When the face and liner panels are stiffened by inserting a subgirt between them, the allowable span will increase to 15' for 22 gauge and to 30' for 18 gauge (and a deeper profile). The structural girt may be eliminated altogether if the liner panel is given a deeper profile or a structural stud is integrated into the liner design. This insulated panel and liner system can be factory assembled to save installation time in the field. The liner panel may be perforated to provide sound absorption. A multi-leaf gypsum wallboard layer can also be enclosed in a double-subgirt system with the face and liner panel to provide a fire-rated wall panel. Some factory assembled panels have a foamed-in-place insulation that is bonded to the face and liner sheet to provide composite structural action.

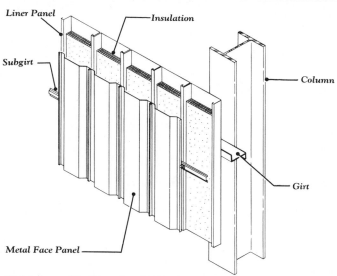

Field Assembled Insulated Metal Wall

Man-hours

Description	m/hr	Unit
Aluminum Siding		
On Steel Frame	.041	sf
On Wood Frame	.030	sf
Closure Strips and Flashing	.040	lf
Steel Siding, on Steel Frame	.040	sf
Metal Siding Panels, Insulated with Liner		
Field Assembled	.164	sf
Factory Assembled	.084	sf
Metal Fascia, No Furring or Framing Included		
Long Panels	.055	sf
Short Panels	.069	sf
Mansard Roofing,		
With Battens, Custom		
Multi-Sloped Surface	.145	sf
Projected Surface	.106	sf
Vertical Surface	.069	sf
Stock, All Surfaces	.069	sf
Framing, to 5' High	.069	lf
Soffits, Metal Panel	.064	sf
Furring		
Metal, 3/4" Channel		
16" On Center	.030	sf
24" On Center	.022	sf
Wood Strips		
On Masonry	.016	lf
On Concrete	.030	lf
Channel Framing, Overhead 24" On Center	.017	sf
Stud Framing		
Metal, Overhead,		
16" On Center	.026	sf
24" On Center	.017	sf
Wood, Overhead		
2" x 4"	.026	lf
2" x 8"	.070	lf
Sheathing		
Drywall	.022	sf
Plywood	.011	sf
Channel Girts, Steel		
Less than 8"	.016	lb
Greater than 8"	.009	lb
Slotted Channel Framing System		
Minimum	.006	lb
Maximum	.010	lb

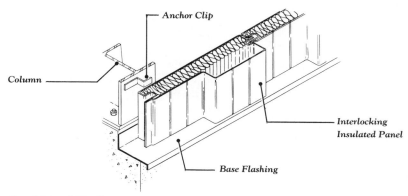

Factory Assembled Insulated Metal Wall

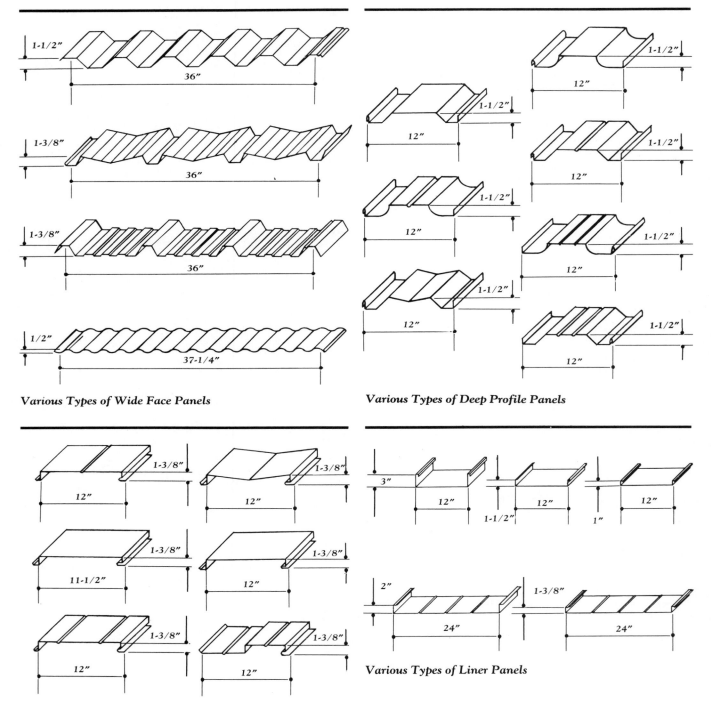

Various Types of Wide Face Panels

Various Types of Deep Profile Panels

Various Types of Flat and Fluted Wide Panels

Various Types of Liner Panels

4 EXTERIOR CLOSURE
CURTAIN WALL

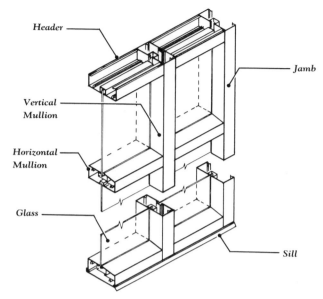

Header

Jamb

Vertical Mullion

Horizontal Mullion

Glass

Sill

Curtain Wall

A curtain wall is a non-structural facade that encloses a building. It consists of panels in a wide variety of materials and constructions. A curtain wall is held in place in a metal frame by caulking, gaskets, and sealants. The extent to which the curtain wall can be prefabricated varies by type: stick, in which all components are field assembled; panel and mullion, in which the panels are prefabricated into frames and field connected to mullions; and total panel systems, in which the mullions are preassembled into the panels.

A curtain wall must be designed to maintain its integrity under a variety of working conditions, including wind, rain, and extreme heat and cold. The mullions are extruded or fabricated I or tubular sections which supply the strength for lateral loads. The mullions may be one- or two-piece sections with "pockets" to receive cover plate assemblies, gaskets and accent strips, and flanges or recesses to receive panel frames and window sashes. Aluminum is an ideal material for complex sections, due to its extrudability, but bronze and steel (stainless or weathering) can be rolled into less complicated sections. Mullions can also be designed as tracks for window washing equipment.

Sills and head pieces are fabricated similar to mullions, but with less substantial cross sections. Horizontal members will support panel frames or sashes, mechanical enclosures at the sills, and drapery pockets at the heads. Gutters must be fabricated into sill and head pieces with weep slots and baffles, and legs must be provided for overlapping with flashing. Floor sills must permit adjustment during installation, as well as provide anchorage to the floor.

Curtain wall panels can provide a wide variety of appearances for the building exterior. For a monolithic exterior, the vision and spandrel panels can be equally spaced, using the same type of glass. The vision panel can

be glazed by means of structural lock-strip gaskets attached directly to the mullion, sill, and head extrusions. The spandrel can be identical, except that it would be backed with an insulated metal panel set behind the glass. A slight variation of the above involves the use of extruded, compression glazing gaskets around the edges of vision or spandrel panels. A closed-cell, neoprene gasket is placed on the outside with a dense "wedge" gasket on the inside; each is inserted into pockets in the mullion and sill extrusions. Pressure plates on the sills and the mullions may be used to further compress the seals after the glass is set.

A window wall system involves the support of only the window panels. Mullions may be set back and extend only from the sill to the head, which may in turn be supported directly by the exterior wall framing. Gaskets may be necessary only at the sill and head with silicone sealant applied at the mullions where the glass panels may be butt-jointed.

Creating a curtain wall is possible with window and spandrel panels that look and function differently. Window panels may be alternated with either a fixed or operable sash, or with fixed and operable sashes in the same panel. Spandrel panels may be constructed of metal, plastic, or hardboard facing sheets, with a variety of finishes, and filled with a foamed-in-place or adhered insulation layer. A system of this type often is manufactured to be field set into a common frame. Like the monolithic system, the spandrel panel may be made of an insulated backer panel mounted independently of the face panel.

Designs have been perfected which eliminate panel framing network of mullions and sills. Seam systems are used at panel joints (inside and outside) and include gaskets, battens, splines, and frames. The panel edges can be interlocking, in addition to being reinforced with extrusions that provide the required stiffness.

Man-hours

Description	m/hr	Unit
Aluminum Sash, Stock	.080	sf Opng.
Custom		
Minimum	.080	sf Opng.
Maximum	.188	sf Opng.
Aluminum Mullions	.066	lf
Steel Sash, Custom	.080	sf Opng.
Steel Mullions	.066	lf
Spandrel Glass		
Over 2000 sf	.133	sf
Under 2000 sf	.145	sf
Spandrel Insulated Glass Panels	.133	sf
Curtain Walls, Stock, Including Glazing		
Minimum	.156	sf
Average	.171	sf
Maximum	.200	sf
Window Walls, Stock, Including Glazing		
Minimum	.150	sf
Average	.171	sf
Maximum	.218	sf

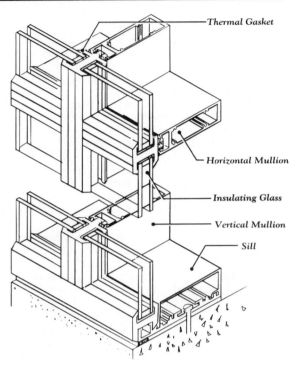

Thermal Gasket — Thermal Gasket

Horizontal Mullion

Insulating Glass

Vertical Mullion

Sill

Thermal Gasket System

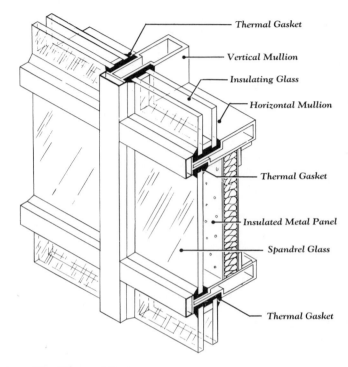

Thermal Gasket

Vertical Mullion

Insulating Glass

Horizontal Mullion

Thermal Gasket

Insulated Metal Panel

Spandrel Glass

Thermal Gasket

Low Rise Thermal System

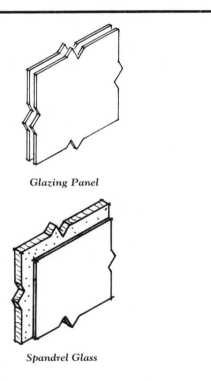

Glazing Panel

Spandrel Glass

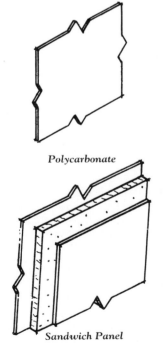

Polycarbonate

Sandwich Panel

Curtain Wall Panel Types

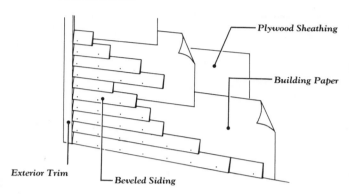

Cedar Clapboards

Wood siding is usually milled of wood species that can withstand extreme variations of weather. Redwood and cedar are two moisture-resistant woods used for board and sheet siding, as well as for shingles. Fir and pine are also used for board and sheet siding, but they must be finished with stain or paint after installation. Man-made materials can also qualify for exterior applications. For example, hardboard and medium-density-overlay products can be used, but they also must be painted or stained. Plywood sheet siding is manufactured with waterproof glue to provide weather protection.

Watertight installation of siding is essential. Cedar and redwood clapboard siding are beveled in widths from 4″ to 10″. Proper installation dictates that upper boards should overlap the lower boards by a minimum of 1″ and be nailed through plywood-sheathing backup to wall studs. Horizontal boards are butted and caulked into vertical corner boards at exterior and interior corners.

Vertical tongue-and-groove board siding is blind-nailed together through the sheathing to horizontal blocking spaced at 24″ on center. It is manufactured in widths from 4″ to 12″. Channel and shiplap board siding are also lapped, but are face-nailed to the blocking. Vertical boards are installed with a 1/2″ joint between the boards and are held in place with a nailed batten strip. All vertical board siding should extend to the corners and be overlapped with corner boards. Vertical siding is usually interrupted at floors with a horizontal wood beltline and flashing strip, or flashing strip alone.

Plywood sheet siding can be installed directly to the stud wall, without sheathing, and nailed along all panel edges and intermediate stud supports. Vertical edges can be lapped, battened, or simply butted and caulked. Horizontal joints are usually flashed.

Clapboard siding comes in the better grades of cedar and redwood, i.e., B, A, and clear grade. Rough sawn and channel cedar are available in no. 3 grade and better.

Man-hours

Description	m/hr	Unit
Siding		
Clapboard		
Cedar, Beveled		
1/2″ x 6″	.032	sf
1/2″ x 8″	.029	sf
3/4″ x 10″	.027	sf
Redwood, Beveled		
1/2″ x 4″	.040	sf
1/2″ x 8″	.032	sf
3/4″ x 10″	.027	sf
Board		
Redwood		
Tongue and Grove		
1″ x 4″	.053	sf
1″ x 8″	.043	sf
Channel		
1″ x 10″	.028	sf
Cedar		
Butted		
1″ x 4″	.033	sf
Channel		
1″ x 8″	.032	sf
Board and Batten		
1″ x 12″	.030	sf
Pine		
Butted		
1″ x 8″	.029	sf
Sheets		
Hardboard, Lapped	.021	sf
Plywood		
MDO		
3/8″ Thick	.021	sf
1/2″ Thick	.022	sf
3/4″ Thick	.024	sf
Texture 1-11		
5/8″ Thick	.023	sf
Sheathing		
Plywood on Walls		
3/8″ Thick	.013	sf
3/4″ Thick	.016	sf
Wood Fiber	.013	sf
Trim Exterior, Up to 1″ x 6″	.040	lf
Fascia		
1″ x 6″	.032	lf
1″ x 8″	.035	lf

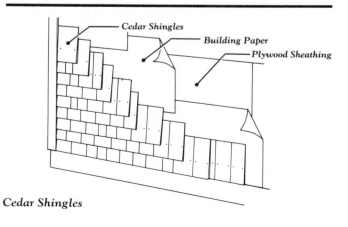

Cedar Shingles

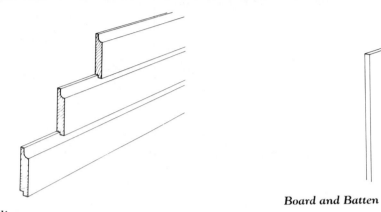

Channel Siding

Board and Batten

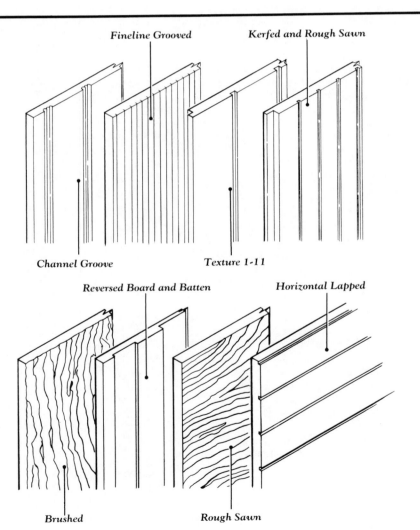

Fineline Grooved

Kerfed and Rough Sawn

Channel Groove

Texture 1-11

Reversed Board and Batten

Horizontal Lapped

Brushed

Rough Sawn

Types of Plywood Paneling

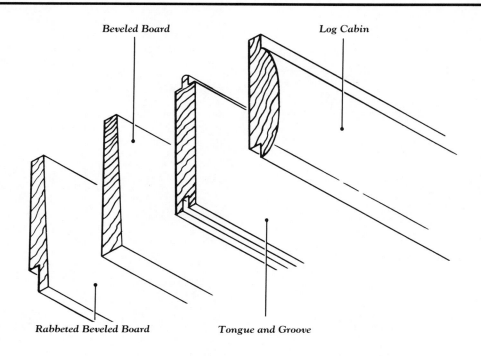

Beveled Board

Log Cabin

Rabbeted Beveled Board

Tongue and Groove

Types of Siding

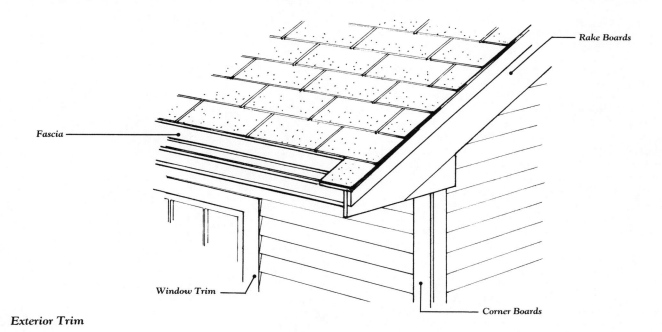

Rake Boards

Fascia

Window Trim

Corner Boards

Exterior Trim

4 EXTERIOR CLOSURE
ENTRANCES

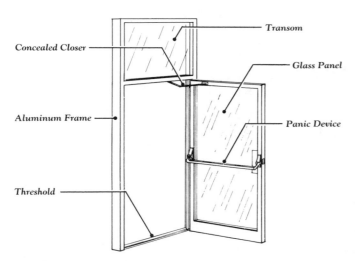

Glass Entrance System with Transom

An entrance system should be designed for appearance, while at the same time maintaining security.

Entrance doors may support the full glass panel in a standard wide stile and rail made of aluminum, solid bronze, or stainless steel. The standard frame is 1-3/4″ x 4″ tubular framing into which the door is center hung with concealed overhead or floor closers, or offset hung with pivot or butt hinges. A balanced door that pivots about its center while sliding to the jamb requires a larger head extrusion for the overhead pivot and roller guide. Sliding door head frames must accommodate rollers or tracks in series or parallel. For the automatic type, motor drives should be provided depending on the specified operation. A revolving door head frame must be of sufficient size to receive the speed control unit. Placing a transom above the entrance door usually will add to the head dimensions also.

The other stile widths are narrow and ultra-narrow, with corresponding thinner frame sections. To compensate for the reduced sections, the extrusion thickness will increase from 1/8″ to 1/4″, and finally will be solid for the ultra-narrow design.

Glass used in the floor-to-ceiling door panels has to be tempered as a safety precaution. Thickness can vary from 1/4″ plain or 1″ insulated for standard doors, but ultra-narrow stile requires a minimum of 1/2″ thickness. Tempered glass doors use 1/2″ and 3/4″ thick glass. This type of door has no stiles and may have top and bottom rails, bottom rails only, or no rails at all.

Flush glazed storefronts consist of single-thickness glass panels, from 1/4″ to 1/2″ thick, mounted in tubular plastic or metal extrusions, measuring up to 1-3/4″ x 4-1/2″, using straight push-in type gaskets on both faces. A seamless version utilizes 1-1/2″ reinforced tubular extrusion on the inside face, with a wedge-type gasket insert and a T-plate extrusion on the outside with pre-shimmed butyl glazing tape for a positive seal. When thicker insulated window units are needed, bigger stick-

type framing extrusions are required, which revert to neoprene exterior and dense "wedge" interior gaskets. The extrusion for a monumental grade storefront will have an I-shaped section whose dimensions will vary from 5-1/2″ to 7-1/2″ deep. A second piece snaps between the flanges to hold the glass and glazing.

An all-glass storefront system is supported top and bottom with exposed or recessed extrusions. Panels can be butt joined with a silicone sealant. For added stiffness, glass mullions can be located at suitable joints and connected with structural sealant.

Man-hours

Description	m/hr	Unit
Glass Entrance Door, Including Frame and Hardware		
Balanced, Including Glass		
3′ x 7′		
Economy	17.780	Ea.
Premium	22.860	Ea.
Hinged, Aluminum		
3′ x 7′	8.000	Ea.
3′ x 7′, 3′ Transom	8.890	Ea.
6′ x 7′	12.310	Ea.
6′ x 10′, 3′ Transom	14.550	Ea.
Stainless Steel, Including Glass		
3′ x 7′		
Minimum	10.000	Ea.
Average	11.430	Ea.
Maximum	13.330	Ea.
Tempered Glass		
3′ x 7′	8.000	
6′ x 7′	11.430	Ea.
Hinged, Automatic, Aluminum		
6′ x 7′	22.860	Ea.
Revolving, Aluminum, 7′-0″		
Diameter x 7′ High		
Minimum	42.667	Ea.
Average	53.333	Ea.
Maximum	71.111	Ea.
Stainless Steel, 7′-0″ Diameter x 7′ High	106.667	Ea.
Bronze, 7′-0″ Diameter x 7′ High	213.333	Ea.
Glass Storefront Systems, Including Frame and Hardware		
Hinged, Aluminum, Including Glass, 400 sf		
w/3′ x 7′ Door		
Commercial Grade	.107	sf
Institutional Grade	.123	sf
Monumental Grade	.139	sf
w/6′ x 7′ Door		
Commercial Grade	.119	sf
Institutional Grade	.139	sf
Monumental Grade	.160	sf
Sliding, Automatic, 12′ x 7′-6″ w/5′ x 7′ door	22.860	Ea.
Mall Front, Manual, Aluminum		
15′ x 9′	12.310	Ea.
24′ x 9′	22.860	Ea.
48′ x 9′ w/Fixed Panels	17.780	Ea.
Tempered All-Glass w/Glass Mullions,		
up to 10′ High	.185	sf
up to 20′ High, Minimum	.218	sf
Average	.240	sf
Maximum	.300	sf

Man-hours (cont.)

Description	m/hr	Unit
Entrance Frames, Aluminum, 3′ x 7′	2.290	Ea.
3′ x 7′, 3′ Transom	2.460	Ea.
6′ x 7′	2.670	Ea.
6′ x 7′, 3′ Transom	2.910	Ea.
Glass, Tempered, 1/4″ Thick	.133	sf
1/2″ Thick	.291	sf
3/4″ Thick	.457	sf
Insulating, 1″ Thick	.213	sf
Overhead Commercial Doors		
Frames not Included		
Stock Sectional Heavy Duty Wood		
1-3/4″ Thick		
8′ x 8′	8.000	Ea.
10′ x 10′	8.889	Ea.
12′ x 12′	10.667	Ea.
Fiberglass and Aluminum Heavy Duty Sectional		
12′ x 12′	10.667	Ea.
20′ x 20′, Chain Hoist	32.000	Ea.
Steel 24′ Gauge Sectional Manual		
8′ x 8′ High	8.000	Ea.
10′ x 10′ High	8.889	Ea.
12′ x 12′ High	10.667	Ea.
20′ x 14′ High, Chain Hoist	22.857	Ea.
For Electric Trolley Operator to 14′ x 14′	4.000	Ea.
Over 14′ x 14′	8.000	Ea.

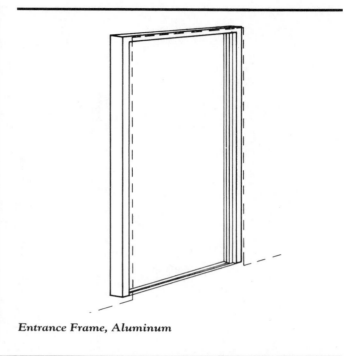

Entrance Frame, Aluminum

Sliding, Automatic, Mall Entrance

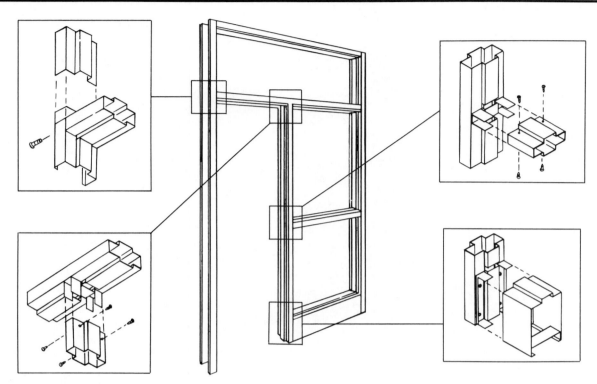

Pre-Engineered, "Stick" System Entrance

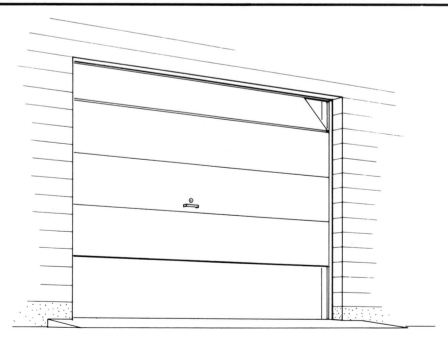

Commercial Overhead Door

4 EXTERIOR CLOSURE
EXTERIOR WINDOWS/SASH

Window systems enhance the living and/or work space by allowing the occupants of a building to enjoy the exterior environment. If window systems are carefully designed and appropriately selected, they can also provide many other advantages, including efficient use of daylight and passive solar energy, as well as the exclusion of exterior noise, glare, and excessive heat and cold. Window units are available in many different sizes, methods of operation, frame configurations, and types of tinting and insulating features to meet the specific needs of each installation situation. Window units are comprised of two basic components: the frame sash, and the glass and glazing materials necessary to support the glass panels within the sash.

The window frame, which supports the sash and provides for attachment to the building structure, and the window sash, which supports the glass within the frame are constructed of three basic materials. Frames and sashes can be constructed of steel, aluminum, or wood. Steel or hollow metal requires finish maintenance, aluminum typically does not. Wood frames and sashes can require maintenance or be maintenance-free. Maintenance-free wood windows are achieved by cladding the exposed wood with vinyl or aluminum.

The varied frame and sash designs of window units provide several operative options besides fixed, or permanently closed, sash. Pivoted sashes come in two configurations; projected sashes swing from horizontal hinges, while casement sashes swing from vertical hinges. Units which are designed to slide in a horizontal track are called sliding sashes; those that slide in vertical tracks are called hung sashes. All types of sashes can be manufactured with thermal breaks to reduce the heat flow through the sash construction material. Depending on the type of sash operation, requirements for mullions or intermediate sub-framing, and the incorporation of thermal breaks, the frame sash may be factory-shipped as complete units or in sections. During the installation of all window units, care must be taken to incorporate weep holes in the sill and to weatherstrip the sash.

The glass itself offers a variety of options which are determined by the location and function of the window unit and the architectural design of the structure. Monolithic glass ranges in thickness from 1/8" to 7/8"; insulating glass ranges in thickness from 1/2" to 1", including the sealed air space between the layers of glass. A general rule is that the thickness of any glass installation is determined by the design wind load. Tinted glass, which is available in bronze, gray, or green color, can reduce solar transmittance by as much as 83%. Glare and heat gain from the outside may be reduced with tinted or untinted reflective glass, which contains a thin coating of metallic oxide. Reflective glass, when installed appropriately, greatly affects a building's heating and cooling requirements. Tempered glass, because it has a high bending strength and breaks into small cubical fragments when shattered, serves as an excellent choice for window installations which require safety glass. Laminated glass may also function as a safety glass because of its shatter resistant construction. It includes a layer of polyvinyl butyral between the glass plies.

However, this type of glass is more commonly empolyed in bullet resistant and sound reduction window installations. Various methods of support for the glass panel are employed depending on the window type. In wood windows, the glass can come in contact with the sash as wood is an insulating material. In metal windows it is necessary to ensure that the glass floats in, but never touches the metal sash of the window unit. The weight of the glass sheet is carried by two special rubber setting blocks which are located between 1/4 and 1/8 of the panel's width from each bottom corner. Soft rubber edge blocking, which is placed in the vertical sash channels and head, prevents horizontal movement of the panel. The location of necessary edge blocking may be adjusted according to the design of the sash's operation and appropriately placed so as to prevent movement of the panel in the plane of the sash.

Movement perpendicular to the plane of glass panels is prevented by several wet and dry glazing techniques which may employ conventional glazing materials as well as newer elastomeric sealants and specialized gaskets. Wet glazing employs shims which are intermittently spaced around the inside and outside of the panel and set in a continuous heel and cap bead of elastomeric sealant or glazing compound. In another wet glazing method, preshimmed tape is used in place of the shims and the sealant on the exterior face of the panel. In both methods, front and back stops maintain pressure against the glazing shims. Dry glazing techniques require no sealants, as extruded rubber compression gaskets perform both the glazing and sealing functions. With this system, a closed-cell neoprene gasket is installed on the outside of the panel and this is compressed when a dense rubber wedge gasket is inserted on the inside. Additional pressure, when needed, may be attained by incorporating into the frame an interior or exterior pressure plate which is tightened with a screwdriver. Another dry glazing system utilizes a structural, or lock-strip, gasket which is a preformed elastomeric seal that also functions as a combined interior-exterior stop. A continuous strip of this gasket material is removed from the frame during installation and then reinstalled with the glass panel to form a permanent, immovable seal.

Combination Storm and Screen Window

Projected Window - Aluminum

Single Hung Window - Aluminum

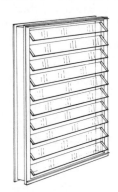

Jalousie Window - Aluminum

Fixed Steel Sash - Custom

Awning Window - Wood

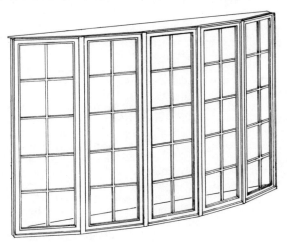

Casement Bow Window - Wood

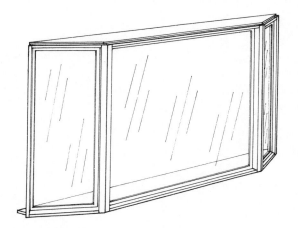

Casement Bay Window - Wood

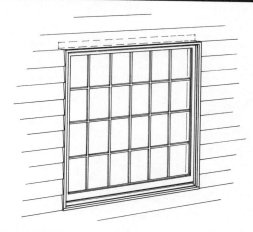

Fixed Picture Window - Wood

Casement Window - Wood

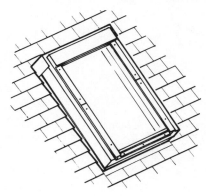

Roof Window

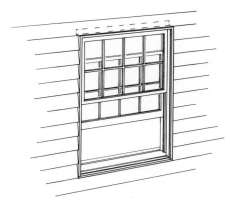

Double Hung Window - Wood

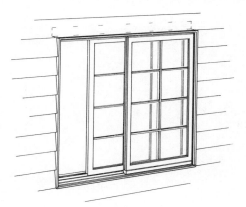

Sliding Window - Wood

163

Man-hours

Description	m/hr	Unit
Custom Aluminum Sash, Glazing Not Included	.080	sf
Stock Aluminum Windows, Frame and Glazing		
Including Casement, 3'-1" x 3'-2",		
Wood Opening	1.600	Ea.
Masonry Opening	2.667	Ea.
Combination Storm and Screen,		
2'-4" x 3'-2" Wood Opening	.533	Ea.
Masonry Openings	.941	Ea.
3'-4" x 5'-6" Wood Opening	.615	Ea.
Masonry Opening	1.143	Ea.
Projected, with Screen,		
3'-1" x 3'-2" Wood Opening	1.600	Ea.
Masonry Opening	2.667	Ea.
4'-5' x 5'-3" Wood Opening	2.000	Ea.
Masonry Opening	4.000	Ea.
Single Hung, 2'x3' Wood Opening	1.600	Ea.
Masonry Opening	2.667	Ea.
3'-4" x 5'-0" Wood Opening	1.778	Ea.
Masonry Opening	3.200	Ea.
Sliding, 3' x 2" Wood Opening	1.600	Ea.
Masonry Opening	2.667	Ea.
5'x3" Wood Opening	1.778	Ea.
Masonry Opening	3.200	Ea.
8'x4' Wood Opening	2.667	Ea.
Masonry Opening	5.333	Ea.
Jalousies, 2'-3" x 4'-0" Opening	1.600	Ea.
3'-1' x 5'-3" Opening	1.600	Ea.
Custom Steel Sash Units, Glazing and		
Trim Not Included	.080	sf
Stock Steel Windows, Frame, Trim and		
Glass Included		
Double Hung, 2'-8" x 4'-6" Opening	1.333	Ea.
Commercial Projected, 3'-9" x 5'-5" Opening	1.600	Ea.
6'-9" x 4'-1" Opening	2.286	Ea.
Intermediate Projected, 2'-9" x 4'-1" Opening	1.333	Ea.
Custom Wood Sash, Including Double/Triple		
Glazing but Not Trim		
5'-4" Opening	3.200	Ea.
7'x4'-6" Opening	3.721	Ea.
8'-6" x 5' Opening	4.571	Ea.
Stock Wood Windows, Frame,		
Trim and Glass Included		
Awning Type, 2'-10" x 1'-9" Opening	1.333	Ea.
4'-4" x 6'-0" Opening	2.286	Ea.
Bow Bay, Fixed Lights, 8' x 5' Opening	5.333	Ea.
9'-9" x 5'-4" Opening	8.000	Ea.
End Units Vent., 8' x 5' Opening	6.400	Ea.
Casement, 1'-10" x 3'x 2" Opening	1.455	Ea.
4'-2" x 4'-2" Opening, 2 Leaf	1.778	Ea.
5'-11" x 5'-2" Opening, 3 Leaf	2.286	Ea.
9'-11" x 6'-3" Opening, 5 Leaf	3.200	Ea.
Double Hung, 2'-8" x 4'-6" Opening	2.353	Ea.
3'-0" x 5'-6" Opening	2.667	Ea.
Picture Window, 4'-6" x 4'-6" Opening	3.200	Ea.
Roof Window, Complete,		
2'-9" x 3'-8" Opening	5.000	Ea.
Sliding, 3'-4" x 2'-7" Opening	2.462	Ea.
5'-4" x 6'-0" Opening	2.667	Ea.

Residential fireplaces are available in three basic types: prefabricated freestanding, prefabricated built-in metal, and traditional masonry. Regardless of the type, fireplaces have the same basic functions, aesthetics and heating.

Prefabricated freestanding fireplaces are of metal construction and come in various designs, both contemporary and traditional. Freestanding fireplace systems consist of the fireplace unit, a fire-resistant hearth (typically brick, stone or tile), and the chimney to exhaust the combustion gases. If the unit is placed close to a combustible wall, a non-combustible wall coating may also be required.

Prefabricated built-in, metal fireplaces are constructed of materials similar to those of freestanding fireplaces. The difference is that they are recessed into the building wall. The system components are the same as for freestanding models, with the addition of a cavity to accept the unit. The space and material requirements for the cavity (as established by the manufacturers) should be closely followed.

Masonry fireplaces are typically constructed of brick and block masonry units. Because of the weight of the fireplace and chimney, a foundation of concrete block and/or cast-in-place concrete is required for support. The fire box consists of fire-resistant brick, while the hearth and face are of standard brick. Stone and tile are also used for these exterior surfaces. Accessories, such as clean-out doors, air vents and dampers complete the typical system.

Chimney types fall into two basic categories: masonry and prefabricated metal. Besides their use with fireplaces, chimneys are also required to vent exhaust gases from water heaters, furnaces, boilers and incinerators. Chimneys can be constructed within the building, on an outside wall, or as a freestanding structure. Foundation and intermediate support requirements will vary accordingly.

Masonry chimneys typically consist of a flue constructed from heat-resistant, refractory material, surrounded by a brick-framed exterior. Fire bricks and clay tiles are commonly used as refractory material. The weight of masonry chimneys is such that foundation support structures are also necessary. Both the chimney and the foundation should be sized according to the requirements of the specific installation.

Metal chimneys consist of sections of insulated or uninsulated pipe, connected with various types of standard fittings to suit the particular installation. Fittings for elbows, bends, support, draft control, and roof and ceiling penetration are typical. The pipe and fittings are most commonly fabricated from galvanized steel, though stainless steel is also used.

Whether metal or masonry, all chimneys should extend at least 3' above the highest point at which they pass through or by the roof. Chimneys should also be at least 2' higher than any roof ridge within a 10' radius. For all chimney installations, it is important to use the appropriate material and to maintain the proper space requirements for installations.

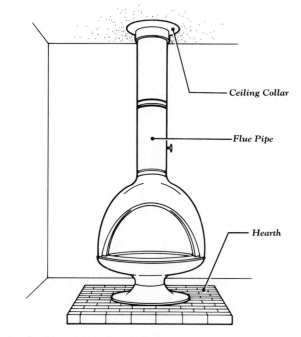

Freestanding Prefabricated Fireplace

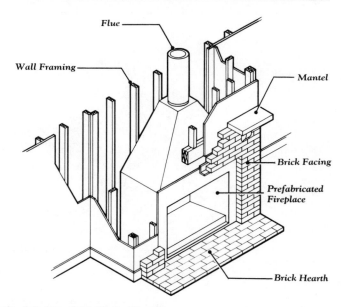

Prefabricated, Built-in Fireplace

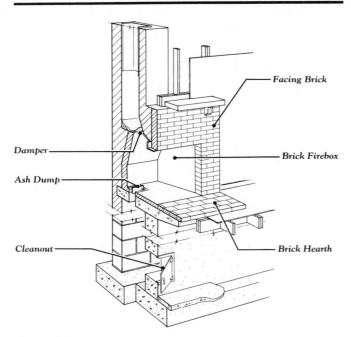

Facing Brick

Damper

Ash Dump

Brick Firebox

Cleanout

Brick Hearth

Masonry Fireplace - Exterior

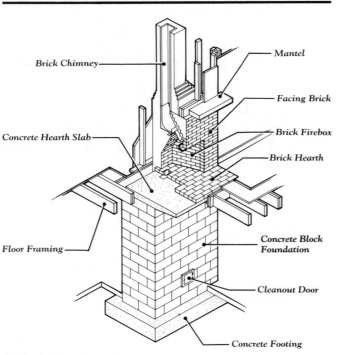

Brick Chimney

Mantel

Facing Brick

Brick Firebox

Concrete Hearth Slab

Brick Hearth

Floor Framing

Concrete Block
Foundation

Cleanout Door

Concrete Footing

Masonry Fireplace - Interior

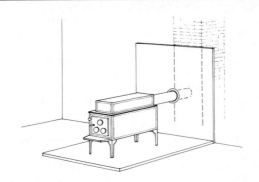

Cast Iron Woodburning Stove

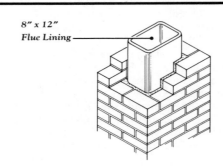

8" x 12"
Flue Lining

Masonry Chimney

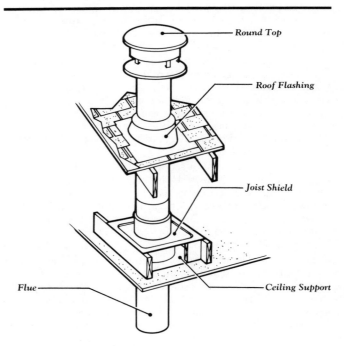

Round Top

Roof Flashing

Joist Shield

Flue

Ceiling Support

Prefabricated Metal Chimney

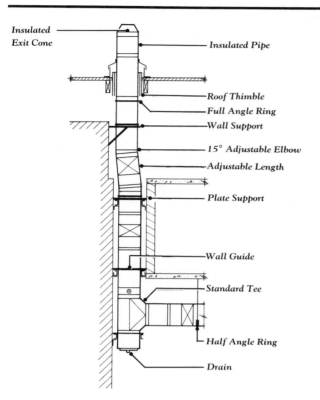

Insulated Exit Cone

Insulated Pipe

Roof Thimble

Full Angle Ring

Wall Support

15° Adjustable Elbow

Adjustable Length

Plate Support

Wall Guide

Standard Tee

Half Angle Ring

Drain

Prefabricated Metal Chimney

Man-hours

Description	m/hr	Unit
Masonry Fireplace, 30" x 24" Opening, Plain Brickwork, Including Hearth, Not Including Chimney or Foundation	48.000	Ea.
Prefabricated Freestanding Fireplace, Not Including Chimney	8.000	Ea.
Prefabricated Built-in Fireplace, Plain Brick Face and Hearth, Not Including Chimney or Foundation	40.000	Ea.
Woodburning Stove, Cast Iron, Not Including Chimney or Hearth	16.000	Ea.
Fireplace Accessories		
Cleanout Door and Frame, 12" x 12"	.800	Ea.
Rotary Control Damper, 30" Opening	1.333	Ea.
72" Opening	1.600	Ea.
Chimney, for Fireplaces		
Masonry, Standard Bricks,		
16" x 16" with one 8" x 8" Flue	.889	vlf
16" x 20" with one 8" x 12" Flue	1.000	vlf
16" x 24" with two 8" x 8" Flues	1.143	vlf
20" x 20" with one 12" x 12" Flue	1.143	vlf
20" x 32" with two 12" x 12" Flues	1.600	vlf
Prefabricated Metal Chimneys, All Fuel,		
Double Wall, Stainless Steel, 6" Diameter	.267	vlf
8" Diameter	.308	vlf
12" Diameter	.364	vlf
14" Diameter	.381	vlf

Man-hours (cont.)

Description	m/hr	Unit
All Fuel, Double Wall, Stainless Steel Fittings		
Ceiling Support 6" Diameter	.533	Ea.
8" Diameter	.615	Ea.
12" Diameter	.727	Ea.
14" Diameter	.762	Ea.
Elbow 15°, 6" Diameter	.533	Ea.
8" Diameter	.615	Ea.
12" Diameter	.727	Ea.
14" Diameter	.762	Ea.
Insulated Tee with Insulated Tee Cap,		
6" Diameter	.533	Ea.
8" Diameter	.615	Ea.
12" Diameter	.727	Ea.
14" Diameter	.762	Ea.
Joist Shield, 6" Diameter	.533	Ea.
8" Diameter	.615	Ea.
12" Diameter	.727	Ea.
14" Diameter	.762	Ea.
Round Top, 6" Diameter	.533	Ea.
8" Diameter	.615	Ea.
12" Diameter	.727	Ea.
14" Diameter	.762	Ea.
Adjustable Roof Flashing, 6" Diameter	.533	Ea.
8" Diameter	.615	Ea.
12" Diameter	.727	Ea.
14" Diameter	.762	Ea.
Simulated Brick Chimney Top, 24" x 24"	1.143	vlf
Chimneys, for Gas Fired Appliances, Prefabricated Metal		
Double-wall Pipe, 3" Diameter	.222	vlf
6" Diameter	.267	vlf
10" Diameter	.333	vlf
14" Diameter	.381	vlf
20" Diameter	.667	vlf
Fittings, 45° Elbow, 3" Diameter	.444	Ea.
6" Diameter	.533	Ea.
10" Diameter	.667	Ea.
14" Diameter	.762	Ea.
20" Diameter	1.333	Ea.
Roof Flashing, 3" Diameter	.444	Ea.
6" Diameter	.533	Ea.
10" Diameter	.667	Ea.
14" Diameter	.800	Ea.
20" Diameter	1.333	Ea.
Tee, 3" Diameter	.593	Ea.
6" Diameter	.667	Ea.
10" Diameter	.762	Ea.
14" Diameter	.889	Ea.
20" Diameter	1.412	Ea.
Top, 3" Diameter	.348	Ea.
6" Diameter	.400	Ea.
10" Diameter	.471	Ea.
14" Diameter	.533	Ea.
20" Diameter	.857	Ea.
Positive Pressure Vent Piping System		
Exit Cone, 6" Diameter	.348	Ea.
12" Diameter	.420	Ea.
24" Diameter	.923	Ea.
36" Diameter	1.090	Ea.
48" Diameter	1.200	Ea.

Man-hours (cont.)

Description	m/hr	Unit
Insulated Pipe, 6″ Diameter	.267	lf
12″ Diameter	.364	lf
24″ Diameter	.750	lf
36″ Diameter	.960	lf
48″ Diameter	1.263	lf
Ventilated Roof Thimble 6″, Diameter	.615	Ea.
12″ Diameter	1.330	Ea.
24″ Diameter	1.714	Ea.
36″ Diameter	2.400	Ea.
48″ Diameter	3.000	Ea.
Full Angle Ring, 6″ Diameter	.348	Ea.
12″ Diameter	.420	Ea.
24″ Diameter	.923	Ea.
36″ Diameter	1.090	Ea.
48″ Diameter	1.200	Ea.
Wall Support Assembly, 6″ Diameter	.727	Ea.
12″ Diameter	1.140	Ea.
24″ Diameter	2.000	Ea.
36″ Diameter	3.000	Ea.
48″ Diameter	4.000	Ea.
15° Adjustable Elbow, 6″ Diameter	.533	Ea.
12″ Diameter	.727	Ea.
24″ Diameter	1.500	Ea.
36″ Diameter	2.000	Ea.
48″ Diameter	2.400	Ea.
Adjustable Length, 6″ Diameter	.533	Ea.
12″ Diameter	.727	Ea.
24″ Diameter	1.500	Ea.
36″ Diameter	2.000	Ea.
48″ Diameter	2.400	Ea.
Wall Guide Assembly, 6″ Diameter	.432	Ea.
12″ Diameter	.533	Ea.
24″ Diameter	1.143	Ea.
36″ Diameter	1.600	Ea.
48″ Diameter	2.182	Ea.
Drain Section, 6″ Diameter	.533	Ea.
12″ Diameter	.727	Ea.
24″ Diameter	1.500	Ea.
36″ Diameter	2.000	Ea.
48″ Diameter	2.400	Ea.
Standard Tee, 6″ Diameter	.667	Ea.
12″ Diameter	.800	Ea.
24″ Diameter	2.000	Ea.
36″ Diameter	2.667	Ea.
48″ Diameter	4.800	Ea.
Plate Support Assembly, 6″ Diameter	.615	Ea.
12″ Diameter	1.330	Ea.
24″ Diameter	1.714	Ea.
36″ Diameter	2.400	Ea.
48″ Diameter	3.000	Ea.
Drained Tee Cap, 6″ Diameter	.432	Ea.
12″ Diameter	.533	Ea.
24″ Diameter	1.143	Ea.
36″ Diameter	1.600	Ea.
48″ Diameter	2.182	Ea.
Half Angle Ring, 6″ Diameter	.174	Ea.
12″ Diameter	.210	Ea.
24″ Diameter	.545	Ea.
36″ Diameter	.665	Ea.
48″ Diameter	.750	Ea.

DIVISION
5
ROOFING

5 ROOFING

GENERAL

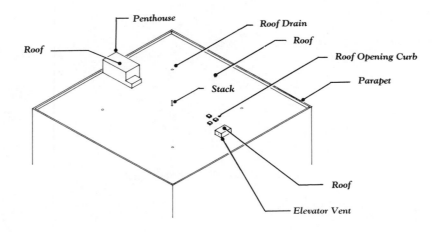

Most often, the type of roofing system selected for a project is predicated on the knowledge and experience of the designer. There are other times when the owner has had contact with someone in the roofing industry who encourages the use of newer and perhaps unproven roofing systems. Also, local fire codes or the community's or owner's architectural guidelines for appearance may dictate one roofing system over another. The system selected should be one that has demonstrated performance in the area where the project will be constructed. There should also exist in the construction location at least one qualified roofing contractor who can, if need be, make any necessary future repairs.

Presently, there are four basic groups of roof covering which are easily discernable one from the other:

. Built-up
. Single-Ply
. Metal
. Shingles and Tiles

The roofing system selected should last the life of the building, provided it is specified, detailed, installed, as per the manufacturer's instructions and cared for properly. It is important to remember that these are systems; a set or arrangement of items so closely related that they form a composite whole. Most often these systems consist of many different items made by different manufacturers; items which must be put together to form either a waterproofing membrane or a water shed. There are also related items that must be accounted for which are not roofing materials, but nevertheless are required for the success of the roofing system. These items include: flashing materials, roof deck insulation, roof hatches, skylights, gutters and downspouts.

Although the roofing system is a low percentage cost item compared to the square foot cost of a building, the system's cost and life expectancy should be carefully evaluated. Water penetration is usually the major problem in building construction and maintenance. If care is taken in the selection of a proper roofing system, before construction, much time, money, and aggravation will be spared later on.

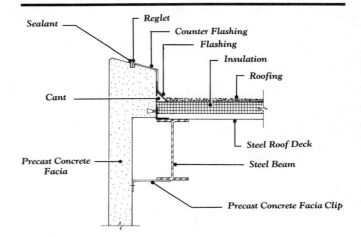

Roofing: *Parapet Detail*

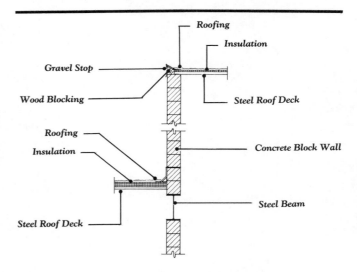

Roofing: *Intersection of Low and High Roof*

170

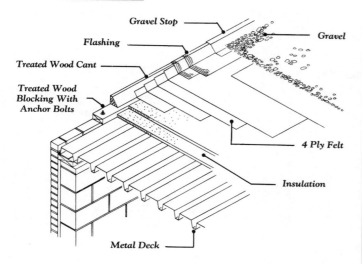

Man-hours

Description	m/hr	Unit
Built-Up Roofing		
Asphalt Flood Coat with Gravel or Slag		
Asbestos Base Sheet		
3 Plies Felt Mopped	2.545	sq
On Nailable Decks	2.667	sq
4 Plies Felt Mopped	2.800	sq
On Nailable Decks	2.947	sq
Coated Glass Fiber Base Sheet		
3 Plies Felt Mopped	2.545	sq
On Nailable Decks	2.667	sq
Organic Base Sheet and 3 Plies Felt	2.333	sq
On Nailable Decks	2.435	sq
Coal Tar Pitch with Gravel or Slag		
Coated Glass Fiber Base Sheet and		
2 Plies Glass Fiber Felt	2.947	sq
On Nailable Decks	3.111	sq
Asphalt Mineral Surface Roll Roofing	2.074	sq
Walkway		
Asphalt Impregnated	.020	sf
Patio Blocks 2″ Thick	.070	sf
Expansion Joints Covers	.048	lf

A built-up roof is composed of three different and distinct elements: felt, bitumen, and surfacing. The felts, which are made of glass, vegetable, or asbestos fibers serve much the same purpose as reinforcing steel in concrete. The felts are necessary as tensile reinforcement to resist the extreme pulling forces in the roofing material. Felts installed in layer fashion also allow more bitumen to be applied in the whole system. Bitumen, either coal-tar pitch or asphalt, is the "glue" that holds the felts together and is also the waterproofing material in the system. Felts do not waterproof; the layers of bitumen provide the waterproofing function.

The surfacings normally applied to built-up roofs are smooth gravel or slag, mineral granules, or a mineral-coated cap sheet. Gravel, slag, and mineral granules may be embedded into the still-fluid flood coat. Gravel and slag serve as an excellent wearing surface to protect the membrane from mechanical damage. On some systems, a mineral-coated cap sheet is applied on top of the plies of felt. The mineral-coated cap sheet is nothing more than a thicker or heavier ply of felt with a mineral granule surface.

The most common built-up systems available contain 2, 3, or 4 piles of felt with either asphalt bitumen or coal tar pitch. Almost all systems are available for application to either nailable or non-nailable decks. All systems may be applied to rigid deck insulation or directly to the structural deck.

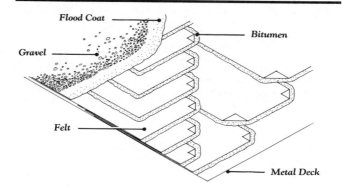

Applied to Nailable Deck

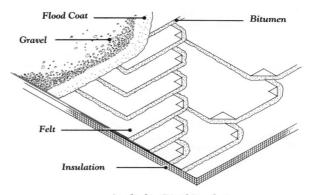

Applied to Rigid Insulation

Built-Up Roof

Single-ply or elastomeric roofing (as it is sometimes called) falls into three categories: thermosetting, thermoplastic, and composites. Various materials and their characteristics are outlined in the following table:

	SINGLE-PLY ROOFING MEMBRANE INSTALLATION GUIDE															
	Generic Materials (Classification)	**Compatible Base**						**Attachment Method**				**Sealing Method**				
		Slip-Sheet Req'd.	Concrete	Exist. Asphalt Memb.	Insulation Board	Plywood	Spray Urethane Foam	Adhesive	Fully Adhered	Loose Laid/Ballast	Partially-Adhered	Adhesive	Hot Air Gun	Self-Sealing	Solvent	Torch Heating
Thermo Setting	EPDM (Ethylene, propylene)	X	X	X	X	X	X	X	X	X	X	X		X	X	
	Neoprene (Synthetic rubber)	X	X		X	X		X	X	X		X				
	PIB (Polyisobutylene)	X	X	X	X	X	X	X			X	X	X		X	
Thermo Plastic	CSPE (Chlorosulfonated polyethyene)	X	X		X	X	X	X	X	X	X	X	X			
	CPE (Chlorinated polyethylene)	X	X		X	X			X	X	X	X				
	PVC (Polyvinyl chloride)	X	X		X	X	X			X	X	X			X	
Composites	Glass reinforced EPDM/neoprene	X	X		X	X	X		X			X				
	Modified bitumen/polyester	X		X	X	X			X			X	X			X
	Modified bitumen/polyethylene & aluminum	X	X		X	X		X	X			X				X
	Modified bitumen/polyethylene sheet	X	X		X	X			X				X			X
	Modified CPE				X	X			X			X				
	Non-woven glass reinforced PVC							X	X	X		X				
	Nylon reinforced PVC		X		X	X			X				X		X	
	Nylon reinforced butyl or neoprene	X							X				X		X	
	Polyester reinforced CPE	X	X	X	X	X	X			X	X	X	X			
	Polyester reinforced PVC	X	X		X	X	X			X	X	X	X		X	
	Rubber asphalt/plastic sheet	X	X	X	X	X			X					X		

Single-ply roofing can be applied loose-laid and ballasted, partially adhered, and fully adhered. The loose-laid and ballasted applications are the most economical. They involve fusing or gluing the side and end laps of the membrane to form a continuous nonadhered sheet held in place with 1/4″ to 1/2″ of stone. Partially adhered, single-ply membrane is attached with a series of strips or plate fasteners to the supporting structure. Because the system allows movement, no ballast is usually required. Fully adhered systems are uniformly, continuously adhered to the manufacturer's approved base. The system is time-consuming to install, but no ballast is required.

Man-hours

Description	m/hr	Unit
Single Ply Membrane, General		
Loose-Laid and Ballasted	.006	sf
Partially Adhered, All Types	.008	sf
Fully Adhered, All Types	.011	sf
Modified Bitumen		
Loose-Laid and Ballasted	.013	sf
Partially Adhered Torch Welding	.016	sf
Fully Adhered Torch Welding or Hot		
Asphalt Attachment	.020	sf
Uncured Neoprene for Flashing	.013	sf
Separator Sheet	.010	sf

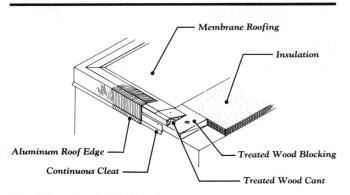

Roof Edge, Single Ply Roofing

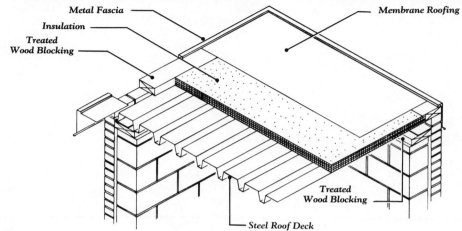

Single Ply Roofing

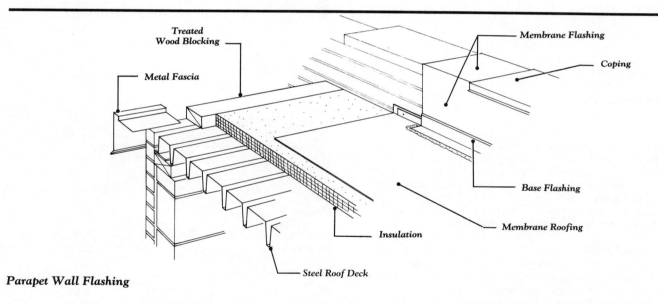

Parapet Wall Flashing

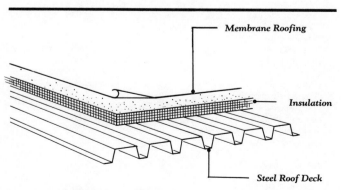

Membrane Roofing-Adhered

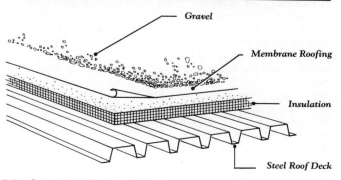

Membrane Roofing-Ballasted

173

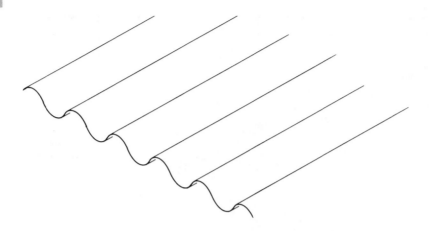

Preformed Roofing

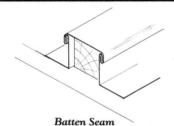

Batten Seam

Formed Metal Roofing

Standing Seam

Flat Seam

Metal roofing systems may be divided into two groups: preformed metal and formed metal. Preformed metal roofs, available in long lengths of varying widths and shapes, are constructed from aluminum, steel, and composition materials, such as asbestos cement and fiberglass. Aluminum roofs are usually prepainted or left natural, while steel roofs are usually galvanized, painted, or asphalt coated.

Preformed metal roofing is installed on pitched roofs according to the manufacturer's recommendations as to minimum required pitch. Lapped ends may be sealed with a preformed sealant (usually available from the manufacturer) to match deck configuration.

Preformed deck is usually installed on purlins, or supporting members, and then fastened with self-tapping screws with an attached neoprene washer to prevent leakage. Span lengths depend on roof pitch, deck thickness, configuration, and geographical area.

Formed metal roofing is used on sloped roofs that have been covered with a base material (plywood, concrete, etc.). Its use is more aesthetic than economical. Typical materials include copper, lead, and zinc alloy. Flat sheets are joined by tool-formed batten-seam, flat-seam, and standing-seam joints. Solder or adhesive is subsequently applied.

Man-hours

Description	m/hr	Unit
Preformed Metal Corrugated Roofing		
Aluminum Mill Finish	.032	sf
Asphalt Panels Smooth or Granulated Finish	.024	sf
Fiberglass Panels	.032	sf
Protected Metal Colored	.065	sf
Steel Galvanized		
29 Gauge	.029	sf
26 Gauge	.030	sf
24 Gauge	.032	sf
22 Gauge	.034	sf
Formed Metal Roofing		
Copper 16 oz		
Batten Seam	7.273	sq
Flat Seam	6.667	sq
Standing Seam	6.154	sq
Copper 18 oz		
Batten Seam	8.000	sq
Flat Seam	6.957	sq
Standing Seam	6.667	sq
Copper 20 oz		
Batten Seam	8.000	sq
Flat Seam or Standing Seam	7.273	sq
Lead 3 lb per sf		
Batten Seam	6.667	sq
Flat Seam or Standing Seam	6.154	sq
Zinc Copper Alloy		
.020" Thick	6.667	sq
.027" Thick	6.957	sq
.032" Thick	7.273	sq
.040" Thick	7.619	sq
Asphalt Coated Felt	.138	sq
Insulation	.010	sf

5 ROOFING
SHINGLE AND TILE ROOFS

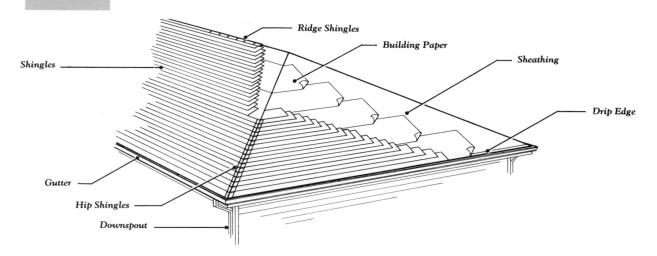

Shingle and tiles are popular materials for cladding sloped roofs. Both are watershed materials, which means that they are not designed to retain water, but to direct the water by means of the sloping or pitching of the roof. Most shingled roofs require a pitch of 3" or more per foot to perform correctly. Shingles may be installed in layer fashion with staggered joints over roofing-felt underlayment. Nails or fasteners are concealed by the exposed area of the shingle course above.

Shingle materials include asphalt, wood, metal, and masonry (tile). Asphalt shingles are available in various weights and styles, with three-tab 240 lb being the most economical. Wood shingles may be either "shingle or shake" grade, with "shakes" being the more expensive of the two systems but generally more asthetically appealing. Metal shingles are either aluminum or steel and can be coated or uncoated.

Slate shingles and tiles of clay and concrete require stronger structural systems to support the added load imposed by the heavier roofing materials. The initial added cost for slate and tile roofs may be offset by replacement and maintenance costs usually associated with other roofing systems.

Man-hours

Description	m/hr	Unit
Shingles		
Aluminum	3.200	sq
Ridge Cap or Valley	.047	lf
Shakes	3.478	sq
Asbestos		
325 lb per sq	2.000	sq
167 lb per sq	2.286	sq
Hip and Ridge	8.000	CLF
Asphalt Standard Strip		
Class A 210 to 235 lb per sq	1.455	sq
Class C 235 to 240 lb per sq	1.600	sq
Standard Laminated		
Class A 240 to 260 lb per sq	1.778	sq
Class C 260 to 300 lb per sq	2.000	sq
Premium Laminated		
Class A 260 to 300 lb per sq	2.286	sq
Class C 300 to 385 lb per sq	2.667	sq
Hip and Ridge Roll	.020	lf
Slate Including Felt Underlay	4.571	sq
Steel	3.636	sq
Wood		
5" Exposure	3.325	sq
5-1/2" Exposure	3.034	sq
Panelized 8' Strips 7" Exposure	2.667	sq
Laps Rakes or Valleys	.040	lf

Man-hours

Description	m/hr	Unit
Tiles		
Aluminum		
Mission	3.200	sq
Spanish	2.667	sq
Clay 8-1/4" x 11"	4.848	sq
Spanish	4.444	sq
Mission	6.957	sq
French	5.926	sq
Norman	8.000	sq
Williamsburg	5.926	sq
Concrete 13" x 16-1/2"	5.926	sq
Steel	3.200	sq

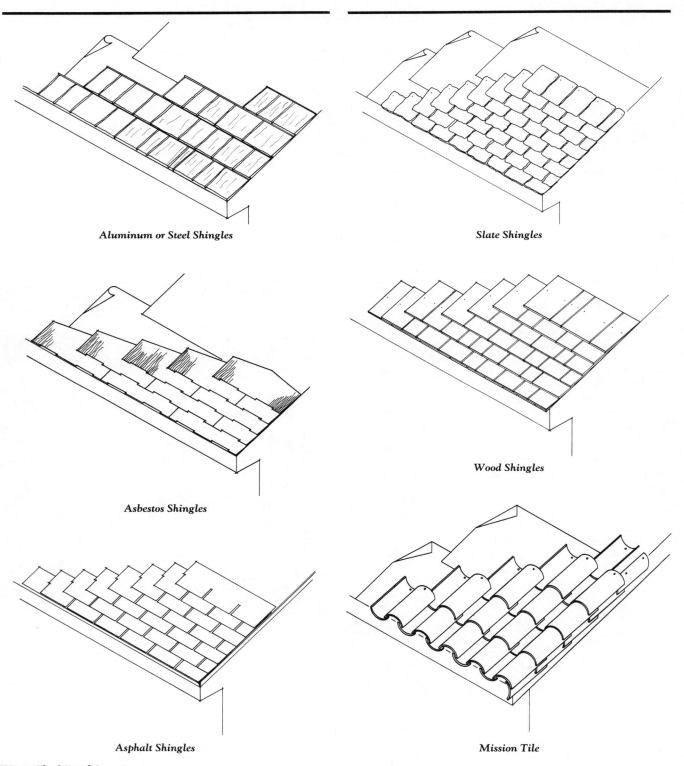

Aluminum or Steel Shingles

Slate Shingles

Asbestos Shingles

Wood Shingles

Asphalt Shingles

Mission Tile

Water Shed Roof Coverings

176

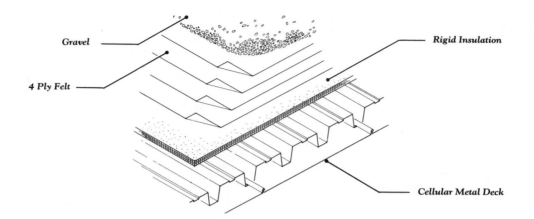

Gravel

4 Ply Felt

Rigid Insulation

Cellular Metal Deck

Roof deck insulation is used to reduce heat transfer through the roof. It may also be used to form an acceptable surface to receive roofing materials and to form slopes to lead water to roof drains.

Major insulation types used as a base for roofing materials include mineral fiberboard, fiberglass boards, fiberglass and urethane composite board, foamglass sheets, perlite sheets, perlite urethane composite sheets, phenolic foam sheets, polystyrene sheets, urethane with felt faces, urethane and gypsum board composite sheets, and poured-in-place gypsum concrete. The roofing system must attain specific R values (resistance to heat transfer) to comply with energy code requirements in most areas.

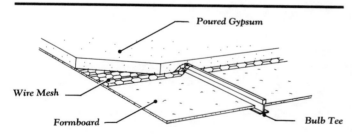

Poured Gypsum

Wire Mesh

Formboard

Bulb Tee

Poured-in-Place Gypsum Concrete and Formboard

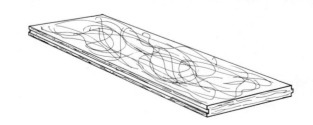

Tongue and Groove Fiberboard Insulation

Man-hours

Description	m/hr	Unit
Roof Deck Insulation		
Fiberboard to 2″ Thick	.010	sf
Fiberglass	.008	sf
Fiberglass and Urethane Composite		
1-11/16″ Thick	.008	sf
2″ and 2-5/8″ Thick	.010	sf
Foamglass		
1-1/2″ and 2″ Thick	.010	sf
3″ and 4″ Thick	.011	sf
Tapered	.013	sf
Perlite		
To 1-1/2″ Thick	.010	sf
2″ Thick	.011	sf
Perlite Urethane Composite		
To 1-3/4″ Thick	.008	sf
2″ Thick	.010	sf
2-1/2″ and 3″ Thick	.011	sf
Phenolic Foam		
To 1-3/4″ Thick	.008	sf
2″ to 3″ Thick	.010	sf
Polystyrene		
1″ Thick	.005	sf
2″ Thick	.006	sf
3″ Thick	.008	sf
Urethane Felt Both Sides		
1″ and 1-1/2″ Thick	.008	sf
2″ to 3″ Thick	.010	sf
Urethane and Gypsum Board Composite		
1-5/8″ Thick	.008	sf
2″ to 3″ Thick	.010	sf
Sprayed Polystyrene or Urethane		
1″ Thick	.031	sf
2″ Thick	.051	sf
Lightweight Cellular Fill		
Portland Cement and Foaming Agent	1.120	cu yd
Vermiculite or Perlite	1.120	cu yd
Ready Mix		
2″ Thick	.006	sf
3″ Thick	.007	sf

5 ROOFING
FLASHING, EXPANSION JOINTS, AND GRAVEL STOPS

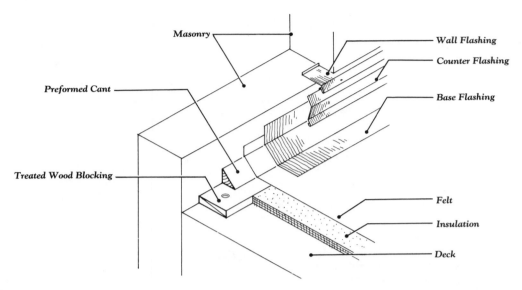

Base Flashing

Flashings are required to ensure that discontinuities in the roofing system are protected from water penetration. Typical flashing locations include the roof perimeter, around penetrations such as vents, hatches, skylights, and equipment curbs, along roof expansion joints, and sometimes at changes in slope.

Typical flashing materials include aluminum, copper, lead-coated copper, lead, polyvinyl chloride, butyl rubber, copper-clad stainless steel, stainless steel, zinc and copper alloy, and galvanized metal. Flashing materials should be compatible with the roofing system and all adjacent materials.

Man-hours

Description	m/hr	Unit
Flashing Aluminum Mill Finish	.055	sf
Fabric-Backed or Mastic-Coated 2 Sides	.024	sf
Copper Sheets Under 6000 lbs		
16 oz Sheets	.070	sf
20 oz Sheets	.073	sf
24 oz Sheets	.076	sf
32 oz Sheets	.080	sf
Paperbacked Fabric-Backed or Mastic-Backed		
2 Sides	.024	sf
Lead-Coated Copper Paperbacked Fabric-		
Backed or Mastic-Backed	.024	sf
Lead 2.5 lb per sf	.059	sf
Polyvinyl Chloride or Butyl Rubber	.028	sf
Copper-Clad Stainless Steel Under 500 lbs		
.015" Thick	.070	sf
.018" Thick	.080	sf
Stainless Steel Sheets or		
Terne Coated Stainless Steel	.052	sf
Paperbacked 2 Sides	.024	sf
Zinc and Copper Alloy	.052	sf

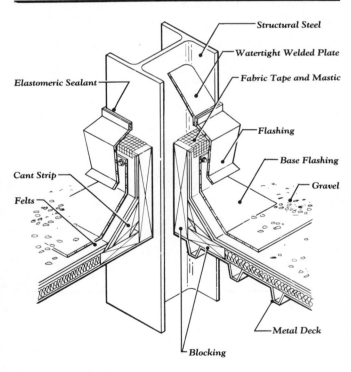

Structural Steel Flashing

Expansion joints are required to compensate for the change in dimensions or volume of building materials due to thermal variations, moisture, or other environmental conditions. In the roof structure, a joint or gap is placed at appropriate intervals to allow for expansion and contraction of the building parts.

Expansion joints typically consist of job-built or prefabricated blocking material designed to raise the joint covering above the roofing material. The joint may be filled with a compressible material, such as felt, rubber, or neoprene to keep it clean and dry. A prefabricated protective covering of rubber, neoprene, or metal inhibits moisture penetration while allowing for movement. This covering, together with the necessary flashings, completes the joint system.

Man-hours

Description	m/hr	Unit
Expansion Joint, Butyl, 1/16″ Thick, 29″ Wide, Metal Flanges	.048	lf
Neoprene, Double-seal Type with Thick Center, 4-1/2″ Wide	.064	lf
Polyethylene Bellows with Galvanized Flanges	.080	lf
Roof Expansion, Joint with Extruded Aluminum Cover, 2″	.070	lf
Roof Expansion Joint, Plastic Curbs, Foam Center	.080	lf
Transitions, Regular, Minimum	.800	Ea.
Maximum	2.000	Ea.
Large, Minimum	.889	Ea.
Maximum	2.667	Ea.
Roof to Wall Expansion Joint with Extruded Aluminum Cover	.070	lf
Wall Expansion Joint, Closed Cell Foam on PVC Cover, 9″ Wide	.064	lf
12″ Wide	.070	lf

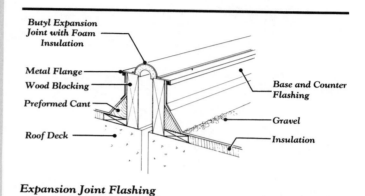

Expansion Joint Flashing

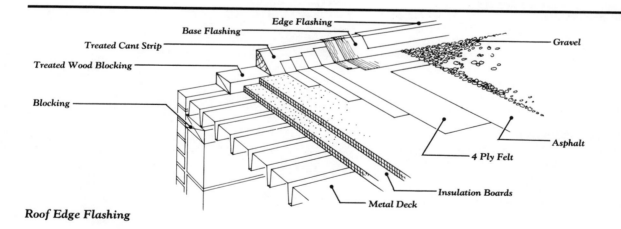

Roof Edge Flashing

Gravel stop is used at the edges of flat or nearly flat roofs to contain the gravel on the roof, as a counter flashing, and as a decorative edge strip. It is usually used in conjunction with treated-lumber blocking and a cant strip to protect the flashing at the roof edge. The exposed face height varies from 4″ to 12″ or more and the flashing return is usually fabricated to suit the roof edge conditions.

Gravel stop may be fabricated from aluminum, copper, lead-coated copper, polyvinyl chloride, galvanized steel, or stainless steel. The finish may be natural or painted. A duranodic finish is commonly used with aluminum gravel stops.

Man-hours

Description	m/hr	Unit
Gravel Stop		
4″ Face Height	.055	lf
6″ Face Height	.059	lf
8″ Face Height	.064	lf
12″ Face Height	.080	lf

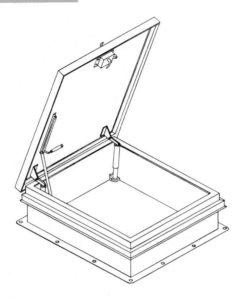

Roof Hatch

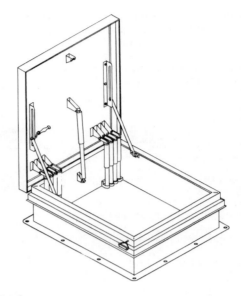

Smoke Hatch

Roof hatches are required by many building codes to allow access to the roof from the building interior. Access to the roof hatch is usually provided by a ladder permanently attached to an adjacent wall. Hatches are generally prefabricated from aluminum or steel and are assembled with a preformed curb with counter flashing. The covers are typically hinged and have operating levers, latches, and sometimes locks. Some models feature a plexiglass cover as an option.

Smoke hatches are required in some buildings with large, open areas to clear the building of smoke in case of fire. They are usually spring-loaded and open automatically when activated by a fusible link. Smoke hatches often are prefabricated with a preformed metal curb.

Skylights and plastic domes are manufactured in many sizes, shapes, and configurations. They may be custom fabricated to satisfy most requirements and are manufactured of acrylic or polycarbonate plastic, or laminated, tempered, or wire glass. Units may be insulated or double glazed. Skylights are used to light interior areas and, if operable, to provide ventilation.

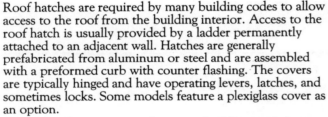

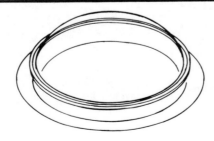

Circular Dome Skylight

Man-hours

Description	m/hr	Unit
Roof Hatches with Curb		
2'-6" x 3'-0"	3.200	Ea.
2'-6" x 4'-6"	3.556	Ea.
2'-6" x 8'-0"	4.848	Ea.
Smoke Hatches		
4'-0" x 4'-0"	2.462	Ea.
4'-0" x 8'-0"	4.000	Ea.
Plastic Roof Domes Flush or Curb Mounted		
Under 10 sf		
Single	.200	sf
Double	.246	sf
10 sf to 20 sf		
Single	.081	sf
Double	.102	sf
20 sf to 30 sf		
Single	.069	sf
Double	.081	sf
30 sf to 65 sf		
Single	.052	sf
Double	.069	sf
Ventilation, Insulated Plexiglass Dome		
Curb Mounted		
30" x 32"	2.667	Ea.
44" x 45"	3.200	Ea.
Skyroofs		
Translucent Panels 2-3/4" Thick	.081	sf horiz.
Continuous Vaulted to 8' Wide		
Single Glazed	.200	sf horiz.
Double Glazed	.221	sf horiz.
To 20' Wide Single Glazed	.183	sf horiz.
Over 20' Wide Single Glazed	.160	sf horiz.
Pyramid Type to 30' Clear Opening	.194	sf horiz.
Grid Type 4' x 10' Modules	.200	sf horiz.
Ridge Units Continuous to 8' Wide		
Double	.246	sf horiz.
Single	.160	sf horiz.

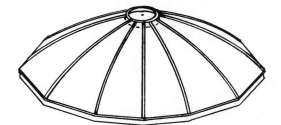

Domed Skylight

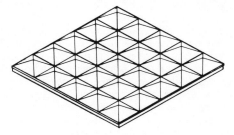

Pyramid Skylights in Grid Form

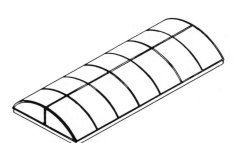

Vaulted Skylight

Double Pitch Skylight

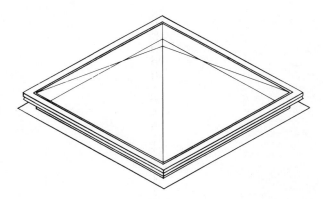

Pyramid Skylight

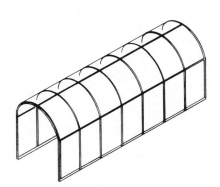

Covered Walkway

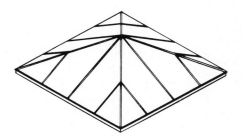

Pyramid Skylight

Skyroofs

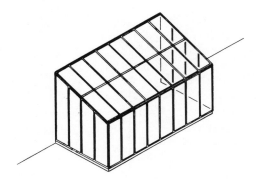

Greenhouse/Solaruim

Gutter and downspout systems are used to collect and distribute water from roof edges. Metal gutters of aluminum, copper, lead-coated copper, galvanized steel, and stainless steel are available in stock lengths with accessories such as mounting brackets, connectors, corners, end caps, downspout connectors, and leaf guards. They may be box, round, or ogee in configuration and are often prepainted. Plastic or vinyl gutters in stock lengths and various cross-sectional shapes are manufactured with matching accessories. These types of gutters have the advantage that they deteriorate less from the effects of weather. Treated wood gutters are premilled in quarter round or ogee patterns. They may be job fabricated in box or V configurations. Downspouts, round or rectangular in cross section, are available with accessories to match the gutter systems. Pipe of black steel or cast iron may be used to provide a more durable downspout.

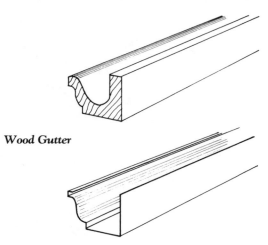

Wood Gutter

Metal or Vinyl Gutter

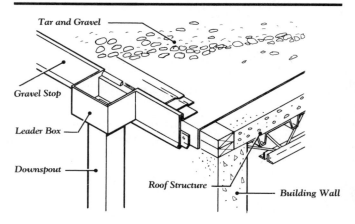

Flat Roof With Leader Box

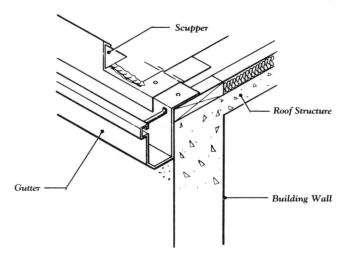

Flat Roof With Gutter

Man-hours

Description	m/hr	Unit
Gutters		
Aluminum	.067	lf
Copper Stock Units		
4″ Wide	.067	lf
6″ Wide	.070	lf
Steel Galvanized or Stainless	.067	lf
Vinyl	.073	lf
Wood	.080	lf
Downspouts		
Aluminum		
2″ x 3″	.042	lf
3″ Diameter	.042	lf
4″ Diameter	.057	lf
Copper		
2″ or 3″ Diameter	.042	lf
4″ Diameter	.055	lf
5″ Diameter	.062	lf
2″ x 3″	.042	lf
3″ x 4″	.055	lf
Steel Galvanized		
2″ or 3″ Diameter	.042	lf
4″ Diameter	.055	lf
5″ Diameter	.062	lf
6″ Diameter	.076	lf
2″ x 3″	.042	lf
3″ x 4″	.055	lf
Epoxy Painted		
2″ x 3″	.042	lf
3″ x 4″	.055	lf
Steel Pipe		
4″ Diameter	.400	lf
6″ Diameter	.444	lf
Stainless Steel		
2″ x 3″ or 3″ Diameter	.042	lf
3″ x 4″ or 4″ Diameter	.055	lf
4″ x 5″ or 5″ Diameter	.059	lf
Vinyl		
2″ x 3″	.038	lf
2-1/2″ Diameter	.036	lf

DIVISION 6

INTERIOR CONSTRUCTION

Concrete block partitions may be erected of nominal 4″, 6″, 8″, 10″ or 12″ thick concrete masonry units. The units may be regular weight or lightweight and solid or 75% solid in horizontal profile. Normal units are nominal 16″ long but some manufacturers produce a nominal 24″ unit. They may be left exposed, painted, or epoxy-coated; or they may be furred and then covered with gypsum board or paneling. Partitions may also be plastered (directly on the block) or covered with self-furring lath and then plastered. Block walls may have joint reinforcing; or they may be reinforced, both horizontally and vertically, and then fully or partially grouted to act as bearing or shear walls.

Special blocks are also available. These include corner, jamb, head and lintel blocks. Heights are usually available in 2″ soaps, 4″ half-block, 8″ full block, and 16″ bond and lintel block. Decorative blocks are available in several shapes and facings for exposed work. Some manufacturers produce a self-furring block, sometimes called a "stud block."

Man-hours

Description	m/hr	Unit
Scaffolding Steel Tubular	.387	CSF
Acoustical Slotted Block		
4″ Thick	.127	sf
6″ Thick	.138	sf
8″ Thick	.151	sf
12″ Thick	.163	sf
Lightweight Block		
4″ Thick	.090	sf
6″ Thick	.095	sf
8″ Thick	.100	sf
10″ Thick	.103	sf
12″ Thick	.130	sf
Regular Block		
Hollow		
4″ Thick	.093	sf
6″ Thick	.100	sf
8″ Thick	.107	sf
10″ Thick	.111	sf
12″ Thick	.141	sf
Solid		
4″ Thick	.095	sf
6″ Thick	.105	sf
8″ Thick	.113	sf
12″ Thick	.150	sf
Glazed Concrete Block		
Single Face 8″ x 16″		
2″ Thick	.111	sf
4″ Thick	.116	sf
6″ Thick	.121	sf
8″ Thick	.129	sf
12″ Thick	.171	sf
Double Face		
4″ Thick	.129	sf
6″ Thick	.138	sf

Man-hours (cont.)

Description	m/hr	Unit
8″ Thick	.148	sf
Joint Reinforcing Wire Strips		
4″ and 6″ Wall	.267	CLF
8″ Wall	.320	CLF
10″ and 12″	.400	CLF
Steel Bars Horizontal		
#3 and #4	.018	lb
#5 and #6	.010	lb
Vertical		
#3 and #4	.023	lb
#5 and #6	.012	lb
Grout Cores Solid		
By Hand 6″ Thick	.035	sf
Pumped 8″ Thick	.038	sf
10″ Thick	.039	sf
12″ Thick	.040	sf
Plaster, Gypsum 3 Coat	.460	sq yd
Perlite or Vermiculite 3 Coat	.541	sq yd
Drywall Gypsum Plasterboard Including Taping		
1/2″ or 5/8″ Thick	.018	sq yd
Furring Steel Galvanized		
7/8″ Deep	3.077	CLF
Resilient	3.137	CLF
Furring Wood	.016	sf
Paint Block Brushwork		
Primer + 1 Coat	.005	sf
+ 2 Coats	.010	sf
Paint Plaster or Drywall		
Primer + 1 Coat	.004	sf
+ 2 Coats	.008	sf
Acrylic Glazed Coating Average	.021	sf
Epoxy Coating Average	.031	sf
Ceramic-like Glazed Coating,		
Cementitious Average	.020	sf

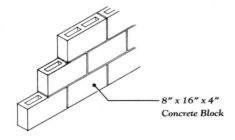

8″ x 16″ x 4″
Concrete Block

Nonbearing Concrete Block Partition

8″ x 16″ x 4″
Glazed Concrete Block

Glazed Concrete Block

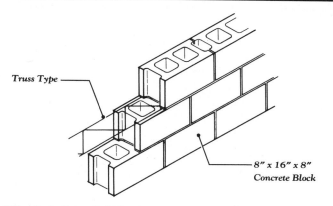

Wire Strip Joint Reinforcing

Truss Type

8" x 16" x 8" Concrete Block

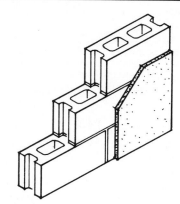

Plaster Direct to Block

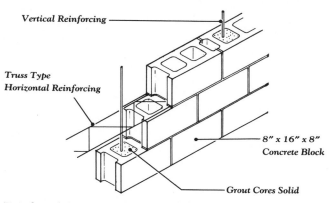

Reinforced Concrete Block Wall

Vertical Reinforcing

Truss Type Horizontal Reinforcing

8" x 16" x 8" Concrete Block

Grout Cores Solid

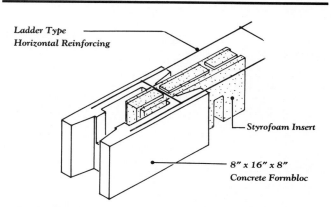

Insulated Concrete Block

Ladder Type Horizontal Reinforcing

Styrofoam Insert

8" x 16" x 8" Concrete Formbloc

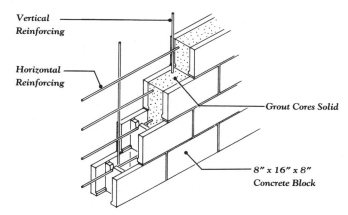

Interlocking Concrete Block

Vertical Reinforcing

Horizontal Reinforcing

Grout Cores Solid

8" x 16" x 8" Concrete Block

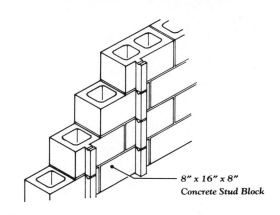

Self-furring Concrete Block

8" x 16" x 8" Concrete Stud Block

Structural glazed facing tile (SGFT) is kiln-fired structural clay with an integral impervious ceramic face. It is also manufactured with an acoustical perforated face. Because it resists stains, marks, impact, abrasion, fading, and crazing, it is ideally suited for use in school corridors, locker rooms, rest rooms, kitchens, and other places where cleanliness and indestructibility are primary considerations.

SGFT is available in a large selection of colors and color combinations in the 6T series, with 5-1/3" x 12" nominal face in 2", 4", 6", and 8" widths and in the 8W series, with 8" x 16" nominal face in 2", 4", 6", and 8" widths. Some manufacturers produce a 4W series with 8" x 8" nominal face in 2" and 4" widths.

Some available tile shapes include stretchers, bullnose jamb or corner, square jamb or corner, coved internal corner, recessed cove base, nonrecessed cove base, bullnose sill, square sill, and universal miter. Walls with openings and returns usually require partition layout drawings to establish quantities of special shapes.

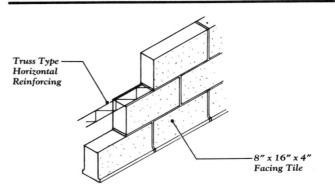

8W Series Structural Glazed Facing Tile

Man-hours

Description	m/hr	Unit
Scaffolding Steel Tubular	.387	CSF
Structural Facing Tile 6T Series		
2" Thick Glazed		
1 Side	.178	sf
4" Thick Glazed		
1 Side	.182	sf
2 Sides	.205	sf
6" Thick Glazed		
1 Side	.190	sf
2 Sides	.216	sf
8" Thick Glazed		
1 Side	.222	sf
8W Series		
2" Thick Glazed		
1 Side	.111	sf
4" Thick Glazed		
1 Side	.116	sf
2 Sides	.123	sf
6" Thick Glazed		
1 Side	.121	sf
8" Thick Glazed		
1 Side	.129	sf

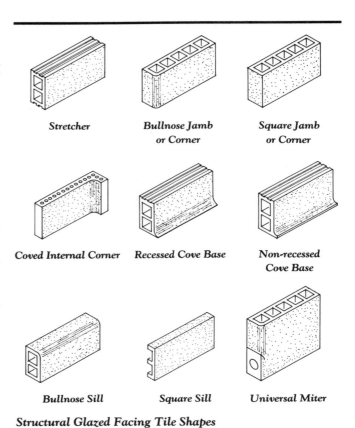

Structural Glazed Facing Tile Shapes

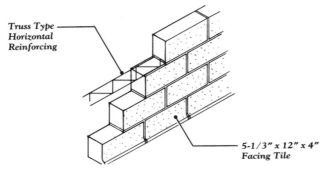

6T Series Structural Glazed Facing Tile

Wood stud partitions may be fabricated of 2″ x 3″, 2″ x 4″, or 2″ x 6″ studs spaced 12″, 16″, or 24″ on center. Bearing partitions normally require 2″ x 4″ or 2″ x 6″ studs 16″ on center; 2″ x 3″ studs are normally used only for separator walls, closet framing, or, sometimes, staggered stud walls.

Wood stud partitions may be clad with drywall, plaster, composition board, or wood paneling and may be used in conjunction with many types of underlayments to provide additional fireproofing or acoustical qualities. Staggered studs, in conjunction with insulation batts or resilient clips to separate the cladding from the studs, may be used to improve sound-deadening qualities.

Wood stud partitions are easily field-fabricated and erected to meet a variety of conditions.

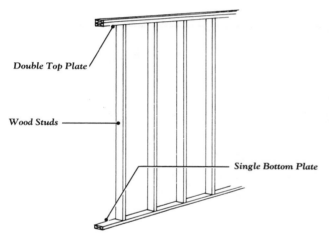

Wood Stud Partition, No Blocking

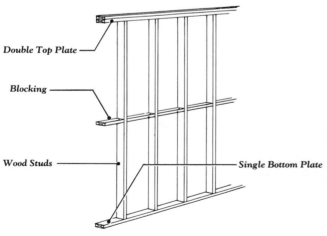

Wood Stud Partition with Blocking

Man-hours

Description	m/hr	Unit
Wood Partitions Studs with Single		
Bottom Plate and Double Top Plate		
2″ x 3″ or 2″ x 4″ Studs		
12″ On Center	.020	sf
16″ On Center	.016	sf
24″ On Center	.013	sf
2″ x 6″ Studs		
12″ On Center	.023	sf
16″ On Center	.018	sf
24″ On Center	.014	sf
Plates		
2″ x 3″	.019	lf
2″ x 4″	.020	lf
2″ x 6″	.021	lf
Studs		
2″ x 3″	.013	lf
2″ x 4″	.012	lf
2″ x 6″	.016	lf
Blocking	.032	lf
Grounds 1″ x 2″		
For Casework	.024	lf
For Plaster	.018	lf
Insulation Fiberglass Batts	.005	sf
Metal Lath Diamond Expanded		
2.5 lb per sq yd	.094	sq yd
3.4 lb per sq yd	.100	sq yd
Gypsum Lath		
3/8″ Thick	.094	sq yd
1/2″ Thick	.100	sq yd
Gypsum Plaster		
2 Coats	.381	sq yd
3 Coats	.460	sq yd
Perlite or Vermiculite Plaster		
2 Coats	.435	sq yd
3 Coats	.541	sq yd
Wood Fiber Plaster		
2 Coats	.556	sq yd
3 Coats	.702	sq yd
Drywall Gypsum Plasterboard Including Taping		
3/8″ Thick	.015	sf
1/2″ or 5/8″ Thick	.017	sf
For Thin Coat Plaster Instead of Taping Add	.013	sf
Prefinished Vinyl Faced Drywall	.015	sf
Sound-deadening Board	.009	sf
Walls in Place		
2″ x 4″ Studs with 5/8″		
Gypsum Drywall Both Sides Taped	.053	sf
2″ x 4″ Studs with 2 Layers Gypsum Drywall		
Both Sides Taped	.078	sf
Paint Brushwork Primer		
+ 1 Coat	.011	sf
+ 2 Coats	.016	sf
+ 3 Coats	.022	sf
Vinyl Wall Covering Medium Weight	.017	sf
Wallpaper High Price Quality Workmanship	.018	sf
Grass Cloth Average	.021	sf
Wood Paneling 4″ x 8″ Sheets		
1/4″ Thick Average	.036	sf
3/4″ Thick Average	.085	sf

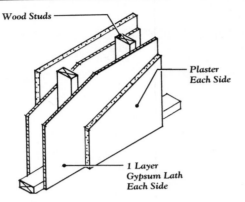

Plaster on Gypsum Lath

Wood Studs

Plaster
Each Side

1 Layer
Gypsum Lath
Each Side

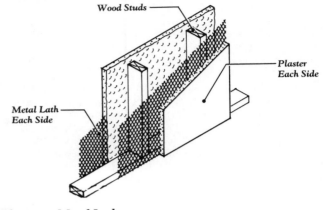

Plaster on Metal Lath

Wood Studs

Plaster
Each Side

Metal Lath
Each Side

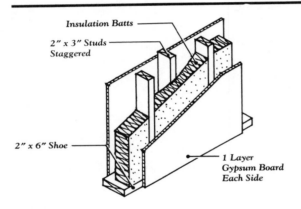

Staggered Stud Wall

Insulation Batts

2" x 3" Studs
Staggered

2" x 6" Shoe

1 Layer
Gypsum Board
Each Side

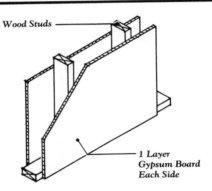

Gypsum Plasterboard

Wood Studs

1 Layer
Gypsum Board
Each Side

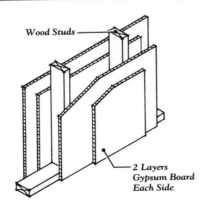

Gypsum Plasterboard, 2 Layers

Wood Studs

2 Layers
Gypsum Board
Each Side

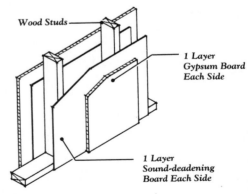

Sound-deadening Board

Wood Studs

1 Layer
Gypsum Board
Each Side

1 Layer
Sound-deadening
Board Each Side

188

Steel studs are formed or fabricated from light-gauge metal in standard lengths and various widths and configurations. They are designated to be load-bearing and nonload-bearing, depending on the gauge of material used in manufacture and the width-versus-height ratio. Non-bearing studs for drywall are available in 25 and 20 gauge, in 1-1/2″, 2″, 2-1/2″, 3″, 3-1/2″, 3-5/8″, 4″, and 6″ depths, with track and bridging to match the studs. C studs are available in 18, 16, and 14 gauge, in depths varying from 1-5/8″ to 12″.

Studs are available factory-painted or galvanized. Bearing studs are available in lengths of 7′ to 37′ ±, in 1″ increments. Drywall studs are available in standard lengths varying from 8′ to 20′.

Studs may be connected to the track with self-tapping screws or self-drilling, hand clinched, or they may be welded. Partition framing may be attached to concrete with expansion bolts, or power-driven fasteners, and to steel framing with power-driven fasteners.

Studs generally are manufactured with openings or knockouts in the web to allow passage for piping or electrical conduit. Studs for partition walls may be covered with gypsum, metal lath and plaster, or gypsum drywall of various thicknesses and laminations. Insulation may be installed to improve acoustical characteristics.

Man-hours

Description	m/hr	Unit
Nonload Bearing Stud, Galv., 25 Gauge,		
1-5/8″ Wide,		
16″ On Center	.019	sf
24″ On Center	.016	sf
2-1/2″ Wide,		
16″ On Center	.020	sf
24″ On Center	.016	sf
20 Gauge, 1-5/8″ Wide,		
16″ On Center	.018	sf
24″ On Center	.016	sf
2-1/2″ Wide,		
16″ On Center	.019	sf
24″ On Center	.016	sf
25 Gauge or 20 Gauge, 3-5/8″ or 4″ Wide		
16″ On Center	.020	sf
24″ On Center	.017	sf
6″ Wide,		
16″ On Center	.022	sf
24″ On Center	.018	sf
Load Bearing Stud, Galv. or Painted, 18 Gauge,		
2-1/2″ Wide,		
16″ On Center	.019	sf
24″ On Center	.016	sf
3-5/8″ or 4″ Wide,		
16″ On Center	.020	sf
24″ On Center	.017	sf
6″ Wide,		
16″ On Center	.022	sf
24″ On Center	.018	sf

Man-hours (cont.)

Description	m/hr	Unit
Load Bearing Stud, Galv. or Painted,		
16 Gauge, 2-1/2″ Wide,		
16″ On Center	.020	sf
24″ On Center	.017	sf
3-5/8″ or 4″ Wide,		
16″ On Center	.021	sf
24″ On Center	.018	sf
6″ Wide,		
16″ On Center	.024	sf
24″ On Center	.019	sf
Insulation Fiberglass Batts	.005	sf
Metal Lath Diamond Expanded		
2.5 lb Screwed to Studs	.100	sq yd
3.4 lb Screwed to Studs or Wired to Framing	.107	sq yd
Rib Lath Wired to Steel		
2.75 lb	.107	sq yd
3.40 lb	.114	sq yd
4.00 lb	.123	sq yd
Gypsum Lath		
3/8″ Thick	.094	sq yd
1/2″ Thick	.100	sq yd
Gypsum Plaster		
2 Coats	.381	sq yd
3 Coats	.460	sq yd
Perlite or Vermiculite Plaster		
2 Coats	.435	sq yd
3 Coats	.541	sq yd
Wood Fiber Plaster		
2 Coats	.556	sq yd
3 Coats	.702	sq yd
Drywall Gypsum Plasterboard Including Taping		
3/8″ Thick	.015	sf
1/2″ or 5/8″ Thick	.017	sf
For Thincoat Plaster Instead of Taping, Add	.013	sf
Prefinished Vinyl Faced Drywall	.015	sf
Sound-deadening Board	.009	sf
Walls in Place		
3-5/8″ Studs, NLB, 25 Gauge,		
16″ On Center with 5/8″ Gypsum Drywall		
Both Sides Taped	.047	sf
24″ On Center	.044	sf
3-5/8″ Studs, NLB, 25 Gauge,		
16″ On Center with 2 Layers Gypsum Drywall		
Both Sides Taped	.065	sf
24″ On Center	.060	sf
Paint Brushwork Primer		
+ 1 Coat	.011	sf
+ 2 Coats	.016	sf
+ 3 Coats	.022	sf
Vinyl Wall Covering Medium Weight	.017	sf
Wallpaper High Price Quality Workmanship	.018	sf
Grass Cloth Average	.021	sf
Wood Paneling 4″ x 8″ Sheets		
1/4″ Thick Average	.036	sf
3/4″ Thick Average	.085	sf

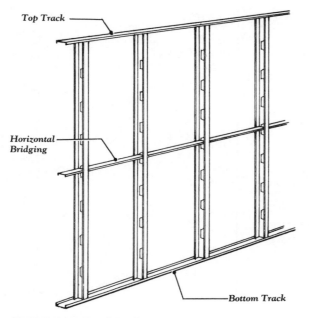

Load Bearing Steel Studs

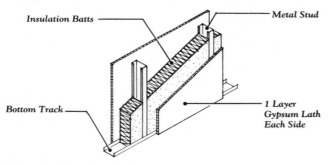

Fiberglass Batt Insulation

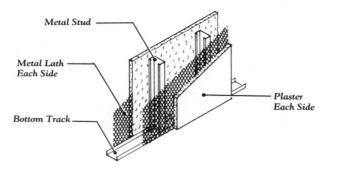

Plaster on Metal Lath

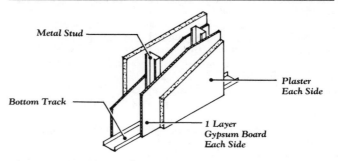

Plaster on Gypsum Lath

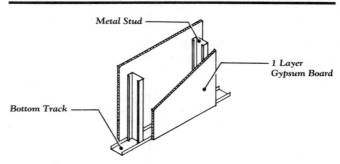

Gypsum Plasterboard

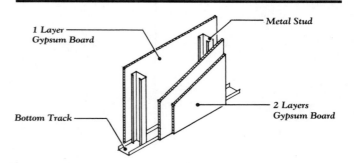

Gypsum Plasterboard, 2 Layers

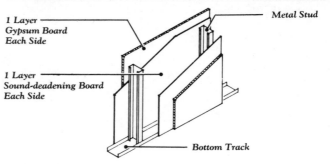

Sound-deadening Board

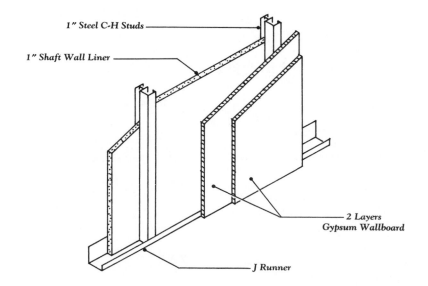

Cavity Shaft Wall

Shaft wall construction is used principally to surround elevator or stair shafts that require specified fire ratings. The walls may be constructed of studs and plaster, or laminated gypsum drywall on studs. The plaster or drywall layers may be used on one or two sides in stud construction. Gypsum stud and solid shaftwall are generally used with shaftliner to increase fire rating capacities. Gypsum drywall specifications usually require staggered joints. For multi-faced walls, one layer of material should be applied horizontally; the other vertically.

Most shaft wall systems are designed for one-side installation to negate the need for staging in the shaftways.

Man-hours

Description	m/hr	Unit
Shaft Wall Cavity Type 1″ Steel C-H Studs With 2 Layers 5/8″ Gypsum Board		
1 Side	.097	sf
Laminated Gypsum Drywall, 2-1/2″ Solid or 3-3/4″ Core With Steel H Sections		
24″ Wide Units	.148	sf
16″ Wide Units	.174	sf
Solid 2″ Thick Steel Edge Gypsum In Channels With 1/2″ Fire Resistant Gypsum Board		
1 Side	.107	sf
2 Sides	.119	sf

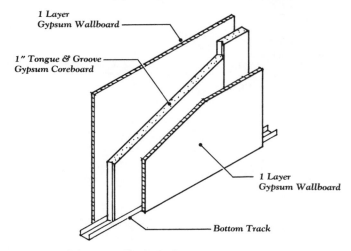

Laminated Gypsum Shaft Wall

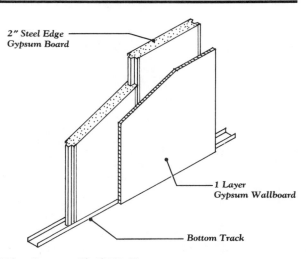

Steel Edge Gypsum Shaft Wall

Wood doors, manufactured in either flush or paneled designs, are separated into three grades: architectural/commercial, residential, and decorator. A wide variety of frames are available for interior and exterior installations in metal, pine, or hardwood, and for various partition thicknesses. Some doors are also available prehung for quick installation.

Architectural or commercial wood doors are the type most often specified in building construction. The stiles are made of hardwood, and the core is dense and of hot-bonded construction. They feature thick face veneers that are exterior-glued and matched in their grain patterns. Because of its durability, this grade of door often carries a lifetime warranty.

Residential wood doors are used in low-use (frequency) situations where economy is a primary consideration. The stiles are manufactured from softwood; the core from low density materials. The face veneers are thin, interior-glued, and broken in their grain patterns.

Decorator wood doors are manufactured from solid wood and are usually hand carved. Because of the choice woods used and the special craftsmanship required in their production, their cost is several times that of similar-size, architectural wood doors.

Wood doors are manufactured in flush or paneled designs. Flush doors are produced with cores of varying density: hollow, particleboard, or solid wood block. Lauan mahogany, birch, or other hardwood veneers are used in their facings. Synthetic veneers, created from a medium-density overlay or high-pressure plastic laminate, may serve as an alternate choice, to natural wood veneers. Flush wood doors may be fire rated.

Paneled wood doors, which are usually manufactured from pine or fir, typically have solid wood stiles and two-, five-, six-, and eight-panel designs. A simulated six-panel door design with a hollow core and moulded hardboard face is also available.

Man-hours

Description	m/hr	Unit
Architectural, Flush, Interior, Hollow Core, Veneer Face		
Up to 3'-0" x 7' x 0"	1.020	Ea.
4'-0" x 7'-0"	1.080	Ea.
High Pressured Plastic Laminate Face		
Up to 2'-6" x 6'-8"	1.000	Ea.
3'-0" x 7'-0"	1.153	Ea.
4'-0" x 7'-0"	1.234	Ea.
Particle Core, Veneer Face		
2'-6" x 6'-8"	1.067	Ea.
3'-0" x 6'-8"	1.143	Ea.
3'-0" x 7'-0"	1.231	Ea.
4'-0" x 7'-0"	1.333	Ea.
M.D.O. on Hardboard Face		
3'-0" x 7'-0"	1.333	Ea.
4'-0" x 7'-0"	1.600	Ea.
High Pressure Plastic Laminate Face		
3'-0" x 7'-0"	1.455	Ea.
4'-0" x 7'-0"	2.000	Ea.
Flush, Exterior, Solid Core, Veneer Face		
2'-6" x 7'-0"	1.067	Ea.
3'-0" x 7'-0"	1.143	Ea.
Decorator, Hand Carved		
Solid Wood		
Up to 3'-0" x 7'-0"	1.143	Ea.
3'-6" x 7'-0"	1.231	Ea.
Fire Door, Flush, Mineral Core		
B Label, 1 Hour, Veneer Face		
2'-6" x 6'-8"	1.143	Ea.
3'-0" x 7'-0"	1.333	Ea.
4'-0" x 7'-0"	1.333	Ea.
High Pressure Plastic Laminate Face		
3'-0" x 7'-0"	1.455	Ea.
4'-0" x 7'-0"	1.600	Ea.
Residential, Interior		
Hollow Core or Panel		
Up to 2'-8" x 6'-8"	.889	Ea.
3'-0" x 6'-8"	.941	Ea.
Bi-Folding Closet		
3'-0" x 6'-8"	1.231	Ea.
5'-0" x 6'-8"	1.455	Ea.
Interior Prehung, Hollow Core or Panel		
Up to 2'-8" x 6'-8"	.800	Ea.
3'-0" x 6'-8"	.842	Ea.
Exterior, Entrance, Solid Core or Panel		
Up to 2'-8" x 6'-8"	1.000	Ea.
3'-0" x 6'-8"	1.067	Ea.
Exterior Prehung, Entrance		
Up to 3'-0" x 7'-0"	1.000	Ea.

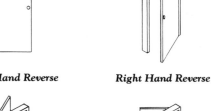

Left Hand Reverse   *Right Hand Reverse*

Left Hand   *Right Hand*

Hand Designations

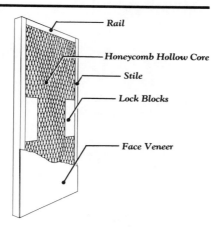

- Rail
- Honeycomb Hollow Core
- Stile
- Lock Blocks
- Face Veneer

Hollow Core Door

- Rail
- Particleboard Core
- Stile
- Face Veneer

Solid Core Door

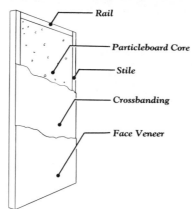

- Rail
- Particleboard Core
- Stile
- Crossbanding
- Face Veneer

Solid Core Door With Crossbanding

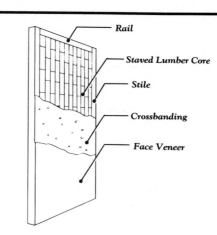

- Rail
- Staved Lumber Core
- Stile
- Crossbanding
- Face Veneer

Staved Wood Core Door

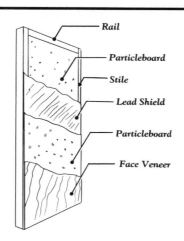

- Rail
- Particleboard
- Stile
- Lead Shield
- Particleboard
- Face Veneer

Lead Core Door

Exterior Entrance, Prehung, Panel Door

Hollow metal doors are available in stock or custom fabrication, flush or embossed, with lights or louvres, labeled or unlabeled, and in various steel gauges and core fills. Stock doors may be supplied for low-, moderate- or high-frequency of use from some manufacturers.

Doors are available in widths of 2′ to 4′ and heights varying from 6′-8″ to 10′ or more. They may be used as single doors, in pairs with both leaves active, or in pairs with one active leaf, including an astragal. Bi-fold hollow metal doors are available for closet applications.

Metal doors may be supplied bonderized and primed, galvanized and primed, factory finished in selected colors, or in stainless steel. Doors with lead lined cores are available for shielded applications.

Hollow metal doors are reinforced at the stress points and premortised for the hardware required for the door application. Hollow metal, labeled fire doors can be supplied stock or custom manufactured with A, B, C, D, or E labels, with 3/4 to 3 hour ratings, depending on the glass area, height and width restrictions, and maximum expected temperature rise. The following door types may be used as a guide for fire door selection.

Man-hours

Description	m/hr	Unit
Hollow Metal Doors Flush		
Full Panel, Commercial		
20 Gauge		
2′-0″ x 6′-8″	.800	Ea.
2′-6″ x 6′-8″	.888	Ea.
3′-0″ x 6′-8″ or 3′-0″ x 7′-0″	.941	Ea.
4′-0″ x 7′-0″	1.066	Ea.
18 Gauge		
2′-6″ x 6′-8″ or 2′-6″ x 7′-0″	.941	Ea.
3′-0″ x 6′-8″ or 3′-0″ x 7′-0″	1.000	Ea.
4′-0″ x 7′-0″	1.066	Ea.
Residential		
24 Gauge		
2′-8″ x 6′-8″	1.000	Ea.
3′-0″ x 7′-0″	1.066	Ea.
Bifolding		
3′-0″ x 6′- 8″	1.000	
5′-0″ x 6′-8″	1.143	Ea.

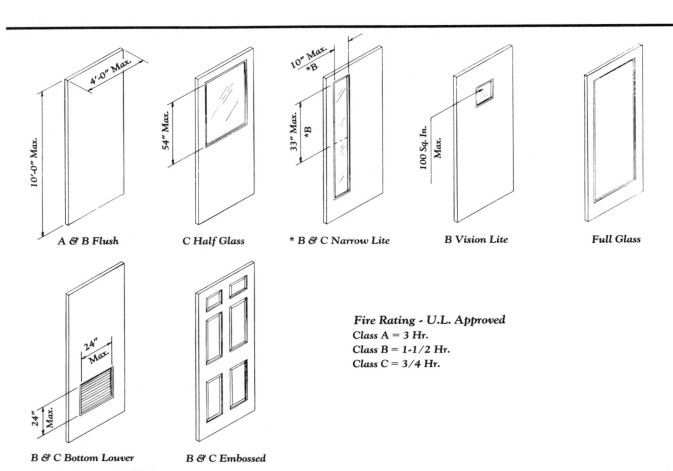

A & B Flush   **C Half Glass**   *** B & C Narrow Lite**   **B Vision Lite**   **Full Glass**

B & C Bottom Louver   **B & C Embossed**

Fire Rated Metal Door Types

Fire Rating - U.L. Approved
Class A = 3 Hr.
Class B = 1-1/2 Hr.
Class C = 3/4 Hr.

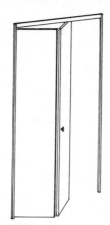

Bi-folding Closet Door and Frame

Double Egress Door and Frame

Hollow metal frames may be supplied in 14, 16, or 18 gauge galvanized or plain steel in knockdown standard frames or welded customized frames that can be fabricated to satisfy most design conditions. Frames with borrowed lights, transoms, or cased openings are available in stick components from some manufacturers. Frames may be wraparound (enclosing the wall) or butt up against the opening. A wraparound frame may terminate into the enclosed wall when it is covered by a finish such as plaster, or the frame may return along the enclosed wall when it is exposed, such as in drywall construction. They are sometimes supplied in two pieces to suit varied wall thicknesses, or in one piece to satisfy standard wall thicknesses. Frames are normally reinforced at stress points and are prepared for hinges and strikes. Anchors to attach the frame to the wall are supplied to suit wall construction requirements. Custom frames normally require a hardware schedule and templates to produce required shop drawings and to accomplish fabrication.

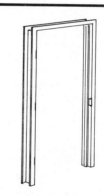

Hollow Metal Frame - Butt or Wraparound

Man-hours

Description	m/hr	Unit
Steel Frames		
18 Gauge		
3'-0" Wide	1.000	Ea.
6'-0" Wide	1.142	Ea.
16 Gauge		
4'-0" Wide	1.066	Ea.
8'-0" Wide	1.333	Ea.
Transom Lite Frames		
Fixed Add	.103	sf
Movable Add	.123	sf

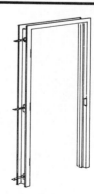

Hollow Metal Frame with Anchors

Folding partitions, operable walls, and relocatable partitions are manufactured in a variety of sizes, shapes, and finishes. Operating partitions include folding accordian, folding leaf, or individual panel systems. These units may be operated by hand or power. Relocatable partitions include the portable type, which is designed for frequent relocation, and the demountable type, which is designed for infrequent relocation.

Operating partitions are supported by aluminum, steel, or wood framing members. The panels are usually filled with a sound-insulation material, as most partitions are rated by their sound reduction qualities. Panel skin materials include aluminum, composition board, fabric, or wood. The panels may be painted or covered with carpeting, fabric, plastic laminate, vinyl, or wood paneling. Chalkboards and tackboards may also be mounted on the skin of the paneling to add another dimension to their use. Panels may also be custom faced to designer's specifications. Large operating partitions are generally installed by factory specialists after the supporting members and framing have been supplied and erected by the building contractor.

Depending on the particular use, the amount of available space, and the design of the area to be partitioned, many different stacking methods and track configurations can be employed. Operating partitions may adopt a recessed-stacked, center-stacked, parallel-stacked, or exposed-stacked format. Track systems are available in straight lengths or curves, or with right-angle layouts and switches. Operating partitions are usually top-track supported, but some models are available with floor supports.

Relocatable and portable panels are manufactured of the same materials used in operable walls. Vinyl-covered gypsum board may also be used as a face material. Partitions may be installed to ceiling height with adjustable top or bottom seals, or to varied heights with end-hinged supports. Manufacturers' details vary widely with various panel systems.

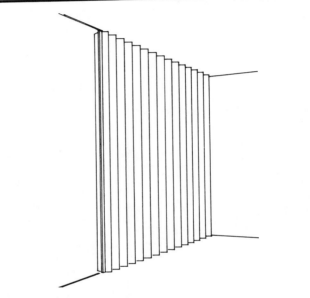

Folding Accordion Partition

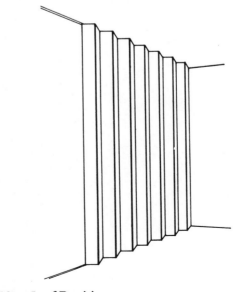

Folding Leaf Partition

Man-hours

Description	m/hr	Unit
Operable Partitions		
Folding Accordian		
Vinyl Covered Average	.107	sf
Acoustical Average	.168	sf
Vinyl Electric Operation Average	.100	sf
Wood to 10' High Average	.053	sf
Folding Leaf		
Wood Acoustic Vinyl Faced, Formica or Hardwood Finish, Steel Acoustical, or Aluminum Acoustical Average	.356	sf
Acoustic Air Wall Average	.044	sf
Portable Partitions, Free Standing 4'-6" High	.100	lf
5'-0" High	.107	lf
6'-0" High	.128	lf
Movable Partitions		
Asbestos Cement Full Height	.400	lf
Gypsum Laminated	.400	lf
3" Thick Acoustical	.400	lf
Vinyl Clad Drywall	.267	lf
Vinyl or Steel Clad Gypsum	.267	lf
Hardboard Vinyl Faced	.356	lf
Metal Enameled Steel to 9'-6" High	.400	lf
Frame All Glass	.400	lf
With 52% Glass	.400	lf
Free Standing 4'-8" High	.160	lf
Acoustical	.160	lf

Relocatable Partition - Demountable and Portable

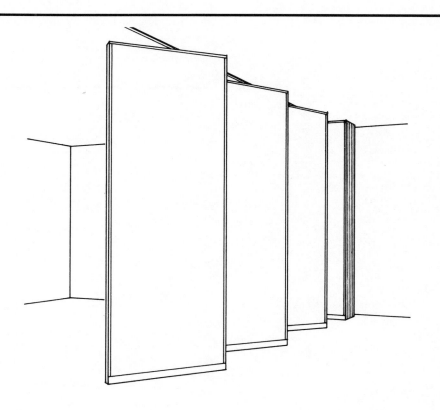

Operable Partition

Toilet partitions, dressing compartments, and privacy screens are manufactured in a variety of materials, finishes, and colors. They are available in many stock sizes for both regular and handicapped-equipped water closets. These partitions may be custom fabricated to fit special size or use requirements, and may be supported from the floor, braced overhead, or hung from the ceiling or wall. Available finish materials include marble, painted metal, plastic laminate, porcelain enamel, and stainless steel.

Toilet Partitions - Floor Mounted

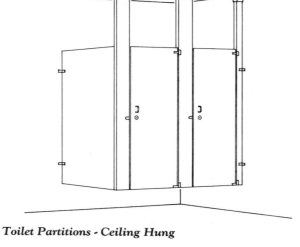

Toilet Partitions - Ceiling Hung

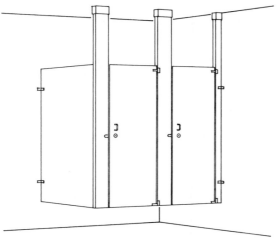

Entrance Screen - Floor Mounted

Man-hours

Description	m/hr	Unit
Toilet Partitions Ceiling Hung Cubicles Marble	8.000	Ea.
Painted Metal, Plastic Laminate, Porcelain		
Enamel or Stainless Steel	14.000	Ea.
Floor and Ceiling Anchored Marble	6.400	Ea.
Painted Metal, Plastic Laminate, Porcelain		
Enamel or Stainless Steel	3.200	Ea.
Floor Mounted Marble	5.333	Ea.
Painted Metal, Plastic Laminate, Porcelain		
Enamel or Stainless Steel		
Floor Mounted Marble	2.286	Ea.
Floor Mounted, Headrail Braced, Marble	5.333	Ea.
Painted Metal, Plastic Laminate, Porcelain		
Enamel or Stainless Steel		
Floor Mounted Marble	2.667	Ea.
Wall Hung Painted Metal, Porcelain Enamel,		
Stainless Steel	2.286	Ea.
Screens, Ceiling Braced and Head Rail		
Braced, Marble	2.667	Ea.
Painted Metal, Plastic Laminate, Porcelain		
Enamel or Stainless Steel	2.000	Ea.
Plaster and Post Braced Marble	1.778	Ea.
Post Braced, Wall Hung Bracket Supported, or		
Flange Supported Painted Metal,		
Plastic Laminate, Porcelain Enamel		
or Stainless Steel	1.600	Ea.
Wedge Type, Painted Metal, Porcelain Enamel,		
Stainless Steel	1.600	Ea.

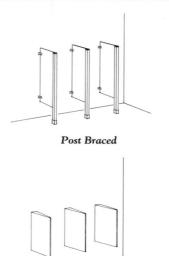

Post Braced

Wedge Type

Urinal Screens

INTERIOR CONSTRUCTION
SPECIAL DOORS, ACCESS PANELS, PASS WINDOWS AND ROLL UP SHUTTERS

Metal access panels and doors are available in steel or stainless steel for fire-rated or non-fire-rated applications. Panels are fabricated for flush installations in drywall, both skim coated or taped, for masonry and tile applications, for plastered walls and ceilings, and for acoustical ceilings. Doors are available in stock sizes and types to suit most applications.

Blast doors are available in standard designs. They may be custom designed to withstand specified pressures and to resist penetration.

Cold-storage doors are available in standard designs in wood, steel, fiberglass, plastic, and stainless steel for all types of cold-storage requirements. Doors are manufactured to provide insulation requirements for cool zones, coolers, and freezers for manual, air, electric, or hydraulic operations. Door operation types include sliding, vertical lift, bi-parting overhead, and single- and double-swing.

Pass windows (vertical and horizontal sliding), rotating shelf windows, ticket windows, cashier doors, and coin and cash trays are available in aluminum, steel, or stainless steel. Pass windows, roll up shutters, and projection booth shutters are available, labeled or non-labeled, with fusible links to suit most applications and requirements.

Darkroom revolving doors and pass throughs are modeled to fit an existing doorframe. They provide protection for multiple darkroom installations and applications.

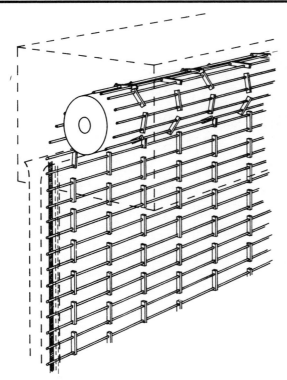

Roll Up Gate

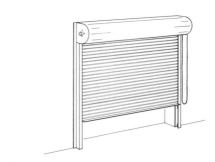

Roll Up Shutter

Man-hours

Description	m/hr	Unit
Access Panels Metal		
12" x 12"	.400	Ea.
12" x 24"	.533	Ea.
18" x 18"	.666	Ea.
24" x 36"	1.000	Ea.
36" x 48"	1.333	Ea.
Cold Storage Doors		
Single Galvanized Steel Horizontal Sliding		
5' x 7'		
Manual Operation	8.000	Ea.
Power Operation	8.421	Ea.
9' x 10'		
Manual Operation	9.411	Ea.
Power Operation	10.000	Ea.
Hinged Lightweight 3' x 6'-6"		
2" Thick	8.000	Ea.
4" Thick	8.421	Ea.
6" Thick	11.428	Ea.
Bi-parting Electric Operated 6' x 8'	20.000	Opng.
For Door Buck Framing and Door Protection Add	6.400	Opng.
Galvanized Batten Door		
4' x 7'	8.000	Opng.
6' x 8'	8.888	Opng.
Fire Door 6' x 8'		
Single Slide	20.000	Opng.
Double Bi-parting	22.857	Opng.
Darkroom Doors Revolving		
2 Way 30" Diameter	4.571	Opng.
3 Way 43" Diameter	5.333	Opng.
4 Way 68" Diameter	8.000	Opng.
Hinged Safety		
2 Way 30" Diameter	5.000	Opng.
3 Way 43" Diameter	5.517	Opng.
Pop Out Safety		
2 Way 30" Diameter	5.161	Opng.
3 Way 43" Diameter	5.714	Opng.
Pass Window Painted Steel		
24" x 36"	10.000	Ea.
48" x 48"	13.333	Ea.
72" x 40"	20.000	Ea.
Vault Front Door and Frame		
32" x 78", 1 Hour, 750 lbs	10.670	Opng.
40" x 78", 2 Hours, 1130 lbs	16.000	Opng.
Day Gate		
32" Wide	10.670	Opng.
42" Wide	11.430	Opng.

Painting and finishing are required to protect interior and exterior surfaces against wear and corrosion and to provide a coordinated finished appearance on the protected material. Paints can be normally classified by their binders or vehicles. Alkyds are oil-modified resin used to manufacture fast-drying enamels. Chlorinated rubber produces coatings that are resistant to alkalies, acids, chemicals, and water. Catalyzed epoxies are two-part coatings that produce a hard film resistant to abrasion, traffic, chemicals, and cleaning. Epoxy esters are epoxies (modified with oil) that dry by oxidation, but are not as hard as catalyzed epoxies. Latex binders, commonly polyvinylacetate, acrylics, or vinyl acetate-acrylics, are binders mixed with a water base. Silicone alkyd binders are oil modified to produce coatings with heat resistance and high-gloss retention. Urethanes are isocynate polymers that are modified with drying oils or alkyds. Vinyl coating solutions are plasticized copolymers of vinyl chloride and vinyl acetate dissolved in strong solvents. Zinc coatings are primers high in zinc dust content dispersed in various vehicles to provide coatings to protect steel.

In order to achieve successful painting, the surface to be painted must first be cleaned. Preparation methods for painting include solvent cleaning, hand-tool cleaning (including wire brushing, scraping, chipping, and sanding), power-tool cleaning, white metal blasting (including removal of all rust and scale), commercial blasting (to remove oil, grease, dirt, rust), and scale or brush off blast to remove oil, grease, dirt, and loose rust and scale.

Fillers, primers, and sealers are manufactured for use with paint or clear coatings for metal, wood, plaster, drywall, and masonry. Sanding fillers are available for clear coatings and block fillers for masonry construction.

Topcoats or finish coats should be specified and applied to meet service requirements. Finishes can be supplied for normal use and atmospheric conditions, hard service areas, and critical areas. Finishes can be furnished in gloss, semi-gloss, low lustre, eggshell, or flat. Intumescent coatings (expandable paints) are available to meet fire-retardant requirements.

Finishes for wood include stains, varnish, and urethanes, as well as sealers and paint. Paint, varnish, and stain are normally applied by brush, roller, or spray gun. Masking and protection is included in the production man-hours shown below where required.

Wall coverings are manufactured, printed, or woven in burlaps, jutes, weaves, grasses, paper, leather, vinyl, silks, stipples, corks, aluminum foil, copper sheets, cork tiles, flexible wood veneers, and flexible mirrors. Wallpaper, vinyl wall coverings, and woven coverings are usually available in different weights, backings, and quality. Surface preparation and adhesive selection are important considerations in placing wall coverings.

Painting

Item	Coat	One Gallon Covers			In 8 Hours Man Covers			Man-hours per 100 SF		
		Brush	Roller	Spray	Brush	Roller	Spray	Brush	Roller	Spray
Paint wood siding	prime	275 sf	250 sf	325 sf	1150 sf	1400 sf	4000 sf	.695	.571	.200
	others	300	275	325	1600	2200	4000	.500	.364	.200
Paint exterior trim	prime	450	—	—	650	—	—	1.230	—	—
	1st	525	—	—	700	—	—	1.143	—	—
	2nd	575	—	—	750	—	—	1.067	—	—
Paint shingle siding	prime	300	285	335	1050	1700	2800	.763	.470	.286
	others	400	375	425	1200	2000	3200	.667	.400	.250
Stain shingle siding	1st	200	190	220	1200	1400	3200	.667	.571	.250
	2nd	300	275	325	1300	1700	4000	.615	.471	.200
Paint brick masonry	prime	200	150	175	850	1700	4000	.941	.471	.200
	1st	300	250	320	1200	2200	4400	.364	.364	.182
	2nd	375	340	400	1300	2400	4400	.615	.333	.182
Paint interior plaster or drywall	prime	450	425	550	1600	2500	4000	.500	.320	.200
	others	500	475	550	1400	3000	4000	.571	.267	.200
Paint interior doors and windows	prime	450	—	—	1300	—	—	.333	—	—
	1st	475	—	—	1150	—	—	.696	—	—
	2nd	500	—	—	1000	—	—	.800	—	—

Man-hours

Description	m/hr	Unit
Cabinets and Casework		
Labor Cost Includes Protection of Adjacent		
Items Not Painted		
Primer Coat, Oil Base, Brushwork	.020	sf
Paint, Oil Base, Brushwork		
1 Coat	.021	sf
2 Coats	.040	sf
Stain, Brushwork, Wipe Off	.022	sf
Shellac, 1 Coat, Brushwork	.021	sf
Varnish, 3 Coats, Brushwork	.034	sf
Doors and Windows		
Labor Cost Includes Protection of Adjacent		
Items Not Painted		
Flush Door and Frame, per Side, Oil Base,		
Primer Coat, Brushwork	.571	Ea.
Paint		
1 Coat	.320	Ea.
2 Coats	.640	Ea.
3 Coats	.889	Ea.
Stain, Brushwork, Wipe Off	.800	Ea.
Shellac, 1 Coat, Brushwork	.667	Ea.
Varnish, 3 Coats, Brushwork	1.600	Ea.
Panel Door and Frame, per Side, Oil Base,		
Primer Coat, Brushwork	.615	Ea.
Paint		
1 Coat	.667	Ea.
2 Coats	1.330	Ea.
3 Coats	2.000	Ea.
Stain, Brushwork, Wipeoff	1.330	Ea.
Shellac, 1 Coat, Brushwork	1.000	Ea.
Varnish, 3 Coats, Brushwork	2.670	Ea.
Windows, Including Frame and Trim, per Side		
Colonial Type, 2' x 3', Oil Base, Primer Coat		
Brushwork	.333	Ea.
Paint		
1 Coat	.364	Ea.
2 Coats	.615	Ea.
3 Coats	.800	Ea.
3' x 5' Opening, Primer Coat, Brushwork	.533	Ea.
Paint		
1 Coat	.615	Ea.
2 Coats	1.000	Ea.
3 Coats	1.380	
4' x 8' Opening, Primer Coat, Brushwork	.667	Ea.
Paint		
1 Coat	.800	Ea.
2 Coats	1.330	Ea.
3 Coats	1.860	Ea.
Standard, 6 to 8 Lites, 2' x 3', Primer	.308	Ea.
Paint		
1 Coat	.333	Ea.
2 Coats	.615	Ea.
3 Coats	.800	Ea.
3' x 5', Primer	.471	Ea.
Paint		
1 Coat	.533	Ea.
2 Coats	.800	Ea.
3 Coats	1.000	Ea.
4' x 8', Primer	.571	Ea.
Paint		
1 Coat	.667	Ea.
2 Coats	1.000	Ea.
3 Coats	1.330	Ea.

Man-hours (cont.)

Description	m/hr	Unit
Single Lite Type, 2' x 3', Oil Base, Primer Coat,		
Brushwork	.200	Ea.
Paint		
1 Coat	.216	Ea.
2 Coats	.400	Ea.
3 Coats	.571	Ea.
3' x 5' Opening, Primer Coat, Brushwork	.296	Ea.
Paint		
1 Coat	.320	Ea.
2 Coats	.571	Ea.
3 Coats	.800	Ea.
4' x 8' Opening, Primer Coat, Brushwork	.571	Ea.
Paint		
1 Coat	.615	Ea.
2 Coats	1.000	Ea.
3 Coats	1.230	Ea.
Miscellaneous		
Fence, Chain Link, Per Side, Oil Base,		
Primer Coat, Brushwork	.013	sf
Spray	.010	sf
Paint 1 Coat, Brushwork	.014	sf
Spray	.010	sf
Picket, Wood, Primer Coat, Brushwork	.039	sf
Spray	.031	sf
Paint 1 Coat, Brushwork	.040	sf
Spray	.031	sf
Floors, Concrete or Wood, Oil Base, Primer or		
Sealer Coat, Brushwork	.007	sf
Roller	.006	sf
Spray	.006	sf
Paint 1 Coat, Brushwork	.008	sf
Roller	.007	sf
Spray	.006	sf
Stain, Wood Floor, Brushwork	.007	sf
Roller	.006	sf
Spray	.006	sf
Varnish, Wood Floor, Brushwork	.008	sf
Roller	.007	sf
Spray	.006	sf
Grilles, per Side, Oil Base, Primer Coat,		
Brushwork	.020	Ea.
Spray	.016	Ea.
Paint 1 Coat, Brushwork	.021	Ea.
Spray	.016	Ea.
Paint 2 Coats, Brushwork	.040	Ea.
Spray	.032	Ea.
Gutters and Downspouts, Oil Base, Primer Coat,		
Brushwork	.025	lf
Paint 1 Coat, Brushwork	.027	lf
Paint 2 Coats, Brushwork	.049	lf
Pipe, to 4" Diameter, Primer or Sealer Coat,		
Oil Base, Brushwork	.020	lf
Spray	.015	lf
Paint 1 Coat, Brushwork	.021	lf
Spray	.015	lf
Paint 2 Coats, Brushwork	.040	lf
Spray	.029	lf
To 8" Diameter, Primer or Sealer Coat,		
Brushwork	.040	lf
Spray	.025	lf
Paint 1 Coat, Brushwork	.046	lf
Spray	.025	lf
Paint 2 Coats, Brushwork	.080	lf
Spray	.043	lf

Man-hours (cont.)

Description	m/hr	Unit
To 12″ Diameter, Primer or Sealer Coat,		
Brushwork	.067	lf
Spray	.050	lf
Paint 1 Coat, Brushwork	.073	lf
Spray	.050	lf
Paint 2 Coats, Brushwork	.133	lf
Spray	.100	lf
To 16″ Diameter, Primer or Sealer Coat,		
Brushwork	.083	lf
Spray	.067	lf
Paint 1 Coat, Brushwork	.089	lf
Spray	.067	lf
Paint 2 Coats, Brushwork	.160	lf
Spray	.123	lf
Trim, Wood, Including Puttying		
Under 6″ Wide		
Primer Coat, Oil Base, Brushwork	.009	lf
Paint, Brushwork		
1 Coat	.009	lf
2 Coats	.016	lf
3 Coats	.027	lf
Over 6″ Wide, Primer Coat, Brushwork	.013	lf
Paint, Brushwork		
1 Coat	.018	lf
2 Coats	.027	lf
3 Coats	.042	lf
Cornice, Simple Design, Primer Coat,		
Oil Base, Brushwork	.029	sf
Paint, Brushwork		
1 Coat	.032	sf
2 Coats	.050	sf
Ornate Design, Primer Coat	.053	sf
Paint		
1 Coat	.057	sf
2 Coats	.089	sf
Balustrades, per Side, Primer Coat, Oil Base,		
Brushwork	.027	sf
Paint		
1 Coat	.028	sf
2 Coats	.047	sf
Trusses and Exposed Wood Frames		
Primer Coat Oil Base, Brushwork	.010	sf
Spray	.007	sf
Paint 1 Coat, Brushwork	.011	sf
Spray	.007	sf
Paint 2 Coats, Brushwork	.016	sf
Spray	.013	sf
Stain, Brushwork, Wipe Off	.013	sf
Varnish, 3 Coats, Brushwork	.029	sf
Siding Exterior		
Steel Siding, Oil Base, Primer or Sealer Coat,		
Brushwork	.009	sf
Spray	.005	sf
Paint 2 Coats, Brushwork	.012	sf
Spray	.006	sf
Stucco, Rough, Oil Base, Paint 2 Coats,		
Brushwork	.012	sf
Roller	.008	sf
Spray	.006	sf
Texture 1-11 or Clapboard, Oil Base		
Primer Coat, Brushwork	.006	sf
Spray	.004	sf
Paint 1 Coat, Brushwork	.006	sf
Spray	.004	sf

Man-hours (cont.)

Description	m/hr	Unit
Paint 2 Coats, Brushwork	.013	sf
Spray	.008	sf
Stain 1 Coat, Brushwork	.006	sf
Spray	.004	sf
Stain 2 Coats, Brushwork	.013	sf
Spray	.008	sf
Wood Shingles, Oil Base Primer Coat, Brushwork	.006	sf
Spray	.004	sf
Paint 1 Coat, Brushwork	.007	sf
Spray	.004	sf
Paint 2 Coats, Brushwork	.014	sf
Spray	.008	sf
Stain 1 Coat, Brushwork	.006	sf
Spray	.004	sf
Stain 2 Coats, Brushwork	.014	sf
Spray	.008	sf
Wall and Ceilings		
Concrete, Dry Wall or Plaster, Oil Base, Primer		
or Sealer Coat		
Smooth Finish, Brushwork	.004	sf
Roller	.004	sf
Spray	.002	sf
Sand Finish, Brushwork	.005	sf
Roller	.004	sf
Spray	.002	sf
Paint 1 Coat		
Smooth Finish, Brushwork	.004	sf
Roller	.004	sf
Spray	.002	sf
Sand Finish, Brushwork	.005	sf
Roller	.004	sf
Spray	.002	sf
Paint 2 Coats		
Smooth Finish, Brushwork	.008	sf
Roller	.007	sf
Spray	.004	sf
Sand Finish, Brushwork	.010	sf
Roller	.008	sf
Spray	.004	sf
Paint 3 Coats		
Smooth Finish, Brushwork	.012	sf
Roller	.010	sf
Spray	.005	sf
Sand Finish, Brushwork	.014	sf
Roller	.011	sf
Spray	.005	sf
Glaze Coating, 5 Coats, Spray		
Clear	.009	sf
Multicolor	.009	sf
Masonry or Concrete Block, Oil Base, Primer		
or Sealer Coat		
Smooth Finish, Brushwork	.005	sf
Spray	.002	sf
Sand Finish, Brushwork	.006	sf
Spray	.002	sf
Paint 1 Coat		
Smooth Finish, Brushwork	.005	sf
Spray	.002	sf
Sand Finish, Brushwork	.006	sf
Spray	.002	sf
Paint 2 Coats		
Smooth Finish, Brushwork	.010	sf
Spray	.004	sf

Man-hours (cont.)

Description	m/hr	Unit
Sand Finish, Brushwork	.011	sf
Spray	.004	sf
Paint 3 Coats		
Smooth Finish, Brushwork	.013	sf
Spray	.005	sf
Sand Finish, Brushwork	.023	sf
Spray	.005	sf
Glaze Coating, 5 Coats, Spray		
Clear	.009	sf
Multicolor	.009	sf
Block Filler, 1 Coat, Brushwork	.006	sf
Silicone, Water Repellent, 2 Coats, Spray	.009	sf
Varnish 1 Coat + Sealer, on Wood Trim, No Sanding Included	.009	sf
Hardwood Floors, 2 Coats, No Sanding Included	.010	sf
Wall Coatings Acrylic Glazed Coatings		
Minimum	.015	sf
Maximum	.026	sf
Epoxy Coatings		
Minimum	.015	sf
Maximum	.047	sf
Exposed Aggregate, Troweled on		
1/16" to 1/4"		
Minimum	.034	sf
Maximum (Epoxy or Polyacrylate)	.062	sf
1/2" to 5/8" Aggregate,		
Minimum	.062	sf
Maximum	.100	sf
1" Aggregate Size		
Minimum	.089	sf
Maximum	.145	sf
Exposed Aggregate, Sprayed on		
1/8" Aggregate		
Minimum	.027	sf
Maximum	.055	sf
High Build Epoxy, 50 Mil		
Minimum	.021	sf
Maximum	.084	sf
Laminated Epoxy with Fiberglass		
Minimum	.027	sf
Maximum	.055	sf
Sprayed Perlite or Vermiculite		
1/16" Thick		
Minimum	.003	sf
Maximum	.013	sf
Vinyl Plastic Wall Coating		
Minimum	.011	sf
Maximum	.033	sf
Urethane on Smooth Surface		
2 Coats		
Minimum	.007	sf
Maximum	.012	sf
3 Coats		
Minimum	.010	sf
Maximum	.017	sf
Ceramic-like Glazed Coating, Cementitious		
Minimum	.018	sf
Maximum	.023	sf
Resin Base		
Minimum	.013	sf
Maximum	.024	sf

Man-hours (cont.)

Description	m/hr	Unit
Wall Covering		
Aluminum Foil	.029	sf
Copper Sheets, .025" Thick		
Vinyl Backing	.033	sf
Phenolic Backing	.033	sf
Cork Tiles, Light or Dark, 12" x 12"		
3/16" Thick	.033	sf
5/16" Thick	.034	sf
1/4" Basket weave	.033	sf
1/2" Natural, Non-directional Pattern	.033	sf
Granular Surface, 12" x 36"		
1/2" Thick	.021	sf
1" Thick	.022	sf
Polyurethane Coated, 12" x 12"		
3/16" Thick	.033	sf
5/16" Thick	.034	sf
Cork Wallpaper, Paperbacked		
Natural	.017	sf
Colors	.017	sf
Flexible Wood Veneer, 1/32" Thick		
Plain Woods	.080	sf
Exotic Woods	.084	sf
Gypsum-based, Fabric-backed, Fire Resistant for Masonry Walls		
Minimum	.020	sf
Average	.023	sf
Maximum	.027	sf
Acrylic, Modified, Semi-rigid PVC		
.028" Thick	.048	sf
.040" Thick	.050	sf
Vinyl Wall Covering, Fabric-backed		
Lightweight	.013	sf
Medium Weight	.017	sf
Heavy Weight	.018	sf
Wallpaper at $8.00 per Double Roll		
Average Workmanship	.013	sf
Paper at $17 per Double Roll		
Average Workmanship	.015	sf
Paper at $40 per Double Roll		
Quality Workmanship	.018	sf
Grass Cloths with Lining Paper		
Minimum	.020	sf
Maximum	.023	sf

6 INTERIOR CONSTRUCTION
CEILINGS

Ceilings may be of three types: exposed structural systems that are painted or stained, directly applied acoustical gypsum board or plaster, or suspended systems, hung below the supporting superstructure to allow space for passage of ductwork, piping and lighting fixtures.

Suspension systems usually consist of aluminum or steel main runners of light, intermediate or heavy duty classification, spaced 2, 3, 4 or 5 feet on center and snap in cross tees usually available in 1 to 5 foot lengths. Main runners are usually hung from the supporting structure with tie wire or metal straps spaced as required by the system. The runners and cross tees may be exposed, natural or painted, or concealed, enclosed by the ceiling tiles. Drywall suspension systems consist of main and cross tee members with flanges to allow the gypsum board to be screwed to the supports.

Acoustical ceilings are available in mineral fiber tiles with many patterns and textures. Sizes vary from 1 foot square to 2′ x 5′ rectangles. The face may be perforated, fissured, textured or plastic covered. The inserts are also available in perforated steel, stainless steel or aluminum to allow for easy cleaning in high humidity or kitchen areas. To retain the acoustical characteristics, these panels are filled with sound absorbing pads. Metal panels are also available as linear ceilings in varying widths in many finishes and configurations.

Integrated ceiling systems combine suspension, air handling, lighting and ceiling tiles into one modular system and have the benefit of a single source of responsibility.

Drywall ceilings may be applied directly to furring strips attached to the supporting structure or attached to a compatible suspension system. They may be painted, sprayed with acoustical material or covered with acoustical tiles.

Plaster ceilings can be applied directly on self-furring or gypsum lath. Suspended plaster ceilings attach to cold rolled channels hung from the supporting structure and covered with metal or gypsum lath.

Fire resistance in hours for the various ceiling systems are evaluated as part of the floor or roof structural and ceiling assembly. To achieve a rating in hours, fire dampers, light fixtures, grilles and diffusers are evaluated along with the structure, suspension system and ceiling.

Man-hours

Description	m/hr	Unit
Ceiling Tile Stapled, Cemented or Installed on Suspension System, 12″ x 12″ or 12″ x 24″, Not Including Furring		
Mineral Fiber, Plastic Coated	.020	sf
Fire Rated, 3/4″ Thick, Plain Faced	.020	sf
Plastic Coated Face	.021	sf
Aluminum Faced, 5/8″ Thick, Plain	.021	sf
Metal Pan Units, 24 ga. Steel, Not Incl.		
Pads, Painted, 12″ x 12″	.023	sf
12″ x 36″ or 12″ x 24″, 7% Open Area	.024	sf
Aluminum, 12″ x 12″	.023	sf
12″ x 24″	.022	sf
Stainless Steel, 12″ x 24″, 26 ga., Solid	.023	sf
5.2% Open Area	.024	sf
Suspended Acoustic Ceiling Boards Not Including Suspension System		
Fiberglass Boards, Film Faced, 2′ x 2′ or 2′ x 4′, 5/8″ Thick	.012	sf
3/4″ Thick	.016	sf
3″ Thick, Thermal, R11	.018	sf
Glass Cloth Faced Fiberglass, 3/4″ Thick	.016	sf
1″ Thick	.016	sf
1-1/2″ Thick, Nubby Face	.017	sf
Mineral Fiber Boards, 5/8″ Thick, Aluminum Faced, 24″ x 24″	.013	sf
24″ x 48″	.012	sf
Plastic Coated Face	.020	sf
Mineral Fiber, 2 Hour Rating, 5/8″ Thick	.012	sf
Mirror Faced Panels, 15/16″ Thick	.016	sf
Air Distributing Ceilings, 5/8″ Thick, F.R.D. Water Felted Board	.020	sf
Eggcrate, Acrylic, 1/2″ x 1/2″ x 1/2″ Cubes	.016	sf
Polystyrene Eggcrate	.016	sf
Luminous Panels, Prismatic	.020	sf
Perforated Aluminum Sheets, .024″ Thick, Corrugated, Painted	.016	sf

Man-hours (cont.)

Description	m/hr	Unit
Mineral Fiber, 24″ x 24″ or 48″, reveal edge, Painted, 5/8″ Thick	.013	sf
3/4″ Thick	.014	sf
Wood Fiber in Cementitious Binder, 2′ x 2′ or 4′,		
Painted, 1″ Thick	.013	sf
2″ Thick	.015	sf
2-1/2″ Thick	.016	sf
3″ Thick	.018	sf
Access Panels, Metal, 12″ x 12″	.400	Ea.
24″ x 24″	.800	Ea.
Suspension Systems		
Class A Suspension System, T Bar, 2′ x 4′ Grid	.010	sf
2′ x 2′ Grid	.012	sf
Concealed Z Bar Suspension System, 12″ Module	015	sf
1-1/2″ Carrier Channels, 4′ O.C., Add	.017	sf
Carrier Channels for Ceiling With Recessed Lighting Fixtures, Add	.017	sf
Hanging Wire, 12 Ga.	.002	sf
Suspended Ceilings, Complete Including Standard Suspension System but Not Incl. 1-1/2″ Carrier Channels		
Air Distributing Ceilings, Incl. Barriers, 2′ x 2′ Board	.027	sf
12″ x 12″ Tile	.033	sf
Ceiling Board System, 2′ x 4′, Plain Faced, Supermarkets	.016	sf
Offices	.021	sf
Wood Fiber, Cementitious Binder, T Bar Susp. 2′ x 2′ x 1″ Board	.023	sf
2′ x 4′ x 1″ Board	.021	sf
Luminous Panels, Flat or Ribbed	.031	sf
Metal Pan with Acoustic Pad	.039	sf
Tile, Z Bar Suspension, 5/8″ Mineral Fiber Tile	.034	sf

Man-hours (cont)

Description	m/hr	Unit
3/4″ Mineral Fiber Tile	.035	sf
Reveal Tile With Drop, 2′ x 2′ Grid		
With Colored Suspension System	.023	sf
Gypsum Plaster Ceilings 2 Coats, No		
Lath Included	.435	sq yd
2 Coats on and Incl. 3/8″ Gypsum Lath		
on Steel	.578	sq yd
3 Coats, No Lath Included	.513	sq yd
3 Coats on and Including Painted Metal		
Lath	.627	sq yd
Metal Lath 2.5 lb Diamond Painted, on		
Wood Framing	.107	sq yd
On Ceilings, 3.4 lb Diamond Painted, on		
Wood Framing	.114	sq yd
3.4 lb. Diamond Painted, Wired to Steel		
Framing	.133	sq yd
Suspended Ceiling System, Incl. 3.4 lb		
Diamond Lath	.533	sq yd
Drywall Ceilings Gypsum Drywall, Fire		
Rated, Finished		
Screwed to Grid, Channel of Joists	.021	sf
Over 8′ High, 1/2″ Thick	.022	sf
5/8″ Thick	.023	sf
Grid Suspension System, Direct Hung		
1-1/2″ C.R.C., With 7/8″ Hi Hat Furring		
Channel, 16″ O.C.	.027	sf
24″ O.C.	.018	sf
3-5/8″ Channel, 25 ga., With Track,		
16″ O.C.	.027	sf
24″ O.C.	.018	sf

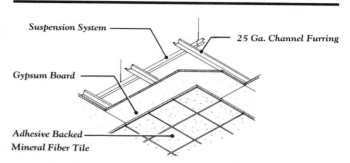

Acoustical Mineral Fiber Tile on Gypsum Board

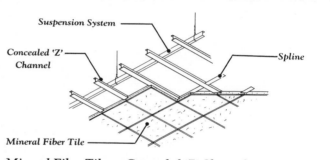

Mineral Fiber Tile on Concealed 'Z' Channel

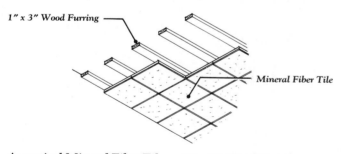

Acoustical Mineral Fiber Tile on 1″ x 3″ Wood Furring

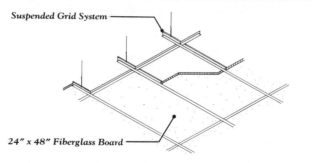

Fiberglass Board on Suspended Grid System

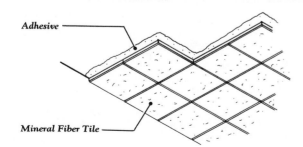

Mineral Fiber Tile Applied with Adhesive

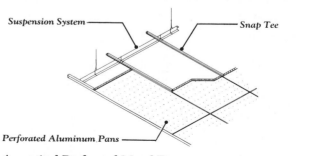

Acoustical Perforated Metal Pans

205

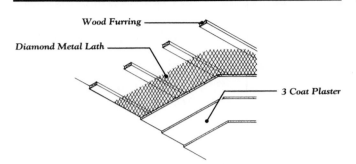

Plaster on Metal Lath and Wood Furring

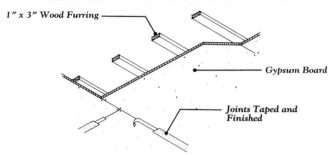

Gypsum Board on 1″ x 3″ Wood Furring

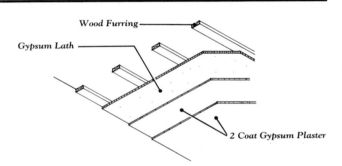

Plaster on Wood Furring

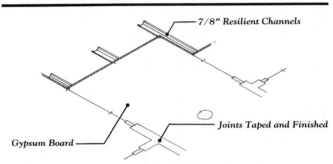

Gypsum Board on 7/8″ Resilient Channel Furring

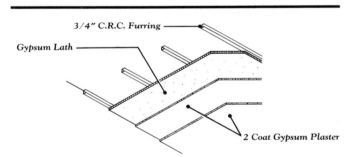

Plaster on Metal Furring

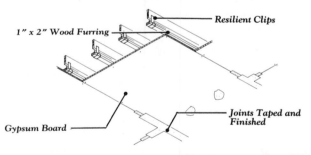

Gypsum Board on 1″ x 2″, Suspended, with Resilient Clips

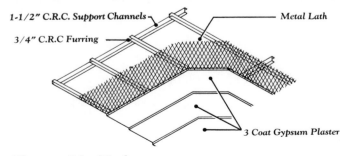

Plaster on Metal Lath

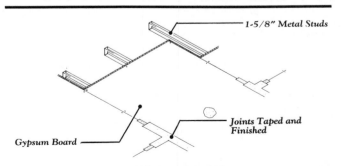

Gypsum Board on 1-5/8″ Metal Stud Furring

6

INTERIOR CONSTRUCTION
FLOORING — CARPET

Carpet is often installed as a flooring material in office, commercial, and residential buildings because it provides an attractive comfortable, and sound-softening covering. Carpeting may be placed in large or small areas with relatively quick and clean installation. Some of the available carpet surface materials include acrylic, nylon, polypropylene, and wood. A variety of backing materials, including polypropylene, jute, cotton, rayon, and polyester, provide support for the surface material. The methods commonly used for integrating the surface materials and attaching them to the backing include tufting, weaving, fusion bonding, knitting, and needle punching. Tufted carpets are usually available in 12- or 15-foot widths, and woven carpets are manufactured in variable widths. A second backing material of foam, urethane, or sponge rubber may be applied to the standard carpet backing to provide an attached pad. Carpet material may be specified or measured by its face weight in ounces per square yard, its pile height, its density, or stitches per inch. A commonly used formula for determining a carpet's density, for example, is

$$\text{Density (oz/sq yd)} = \frac{\text{Face Weight, oz/sq yd} \times 36}{\text{Pile Height, in.}}$$

The method of installation of carpeting material is usually determined by such variables as the size of the area to be covered, the amount and type of traffic that it will carry, and the style of the carpeting. Depending on these conditions, carpet may be installed over separate felt, sponge rubber, or urethane foam pads, or it may be directly adhered to the sub-floor. Wide expanses and carpet locations which carry heavy or wheeled traffic usually require direct glue-down installation to prevent wrinkling, bulging, and movement of the carpet. If padding is required in a glue-down installation, it must consist of an attached pad of synthetic latex, polyurethane foam, or similar material. Carpet may also be installed over a separate pad and then stretched and edge fastened. This method of installation is not recommended for large areas or heavy traffic situations.

With the development of flat undercarpet cable systems for power, telephone, and data line wiring, the use of modular carpet tiles has increased greatly. Modular carpet tiles, when placed directly over the undercarpet cable run, provide easy access to the wiring system. Because taps can be made at any point along the route of the undercarpet system, facilities relocation can be accomplished conveniently, quickly, and economically.

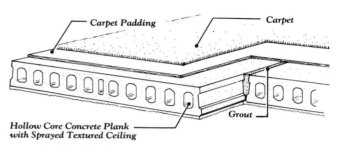

Hollow Core Concrete Plank with Sprayed Textured Ceiling

Carpet on Hollow Core Concrete Plank

Man-hours

Description	m/hr	Unit
Carpet Commercial Grades, Cemented		
Acrylic, 26 oz, Light to Medium Traffic	.216	sf
28 oz, Medium Traffic	.229	sf
35 oz, Medium to Heavy Traffic	.242	sf
Nylon, Non Anti-Static, 15 oz, Light Traffic	.229	sf
Nylon, With Anti-Static, 17 oz, Light to		
Medium Traffic	.211	sf
20 oz, Medium Traffic	.216	sf
22 oz, Medium Traffic	.216	sf
24 oz, Medium to Heavy Traffic	.229	sf
26 oz, Medium to Heavy Traffic	.229	sf
28 oz, Heavy Traffic	.229	sf
32 oz, Heavy Traffic	.242	sf
42 oz, Heavy Traffic	.258	sf
Needle Bonded, 20 oz, No Padding	.143	sf
Polypropylene, 15 oz, Light Traffic	.143	sf
22 oz, Medium Traffic	.182	sf
24 oz, Medium to Heavy Traffic	.182	sf
26 oz, Medium to Heavy Traffic	.182	sf
28 oz, Heavy Traffic	.205	sf
32 oz, Heavy Traffic	.205	sf
42 oz, Heavy Traffic	.216	sf
Scrim Installed, Nylon Sponge Back		
Carpet, 20 oz	.242	sq yd
60 oz	.267	sq yd
Tile, Foam-Backed, Needle Punch	.014	sf
Tufted Loop or Shag	.014	sq yd
Wool, 30 oz, Medium Traffic	.229	sq yd
Wool, 36 oz, Medium to Heavy Traffic	.242	sq yd
Sponge Back, Wool 36 oz, Medium to		
Heavy Traffic	.143	sq yd
42 oz, Heavy Traffic	.229	sq yd
Padding, Sponge Rubber Cushion, Minimum	.108	sq yd
Maximum	.123	sq yd
Felt, 32 oz to 56 oz, Minimum	.108	sq yd
Maximum	.123	sq yd
Bonded Urethane, 3/ 8" Thick, Minimum	.094	sq yd
Maximum	.107	sq yd
Prime Urethane, 1/4" Thick, Minimum	.094	sq yd
Maximum	.107	sq yd

6 INTERIOR CONSTRUCTION
COMPOSITION FLOORING

Composition floors are seamless coverings which are based on epoxy, polyester, acrylic, and polyurethane resins. Various aggregates or color chips may be combined with the resins to produce decorative effects, or special coatings may be applied to enhance the functional effectiveness of a floor's surface. Composition floors are normally installed in special-use situations where other flooring materials do not meet sanitary, safety, or durability standards. Some of these special-use applications of composition flooring include: manufacturing areas, food processing areas, kitchens, beverage bottling plants, parking facilities, multi-purpose rooms, gymnasiums, locker rooms, explosion hazard areas, and other similar flooring installations.

In most cases, the materials which are combined to form the composition floor are determined by and designed for the specific application. Some of the available composition flooring systems include: terrazzo (for decorative floors), wood toppings and sealers, concrete topping and sealers, waterproof and chemical resistant floors, gymnasium sports surfacing, interior and exterior sports surfacing, and conductive floors. The materials employed in these systems are applied by trowel, roller, squeegee, notched trowel, spray gun, brush, or similar tools. After the flooring materials have been applied and allowed to cure, they may be sanded and sealed to produce the desired texture and surface finish. Composition floors are normally installed by trained and approved specialty contractors.

Man-hours

Description	m/hr	Unit
Composition Flooring, Acrylic, 1/4″ Thick	.092	sf
3/8″ Thick	.107	sf
Cupric Oxychloride, on Bond Coat, Minimum	.100	sf
Maximum	.114	sf
Epoxy, with Colored Quartz Chips,		
Broadcast, Minimum	.071	sf
Maximum	.098	sf
Trowelled, Minimum	.086	sf
Maximum	.100	sf
Heavy Duty Epoxy Topping, 1/4″ Thick		
500 to 1,000 sf	.114	sf
1,000 to 2,000 sf	.107	sf
Over 10,000 sf	.100	sf
Epoxy Terrazzo, 1/4″ Thick, Chemical		
Resistant, Minimum	.128	sf
Maximum	.171	sf
Conductive, Minimum	.135	sf
Maximum	.178	sf
Mastic, Hot Laid, 2 Coat, 1-1/2″ Thick		
Standard, Minimum	.070	sf
Maximum	.092	sf
Acidproof, Minimum	.079	sf
Maximum	.137	sf
Neoprene, Trowelled on, 1/4″ Thick,		
Minimum	.088	sf
Maximum	.112	sf
Polyacrylate Terrazzo, 1/4″ Thick, Minimum	.065	sf
Maximum	.100	sf
3/8″ Thick, Minimum	.077	sf
Maximum	.100	sf
Conductive Terrazzo, 1/4″ Thick, Minimum	.107	sf
Maximum	.157	sf
3/8″ Thick, Minimum	.132	sf
Maximum	.188	sf
Granite, Conductive, 1/4″ Thick, Minimum	.069	sf
Maximum	.114	sf
3/8″ Thick, Minimum	.107	sf
Maximum	.126	sf
Polyester, with Colored Quartz Chips,		
1/16″ Thick, Minimum	.045	sf
Maximum	.086	sf
1/8″ Thick, Minimum	.059	sf
Maximum	.071	sf
Polyester, Heavy Duty, Compared to		
Epoxy, Add	.019	sf
Polyurethane, with Suspended Vinyl Chips,		
Minimum	.045	sf
Maximum	.056	sf

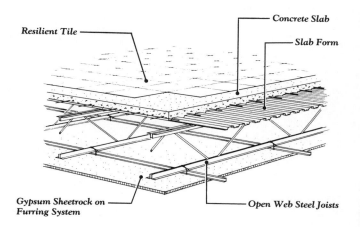

Resilient Tile

Concrete Slab

Slab Form

Gypsum Sheetrock on Furring System

Open Web Steel Joists

Resilient floors are designed for situations where durability and low maintenance are primary considerations. The various resilient flooring materials and their installation formats include: asphalt tiles, cork tiles, polyethylene in rolls, polyvinyl chloride in sheets, rubber in tiles and sheets, vinyl composition tiles, and vinyl in sheets and tiles. All of these materials may be manufactured with or without resilient backing and, except for the polyethylene rolls, they are available in a wide range of colors, designs, textures, compositions, and styles. Rubber and vinyl accessories, which are designed to complement any type of resilient flooring material, include: bases, beveled edges, thresholds, and corner guards, as well as stair treads, risers, nosings, and stringer covers.

The manufacturer's recommendations should be carefully followed for the detailed aspects of the installation of any resilient floor. Generally, any concrete floor surface should be dry, clean, and free from depressions and adhered droppings. Curing and separating compounds should also be thoroughly removed from the surface before the resilient floor is placed. Special consideration is required for installations on slabs or wood subfloors that are located below grade level or above low crawl spaces.

Man-hours

Description	m/hr	Unit
Resilient Asphalt Tile, on Concrete, 1/8" Thick		
Color Group B	.015	sf
Color Group C and D	.015	sf
For Less Than 500 sf, Add	.002	sf
For Over 5000 sf, Deduct	.007	sf
Base, Cove, Rubber or Vinyl, .080" Thick		
Standard Colors, 2-1/2" High	.026	lf
4" High	.027	lf
6" High	.027	lf
1/8" Thick, Standard Colors, 2-1/2" High	.026	lf
4" High	.027	lf
6" High	.027	lf
Corners, 2-1/2" High	.026	Ea.
4" High	.027	Ea.
6" High	.027	Ea.
Conductive Flooring, Rubber Tile, 1/8" Thick	.025	sf
Homogeneous Vinyl Tile, 1/8" Thick	.025	sf
Cork Tile, Standard Finish, 1/8" Thick	.025	sf
3/16" Thick	.025	sf
5/16" Thick	.025	sf
1/2" Thick	.025	sf
Urethane Finish, 1/8" Thick	.025	sf
3/16" Thick	.025	sf
5/16" Thick	.025	sf
1/2" Thick	.025	sf
Polyethylene, in Rolls, No Base Incl., Landscape Surfaces	.029	sf
Nylon Action Surface, 1/8" Thick	.029	sf
1/4" Thick	.029	sf
3/8" Thick	.029	sf
Golf Tee Surface with Foam Back	.033	sf
Practice Putting, Knitted Nylon Surface	.033	sf
Polyurethane, Thermoset, Prefabricated in Place, Indoor		
3/8" Thick for Basketball, Gyms, Etc.	.080	sf
Stair Treads and Risers		
Rubber, Molded Tread, 12" Wide, 5/16" Thick, Black	.070	lf
Colors	.070	lf
1/4" Thick, Black	.070	lf
Colors	.070	lf
Grit Strip Safety Tread, Colors 5/16" Thick	.070	lf
3/16" Thick	.067	lf
Landings, Smooth Sheet Rubber, 1/8" Thick	.029	sf
3/16" Thick	.030	sf
Nosings, 1-1/2" Deep, 3" Wide, Residential	.057	lf
Commercial	.057	lf
Risers, 7" High, 1/8" Thick, Flat	.046	lf
Coved	.046	lf
Vinyl, Molded Tread, 12" Wide, Colors, 1/8" Thick	.070	lf
1/4" Thick	.070	lf
Landing Material, 1/8" Thick	.040	sf
Riser, 7" High, 1/8" Thick, Coved	.046	lf
Threshold, 5-1/2" Wide	.080	lf
Tread and Riser Combined, 1/8" Thick	.100	lf

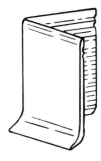

Rubber Cove Base - Corner

Butt Type

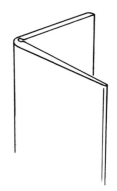

Rubber or Vinyl Corner Guards

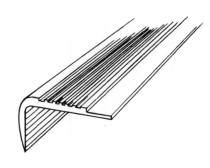

Lap Type

Rubber Nosings, Safety Rib

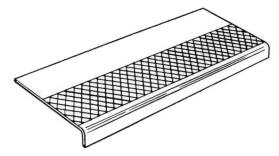

Half Diamond Molded Rubber Stair Tread

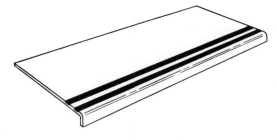

Grit - Strip Molded Rubber Safety Stair Tread

Rubber Stair Treads

210

6 INTERIOR CONSTRUCTION
TILE

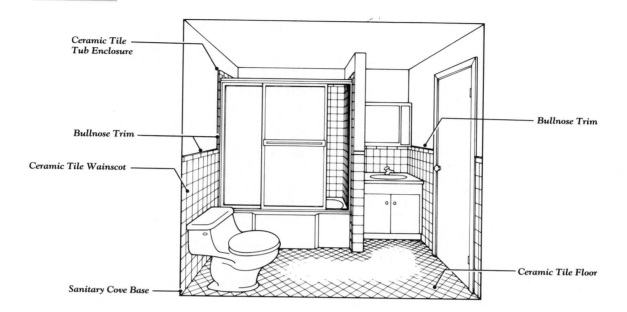

Ceramic Tile Tub Enclosure

Bullnose Trim

Ceramic Tile Wainscot

Sanitary Cove Base

Bullnose Trim

Ceramic Tile Floor

Because of their modular installation and multi-purpose capabilities, ceramic tiles and tiles manufactured from other materials provide an endless source of interior and exterior wall, floor, and counter top coverings. Standard tiles and the methods employed in setting them are designed and manufactured for the intended use of the covering. Many manufacturers can supply special shapes, complementary items, and, especially, bathroom accessories to complete the tile installation. Some of the more commonly applied uses of ceramic and other tiles include: toilet rooms, tubs, steam rooms, swimming pools, and other related installations where easily cleaned, water repellent, and durable surfaces are required.

Ceramic and other tiles are manufactured from several basic materials in a wide range of shapes, sizes, and finishes. Ceramic tiles are manufactured from clay, porcelain, or cement; metal and plastic are the most commonly used materials for tiles which are not labeled as ceramic. In addition to standard ceramic tile materials, floor tiles may also include split brick pavers, quarry tiles, and terra cotta tiles. Tile shapes and sizes vary from 1″ squares to variously sized rectangles, mosaics, patterned combinations, hexagons, octagons, valencia, wedges, and circles. They are available in multi-colored designs and mural sets or with individual pieces that are embossed with designs or pictures. Tiles with special designs and logos may also be custom manufactured. Standard and custom ceramic tiles are produced with glazed and unglazed surfaces with bright, matte, nonslip, and textured finishes.

Proper installation methods for ceramic tiles are determined by the location and the type of backing surface, which may vary from stud or masonry walls to wood or concrete floors. The tiling material may be placed in individual pieces, or it may be installed in factory-prepared back-mounted and ungrouted sections which cover 2

square feet or more as a unit. Tiling material is also available in factory-prepared sections and patterns in which the tiles have been preset and pregrouted with silicone rubber.

The two recognized methods of installation are the thick set, or mud set, method and the thin set method, in which the tile is directly adhered to the base, sub-base, or wall material. In thick set floor installations, Portland cement and sand are placed and screeded to a thickness of 3/4″ to 1-1/4″. For thick set wall installations Portland cement, sand, and lime are placed on the backing surface and troweled to a thickness of 3/4″ to 1″. With both floor and wall placement, the mortars may be reinforced with metal lath or mesh and can be backed with membranes. The tiles may be placed on and adhered to the mortar bed while it remains plastic, or they may be placed after the bed has cured and adhered with a thin bond coat of Portland cement with sand additives. The thin set method of tile installation requires specially prepared mortars and adhesives on properly prepared surfaces. Latex Portland cement, epoxy mortar, epoxy emulsion mortar, epoxy adhesive, and organic adhesives are some of these specialized thin set preparations.

The grouting of the placed and adhered tile material is the final critical step in installation process, as it ensures the sealing of the joints between the tiles and affects the appearance of the installation. The choice of grouting material to be used depends on the type of tile material and the conditions of its exposure. Some of the available grouting mixtures include Portland cement grouts with additives, mastic grout, furan resin grout, epoxy grout, and silicone rubber grout. The recommendations of the manufacturer should be carefully followed in all aspects of the grouting operation.

Man-hours

Description	m/hr	Unit
Flooring Cast Ceramic 4" x 8" x 3/4" Pressed	.160	sf
Hand Molded	.168	sf
8" x 3/4" Hexagonal	.188	sf
Heavy Duty Industrial Cement Mortar Bed	.200	sf
Ceramic Pavers 8" x 4"	.168	sf
Ceramic Tile Base, Using 1" x 1" Tiles, 4" High, Mud Set	.195	lf
Thin Set	.125	lf
Cove Base, 4-1/4" x 4-1/4" High, Mud Set	.176	lf
Thin Set	.125	lf
6" x 4-1/4" High, Mud Set	.160	lf
Thin Set	.117	lf
Sanitary Cove Base, 6" x 4-1/4" High, Mud Set	.172	lf
Thin Set	.129	lf
6" x 6" High, Mud Set	.190	lf
Thin Set	.137	lf
Bullnose Trim, 4-1/4" x 4-1/4", Mud Set	.195	lf
Thin Set	.125	lf
6" x 4-1/4" Bullnose Trim, Mud Set	.190	lf
Thin Set	.129	lf
Floors, Natural Clay, Random or Uniform, Thin Set, Color Group 1	.087	sf
Color Group 2	.087	sf
Porcelain Type, 1 Color, Color Group 2, 1" x 1"	.087	sf
2" x 2" or 2" x 1", Thin Set	.084	sf
Conductive Tile, 1" Squares, Black	.147	sf
4" x 8" or 4" x 4", 3/8" Thick	.133	sf
Trim, Bullnose, Etc.	.080	lf
Specialty Tile, 3" x 6" x 1/2", Decorator Finish	.087	sf
Add For Epoxy Grout, 1/16" Joint, 1" x 1" Tile	.020	sf
2" x 2" Tile	.020	sf
Pregrouted Sheets, Walls, 4-1/4", 6" x 4-1/4", and 8-1/2" x 4-1/4", sf Sheets, Silicone Grout	.067	sf
Floors, Unglazed, 2 sf Sheets Urethane Adhesive	.089	sf
Walls, Interior, Thin Set, 4-1/4" x 4-1/4" Tile	.084	sf
6" x 4-1/4" Tile	.084	sf
8-1/2" x 4-1/4" Tile	.084	sf
6" x 6" Tile	.080	sf
Decorated Wall Tile, 4-1/4" x 4-1/4", Minimum	.018	Ea.
Maximum	.028	Ea.
Exterior Walls, Frostproof, Mud Set, 4-1/4" x 4-1/4"	.157	sf
1-3/8" x 1-3/8"	.172	sf
Crystalline Glazed, 4-1/4" x 4-1/4", Mud Set, Plain	.160	sf
4-1/4" x 4-1/4", Scored Tile	.160	sf
1-3/8" Squares	.172	sf
For Epoxy Grout, 1/16" Joints, 4-1/4" Tile, Add	020	sf

Man-hours (cont.)

Description	m/hr	Unit
For Tile Set in Dry Mortar, Add	.009	sf
For Tile Set in Portland Cement Mortar, Add	.055	sf
Regrout Tile 4-1/2 x 4-1/2, or Larger, Wall	.080	sf
Floor	.073	sf
Ceramic Tile Panels Insulated, Over 1000 Square Feet, 1-1/2" Thick	.073	sf
2-1/2" Thick	.073	sf
Glass Mosaics 3/4" Tile on 12" Sheets, Color Group 1 and 2 Minimum	.195	sf
Maximum (Latex Set)	.219	sf
Color Group 3	.219	sf
Color Group 4	.219	sf
Color Group 5	.219	sf
Color Group 6	.219	sf
Color Group 7	.219	sf
Color Group 8, Gold, Silvers and Specialties	.250	sf
Marble Thin Gauge Tile, 12" x 6", 9/32", White Carara	.250	sf
Filled Travertine	.250	sf
Synthetic Tiles, 12" x 12" x 5/8", Thin Set, Floors	.250	sf
On Walls	.291	sf
Metal Tile Cove Base, Standard Colors, 4-1/4" Square	.053	lf
4-1/8" x 8-1/2"	.040	lf
Walls, Aluminum, 4-1/4" Square, Thin Set, Plain	.100	sf
Epoxy Enameled	.107	sf
Leather on Aluminum, Colors	.123	sf
Stainless Steel	.107	sf
Suede on Aluminum	.123	sf
Plastic Tile Walls, 4-1/4" x 4-1/4", .050" Thick	.064	sf
.110" Thick	.067	sf
Quarry Tile Base, Cove or Sanitary, 2" or 5" High, Mud Set 1/2" Thick	.145	lf
Bullnose Trim, Red, Mud Set, 6" x 6" x 1/2" Thick	.133	lf
4" x 4" x 1/2" Thick	.145	lf
4" x 8" x 1/2" Thick, Using 8" as Edge	.123	lf
Floors, Mud Set, 1000 sf Lots, Red, 4" x 4" x 1/2" Thick	.133	sf
6" x 6" x 1/2" Thick	.114	sf
4" x 8" x 1/2" Thick	.123	sf
Brown Tile, Imported, 6" x 6" x 7/8"	.133	sf
9" x 9" x 1-1/4"	.145	sf
For Thin Set Mortar Application, Deduct	.023	sf
Stair Tread and Riser, 6" x 6" x 3/4", Plain	.320	sf
Abrasive	.340	sf
Wainscot, 6" x 6" x 1/2", Thin Set, Red	.152	sf
Colors Other Than Green	.152	sf
Window Sill, 6" Wide, 3/4" Thick	.178	lf
Corners	.200	Ea.
Terra Cotta Tile on Walls, Dry Set, 1/2" Thick Square, Hexagonal or Lattice Shapes, Unglazed	.059	sf
Glazed, Plain Colors	.062	sf
Intense Colors	.064	sf

Stone floors may be constructed from any type of stone that meets the durability standards and aesthetic requirements of their locations. Some of the commonly used stone flooring materials include slate, flagstone, granite, and marble. The floors may adopt patterned or random designs and use stones that are randomly cut, uniform in size and shape, or designed in patterned sets. The stones may feature neat sawn edges or the irregular shapes of field-cut edges. The various possible exposed face finishes for the stones include: natural cleft, sawn, sawn and polished, and any specialized finish which is normal to the type of stone and functionally and aesthetically appropriate.

Stone floors may be placed on mortar beds or applied on mastic adhesive. When the floor is installed on a mortar bed, the stone may vary slightly in thickness, with the total floor usually measuring 1-1/2″ to 2″ thick from subfloor to finish. When the stone floor is thin set, or laid in mastic, gauged stones with a constant thickness must be used.

After the stone flooring material has been placed, premixed grouting material in various colors or mortar is used to fill the joints between the stones. Some stone flooring materials with consistent sizes and regular edges may not require grouted joints. Many different sealants for the various stone types are available for the final step in the installation process.

Man-hours

Description	m/hr	Unit
Bluestone Flooring, Natural Cleft, Smooth		
or Thermal Finish 1″ Thick	.267	sf
1-1/2″ Thick	.276	sf
Granite Flooring 3/4″ to 1-1/2″ Thick	.276	sf
Granite Pavers 4″ x 4″ x 4″ Blocks	.369	sf
Marble Flooring Tiles, 3/8″ Thick, Thinset	.267	sf
Mortar Bed	.369	sf
Marble Travertine, 1-1/4″ Thick, Mortar Bed	.308	sf
Slate 1/2″ Thick 6 x 6 x 1/2″, Thinset	.267	sf
Mortar Bed	.320	sf
24″ x 24″ x 1/2″ Mortar Bed	.267	sf

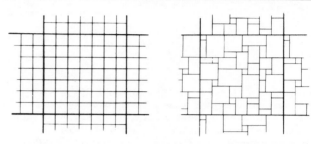

Square Pattern   *Random Rectangular Pattern*

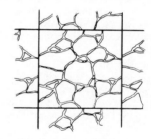

Random Irregular Pattern

Marble Flooring

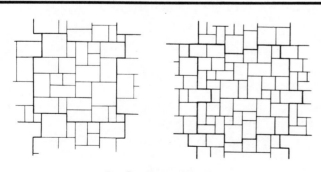

Random Repeat Patterns

Bluestone Flooring

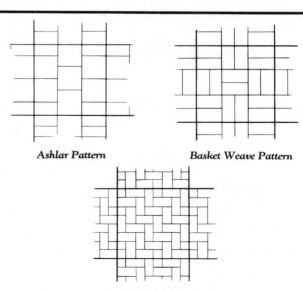

Ashlar Pattern   *Basket Weave Pattern*

Herringbone Pattern

Slate Flooring

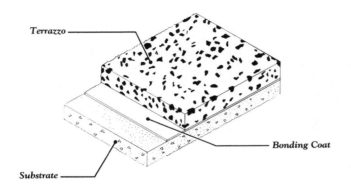

Terrazzo

Bonding Coat

Substrate

Man-hours

Description	m/hr	Unit
Terrazzo Cast in Place Cove Base, 6" High	.348	lf
Curb, 6" High and 6" Wide	.533	lf
Floor, Bonded to Concrete, 1-3/4" Thick,		
Gray Cement	.123	sf
White Cement	.128	sf
Not Bonded, 3" Total Thickness,		
Gray Cement	.160	sf
White Cement	.168	sf
Bonded Conductive Floor for Hospitals	.145	sf
Epoxy Terrazzo, 1/4" Thick, Minimum	.094	sf
Average	.123	sf
Maximum	.160	sf
Monolithic Terrazzo, 5/8" Thick, Including		
3-1/2" Base Slab, 10' Panels, Mesh and Felt	.070	sf
Stairs, Cast in Place, Pan Filled Treads	.291	lf
Treads and Risers	.800	lf
Stair Landings, Add to Floor Prices	.258	sf
Stair Stringers and Fascia	.291	sf
For Abrasive Metal Nosings on Stairs, Add	.056	lf
For Abrasive Surface Finish, Add	.027	sf
For Flush Abrasive Strips, Add	.026	lf
Wainscot, Bonded, 1-1/2" Thick	.400	sf
Epoxy Terrazzo, 1/4" Thick	.229	sf
Terrazzo Precast Base, 6" High, Straight	.067	lf
Cove	.080	lf
8" High Base, Straight	.073	lf
Cove	.089	lf
Curbs, 4" x 4" High	.145	lf
8" x 8" High	.178	lf
Floor Tiles, Non-slip, 1" Thick, 12" x 12"	.267	sf
1-1/4" Thick, 12" x 12"	.267	sf
16" x 16"	.291	sf
1-1/2" Thick, 16" x 16"	.320	sf
Floor Tiles, 12" x 12", 3/16" Thick	.062	sf
Stair Treads, 1-1/2" Thick, Non-slip,		
Diamond Pattern	.168	lf
Line Pattern	.178	lf
2" Thick Treads, Straight	.178	lf
Curved	.188	lf
Stair Risers, 1" Thick, to 6" High,		
Straight Sections	.100	lf
Cove	.107	lf
Curved, 1" Thick, to 6" High, Vertical	.119	lf
Cove	.123	lf
Stair Tread and Riser, Single Piece, Straight,		
Minimum	.246	lf
Maximum	.267	lf
Curved Tread and Riser, Minimum	.267	lf
Maximum	.291	lf
Stair Stringers, Notched, 1" Thick	.152	lf
2" Thick	.267	lf
Stair Landings, Structural, Non-slip,		
1-1/2" Thick	.229	sf
3" Thick	.168	sf
Wainscot, 12" x 12" x 1" Tiles	.229	sf
16" x 16" x 1-1/2" Tiles	.267	sf

Terrazzo flooring materials provide many options to produce colorful, durable, and easily cleaned floors. Conventional ground and polished terrazzo floors employ granite, marble, glass, and onyx chips in a choice of specialized matrices, but well-graded gravel and other stone materials may be used to create different textures when added to the mix. Precast terrazzo tiles and bases, which are normally installed finished and polished on a cement sand base, are also available in many color combinations and aggregates.

Conventional installation of terrazzo flooring involves the mixing of the aggregate with one of three commonly used bonding matrices and the placing of the mix in sections which are defined by divider strips. The matrices employed to bond the aggregate include: cementitious matrices, which consist of natural or colored Portland cement; cement with an acrylic additive; and resinous matrices, which consist of epoxy and polyester. The divider strips, which are manufactured from zinc, brass, or colored plastic, provide for expansion and are therefore positioned over breaks in the substrate and at critical locations where movement is expected in large sections of flooring. The divider strips are also used to terminate pours, to act as leveling guides, and to permit changes in the aggregate mix, section-to section, to create designs or logos.

The terrazzo flooring material may be installed with various methods on a cement sand underbed, or it may be applied in the thin set method directly to the concrete slab. Sand-cushioned terrazzo employs three layers of material: a 1/4" thick sand cushion placed on the slab and covered by an isolation membrane; a mesh-reinforced underbed of approximately 1-3/4" in thickness; and a 1/2" thick terrazzo topping. Bonded terrazzo consists of two layers of material: a 1-3/4" thick underbed placed on the slab and a 1/2" thick terrazzo topping. Monolithic terrazzo flooring is comprised of a single 1/2" thick layer of terrazzo topping applied directly to the slab after control joints have been saw cut into the slab and the divider strips grouted in place. With the thin set method, the terrazzo mix is applied in a single 1/4" thick layer directly on the concrete slab. The surfaces of stair treads and risers may be poured in place with a 1/2" thick layer of terrazzo topping on a 3/4" underbed, or they may be installed as precast terrazzo pieces on a 3/4" thick underbed.

6 INTERIOR CONSTRUCTION
WOOD FLOORS

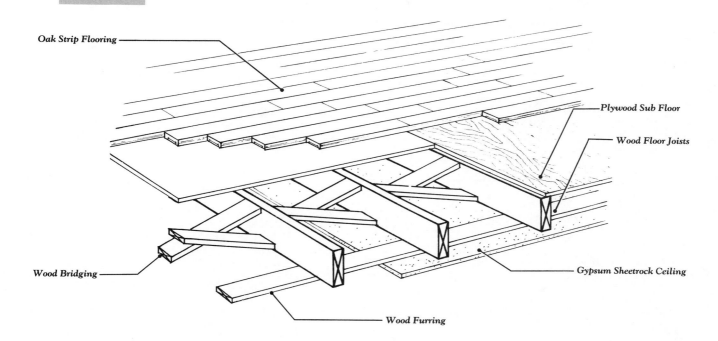

Oak Strip Flooring

Plywood Sub Floor

Wood Floor Joists

Wood Bridging

Gypsum Sheetrock Ceiling

Wood Furring

Wood flooring material is manufactured in three common formats: solid or laminated planks; solid or laminated parquet; and end-grain blocks, which are available in individual pieces or in preassembled strips. Commonly employed hardwood flooring materials include ash, beech, cherry, mahogany, oak, pecan, teak, and walnut, as well as exotic hardwood species, such as ebony, karpa wood, rosewood, and zebra wood. Cedar, fir, pine, and spruce are among the popular softwood flooring materials. End-grain block flooring is manufactured from alder, fir, hemlock, mesquite, oak, and yellow pine.

Strip flooring material is supplied in several different milling formats and combinations, including tongue-and-grooved and matched, square-edged, and jointed with square edges and splines. The installation of tongue-and-groove flooring usually requires blind nailing or fastening by means of metal attachment clips, while square-edged flooring is usually face fastened. Strip floors may be installed over wood-framed subfloors of planks or plywood sheets which have been fastened to the floor joists. A layer of building paper should be laid between the subfloor and the finish strip floor to provide a comfort cushion and to reduce noise. If the strip floor is placed over a concrete surface, a subfloor of exterior plywood should be fastened directly to the slab and protected from moisture by a polyethylene vapor barrier between the concrete and the subfloor. The plywood subfloor may also be fastened to lapped sleepers which have been imbedded in asphalt floor mastic on the slab and draped with the polyethylene vapor barrier before the subfloor is placed.

Parquet floors are prefabricated in panels of various sizes which are milled with square edges, tongue-and-groove edges, or splines. The panels may also contain optional adhered backings which protect the flooring material from moisture, add comfort to the walking area, provide insulation, and deaden sound. Adhesives are normally employed to attach the parquet flooring panels to firm, level subfloors of concrete, wood, plywood, particleboard, resilient tile, or terrazzo. Some parquet floor manufacturers also supply feature strips and factory-fashioned moldings to cover the required expansion space at the edges of the floor.

Wood block floors for industrial application are manufactured in individual blocks and preassembled strips. The surface dimensions of individual blocks are 3″ by 6″, 4″ by 6″, and 4″ by 8″, with nominal thicknesses ranging from 1-1/2″ to 4″. The blocks are normally installed in a layer of pitch applied to a concrete floor. The finish coating of pitch or similar material is then squeezed into the joints between the blocks to provide additional fastening strength. A sealer is applied to provide surface finish. End-grain strip block flooring is placed in mastic adhesive which has been troweled onto a dampproofing membrane that covers the concrete subfloor. After the strip block flooring material has been laid, it is sanded, filled, and finished with penetrating oil.

Wood gymnasium or sports floors usually require specialized installation methods because of their unique function and large size. They may be placed over sleepers that are installed on cushions or pads, or they may be laid directly over a resilient base material or plywood sub-base on cushioned pads. If the flooring material is placed on a sleeper support system, metal clips may be employed for fastening; if it is placed on a plywood subfloor, direct nailing is normally used for fastening. Because large floor areas, such as gymnasiums, require a wide expansion space at the edges, the placement and type of closure strip installed to cover the space deserve special consideration.

215

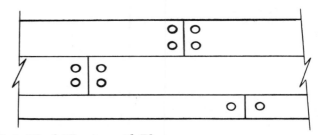

Strip Plank Flooring with Plugs

Random Patterns of Parquet Flooring

Man-hours (cont.)

Description	m/hr	Unit
Running Tracks, Sitka Spruce Surface	.129	sf
3/4" Plywood Surface	.080	sf
Maple, Strip, Not Including Finish	.047	sf
Oak Strip, White or Red, Not Including Finish	.047	sf
Parquetry, Standard, 5/16" Thick, Not		
Including Finish, Minimum	.050	sf
13/16" Thick, Select Grade, Minimum	.050	sf
Maximum	.080	sf
Custom Parquetry, Including Finish,		
Minimum	.080	sf
Maximum	.160	sf
Prefinished White Oak, Prime Grade, 2-1/4"		
Wide	047	sf
3-1/4" Wide	.043	sf
Ranch Plank	.055	sf
Hardwood Blocks, 9" x 9", 25/32" Thick	.050	sf
Acrylic Wood Parquet Blocks,		
12" x 12" x 5/16", Irradiated, Set in Epoxy	.050	sf
Yellow Pine, 3/4" x 3-1/8", T & G, C and		
Better, Not Including Finish	.040	sf
Refinish Old Floors, Minimum	.020	sf
Maximum	.062	sf
Sanding and Finishing, Fill, Shellac, Wax	.027	sf
Wood Block End Grain Flooring,		
Creosoted, 2" Thick	.027	sf
Natural Finish, 1" Thick	.029	sf
1-1/2" Thick	.031	sf
2" Thick	.033	sf
Wood Composition Gym Floors		
2-1/4" x 6-7/8" x 3/8", on 2" Grout Setting		
Bed	.107	sf
Thin set, on Concrete	.064	sf
Sanding and Finishing, Add	.040	sf

Man-hours

Description	m/hr	Unit
Wood Floors Fir, Vertical Grain, 1" x 4",		
Not Including Finish	.031	sf
Gym Floor, in Mastic, Over 2 Ply Felt, #2		
and Better 25/32" Thick Maple, Including		
Finish	.080	sf
33/32" Thick Maple, Including Finish	.082	sf
For 1/2" Corkboard Underlayment, Add	.011	sf
Maple Flooring, Over Sleepers, #2 and Better		
Including Finish, 25/32" Thick	.094	sf
33/32" Thick	.096	sf
For 3/4" Subfloor, Add	.023	sf
With Two 1/2" Subfloors, 25/32" Thick	.116	sf
Maple, Including Finish, #2 and Better,		
25/32" Thick, on Rubber		
Sleepers, with Two 1/2" Subfloors	.105	sf
With Steel Spline, Double Connection		
to Channels	.110	sf
Portable Hardwood, Prefinished Panels	.096	sf
Insulated with Polystyrene, Add	.048	sf

6 INTERIOR CONSTRUCTION
FINISH HARDWARE

Finish hardware is the construction industry term for the devices used to operate doors, windows, drawers, shutters, closets, and cabinets. This category includes such items as hinges, latches, locks, panic devices, security and detection systems, astragals, and weather stripping. In a typical building, the finish hardware accounts for between 2% and 3% of the total job cost. Consequently, a 1% difference in the total building cost can determine the difference between economy and quality hardware.

A hardware specialist will often prepare a schedule and specify the hardware that is to be used for each opening. There are two general classifications of hardware: "Builder's", and "Commercial". "Builder's" is generally used for residential construction.

Another hardware specification is frequency of use, such as heavy, light, or medium. The location where the hardware will be used is one more consideration. The size, weight, and material of a door and its frame, for example, will dictate the size and number of hinges.

Building codes and security requirements will determine the selection of the proper fire barrier and electronic hardware in a modern building.

Most finish hardware is made of cast, wrought, or forged metal, such as iron, brass, bronze, white metal, or aluminum. These metals can be natural or plated and can be finished in dull, polished, Japaned, or antique finishes.

Spring Hinged Security Seal

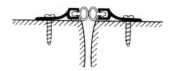

Two Piece Overlapping

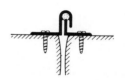

Interlocking with Bulb Insert

Aluminum Extrusion with Sponge Insert

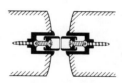

Two Piece Adjustable

Spring Loaded Locking Bolt

Spring Bronze Strip

Man-hours

Description	m/hr	Unit
Astragals, 1/8″ x 3″	.089	lf
Spring Hinged Security Seal with Cam	.107	lf
Two Piece Overlapping	.133	lf
Automatic Openers, Swing Doors, Single	20.000	Ea.
Single Operating, Pair	32.000	Pair
Sliding Doors, 3′ Wide Including Track and		
Hanger, Single	26.667	Opng.
Bi-parting	32.000	Opng.
Handicap Opener Button, Operating	2.000	Ea.
Bolts, Flush, Standard, Concealed	1.143	Ea.
Bumper Plates, 1-1/2″ x 3/4″ U Channel	.200	lf
Door Closer, Adjustable Backcheck,		
3 Way Mount	1.333	Ea.
Doorstops	.333	Ea.
Kickplate	533	Ea.
Lockset, Non Keyed	.667	Ea.
Keyed	.800	Ea.
Dead Locks, Heavy Duty	.889	Ea.
Entrance Locks, Deadbolt	1.000	Ea.
Mortise Lockset, Non Keyed	.889	Ea.
Keyed	1.000	Ea.
Panic Device for Rim Locks		
Single Door, Exit Only	1.333	Ea.
Outside Key and Pull	1.600	Ea.
Bar and Vertical Rod, Exit Only	1.600	Ea.
Outside Key and Pull	2.000	Ea.
Push Pull Plate	.667	Ea.
Weatherstripping, Window, Double Hung	1.111	Opng.
Doors, Wood Frame, 3′ x 7′	1.053	Opng.
6′ x 7′	1.143	Opng.
Metal Frame, 3′ x 7′	2.667	Opng.
6′ x 7′	3.200	Opng.

Astragals

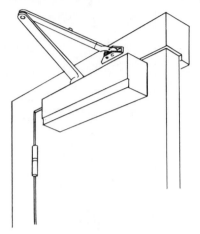

Door Closer

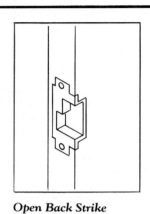

Flush Bolt, Concealed　　　**Open Back Strike**

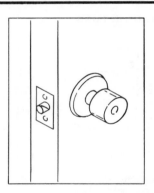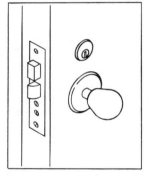

Cylindrical Lock　　　　　Mortise Lock

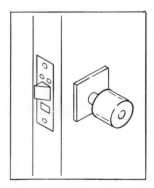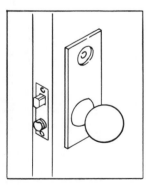

Integral Lock　　　　　"G" Lock

Locksets

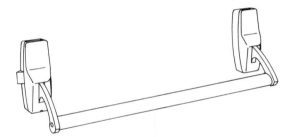

Rim Mounted Panic Bar

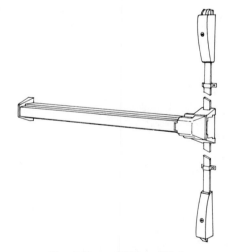

Touch Bar and Vertical Rod

Panic Devices

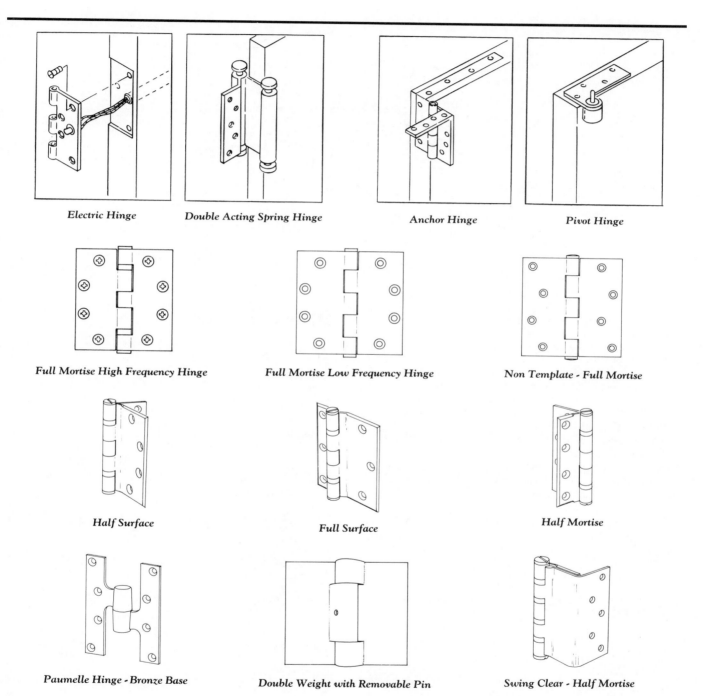

Electric Hinge

Double Acting Spring Hinge

Anchor Hinge

Pivot Hinge

Full Mortise High Frequency Hinge

Full Mortise Low Frequency Hinge

Non Template - Full Mortise

Half Surface

Full Surface

Half Mortise

Paumelle Hinge - Bronze Base

Double Weight with Removable Pin

Swing Clear - Half Mortise

Hinges

Soffit

Wall Cabinets

Counter Top

Base Cabinets

Wall Cabinets

Counter Top

Base Cabinets

Kitchen Cabinets

Millwork refers to finish material made of wood, plastic, and sometimes molded gypsum or polyurethane. Millwork can be custom designed and fabricated, or factory fabricated or milled.

The millwork contractor will commonly furnish the following wood items: doors, windows, factory- and custom-fabricated cabinetry and casework, columns, mantels, grilles, louvers, moldings, paneling, railings, shelving, siding and stairs.

Plastic materials include: laminates, cabinets, moldings, doors and windows.

Factory-molded gypsum and polyurethane, medallions, mantels, stair brackets, door and window features and moldings may also be supplied under the millwork contract.

Paneling may be specified as "matched" or "unmatched" in one or more than 40 wood species with milled moldings, doors, door frames and cabinetry material. Kitchen cabinets may be factory-fabricated or custom designed of wood or plastic laminate.

Man-hours

Description	m/hr	Unit
Beams, Decorative 4″ x 8″	.100	lf
Cabinets, Kitchen Base, 24″ x 24″ x 35″ High	.717	Ea.
Kitchen Wall, 12″ x 24″ x 30″ High	.788	Ea.
Casework Frames		
Base Cabinets, 36″ High, Two Bay, 36″ Wide	3.636	Ea.
Book Cases, 7′ High, Two Bay, 36″ Wide	5.000	Ea.
Coat Racks, 7′ High, Two Bay, 48″ Wide	2.909	Ea.
Wall Mounted Cabinets, 30″ High, Two Bay, 36″ Wide	3.721	Ea.
Wardrobe, 7′ High, 48″ Wide	4.706	Ea.
Cabinet Doors		
Glass Panel, Hardwood Frame, 18″ Wide, 30″ High	.276	Ea.
Hardwood, Raised Panel, 18″ Wide, 30″ High	.571	Ea.

Man-hours (cont.)

Description	m/hr	Unit
Plastic Laminate, 18″ Wide, 30″ High	.364	Ea.
Counter Tops, Stock, Plastic Laminate, 1-1/4″ Thick	.286	lf
Moldings, Base, 9/16″ x 3-1/2″	.033	lf
Casings, Apron, 5/8″ x 3-1/2″	.036	lf
Band, 11/16″ x 1-3/4″	.032	lf
Casing, 11/16″ x 3-1/2″	.037	lf
Ceilings, Bed, 9/16″ x 2″	.033	lf
Cornice, 9/16″ x 2-1/4″	.027	lf
Cove Scotia, 11/16″ x 2-3/4″	.031	lf
Crown, 11/16″ x 4-5/8″	.036	lf
Exterior Cornice Boards, 1″ x 6″	.040	lf
Corner Board, 1″ x 6″	.040	lf
Fascia, 1″ x 6″	.032	lf
Moldings	.032	lf
Verge Board, 1″ x 6″	.040	lf
Trim, Astragal, 1-5/16″ x 2-3/16″	.033	lf
Chair Rail, 5/8″ x 3-1/2″	.033	lf
Handrail, 1-1/2″ x 1-3/4″	.100	lf
Door Moldings	.471	set
Door Trim Including Headers, Stops, and Casings 4-1/2″ Wide	1.509	Opng.
Stool Caps, 1-1/16″ x 3-1/4″	.053	lf
Threshold, 3′ Long, 5/8″ x 3-5/8″	.250	Ea.
Window Trim Sets Including Casings, Header, Stops, Stool and Apron	.800	Opng.
Paneling, 1/4″ Thick	.040	sf
3/4″ Thick, Stock	.050	sf
Architectural Grade	.071	sf
Stairs, Prefabricated		
Box 3′ Wide, 8′ High	5.333	Flight
Open Stairs, 8′ High	5.333	Flight
3 Piece Wood Railings and Baluster, 8′ High	1.333	Ea.
Spiral Oak, 4′-6″ Diameter Including Rail	10.667	Flight

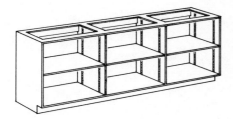

Three Bay Base Cabinet

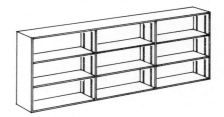

Three Bay Wall Mounted Cabinet

Three Bay Book Case

Two Bay Coat Rack

Wardrobe with Drawers

Wardrobe with Shelves

Casework Frames

Corner China Cabinet

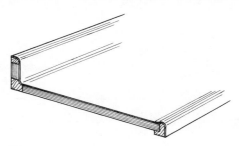

Counter Top

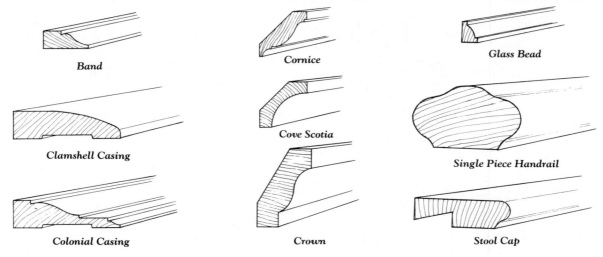

Band Cornice Glass Bead

Clamshell Casing Cove Scotia Single Piece Handrail

Colonial Casing Crown Stool Cap

Mouldings

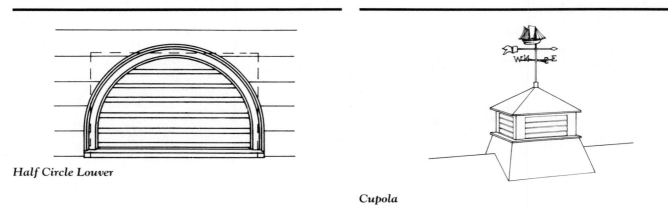

Half Circle Louver

Cupola

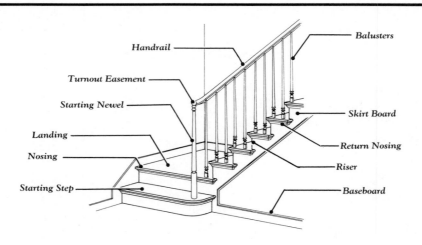

Handrail Balusters

Turnout Easement

Starting Newel Skirt Board

Landing

Nosing Return Nosing

Riser

Starting Step Baseboard

Stairs

DIVISION
7
CONVEYING

Conveying systems are used to carry passengers and materials in or on vertical, horizontal, or inclined transport assemblies. Because they are very specialized and complex pieces of machinery, they are installed by specialty contractors. The list below shows the many types of conveying systems available and the most important points to be considered in the design and/or selection of a system.

Dumbwaiters:
- Capacity
- Floors
- Speed
- Size
- Stops
- Finish

Elevators:
- Hydraulic or electric
- Capacity
- Floors
- Stops
- Finish
- Door type
- Special requirements
- Geared or gearless
- Size
- Number required
- Speed
- Machinery location
- Signals

Escalators:
- Capacity
- Floors
- Story height
- Finish
- Incline angle
- Size
- Number required
- Speed
- Machinery location
- Special requirements

Material Handling Systems:
- Automated
- Non-automated

Moving Stairs and Walks:
- Capacity
- Floors
- Story height
- Finish
- Incline angle
- Size
- Number required
- Speed
- Machinery location
- Special requirements

Pneumatic Tube Systems:
- Automatic
- Size
- Length
- Manual
- Stations
- Special requirements

Vertical Conveyer:
- Automatic
- Non-automatic

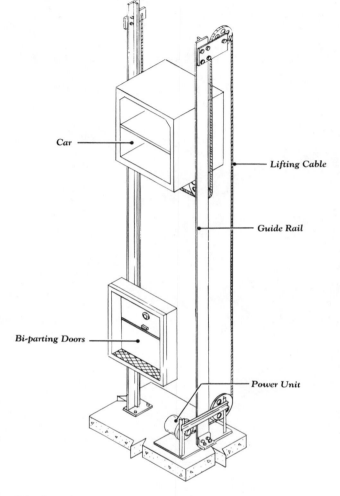

Dumbwaiter

Among the most commonly used types of conveying systems are elevators, which are manufactured for all categories of multi-storied buildings—commercial, residential, institutional, or industrial. They are powered by hydraulic cylinders or electric motors and may be used to transport passengers, freight, and equipment.

Hydraulic elevators are commonly installed in buildings that do not exceed 70′ in height. Because they are relatively slow, with travel speeds ranging between 25′ and 150′ per minute, and because their travel distance is short, they are most often installed in low-rise commercial, industrial, and residential buildings. They are also appropriate elevator systems for motels and hotels and for structures where freight is moved regularly.

The operation of hydraulic elevators involves a hydraulic piston which is encased in a cylinder and mounted to the base of the cab. The cylinder and its casing must usually extend vertically into the ground beneath the elevator shaft to a depth that is equal to the required rise of the system. For example, a hydraulic elevator that travels 55′ in height must be equipped with a piston and casing that are driven 55′ below the shaft's lowest point. Because the piston and casing assembly must be placed by a drilling rig that requires excessive head room, hydraulic elevators are difficult to install in existing buildings. Typically, the cost of installing hydraulic elevators in new low-rise buildings is less than that of electrically powered systems.

Electric elevators are usually installed in mid- to high-rise commercial and institutional buildings of more than five stories. They are much faster than hydraulic elevators, with potential speeds of 1800′ per minute in buildings above 50 stories in height. They are unlimited in their travel length. Although electric elevators are not thought of as freight haulers, they can transport capacities of up to 10,000 pounds.

The lifting operation of electric elevators involves a system of steel hoisting cables and counterweights which is powered by electric motors. One end of the hoisting cables is attached to the cab and the other to the counterweights, which can weigh up to 40% of the load capacity and which help to reduce the power requirements of the motors. The electric motors drive drum assemblies that may be geared or gearless units. The geared drum assemblies are normally used for low-speed applications, while the gearless units are preferred for high-speed installations.

Freight elevators are specifically designed to transport equipment and materials rather than people. Hydraulically powered freight systems are normally installed in low-rise structures, and electrically powered units are more economical for lifts over 50′ in height. Generally, freight elevators operate on the same principles as conventional hydraulic and electric elevators, but they are more rugged in construction and appearance. Also, because they are built to transport heavier loads, they tend to be slower, with standard speeds ranging from 75′ to 200′ per minute. Higher-speed systems are available for low-capacity and light-duty functions. Freight elevators may carry payloads ranging from 2,500 to 25,000 pounds or more, if necessary.

"Minimum Guidelines and Requirements for Accessible Design," by the Architectural and Transportation Barriers Compliance Board, became law on September 3, 1982. The guidelines and requirements provide a basis for consistent and improved accessibility standards for everyone, not just the handicapped. For complete and up-to-date information, write to: U.S. Architectural and Transportation Barriers Compliance Board, 330 C Street, S.W., Room 1010, Switzer Building, Washington, D.C. 20202.

Elevator Hoistway Sizes

Elevator Type	Floors	Building Type	Capacity Lbs	Capacity Passengers	Entry *	Hoistway Width	Hoistway Depth	SF Area Per Floor
Hydraulic	5	Apt./Small Office	1500	10	S	6′-7″	4′-6″	29.6
			2000	13	S	7′-8″	4′-10″	37.4
	7	Av. Office/Hotel	2500	16	S	8′-4″	5′-5″	45.1
			3000	20	S	8′-4″	5′-11″	49.3
		Lg. Office/Store	3500	23	S	8′-4″	6′-11″	57.6
		Freight Light Duty	2500		D	7′-2″	7′-10″	56.1
		Heavy Duty	5000		D	10′-2″	10′-10″	110.1
			7500		D	10′-2″	12′-10″	131
			5000		D	10′-2″	10′-10″	110.1
			7500		D	10′-2″	12′-10″	131
			10000		D	10′-4″	14′-10″	153.3
		Hospital	3500		D	6′-10″	9′-2″	62.7
			4000		D	7′-4″	9′-6″	69.6
Electric Traction, High Speed	High Rise	Apt./Small Office	2000	13	S	7′-8″	5′-10″	44.8
			2500	16	S	8′-4″	6′-5″	54.5
			3000	20	S	8′-4″	6′-11″	57.6
			3500	23	S	8′-4″	7′-7″	63.1
		Store	3500	23	S	9′-5″	6′-10″	64.4
		Large Office	4000	26	S	9′-4″	7′-6″	70
		Hospital	3500		D	7′-6″	9′-2″	69.4
			4000		D	7′-10″	9′-6″	58.8
Geared, Low Speed		Apartment	1200	8	S	6′-4″	5′-3″	33.2
			2000	13	S	7′-8″	5′-8″	43.5
			2500	16	S	8′-4″	6′-3″	52
		Office	3000	20	S	8′-4″	6′-9″	56
		Store	3500	23	S	9′-5″	6′-10″	64.4
						Add 4″ width for multiple units		

* S = Single Door
* D = Double Door

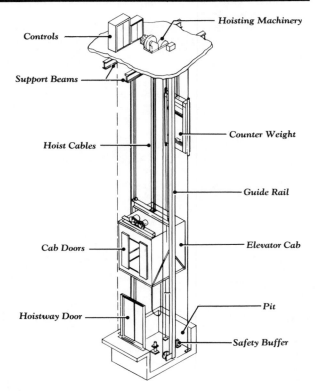

Controls

Hoisting Machinery

Support Beams

Hoist Cables

Counter Weight

Guide Rail

Cab Doors

Elevator Cab

Hoistway Door

Pit

Safety Buffer

Electric Elevator

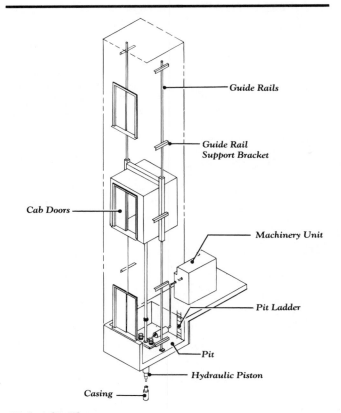

Guide Rails

Guide Rail
Support Bracket

Cab Doors

Machinery Unit

Pit Ladder

Pit

Hydraulic Piston

Casing

Hydraulic Elevator

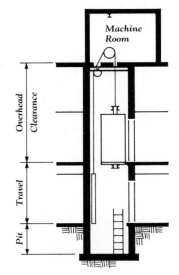

Machine
Room

Overhead
Clearance

Travel

Pit

Electric Elevator

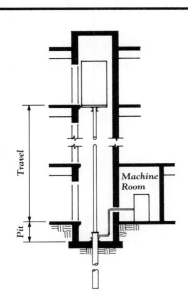

Travel

Machine
Room

Pit

Hydraulic Elevator

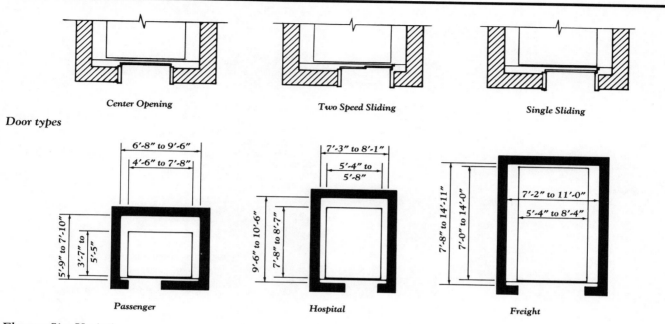

Door types

Center Opening

Two Speed Sliding

Single Sliding

Elevator Size Variations

Passenger

Hospital

Freight

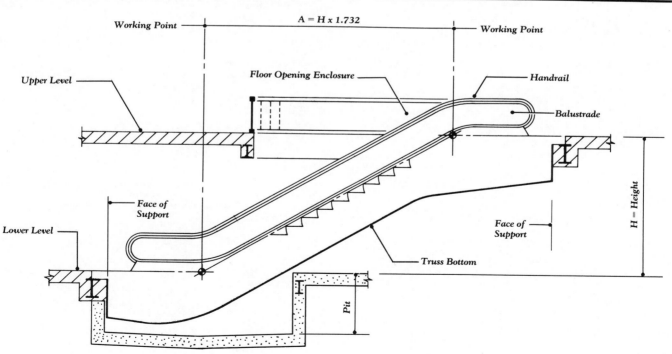

A = H x 1.732

Working Point

Working Point

Upper Level

Floor Opening Enclosure

Handrail

Balustrade

Face of Support

Face of Support

Lower Level

Truss Bottom

H = Height

Pit

Escalator

Man-hours

Description	m/hr	Unit
Conveying Systems		
Dumbwaiters 2 Stop Electric Average	134.000	Ea.
For Each Additional Stop Add	29.630	Stop
Hand Average	76.888	Ea.
For Each Additional Stop Add	26.667	Stop
Elevators for Multi-Story Buildings		
Freight 2 Story Hydraulic		
4,000 lb Capacity Average	498.000	Ea.
10,000 lb Capacity Average	549.000	Ea.
6 Story 4,000 lb or 10,000 lb Capacity	1067.000	Ea.
Geared Electric		
4,000 lb Capacity	800.000	Ea.
10,000 lb Capacity	1067.000	Ea.
Passenger Pre-Engineered		
5 Story Hydraulic 2,500 lb Capacity	800.000	Ea.
For Less Than 5 Stop Deduct	110.000	Stop
10 Story Geared Traction 200'/min		
2,500 lb Capacity	1600.000	Ea.
For Less Than 10 Stop Deduct	94.118	Stop
For 4,500 lb Capacity General Purpose	1600.000	Ea.
Residential Cab Type		
1 Floor 2 Stop Average	80.000	Ea.
2 Floor 3 Stop Average	160.000	Ea.
Stair Climber (Chair Lift)		
Single Seat Average	48.000	Ea.
Wheel Chair Porch Lift Average	24.000	Ea.
Stair Lift Average	48.000	Ea.
Escalators per Single Unit Average	667.000	Ea.
Material Handling		
Conveyers Gravity Type 2" Rollers 3"		
On Center Horizontal Belt Center Drive		
and Takeup 45'/min		
16" Belt 26.5' Length	32.000	Ea.
24" Belt		
41.5' Length	40.000	Ea.
61.5' Length	53.333	Ea.
Inclined Belt 25° Incline with Horizontal		
Loader and End Idler Assembly 34' Length		
12" Belt	53.333	Ea.
16" Belt	80.000	Ea.
24" Belt	106.667	Ea.

Man-hours (cont.)

Description	m/hr	Unit
Conveyer Overhead Automatic Powered		
Chain Conveyer		
130 lb/lf Capacity	.941	lf
Monorail Overhead Manual Channel Type		
125 lb/lf	.308	lf
500 lb/lf	.381	lf
Motorized Car Distribution System Single		
Track 20 lb per Car Capacity Average	190.877	station
Larger System Including Hospital		
Transport Track Type Fully		
Automated Average	1600.000	station
Moving Walk		
24" Tread Width Average	6.073	lf
40" Tread Width Average	7.800	lf
Moving Ramp 12° Incline		
32" Tread Width Average	7.225	lf
40" Tread Width Average	9.981	lf
Parcel Lift 20" x 20" 100 lb Capacity Electric		
Per Floor	64.000	Ea.
Pneumatic Tube System Single Tube 2 Stations		
100' Long Stock		
Economy		
3" Diameter	133.000	total
4" Diameter	177.000	total
Twin Tube Two Stations or More		
Conventional System		
2-1/2" Round	.256	lf
3" Round	.348	lf
4" Round	.323	lf
4" x 7" Oval	.426	lf
Add for Blower	8.000	system
Plus for Each Station Add	2.133	Ea.
Completely Automatic System		
4" Round		
15 to 50 Stations	55.172	station
51 to 144 Stations	50.000	station
6" Round or 4" x 7" Oval		
15 to 50 Stations	66.667	station
51 to 144 Stations	69.565	station
Vertical Conveyer Automatic Selective Central		
Control to 10 Floors Base Price	400.000	total
For Automatic Service Any Floor to Any		
Floor Add	13.913	floor

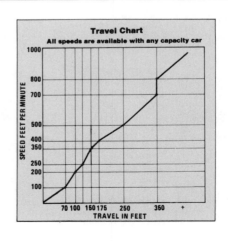

Passenger Capacity During Peak Periods

Travel Chart — All speeds are available with any capacity car

DIVISION
8
MECHANICAL

Because piping in building construction is used for both pressurized and nonpressurized applications, the materials and joining methods employed in a given installation vary. Location, type of fluid being conveyed, and economic restrictions are other variables which must be considered when selecting piping material.

If pressurized piping is to be placed underground to convey gas, oil, pumped sewage, steam or water, many suitable materials are available. These include asbestos cement, brass, concrete (reinforced), copper, cast and ductile iron, lead, plastic, and steel. The methods of joining and sealing these pressurized systems vary with the type of piping material and include caulked, cemented, compressed, flanged, gasketed, soldered, threaded, welded, and mechanical joint connections. The materials used for pressurized piping systems within a building include brass,

copper, plastic, and steel. Joining and sealing methods for these materials are the same as for those used in underground pressurized systems of similar piping material. When nonpressurized piping is to be used underground to convey storm or sanitary drainage, a number of materials are available, including asbestos cement, bituminous fiber, cast iron, clay tile, concrete, corrugated metal, and a variety of plastics. The joining and sealing methods employed for a given system depend on the type of piping material and include caulked, cemented, compression, gasketed, and mechanical joint connections. Nonpressurized piping systems within a building consist of cast iron soil pipe, copper, galvanized steel or plastic pipe. The joining and sealing methods used for these materials are the same as for those used in nonpressurized underground systems of similar piping materials.

Cast Iron Soil Pipe

Cast iron soil pipe is used to drain water and sanitary wastes from the plumbing fixtures and to convey these products to the public sewer main or private septic system. Cast iron may also be used for other elements within the sanitary system, including the soil stack and vent piping, but plastic piping and fittings designed for drainage, waste, and venting are also commonly installed.

Historically, the most common method of joining cast iron pipe has been with a hub and spigot fitting sealed with a lead and oakum joint. Where approved by local codes, the use of no-hub pipe and fittings, sealed with neoprene gaskets and metal couplings, has been popularly applied in recent years as a method of joining cast iron pipe.

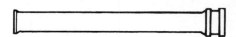

Single Hub Soil Pipe

Eighth Bend

Sanitary Tee

Man-hours for Installation

Description	m/hr	Unit
Cast Iron Soil Pipe Service Weight, Single Hub with Hangers Every Five Feet,		
Lead and Oakum Joints Every Ten Feet		
2" Pipe Size	.254	lf
3" Pipe Size	.267	lf
4" Pipe Size	.291	lf
5" Pipe Size	.316	lf
6" Pipe Size	.329	lf
8" Pipe Size	.542	lf
10" Pipe Size	.593	lf
12" Pipe Size	.667	lf
Cast Iron Soil Pipe Fittings		
Hub and Spigot Service Weight		
Bends or Elbows		
2" Pipe Size	1.000	Ea.
3" Pipe Size	1.140	Ea.
4" Pipe Size	1.230	Ea.
5" Pipe Size	1.330	Ea.
6" Pipe Size	1.410	Ea.
8" Pipe Size	2.910	Ea.
10" Pipe Size	3.200	Ea.
12" Pipe Size	3.560	Ea.
Tees or Wyes		
2" Pipe Size	1.600	Ea.
3" Pipe Size	1.780	Ea.
4" Pipe Size	2.000	Ea.
5" Pipe Size	2.000	Ea.
6" Pipe Size	2.180	Ea.
8" Pipe Size	4.570	Ea.
10" Pipe Size	4.870	Ea.
12" Pipe Size	5.330	Ea.

Man-hours (cont.)

Description	m/hr	Unit
Cleanouts		
Floor Type		
2" Pipe Size	.800	Ea.
3" Pipe Size	1.000	Ea.
4" Pipe Size	1.333	Ea.
5" Pipe Size	2.000	Ea.
6" Pipe Size	2.667	Ea.
8" Pipe Size	4.000	Ea.
Cleanout Tee		
2" Pipe Size	2.000	Ea.
3" Pipe Size	2.222	Ea.
4" Pipe Size	2.424	Ea.
5" Pipe Size	2.909	Ea.
6" Pipe Size	3.200	Ea.
8" Pipe Size	6.400	Ea.
Drains		
Heelproof Floor Drain		
2" to 4" Pipe Size	1.600	Ea.
5" and 6" Pipe Size	1.778	Ea.
8" Pipe Size	2.000	Ea.
Shower Drain		
1-1/2" to 3" Pipe Size	2.000	Ea.
4" Pipe Size	2.286	Ea.
Cast Iron Service Weight Traps		
Deep Seal		
2" Pipe Size	1.143	Ea.
3" Pipe Size	1.333	Ea.
4" Pipe Size	1.455	Ea.
P Trap		
2" Pipe Size	1.000	Ea.
3" Pipe Size	1.143	Ea.
4" Pipe Size	1.231	Ea.
5" Pipe Size	1.333	Ea.
6" Pipe Size	1.412	Ea.
8" Pipe Size	2.909	Ea.
10" Pipe Size	3.200	Ea.
Running Trap with Vent		
3" Pipe Size	1.143	Ea.
4" Pipe Size	1.231	Ea.
5" Pipe Size	2.182	Ea.
6" Pipe Size	3.000	Ea.
8" Pipe Size	3.200	Ea.
S Trap		
2" Pipe Size	1.067	Ea.
3" Pipe Size	1.143	Ea.
4" Pipe Size	1.231	Ea.
No Hub with Couplings Every Ten Feet OC		
1-1/2" Pipe Size	.225	lf
2" Pipe Size	.239	lf
3" Pipe Size	.250	lf
4" Pipe Size	.276	lf
5" Pipe Size	.289	lf
6" Pipe Size	.304	lf
8" Pipe Size	.464	lf
10" Pipe Size	.525	lf

Cleanout, Floor Type

Cleanout Tee

Heelproof Floor Drain

Shower Drain

Deep Seal Trap

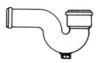

P Trap

Running Trap with Vent

S Trap

Man-hours (cont.)

Description	m/hr	Unit
No Hub Couplings*		
1-1/2" Pipe Size	.333	Ea.
2" Pipe Size	.364	Ea.
3" Pipe Size	.421	Ea.
4" Pipe Size	.485	Ea.
5" Pipe Size	.545	Ea.
6" Pipe Size	.600	Ea.
8" Pipe Size	.970	Ea.
10" Pipe Size	1.230	Ea.

*Note: In estimating labor for no hub fittings, all the labor is included in the no hub couplings. One coupling per joint.

No Hub Coupling

Copper Pipe and Tubing

Copper and brass piping are not commonly used in today's construction industry, having been replaced by the more economical and versatile copper tubing. This tubing is manufactured in various wall thicknesses, which are referred to by type. Types K, L, M, and DWV are the most commonly used sizes for plumbing and heating installations; and type ACR, for air conditioning and refrigeration applications. Generally, all types of copper tubing are available in straight lengths, both hard-drawn and annealed for bending, except for Type DWV, which is manufactured in hard-drawn lengths only. Annealed, or soft-drawn, copper tubing measuring 2" and less in diameter is also available in coils.

The type of copper tubing installed in a given situation depends on its location and function. Type K tubing, because of its relatively thick-walled construction, is normally used for underground applications and for interior systems where more durable tubing is called for. Medium walled Type L tubing is commonly used for interior plumbing, heating, and cooling systems. Light-walled Type M tubing is frequently installed in low-

pressure situations, such as residential heating systems and drainage lines measuring 1" and less in diameter. Type DWV, which is the thinnest-walled copper tubing, is designed for nonpressurized applications, such as drainage waste and vent piping. Type DWV tubing is manufactured in sizes ranging from 1-1/4" to 8" diameter. Type ACR tubing, which is commonly used in air conditioning and refrigeration systems, is identical in size to Type L tubing, but it is usually shipped to the job cleaned and capped. It may also be dehydrated or charged with nitrogen at the factory prior to shipping.

The joining methods and fittings used in copper tubing installations vary with the type of tubing, its location, and its function. Generally, brazing, soldering, flanged-joint, flared-joint, and compression are the most commonly used joining methods. Special thin-walled, solder-type drainage fittings are available for DWV applications. Cast bronze and wrought copper solder-type fittings are utilized for Types K, L, M and ACR tubing, regardless of the amount of pressure in the system. For high pressure and extreme temperature conditions, silver solder or brazing is used.

Man-hours

Description	m/hr	Unit
Man-hours to Install The Several Types of Copper Tubing Based On A Soft Soldered Coupling and A Clevis Hanger Every Ten Feet		
Type K Tubing		
1/2" Pipe Size	.103	lf
3/4" Pipe Size	.108	lf
1" Pipe Size	.121	lf
1-1/4" Pipe Size	.143	lf
1-1/2' Pipe Size	.160	lf
2" Pipe Size	.200	lf
2-1/2" Pipe Size	.267	lf
3" Pipe Size	.296	lf
3-1/2" Pipe Size	.381	lf
4" Pipe Size	.421	lf
5" Pipe Size	.500	lf
6" Pipe Size	.632	lf
8" Pipe Size	.706	lf

Man-hours (cont.)

Description	m/hr	Unit
Type L Tubing		
1/2" Pipe Size	.099	lf
3/4" Pipe Size	.105	lf
1" Pipe Size	.118	lf
1-1/4" Pipe Size	.138	lf
1-1/2" Pipe Size	.154	lf
2" Pipe Size	.190	lf
2-1/2" Pipe Size	.258	lf
3" Pipe Size	.286	lf
3-1/2" Pipe Size	.372	lf
4" Pipe Size	.410	lf
5" Pipe Size	.471	lf
6" Pipe Size	.600	lf
8" Pipe Size	.667	lf

Man-hours (cont.)

Description	m/hr	Unit
Type M Tubing		
1/2″ Pipe Size	.095	lf
3/4″ Pipe Size	.103	lf
1″ Pipe Size	.114	lf
1-1/4″ Pipe Size	.133	lf
1-1/2″ Pipe Size	.148	lf
2″ Pipe Size	.182	lf
2-1/2″ Pipe Size	.250	lf
3″ Pipe Size	.276	lf
3-1/2″ Pipe Size	.356	lf
4″ Pipe Size	.400	lf
5″ Pipe Size	.444	lf
6″ Pipe Size	.571	lf
8″ Pipe Size	.632	lf

Man-hours

Description	m/hr	Unit
Fittings for Copper Tubing, Pressurized Systems, Solder Joint		
Elbows 90° or 45°		
1/2″ Tubing Size	.400	Ea.
3/4″ Tubing Size	.421	Ea.
1″ Tubing Size	.500	Ea.
1-1/4″ Tubing Size	.533	Ea.
1-1/2″ Tubing Size	.615	Ea.
2″ Tubing Size	.727	Ea.
2-1/2″ Tubing Size	1.231	Ea.
3″ Tubing Size	1.455	Ea.
3-1/2″ Tubing Size	1.600	Ea.
4″ Tubing Size	1.778	Ea.
5″ Tubing Size	2.667	Ea.
6″ Tubing Size	2.667	Ea.
8″ Tubing Size	3.000	Ea.
Tees		
1/2″ Tubing Size	.615	Ea.
3/4″ Tubing Size	.667	Ea.
1″ Tubing Size	.800	Ea.
1-1/4″ Tubing Size	.889	Ea.
1-1/2″ Tubing Size	1.000	Ea.
2″ Tubing Size	1.143	Ea.
2-1/2″ Tubing Size	2.000	Ea.
3″ Tubing Size	2.286	Ea.
3-1/2″ Tubing Size	2.667	Ea.
4″ Tubing Size	3.200	Ea.
5″ Tubing Size	4.000	Ea.
6″ Tubing Size	4.000	Ea.
8″ Tubing Size	4.800	Ea.
Drainage Waste and Vent Systems, Solder Joint		
Elbows 90° or 45°		
1-1/4″ Tubing Size	.615	Ea.
1-1/2″ Tubing Size	.667	Ea.
2″ Tubing Size	.800	Ea.
3″ Tubing Size	1.600	Ea.
4″ Tubing Size	1.778	Ea.
5″ Tubing Size	2.667	Ea.
6″ Tubing Size	2.667	Ea.
8″ Tubing Size	3.000	Ea.

Man-hours (cont.)

Description	m/hr	Unit
Type DWV Tubing		
1-1/4″ Pipe Size	.133	lf
1-1/2″ Pipe Size	.148	lf
2″ Pipe Size	.182	lf
2-1/2″ Pipe Size	.225	lf
3″ Pipe Size	.276	lf
4″ Pipe Size	.400	lf
5″ Pipe Size	.444	lf
6″ Pipe Size	.571	lf
8″ Pipe Size	.632	lf

Type ACR tubing, having the same physical properties as Type L tubing, will require the same labor for hanging and coupling together. Refrigerant systems, on the other hand, have special cleaning and testing requirements, and additional labor is needed for these functions.

90° Elbow - Solder Joint

45° Elbow - Solder Joint

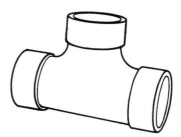

Tee - Solder Joint

Flexible Plastic Tubing

Polybutylene (PB) and polyethylene (PE) are classified as flexible piping. They are both supplied in coils of one hundred feet or more. Because of this flexibility and the availability of long sections, the use of couplings and bends is greatly reduced and installing labor is also much less than using rigid plastic piping.

Polyethylene has a lower material cost than polybutylene but PB has higher pressure and temperature ratings.

Each of these flexible systems may be joined by fusion, insert fittings (plastic or brass), compression or flared fittings. PE is customarily used for water or gas service, whereas PB is used for hot and cold water and is currently gaining limited approval in fire sprinkler systems.

Man-hours

Description	m/hr	Unit
Man-hours to Install Flexible Plastic Piping. Pipe Hangers and Couplings Are Not Included Since They Are Not Always Necessary Such as When This Piping Is Laid in A Trench or Strung Through Joists or Clipped to Structural Members.		
1/2″ Pipe Size		
3/4″ Pipe Size	.030	lf
1″ Pipe Size	.040	lf
1-1/4″ Pipe Size	.060	lf
1-1/2″ Pipe Size	.080	lf
2″ Pipe Size	.100	lf
3″ Pipe Size	.130	lf
Fittings for Polybutylene or Polyethylene Pipe		
Acetal Insert Type with Crimp Rings (Hot and Cold Water)		
Elbows 90°		
1/2″ Pipe Size	.350	Ea.
3/4″ Pipe Size	.360	Ea.
Couplings		
1/2″ Pipe Size	.350	Ea.
3/4″ Pipe Size	.350	Ea.
Tees		
1/2″ Pipe Size	.530	Ea.
3/4″ Pipe Size	.570	Ea.
Nylon Insert Type with Stainless Steel Clamps (Cold Water Only)		
Elbows 90°		
3/4″ Pipe Size	.360	Ea.
1″ Pipe Size	.420	Ea.
1-1/4″ Pipe Size	.440	Ea.
1-1/2″ Pipe Size	.470	Ea.
2″ Pipe Size	.500	Ea.
Couplings		
3/4″ Pipe Size	.360	Ea.
1″ Pipe Size	.420	Ea.
1-1/4″ Pipe Size	.440	Ea.
1-1/2″ Pipe Size	.470	Ea.
2″ Pipe Size	.500	Ea.
Tees		
3/4″ Pipe Size	.570	Ea.
1″ Pipe Size	.620	Ea.
1-1/4″ Pipe Size	.670	Ea.
1-1/2″ Pipe Size	.730	Ea.
2″ Pipe Size	.800	Ea.
Stainless Steel Clamps		
3/4″ Pipe Size	.070	Ea.
1″ Pipe Size	.070	Ea.
1-1/4″ Pipe Size	.080	Ea.
1-1/2″ Pipe Size	.080	Ea.
2″ Pipe Size	.090	Ea.

90° Elbow - Acetal Insert

Coupling - Acetal Insert

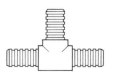

Tee - Acetal Insert

90° Elbow - Nylon Insert

Coupling - Nylon Insert

Tee - Nylon Insert

Stainless Steel Clamp Ring

Man-hours (cont.)

Description	m/hr	Unit
Acetal Flare Type (Hot and Cold Water)		
Elbows 90°		
1/2" Pipe Size	.360	Ea.
3/4" Pipe Size	.380	Ea.
1" Pipe Size	.440	Ea.
Couplings		
1/2" Pipe Size	.360	Ea.
3/4" Pipe Size	.380	Ea.
1" Pipe Size	.440	Ea.
Tees		
1/2" Pipe Size	.570	Ea.
3/4" Pipe Size	.620	Ea.
1" Pipe Size	.670	Ea.
Fusion Type (Hot and Cold Water)		
Elbows 90° or 45°		
1" Pipe Size	.420	Ea.
1-1/4" Pipe Size	.440	Ea.
1-1/2" Pipe Size	.470	Ea.
2" Pipe Size	.500	Ea.
3" Pipe Size	.800	Ea.
Couplings		
1" Pipe Size	.420	Ea.
1-1/4" Pipe Size	.440	Ea.
1-1/2" Pipe Size	.470	Ea.
2" Pipe Size	.500	Ea.
3" Pipe Size	.800	Ea.
Tees		
1" Pipe Size	.620	Ea.
1-1/4" Pipe Size	.670	Ea.
1-1/2" Pipe Size	.730	Ea.
2" Pipe Size	.800	Ea.
3" Pipe Size	1.140	Ea.
Brass Insert Type Fittings		
(Hot and Cold Water)		
Elbows 90°		
1/2" Pipe Size	.350	Ea.
3/4" Pipe Size	.360	Ea.
Couplings		
1/2" Pipe Size	.350	Ea.
3/4" Pipe Size	.360	Ea.
Tees		
1/2" Pipe Size	.530	Ea.
3/4" Pipe Size	.570	Ea.

90° Elbow - Acetal Flare Type

Coupling - Acetal Flare Type

Tee - Acetal Flare Type

90° Elbow - Fusion Type

Coupling - Fusion Type

Tee - Fusion Type

90° Elbow - Brass Insert Type

Coupling - Brass Insert Type

Tee - Brass Insert Type

Man-hours (cont.)

Description	m/hr	Unit
Tees		
1-1/4" Tubing Size	.889	Ea.
1-1/2" Tubing Size	1.000	Ea.
2" Tubing Size	1.143	Ea.
3" Tubing Size	2.286	Ea.
4" Tubing Size	3.200	Ea.
5" Tubing Size	4.000	Ea.
6" Tubing Size	4.000	Ea.
8" Tubing Size	4.800	Ea.

Plastic Piping

Plastic piping is becoming increasingly popular in the building construction industry. Subject to local codes and restrictions, various types of plastic piping are in use for gas, site drainage, water service, fire protection, heating and cooling, as well as plumbing systems.

Among the many advantages of plastic piping are its resistance to corrosion, light weight, and ease of installation. It does have some drawbacks, however, and these should be fully investigated before a decision is made to substitute plastic for metal.

In an area of high temperatures, some plastics become soft and tend to sag. In such cases, more concentrated support is needed, rather than the nominal, one support every ten feet. Also, some plastics are acceptable for cold water, but not for hot. Some types should not be exposed to direct sunlight.

Another point to consider is the fact that a bad joint in solvent-welded piping cannot be repaired. The fitting and adjoining pipe sections must be replaced in such cases, and couplings added to the fit.

Also, certain precautions should be kept in mind concerning the cement, cleaners and primers used in joining plastic pipe and fittings. These materials are combustible and should be used only in well-ventilated areas, away from heat, flames or sparks. Smoking in the immediate vicinity where plastics are being primed or solvent welded is hazardous!

The manufacture of plastic piping is a relatively new industry, still growing and evolving. It has produced such a variety of plastic piping types and joining methods that no one material or method can lay claim to being the perfect replacement in every piping application. On the other hand, brass, cast iron, copper and steel, while they each have limitations, have, over the years, been improved and perfected for their own special applications.

In addition to its use for plumbing, plastic piping is also used for chemicals, electrical conduit, telephone ducts, irrigation, and sewer mains. Plastic is popular as a lining or coating of metal pipe, primarily in the chemical industry or for underground use where the physical strength of the metal is enhanced by the corrosion-resistant qualities of the plastic.

Plastic type DWV pipe is used in the drainage waste and vent piping of plumbing systems. DWV piping may be ABS (Acrylonitrile-Butadiene-Styrene) or PVC (Polyvinyl Chloride). Each of these drainage systems is joined by the solvent weld method. While both require the identical labor man-hours, the ABS material costs are higher.

Man-hours

Description	m/hr	Unit
Man-hours to Install DWV Piping, ABS, or PVC with A Coupling and 3 Pipe Hangers Every Ten Feet		
1-1/4" Pipe Size	.190	lf
1-1/2" Pipe Size	.222	lf
2" Pipe Size	.271	lf
3" Pipe Size	.302	lf
4" Pipe Size	.333	lf
6" Pipe Size	.410	lf
Fittings for Use with Plastic Pipe DWV, ABS or PVC, with Socket Joints		
Bends - Quarter or Eighth		
1-1/4" Pipe Size	.471	Ea.
1-1/2" Pipe Size	.500	Ea.
2" Pipe Size	.571	Ea.
3" Pipe Size	.941	Ea.
4" Pipe Size	1.143	Ea.
6" Pipe Size	2.000	Ea.
Couplings		
1-1/4" Pipe Size	.471	Ea.
1-1/2" Pipe Size	.500	Ea.
2" Pipe Size	.571	Ea.
3" Pipe Size	.727	Ea.
4" Pipe Size	.941	Ea.
6" Pipe Size	1.333	Ea.

Quarter Bend - Socket Joint

Eighth Bend - Socket Joint

Man-hours (cont.)

Description	m/hr	Unit
Tees and Wyes - Sanitary		
1-1/4" Pipe Size	.727	Ea.
1-1/2" Pipe Size	.800	Ea.
2" Pipe Size	.941	Ea.
3" Pipe Size	1.455	Ea.
4" Pipe Size	1.778	Ea.
6" Pipe Size	3.200	Ea.
Man-hours to Install PVC Schedule 40		
Piping with A Coupling and 3 Hangers		
Every Ten Feet. PVC Piping Is Used for Water		
Service, Gas Service, Irrigation and		
Air Conditioning.		
1/2" Pipe Size	.148	lf
3/4" Pipe Size	.157	lf
1" Pipe Size	.174	lf
1-1/4" Pipe Size	.190	lf
1-1/2" Pipe Size	.222	lf
2" Pipe Size	.271	lf
2-1/2" Pipe Size	.286	lf
3" Pipe Size	.302	lf
4" Pipe Size	.333	lf
5" Pipe Size	.372	lf
6" Pipe Size	.410	lf
8" Pipe Size	.500	lf
Fittings for Use with PVC Schedule 40,		
with Socket Joints,		
Elbows, 90° or 45°		
1/2" Pipe Size	.364	Ea.
3/4" Pipe Size	.381	Ea.
1" Pipe Size	.444	Ea.
1-1/4" Pipe Size	.471	Ea.
1-1/2" Pipe Size	.500	Ea.
2" Pipe Size	.571	Ea.
2-1/2" Pipe Size	.727	Ea.
3" Pipe Size	.941	Ea.
4" Pipe Size	1.143	Ea.
5" Pipe Size	1.333	Ea.
6" Pipe Size	2.000	Ea.
8" Pipe Size	2.400	Ea.
Couplings		
1/2" Pipe Size	.364	Ea.
3/4" Pipe Size	.381	Ea.
1" Pipe Size	.444	Ea.
1-1/4" Pipe Size	.471	Ea.
1-1/2" Pipe Size	.500	Ea.
2" Pipe Size	.571	Ea.
2-1/2" Pipe Size	.700	Ea.
3" Pipe Size	.842	Ea.
4" Pipe Size	1.000	Ea.
5" Pipe Size	1.143	Ea.
6" Pipe Size	1.333	Ea.
8" Pipe Size	1.714	Ea.
Tees		
1/2" Pipe Size	.571	Ea.
3/4" Pipe Size	.615	Ea.
1" Pipe Size	.667	Ea.
1-1/4" Pipe Size	.727	Ea.
1-1/2" Pipe Size	.800	Ea.
2" Pipe Size	.941	Ea.
2-1/2" Pipe Size	1.143	Ea.
3" Pipe Size	1.455	Ea.
4" Pipe Size	1.778	Ea.
5" Pipe Size	2.000	Ea.
6" Pipe Size	3.200	Ea.
8" Pipe Size	4.000	Ea.

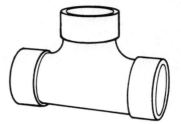

Vent Tee - Socket Joint

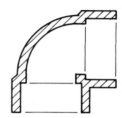

90° Elbow - Socket Joint

Coupling - Socket Joint

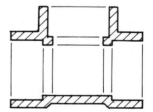

Tee - Socket Joint

Man-hours (cont.)

Description	m/hr	Unit
Man-hours to Install PVC Schedule 80		
Piping with A Coupling and 3		
Hangers Every Ten Feet		
1/2" Pipe Size	.160	lf
3/4" Pipe Size	.170	lf
1" Pipe Size	.186	lf
1-1/4" Pipe Size	.205	lf
1-1/2" Pipe Size	.235	lf
2" Pipe Size	.291	lf
2-1/2" Pipe Size	.308	lf
3" Pipe Size	.320	lf
4" Pipe Size	.348	lf
5" Pipe Size	.381	lf
6" Pipe Size	.421	lf
8" Pipe Size	.511	lf
Fittings for Use with PVC Schedule 80,		
Socket or Threaded Joints,		
Elbows, 90° or 45°		
1/2" Pipe Size	.444	Ea.
3/4" Pipe Size	.471	Ea.
1" Pipe Size	.533	Ea.
1-1/4" Pipe Size	.571	Ea.
1-1/2" Pipe Size	.615	Ea.
2" Pipe Size	.727	Ea.
2-1/2" Pipe Size	.941	Ea.
3" Pipe Size	1.143	Ea.
4" Pipe Size	1.333	Ea.
5" Pipe Size	2.000	Ea.
6" Pipe Size	2.286	Ea.
8" Pipe Size	3.000	Ea.
Couplings		
1/2" Pipe Size	.444	Ea.
3/4" Pipe Size	.471	Ea.
1" Pipe Size	.533	Ea.
1-1/4" Pipe Size	.571	Ea.
1-1/2" Pipe Size	.615	Ea.
2" Pipe Size	.727	Ea.
2-1/2" Pipe Size	.800	Ea.
3" Pipe Size	.842	Ea.
4" Pipe Size	1.000	Ea.
5" Pipe Size	1.143	Ea.
6" Pipe Size	1.333	Ea.
8" Pipe Size	1.714	Ea.
Tees		
1/2" Pipe Size	.667	Ea.
3/4" Pipe Size	.727	Ea.
1" Pipe Size	.800	Ea.
1-1/4" Pipe Size	.889	Ea.
1-1/2" Pipe Size	1.000	Ea.
2" Pipe Size	1.143	Ea.
2-1/2" Pipe Size	1.455	Ea.
3" Pipe Size	1.778	Ea.
4" Pipe Size	2.000	Ea.
5" Pipe Size	3.200	Ea.
6" Pipe Size	3.200	Ea.
8" Pipe Size	4.000	Ea.

Note: The man-hours are the same to install
CPVC schedule 80 piping (chlorinated
polyvinyl chloride) with a coupling and 3
hangers every ten feet. CPVC piping was
developed for hot water use and is used for cold
water also.

45° Elbow - Threaded Joint

Coupling - Threaded Joint

Tee - Threaded Joint

Steel Pipe

Steel pipe is often referred to as "black steel" or "carbon steel" and, occasionally, as "black iron." It also may be manufactured with coated or galvanized surfaces and in a variety of coverings and linings to withstand the effects of corrosive atmospheres and fluids. It is also available in various joint arrangements, wall thicknesses, and schedules.

Joining and sealing arrangements include flanged, grooved, threaded, and welded joints. Generally, grooved-joint piping is the least expensive to install, with threaded, welded, and flanged-joint piping increasing in cost in that order. However, if the piping is 2-1/2" in diameter, or larger, welded-joint piping is less expensive to install than threaded-joint piping.

Steel pipe is also available in various wall thicknesses and schedules. The most commonly used for everyday heating-cooling, gas, fire protection, oil, and plumbing systems is Schedule 40, with 125 lb fittings. Fire protection piping is becoming less expensive in areas where local codes allow thinner-walled steel pipe, such as schedule 10, with grooved-joint fittings and valves.

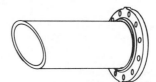

Flanged Steel Pipe

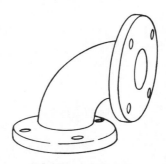

90° Elbow - Flanged

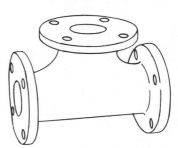

Tee - Flanged

Man-hours

Description	m/hr	Unit
Man-hours to Install Black Schedule #40 Flanged with A Pair of Weld Neck Flanges and a Roll Type Hanger Every Ten Feet. The Pipe Hanger Is Oversized to Allow for Insulation.		
1-1/4" Pipe Size	.250	lf
1-1/2" Pipe Size	.276	lf
2" Pipe Size	.356	lf
2-1/2" Pipe Size	.444	lf
3" Pipe Size	.500	lf
3-1/2" Pipe Size	.552	lf
4" Pipe Size	.615	lf
5" Pipe Size	.762	lf
6" Pipe Size	.960	lf
8" Pipe Size	1.263	lf
10" Pipe Size	1.500	lf
12" Pipe Size	1.714	lf
Fittings for Use with Steel Pipe		
Flanged Elbows, Cast Iron 90° or 45°, 125 lb		
1-1/2" Pipe Size	1.140	Ea.
2" Pipe Size	1.231	Ea.
2-1/2" Pipe Size	1.333	Ea.
3" Pipe Size	1.455	Ea.
3-1/2" Pipe Size	2.000	Ea.
4" Pipe Size	2.000	Ea.
5" Pipe Size	2.286	Ea.
6" Pipe Size	2.667	Ea.
8" Pipe Size	3.000	Ea.
10" Pipe Size	3.429	Ea.
12" Pipe Size	4.000	Ea.
Flanged Tees, Cast Iron 125 lb		
1-1/2" Pipe Size	1.778	Ea.
2" Pipe Size	1.778	Ea.
2-1/2" Pipe Size	2.000	Ea.
3" Pipe Size	2.286	Ea.
3-1/2" Pipe Size	3.200	Ea.
4" Pipe Size	3.200	Ea.
5" Pipe Size	4.000	Ea.
6" Pipe Size	4.000	Ea.
8" Pipe Size	4.800	Ea.
10" Pipe Size	6.000	Ea.
12" Pipe Size	8.000	Ea.

Man-hours (cont.)

Description	m/hr	Unit
Gasket and Bolt Sets Required at Each		
Flanged Joint		
1-1/2″ Pipe Size	.267	Ea.
2″ Pipe Size	.267	Ea.
2-1/2″ Pipe Size	.267	Ea.
3″ Pipe Size	.267	Ea.
3-1/2″ Pipe Size	.286	Ea.
4″ Pipe Size	.296	Ea.
5″ Pipe Size	.308	Ea.
6″ Pipe Size	.333	Ea.
8″ Pipe Size	.400	Ea.
10″ Pipe Size	.444	Ea.
12″ Pipe Size	.500	Ea.
Steel Pipe Man-hours to Install Black,		
Schedule #10, Grooved Joint or Plain		
End with A Mechanical Joint Coupling		
and Pipe Hanger Every Ten Feet		
2″ Pipe Size	.186	lf
2-1/2″ Pipe Size	.262	lf
3″ Pipe Size	.291	lf
3-1/2″ Pipe Size	.302	lf
4″ Pipe Size	.327	lf
5″ Pipe Size	.400	lf
6″ Pipe Size	.522	lf
8″ Pipe Size	.585	lf
10″ Pipe Size	.706	lf
12″ Pipe Size	.800	lf
Black or Galvanized Schedule #40 Grooved		
Joint or Plain End with A Mechanical		
Joint Coupling and Pipe Hanger Every		
Ten Feet		
3/4″ Pipe Size	.113	lf
1″ Pipe Size	.127	lf
1-1/4″ Pipe Size	.138	lf
1-1/2″ Pipe Size	.157	lf
2″ Pipe Size	.200	lf
2-1/2″ Pipe Size	.281	lf
3″ Pipe Size	.320	lf
3-1/2″ Pipe Size	.340	lf
4″ Pipe Size	.356	lf
5″ Pipe Size	.432	lf
6″ Pipe Size	.571	lf
8″ Pipe Size	.649	lf
10″ Pipe Size	.774	lf
12″ Pipe Size	.889	lf
Fittings Are for Use with Grooved Joint		
or Plain End Steel Pipe		
Elbows 90° or 45°		
3/4″ Pipe Size	.160	Ea.
1″ Pipe Size	.160	Ea.
1-1/4″ Pipe Size	.200	Ea.
1-1/2″ Pipe Size	.242	Ea.
2″ Pipe Size	.320	Ea.
2-1/2″ Pipe Size	.400	Ea.
3″ Pipe Size	.485	Ea.
4″ Pipe Size	.640	Ea.
5″ Pipe Size	.800	Ea.
6″ Pipe Size	.960	Ea.
8″ Pipe Size	1.143	Ea.
10″ Pipe Size	1.333	Ea.
12″ Pipe Size	1.600	Ea.

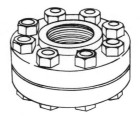

Flanged Joint

Grooved Joint Steel Pipe

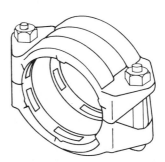

Grooved Joint Coupling

Mechanical Joint Elbow 90° - Plain End Pipe

Man-hours (cont.)

Description	m/hr	Unit
Tees		
3/4" Pipe Size	.211	Ea.
1" Pipe Size	.242	Ea.
1-1/4" Pipe Size	.296	Ea.
1-1/2" Pipe Size	.364	Ea.
2" Pipe Size	.471	Ea.
2-1/2" Pipe Size	.593	Ea.
3" Pipe Size	.727	Ea.
4" Pipe Size	.941	Ea.
5" Pipe Size	1.231	Ea.
6" Pipe Size	1.412	Ea.
8" Pipe Size	1.714	Ea.
10" Pipe Size	2.000	Ea.
12" Pipe Size	2.400	Ea.

Man-hours to Install Black or Galvanized
Schedule #40 Threaded with A Coupling and
Pipe Hanger Every Ten Feet. The Pipe Hanger Is
Oversized to Allow for Insulation.

Description	m/hr	Unit
1/2" Pipe Size	.127	lf
3/4" Pipe Size	.131	lf
1" Pipe Size	.151	lf
1-1/4" Pipe Size	.180	lf
1-1/2" Pipe Size	.200	lf
2" Pipe Size	.250	lf
2-1/2" Pipe Size	.320	lf
3" Pipe Size	.372	lf
3-1/2" Pipe Size	.400	lf
4" Pipe Size	.444	lf
5" Pipe Size	.615	lf
6" Pipe Size	.774	lf
8" Pipe Size	.889	lf
10" Pipe Size	1.043	lf
12" Pipe Size	1.333	lf

Fittings for Use with Steel Pipe. Threaded
Fittings, Cast Iron, 125 lb or
Malleable Iron Rated at 150 lb.
Elbows, 90° or 45°

Description	m/hr	Unit
1/2" Pipe Size	.533	Ea.
3/4" Pipe Size	.571	Ea.
1" Pipe Size	.615	Ea.
1-1/4" Pipe Size	.727	Ea.
1-1/2" Pipe Size	.800	Ea.
2" Pipe Size	.889	Ea.
2-1/2" Pipe Size	1.143	Ea.
3" Pipe Size	1.600	Ea.
3-1/2" Pipe Size	2.000	Ea.
4" Pipe Size	2.667	Ea.
5" Pipe Size	3.200	Ea.
6" Pipe Size	3.429	Ea.
8" Pipe Size	4.000	Ea.
Tees		
1/2" Pipe Size	.889	Ea.
3/4" Pipe Size	.889	Ea.
1" Pipe Size	1.000	Ea.
1-1/4" Pipe Size	1.143	Ea.
1-1/2" Pipe Size	1.231	Ea.
2" Pipe Size	1.455	Ea.
2-1/2" Pipe Size	1.778	Ea.
3" Pipe Size	2.667	Ea.
3-1/2" Pipe Size	3.200	Ea.
4" Pipe Size	4.000	Ea.
5" Pipe Size	5.333	Ea.
6" Pipe Size	6.000	Ea.
8" Pipe Size	8.000	Ea.

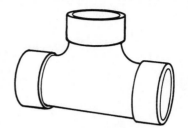

Tee - Plain End Pipe

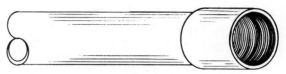

Threaded and Coupled Steel Pipe

45° Elbow - Malleable Iron

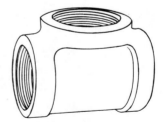

Tee - Cast Iron

Man-hours (cont.)

Description	m/hr	Unit
Unions Malleable Iron 150 lb		
1/2" Pipe Size	.571	Ea.
3/4" Pipe Size	.615	Ea.
1" Pipe Size	.667	Ea.
1-1/4" Pipe Size	.762	Ea.
1-1/2" Pipe Size	.842	Ea.
2" Pipe Size	.941	Ea.
2-1/2" Pipe Size	1.231	Ea.
3" Pipe Size	1.778	Ea.
3-1/2" Pipe Size	3.200	Ea.
4" Pipe Size	3.200	Ea.

Note: For unions larger than 4", flanges must be used.

Man-hours (cont.)

Description	m/hr	Unit
Steel Pipe Man-hours to Install Black, Schedule #40 Welded with A Butt Joint and A Roll Type Hanger Every Ten Feet. The Pipe Hanger Is Oversized to Allow for Insulation.		
1-1/4" Pipe Size	.190	lf
1-1/2" Pipe Size	.211	lf
2" Pipe Size	.262	lf
2-1/2" Pipe Size	.340	lf
3" Pipe Size	.372	lf
3-1/2" Pipe Size	.410	lf
4" Pipe Size	.432	lf
5" Pipe Size	.500	lf
6" Pipe Size	.667	lf
8" Pipe Size	.828	lf
10" Pipe Size	1.000	lf
12" Pipe Size	1.263	lf
Fittings for Use with Steel Pipe		
Butt Weld, Long Radius 90° Elbows 150 lb		
1-1/4" Pipe Size	1.143	Ea.
1-1/2" Pipe Size	1.231	Ea.
2" Pipe Size	1.600	Ea.
2-1/2" Pipe Size	2.000	Ea.
3" Pipe Size	2.286	Ea.
3-1/2" Pipe Size	3.200	Ea.
4" Pipe Size	3.200	Ea.
5" Pipe Size	4.800	Ea.
6" Pipe Size	4.800	Ea.
8" Pipe Size	6.000	Ea.
10" Pipe Size	8.000	Ea.
12" Pipe Size	9.600	Ea.

Union - Malleable Iron

Bevel End Steel Pipe

90° Elbow, Butt Weld, Long Radius - Steel Pipe

Man-hours

Description	m/hr	Unit
Forty-five Degree Elbows Require The Same Man-hours for Installation As Ninety Degree Elbows. Since Forty-five Degree Ells Are Often Double The Price of Ninety Degree Ells, A Cost-conscious Pipe Fitter Will Cut A Ninety in Half to Make Two Forty-fives.		
Butt Weld Tees 150 lb		
1-1/4" Pipe Size	1.778	Ea.
1-1/2" Pipe Size	2.000	Ea.
2" Pipe Size	2.667	Ea.
2-1/2" Pipe Size	3.200	Ea.
3" Pipe Size	4.000	Ea.
3-1/2" Pipe Size	5.333	Ea.
4" Pipe Size	5.333	Ea.
5" Pipe Size	8.000	Ea.
6" Pipe Size	8.000	Ea.
8" Pipe Size	9.600	Ea.
10" Pipe Size	12.000	Ea.
12" Pipe Size	15.000	Ea.
Weld Neck Flanges 150 lb		
1-1/4" Pipe Size	.552	Ea.
1-1/2" Pipe Size	.615	Ea.
2" Pipe Size	.800	Ea.
2-1/2" Pipe Size	1.000	Ea.
3" Pipe Size	1.143	Ea.
3-1/2" Pipe Size	1.333	Ea.
4" Pipe Size	1.600	Ea.
5" Pipe Size	2.000	Ea.
6" Pipe Size	2.400	Ea.
8" Pipe Size	3.429	Ea.
10" Pipe Size	4.000	Ea.
12" Pipe Size	4.800	Ea.
Slip-on Weld Flanges 150 lb		
1-1/4" Pipe Size	1.000	Ea.
1-1/2" Pipe Size	1.067	Ea.
2" Pipe Size	1.333	Ea.
2-1/2" Pipe Size	1.600	Ea.
3" Pipe Size	1.778	Ea.
3-1/2" Pipe Size	2.286	Ea.
4" Pipe Size	2.667	Ea.
5" Pipe Size	3.200	Ea.
6" Pipe Size	4.000	Ea.
8" Pipe Size	4.800	Ea.
10" Pipe Size	6.000	Ea.
12" Pipe Size	8.000	Ea.

Tee, Butt Weld - Steel Pipe

Weld Neck Flange - Steel Pipe

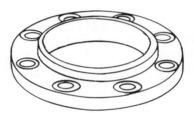

Slip-on Weld Flange - Steel Pipe

The man-hours for the above weld flanges are for aligning and welding only. Labor must be added for bolting up each joint. The above man-hours indicated for slip-on flange welding include an internal weld as well as an external weld at the hub. In low pressure heating and cooling applications, it is not an uncommon practice to tack weld the hub and complete the normal weld internally. This practice, although not recommended, takes considerably less time than the standard procedure.

8 MECHANICAL
VALVES

The flow of fluids in a piping system is controlled or regulated by the use of valves. Valves are used to start, stop, divert, relieve, or regulate the flow, pressure, or temperature in a piping system. Valves are manufactured in several configurations according to use. Some of the types include: gate, globe, angle, check, ball, butterfly, and plug. Valves are further classified by their piping connections, stem position, pressure and temperature limits, as well as by the materials from which they are made.

Stainless steel, or steel alloy, valves can be used effectively in most instances for corrosion protection.

Bronze is one of the oldest materials used to make valves. It is most commonly used in steam, hot- and cold-water systems and other non-corrosive services. Bronze is often used as a seating surface in larger iron-body valves to ensure tight closure. Pressure ratings of 300 psi and temperatures up to 150 degrees F are typical.

Iron valves are normally used in medium to large pipe lines to control non-corrosive fluids and gases. Pressures for these valves should not exceed 250 psi at 450 degrees F, or 500 psi cold working pressures for water, oil, or gas.

Carbon steel is a high-strength material, and the valves made from this metal are therefore used in higher-pressure services, such as steam lines up to 600 psi at 850 degrees F. Many steel valves are available with butt-weld ends for economy and are generally used in high-pressure steam service, as well as other higher-pressure, non-corrosive services.

Forged steel valves are made of tough carbon steel. They are used at pressures up to 2,000 psi and temperatures up to 1,000 degrees F.

Plastic is used for a great variety of valves, generally in high-corrosive service, at low temperatures and low pressures. Plastic lining of metal valves for corrosive service and high-purity applications and temperatures are also available.

Outside screw and yoke valves offer a visual indication of whether the valve is open or closed, according to the position of the screw. This type of valve is recommended where high temperatures, corrosives, and solids in the line might cause damage to inside-valve stem threads. The stem threads are engaged by the yoke bushing, so that the stem rises through the hand wheel as it is turned.

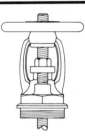

Rising Stem - Outside Screw and Yoke

Rising stem, inside screw valves need adequate clearance above the stem for operation because both the hand wheel and the stem rise during operation. The valve wedge position is indicated by the position of the stem and hand wheel.

Rising Stem - Inside Screw

Non-rising stem, inside screw valves need a minimum clearance for operation. Excessive wear or damage to stem threads inside the valve may be caused by heat, corrosion, and solids. Because the hand wheel and stem do not rise, the wedge position cannot be visually determined.

Non-rising Stem - Inside Screw

Gate valves provide full flow, minute pressure drop, and minimum turbulence. They are normally used where operation is infrequent, such as for equipment isolation.

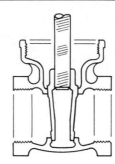

Gate Valve

Globe valves are designed for throttling and/or frequent operation with positive shutoff. Particular attention must be paid to the several types of seating material available to avoid unnecessary wear. The seats must be compatible with the fluid in service and may be composition or metal in construction. The configuration of the globe valve opening causes turbulence, which results in increased flow resistance.

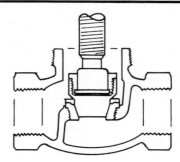

Globe Valve

The fundamental difference in use between the angle valve and the globe valve is the fluid flow through the angle valve. The flow makes a 90 degree turn but offers less resistance than a combination of a globe valve and an elbow. An angle valve thus reduces the number of joints and the installation labor.

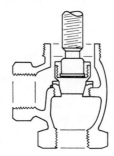

Angle Valve

Check valves are one-way valves and are designed to prevent backflow by automatically seating when the direction of fluid is reversed. Swing check valves are generally installed with gate valves, as they provide comparable full flow, and are usually recommended for lines where flow velocities are low. They should not be used on lines with pulsating flow. They are also recommended for horizontal installation, or in vertical lines where flow is only upward.

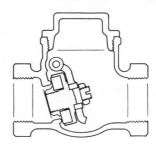

Check Valve

Lift check valves are commonly used with globe and angle valves, since they all have similar diaphragm seating arrangements. Horizontal lift checks should be used for horizontal lines, and vertical lift checks should be used for vertical lines.

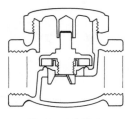

Horizontal Type

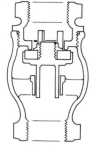

Vertical Type

Lift Check Valve

Ball valves are light and easily installed, yet because of modern elastomeric seats, they provide tight closure. Flow is controlled by rotating 90 degrees a drilled ball that fits tightly against resilient seals. This ball seals with flow in either direction, and the valve handle indicates the degree of opening. This type of valve is recommended for frequent operation, such as for tank filling, and is readily adaptable to automation. Ball valves are ideal for installation where space is limited.

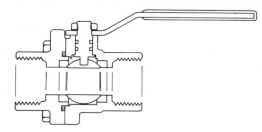

Ball Valve

Butterfly valves provide bubble-tight closure with excellent throttling characteristics. They can be used for full-open, closed, and for throttling applications. The butterfly valve consists of a disc, controlled by a shaft, within the valve body. In its closed position, the valve disc seals against a resilient seat. The disc position throughout the full 90 degree rotation is visually indicated by the position of the operator. Butterfly valves are only a fraction of the weight of a gate valve and require no gaskets between flanges in most cases. They are recommended for frequent operation and are adaptable to automation where space is limited. Wafer- and lug-type bodies, when installed between two pipe flanges, can be easily removed from the line. The pressure of the bolted flanges holds the valve in place. Locating lugs makes installation easier.

Wafer Type

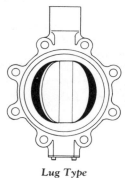

Lug Type

Butterfly Valve

Lubricated plug valves, because of the wide range of service to which they are adapted, may be classified as all-purpose valves. They can be safely used at all pressures and vacuums, and at all temperatures up to the limits of available lubricants. They are the most satisfactory valves for the handling of gritty suspensions and many other destructive, erosive, corrosive, and chemical solutions.

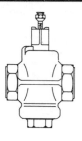

Plug Valve

Backflow preventers are mechanical devices installed to stop contaminated fluids from entering the potable water system. This reversal of flow may be caused by back pressure or back syphonage. The assembly incorporates double check valves, a relief valve and vacuum breaker all contained in one housing.

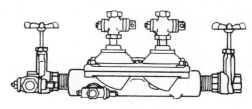

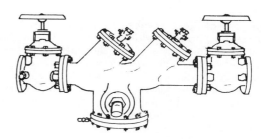

Double Check

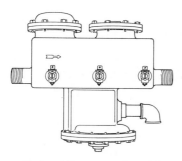

Reduced Pressure - Threaded

Reduced Pressure - Flanged

Backflow Preventers

Man-hours

Description	m/hr	Unit
Valves		
Threaded or Solder End		
1/8" Diameter	.333	Ea.
1/4" Diameter	.333	Ea.
3/8" Diameter	.333	Ea.
1/2" Diameter	.333	Ea.
3/4" Diameter	.400	Ea.
1" Diameter	.421	Ea.
1-1/4" Diameter	.533	Ea.
1-1/2" Diameter	.615	Ea.
2" Diameter	.727	Ea.
2-1/2" Diameter	1.067	Ea.
3" Diameter	1.231	Ea.
Flanged End		
2" Diameter	1.600	Ea.
2-1/2" Diameter	3.200	Ea.
3" Diameter	3.556	Ea.
3-1/2" Diameter	5.333	Ea.
4" Diameter	5.333	Ea.
5" Diameter	7.059	Ea.
6" Diameter	8.000	Ea.
8" Diameter	9.600	Ea.
10" Diameter	10.909	Ea.
12" Diameter	14.118	Ea.
Plastic Threaded or Socket End		
1/4" Diameter	.308	Ea.
1/2" Diameter	.308	Ea.
3/4" Diameter	.320	Ea.
1" Diameter	.348	Ea.
1-1/4" Diameter	.381	Ea.
1-1/2" Diameter	.400	Ea.
2" Diameter	.471	Ea.
2-1/2" Diameter	.615	Ea.
3" Diameter	.667	Ea.

Man-hours (cont.)

Description	m/hr	Unit
Backflow Preventers		
Threaded End		
3/4" Diameter	.500	Ea.
1" Diameter	.571	Ea.
1-1/2" Diameter	.800	Ea.
2" Diameter	1.143	Ea.
Flanged End		
2-1/2" Diameter	3.200	Ea.
3" Diameter	3.556	Ea.
4" Diameter	5.333	Ea.
6" Diameter	8.000	Ea.
8" Diameter	12.000	Ea.
10" Diameter	24.000	Ea.

8 MECHANICAL
PIPING SUPPORT

Piping for mechanical and electrical installations is supported in the same basic manner. The pipe hanger material is usually black or galvanized steel. For appearance or in corrosive atmospheres, chrome- or copper-plated steel, stainless steel, cast iron, or a variety of plastics may also be used. The location of piping supports is an important consideration, and many different building components may be used for anchoring the pipe hangers. Some of the commonly used locations for hanger supports include the roof or floor above, structural members, side walls, another pipe line, machinery, building equipment, pedestals, rolls, or racks.

Another consideration is the method of anchoring the hanger assembly to the building structure. The method selected for anchoring the hanger depends on the type of material to which it is being secured—usually concrete, steel, or wood. If the roof or floor slab is constructed of concrete, formed and placed at the site, then concrete inserts may be nailed in the forms at the required locations prior to the placing of the concrete. These inserts may be manufactured from steel or malleable iron, and they either are tapped to receive the hanger rod machine thread or contain a slot to receive an insert nut. Because the slotted type of insert requires separate insertable nuts for the various rod diameters, only one size of insert need be warehoused. For multiple side-by-side runs, long, slotted insert channels of up to 10' increments are available with several types of adjustable inserts.

When precast slabs are used, the inserts are installed in the joints between slabs, or anchors must be drilled or shot into the slab at the site. Electric and pneumatic drills and hammers are available to drill holes for anchors, shields, or expansion bolts. Another method of installing anchors on site is the gunpowder-actuated stud driver, which partially embeds a threaded stud into the concrete.

If the piping is to be supported from the sidewalls, the methods of drilling, driving, or anchoring mentioned above are used for solid concrete walls. Where hollow-core masonry walls are to be fitted, holes may be drilled for toggle or expansion-type bolts or anchors.

Attaching the pipe supports to steel or wood requires different methods from those used for concrete. When the piping is to be supported from the building's structural steel members, a wide variety of beam clamps, fish plates, and welded attachments can be employed. If the piping is being run in areas where the structural steel is not located directly overhead, then intermediate steel is used to bridge the gap. This intermediate steel is usually erected at the piping contractor's expense. If the building is constructed of wood, then lag screws, drive screws, or nails are used to secure the support assembly.

From the anchoring device a steel hanger rod, threaded on both ends, extends to receive the pipe hanger. One end of this rod is threaded into the anchoring device, and the other is fastened to the hanger itself by a washer and nut. For cost effectiveness and convenience, continuous thread rod may be used. The pipe hanger itself may be a ring, band, roll, or clamp, depending on the function and size of the piping being supported. Spring-type hangers are also used to cushion vibration.

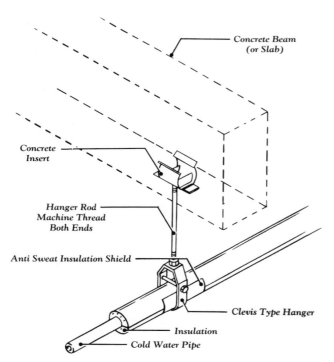

Pipe Supports

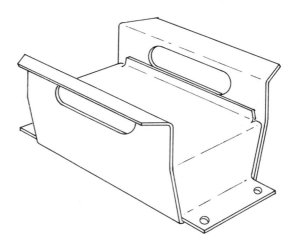

Wedge Type Concrete Insert

Man-hours

Description	m/hr	Unit
Side Beam Brackets	.167	Ea.
Wall Brackets	.235	Ea.
C Clamps	.050	Ea.
I Beam Clamps		
2″ Flange Size	.083	Ea.
3″	.084	Ea.
4″	.086	Ea.
5″	.087	Ea.
6″	.089	Ea.
7″	.091	Ea.
8″	.093	Ea.
Riser Clamps		
3/4″ Pipe Size	.167	Ea.
1″	.170	Ea.
1-1/4″	.174	Ea.
1-1/2″	.178	Ea.
2″	.186	Ea.
2-1/2″	.195	Ea.
3″	.200	Ea.
3-1/2″	.205	Ea.
4″	.211	Ea.
5″	.216	Ea.
6″	.222	Ea.
8″	.235	Ea.
10″	.250	Ea.
12″	.286	Ea.
Split Ring Clamps		
1/2″ Pipe Size	.117	Ea.
3/4″	.119	Ea.
1″	.121	Ea.
1-1/4″	.124	Ea.
1-1/2″	.127	Ea.
2″	.129	Ea.
2-1/2″	.133	Ea.
3″	.137	Ea.
3-1/2″	.140	Ea.
4″	.145	Ea.
5″	.151	Ea.
6″	.154	Ea.
8″	.160	Ea.
High Temperature Clamps		
4″ Pipe Size	.151	Ea.
6″	.151	Ea.
8″	.165	Ea.
10″	.190	Ea.
12″	.222	Ea.
Two Bolt Clamps		
1/2″ Pipe Size	.117	Ea.
3/4″	.119	Ea.
1″	.121	Ea.
1-1/4″	.123	Ea.
1-1/2″	.127	Ea.
2″	.129	Ea.
2-1/2″	.133	Ea.
3″	.137	Ea.
3-1/2″	.140	Ea.
4″	.145	Ea.
5″	.151	Ea.
6″	.154	Ea.
8″	.160	Ea.
10″	.167	Ea.
12″	.180	Ea.

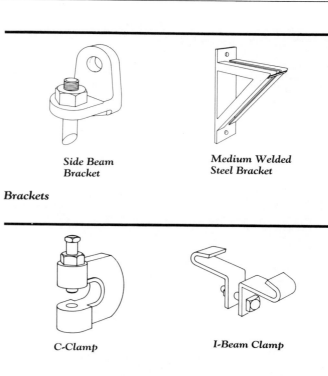

Side Beam
Bracket

Medium Welded
Steel Bracket

Brackets

C-Clamp

I-Beam Clamp

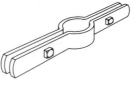
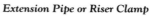

Extension Pipe or Riser Clamp

Split Ring Pipe Clamp

Alloy Steel Pipe Clamp

Medium Pipe Clamp

Clamp Types

Man-hours (cont.)

Description	m/hr	Unit
Pipe Alignment Guides		
1" Pipe Size	.308	Ea.
1-1/4" to 2"	.381	Ea.
2-1/2" to 3-1/2"	.444	Ea.
4" to 5"	.500	Ea.
6"	.762	Ea.
8"	1.000	Ea.
10"	1.333	Ea.
12"	1.412	Ea.
Band Hangers		
1/2" Pipe Size	.113	Ea.
3/4"	.114	Ea.
1"	.117	Ea.
1-1/4"	.119	Ea.
1-1/2"	.122	Ea.
2"	.124	Ea.
2-1/2"	.128	Ea.
3"	.131	Ea.
3-1/2"	.134	Ea.
4"	.140	Ea.
5"	.145	Ea.
6"	.148	Ea.
8"	.154	Ea.
Adjustable Swivel Rings		
1/2" Pipe Size	.117	Ea.
3/4"	.118	Ea.
1"	.121	Ea.
1-1/4"	.124	Ea.
1-1/2"	.127	Ea.
2"	.129	Ea.
2-1/2"	.133	Ea.
3"	.137	Ea.
3-1/2"	.140	Ea.
4"	.145	Ea.
5"	.151	Ea.
6"	.154	Ea.
8"	.160	Ea.
Clevis		
1/2" Pipe Size	.117	Ea.
3/4"	.119	Ea.
1"	.121	Ea.
1-1/4"	.124	Ea.
1-1/2"	.127	Ea.
2"	.129	Ea.
2-1/2"	.133	Ea.
3"	.137	Ea.
3-1/2"	.140	Ea.
4"	.145	Ea.
5"	.151	Ea.
6"	.154	Ea.
8"	.160	Ea.
10"	.167	Ea.
12"	.180	Ea.

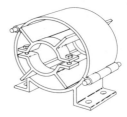

Pipe Alignment Guide

Adjustable Band Hanger

Adjustable Pipe Ring

Adjustable Clevis

Man-hours (cont.)

Description	m/hr	Unit
Insulation Protection Saddles		
3/4" Pipe Size	.118	Ea.
1"	.118	Ea.
1-1/4"	.118	Ea.
1-1/2"	.121	Ea.
2"	.121	Ea.
2-1/2"	.125	Ea.
3"	.125	Ea.
3-1/2"	.129	Ea.
4"	.129	Ea.
5"	.133	Ea.
6"	.133	Ea.
Bird Cage Rolls		
2-1/2" Pipe Size	.117	Ea.
3"	.122	Ea.
3-1/2"	.129	Ea.
4"	.137	Ea.
5"	.145	Ea.
6"	.154	Ea.
8"	.167	Ea.
10"	.200	Ea.
12"	.235	Ea.
Two Rod Rolls		
2-1/2" Pipe Size	.136	Ea.
3"	.139	Ea.
3-1/2"	.142	Ea.
4"	.143	Ea.
5"	.145	Ea.
6"	.158	Ea.
8"	.178	Ea.
10"	.200	Ea.
12"	.235	Ea.
Chair Roll		
2" Pipe Size	.118	Ea.
2-1/2"	.123	Ea.
3"	.129	Ea.
3-1/2"	.133	Ea.
4"	.138	Ea.
5"	.143	Ea.
6"	.151	Ea.
8"	.160	Ea.
10"	.167	Ea.
12"	.174	Ea.
Pipe Straps		
1/2" Pipe Size	.113	Ea.
3/4"	.114	Ea.
1"	.117	Ea.
1-1/4"	.119	Ea.
1-1/2"	.122	Ea.
2"	.124	Ea.
2-1/2"	.128	Ea.
3"	.131	Ea.
3-1/2"	.134	Ea.
4"	.140	Ea.

Pipe Covering Protection Saddle

Adjustable Steel Yoke Pipe Roll

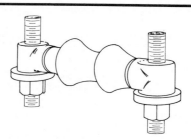

Adjustable Two-Rod Roller Hanger

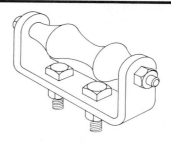

Roller Chair

Pipe Strap

Man-hours (cont.)

Description	m/hr	Unit
U-Bolts		
1/2" Pipe Size	.050	Ea.
3/4"	.051	Ea.
1"	.053	Ea.
1-1/4"	.054	Ea.
1-1/2"	.056	Ea.
2"	.058	Ea.
2-1/2"	.060	Ea.
3"	.063	Ea.
3-1/2"	.066	Ea.
4"	.068	Ea.
5"	.070	Ea.
6"	.072	Ea.
8"	.073	Ea.
10"	.075	Ea.
12"	.077	Ea.
U-Hooks		
3/4" to 2" Pipe Size	.083	Ea.

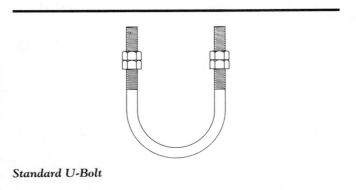

Standard U-Bolt

U-Hook

8 MECHANICAL
PUMPS

A pump is a mechanical device used to convey, raise or variate the pressure of fluids, such as water, oil or sewage.

In the building construction industry, pumps are employed in dewatering the site of excavation, increasing water pressure for potable or fire protection systems, circulating potable, cooling or heating water, transferring heating oil, draining wet basements or ejecting sewage from sub-basement pits.

The most commonly used pumps are centrifugal, driven by an electric motor. In some instances firepumps may be driven by diesel engine as a precaution against electrical failure. Most pump bodies are constructed of cast iron. For potable water pumps, bronze is usually substituted for cast iron at an increase of cost.

To ensure proper operation, a check valve should be installed on the discharge side of the pump for backflow prevention. If a throttling or flow regulating valve is required, it should also be on the discharge side to prevent pump cavitation.

In-Line Centrifugal Pump

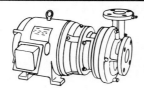

Close Coupled, Centrifugal Pump

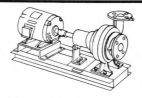

Base Mounted Centrifugal Pump

Submersible Sump Pump

Man-hours

Description	m/hr	Unit
Circulating Pumps, Heating, Cooling, Potable Water		
In Line Type		
3/4" Pipe Size 1/40 hp	1.000	Ea.
3/4" Through 1-1/2" Pipe Size 1/3 hp	2.667	Ea.
2" Pipe Size 1/6 hp	3.200	Ea.
2-1/2" Pipe Size 1/4 hp	3.200	Ea.
3" Pipe Size 1/3 hp through 1 hp	4.000	Ea.
Close Coupled Type		
1-1/2" Pipe Size 1-1/2 hp 50 GPM	5.333	Ea.
2" Pipe Size 3 hp 90 GPM	6.957	Ea.
2-1/2" Pipe Size 3 hp 150 GPM	8.000	Ea.
3" Pipe Size 5 hp 225 GPM	8.889	Ea.
4" Pipe Size 7-1/2 hp 350 GPM	10.000	Ea.
5" Pipe Size 15 hp 1000 GPM	14.118	Ea.
6" Pipe Size 25 hp 1550 GPM	16.000	Ea.
Base Mounted Type		
2-1/2" Pipe Size 3 hp 150 GPM	8.889	Ea.
3" Pipe Size 5 hp 225 GPM	10.000	Ea.
4" Pipe Size 7-1/2 hp 350 GPM	10.667	Ea.
5" Pipe Size 15 hp 1000 GPM	15.000	Ea.
6" Pipe Size 25 hp 1550 GPM	17.143	Ea.
8" Pipe Size 30 hp 1660 GPM	20.000	Ea.
10" Pipe Size 40 hp 1800 GPM	20.000	Ea.
12" Pipe Size 50 hp 2200 GPM	24.000	Ea.
Pumps		
Sewage Ejector with Basin and Cover		
Single		
110 GPM 1/2 hp	6.400	Ea.
173 GPM 3/4 hp	8.000	Ea.
218 GPM 1 hp	10.000	Ea.
285 GPM 2 hp	12.308	Ea.
325 GPM 3 hp	17.143	Ea.
370 GPM 5 hp	24.000	Ea.
Duplex		
110 GPM 1/2 hp	8.000	Ea.
173 GPM 3/4 hp	10.000	Ea.
218 GPM 1 hp	20.000	Ea.
285 GPM 2 hp	24.000	Ea.
325 GPM 3 hp	30.000	Ea.
370 GPM 5 hp	48.000	Ea.
Sump Pumps, Submersible		
Pedestal Type Cellar Drainer with External Float		
42 GPM 1/3 hp	1.600	Ea.
Submersible with Built in Float		
22 GPM 1/4 hp	1.330	Ea.
68 GPM 1/2 hp	1.600	Ea.
94 GPM 1/2 hp	1.600	Ea.
105 GPM 1/2 hp	2.000	Ea.
Well Pumps		
Deep Well Multi-Stage	10.000	Ea.
Shallow Well Jet Type	4.000	Ea.
Fire Pumps with Controller and Fittings		
Electric		
4" Pipe Size 100 hp 500 GPM	51.619	Ea.
6" Pipe Size 250 hp 1000 GPM	88.889	Ea.
8" Pipe Size 300 hp 2000 GPM	114.286	Ea.
10" Pipe Size 450 hp 3500 GPM	133.333	Ea.
Diesel		
4" Pipe Size 111 hp 500 GPM	53.333	Ea.
6" Pipe Size 255 hp 1000 GPM	80.000	Ea.
8" Pipe Size 255 hp 2000 GPM	100.000	Ea.
10" Pipe Size 525 hp 3500 GPM	160.000	Ea.

253

MECHANICAL
AUTOMATIC STORAGE WATER SYSTEMS

Hot water heaters are nominally rated by their storage capacity in residential use and by various other parameters, such as recovery and flow capacity, for commercial use. Depending on the economic advantage for different geographical areas, storage water systems are heated by electricity, gas, or oil. Sizes range from 5 gallons to several thousand gallons. Units may be installed in multiples for increased capacity.

Residential water heaters are sized by the number of bedrooms in the dwelling unit according to the following table:

Man-hours

Description		m/hr	Unit
Number of Bedrooms	H.W. Storage Capacity		
1	20 gal.	3.810	Ea.
2	30 gal.	4.000	Ea.
3	40 gal.	4.211	Ea.
4	50 gal.	4.444	Ea.

(For each additional bedroom go to the next larger size 70, 85, and 100 gallon.)

Electric water heaters can be installed practically anywhere in a building, whereas gas or oil-fired units are confined to placement in well-ventilated areas and require an exhaust flue to carry away the products of combustion.

Storage water systems consist of a tank with a non-corrosive shell or lining; an insulated metal jacket; pressure, temperature, and vacuum safety controls; operating control; and the heat source. A recirculation line from the end of the hot water distribution main back to the heater assures an immediate hot water source at each outlet or fixture.

Long runs of piping should be avoided because of line temperature loss. For isolated outlets not requiring storage, a small electric "point of use" heater may be economically justified.

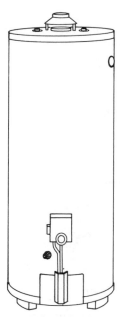

Residential Gas Fired Water Heater

Hot Water Consumption Rates For Commercial Applications

Type of Building	Size Factor	Maximum Hourly Demand	Average Day Demand
Apartment Dwellings	No. of Apartments:		
	Up to 20	12.0 gal. per Apt.	42.0 gal. per Apt.
	21 to 50	10.0 gal. per Apt.	40.0 gal. per Apt.
	51 to 75	8.5 gal. per Apt.	38.0 gal. per Apt.
	76 to 100	7.0 gal. per Apt.	37.0 gal. per Apt
	101 to 200	6.0 gal. per Apt.	36.0 gal. per Apt.
	201 up	5.0 gal. per Apt.	35.0 gal. per Apt.
Dormitories	Men	3.8 gal. per Man	13.1 gal. per Man
	Women	5.0 gal. per Woman	12.3 gal. per Woman
Hospitals	Per Bed	23.0 gal. per Patient	90.0 gal. per Patient
Hotels	Single Room with Bath	17.0 gal. per Unit	50.0 gal. per Unit
	Double Room with Bath	27.0 gal. per Unit	80.0 gal. per Unit
Motels	No. of Units:		
	Up to 20	6.0 gal. per Unit	20.0 gal. per Unit
	21 to 100	5.0 gal. per Unit	14.0 gal. per Unit
	101 Up	4.0 gal. per Unit	10.0 gal. per Unit
Nursing Homes		4.5 gal. per Bed	18.4 gal. per Bed
Office Buildings		0.4 gal. per Person	1.0 gal. per Person
Restaurants	Full Meal Type	1.5 gal./max. Meals/hr	2.4 gal. per Meal
	Drive-in Snack Type	0.7 gal./max. Meals/hr	0.7 gal. per Meal
Schools	Elementary	0.6 gal. per Student	0.6 gal. per Student
	Secondary and High	1.0 gal. per Student	1.8 gal. per Student

For evaluation purposes, recovery rate and storage capacity are inversely proportional. Water heaters should be sized so that the maximum hourly demand anticipated can be met in addition to an allowance for the heat loss from the pipes and storage tank.

When sizing storage capacity, assume that 75% of the tank volume is available during a peak demand of one hour.

Gas Fired Water Heaters-Commercial

Included below is the heater with self-energizing gas controls, safety pilots, insulated jacket, hi-limit aquastat and pressure relief valve. Installation time includes piping and fittings within 10' of heater. Gas heaters require vent piping (not included).

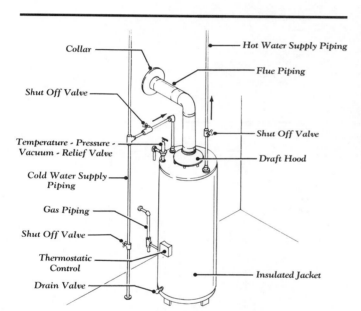

Gas Fired Hot Water System

Man-hours

Description	m/hr	Unit
Gas Fired Water, Commercial, 100°F Rise		
75.5 MBH Input, 63 gal./hr	5.714	Ea.
100 MBH Input, 91 gal./hr	5.714	Ea.
155 MBH Input, 150 gal./hr	10.000	Ea.
200 MBH Input, 192 gal./hr	13.333	Ea.
300 MBH Input, 278 gal./hr	20.000	Ea.
390 MBH Input, 374 gal./hr	20.000	Ea.
500 MBH Input, 480 gal./hr	22.857	Ea.
600 MBH Input, 576 gal./hr	26.667	Ea.
800 MBH Input, 768 gal./hr	32.000	Ea.
1000 MBH Input, 960 gal./hr	32.000	Ea.
1500 MBH Input, 1440 gal./hr	40.000	Ea.
1800 MBH Input, 1730 gal./hr	48.000	Ea.
2450 MBH Input, 2350 gal./hr	60.000	Ea.
3000 MBH Input, 2880 gal./hr	80.000	Ea.
3750 MBH Input, 3600 gal./hr	80.000	Ea.

Components

Description	m/hr	Unit
Water Heater, Commercial, Gas, 75.5 MBH, 63 gal./hr	5.714	Ea.
Copper Tubing, Type L, Solder Joint, Hanger 10' On Center, 1" Diameter	.118	lf
Wrought Copper 90° Elbow for Solder Joints 1" Diameter	.500	Ea.
Wrought Copper Tee for Solder Joints 1" Diameter	.800	Ea.
Wrought Copper Union for Soldered Joints 1" Diameter	.533	Ea.
Valve, Gate, Bronze, 125 lb, NRS, Soldered 1" Diameter	.421	Ea.
Relief Valve, Bronze, Press and Temp, Self Close, 3/4" IPS	.286	Ea.
Copper Tubing, Type L, Solder Joints, 3/4" Diameter	.105	lf
Wrought Copper 90° Elbow for Solder Joints 3/4" Diameter	.421	Ea.
Wrought Copper, Adapter, CTS to MPT, 3/4" IPS	.381	Ea.
Pipe Steel Black, Schedule 40, Threaded, 3/4" Diameter	.131	lf
Pipe, 90° Elbow, Malleable Iron Black, 150 lb Threaded, 3/4" Diameter	.571	Ea.
Pipe, Union with Brass Seat, Malleable Iron Black, 3/4" Diameter	.615	Ea.
Valve, Gas Stop w/o Check, Brass, 3/4" IPS	.364	Ea.

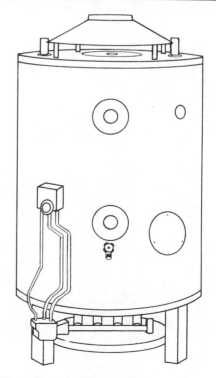

Gas Fired Water Heater - Commercial

Man-hours

Description	m/hr	Unit
Oil Fired Water Heater, Commercial, 100°F Rise		
97 MBH Output, 116 gal./hr	7.273	Ea.
134 MBH Output, 161 gal./hr	10.000	Ea.
161 MBH Output, 192 gal./hr	13.333	Ea.
187 MBH Output, 224 gal./hr	16.000	Ea.
262 MBH Output, 315 gal./hr	22.857	Ea.
341 MBH Output, 409 gal./hr	22.857	Ea.
420 MBH Output, 504 gal./hr	26.667	Ea.
525 MBH Output, 630 gal./hr	32.000	Ea.
630 MBH Output, 756 gal./hr	32.000	Ea.
735 MBH Output, 880 gal./hr	40.000	Ea.
840 MBH Output, 1000 gal./hr	40.000	Ea.
1050 MBH Output, 1260 gal./hr	40.000	Ea.
1365 MBH Output, 1640 gal./hr	48.000	Ea.
1680 MBH Output, 2000 gal./hr	60.000	Ea.
2310 MBH Output, 2780 gal./hr	80.000	Ea.
2835 MBH Output, 3400 gal./hr	80.000	Ea.
3150 MBH Output, 3780 gal./hr	120.000	Ea.

Man-hours

Description	m/hr	Unit
Electric Water Heater, Commercial, 100°F Rise		
50 gal. Tank, 9 kW 37 gal./hr	4.444	Ea.
80 gal., 12 kW 49 gal./hr	5.333	Ea.
36 kW 147 gal./hr	5.333	Ea.
120 gal., 36 kW 147 gal./hr	6.667	Ea.
150 gal., 120 kW 490 gal./hr	8.000	Ea.
200 gal., 120 kW 490 gal./hr	9.412	Ea.
250 gal., 150 kW 615 gal./hr	10.667	Ea.
300 gal., 180 kW 738 gal./hr	12.308	Ea.
350 gal., 30 kW 123 gal./hr	14.545	Ea.
180 kW 738 gal./hr	14.545	Ea.
500 gal., 30 kW 123 gal./hr	20.000	Ea.
240 kW 984 gal./hr	20.000	Ea.
700 gal., 30 kW 123 gal./hr	24.000	Ea.
300 kW 1230 gal./hr	24.000	Ea.
1000 gal., 60 kW 245 gal./hr	34.286	Ea.
480 kW 1970 gal./hr	34.286	Ea.
1500 gal., 60 kW 245 gal./hr	48.000	Ea.
480 kW 1970 gal./hr	48.000	Ea.
2000 gal., 60 kW 245 gal./hr	80.000	Ea.
480 kW 1970 gal./hr	80.000	Ea.

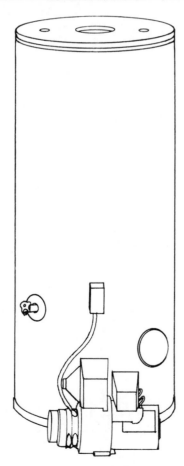

Oil Fired Water Heater - Commercial

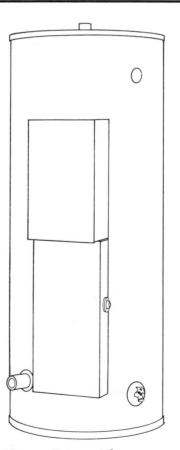

Electric Water Heater - Commercial

8 MECHANICAL
ROOF STORM DRAINS

Storm drainage systems convey rainwater from roofs and upper portions of structures where the water could cause damage to the building or constitute a hazard to the public. The water is discharged into a storm sewer or other approved location where it will not cause damage. The drainage system is usually not connected to a public sanitary sewage line.

Storm drains located in buildings above ground may be fabricated using brass, cast iron, copper, galvanized steel, lead, ABS, or PVC. Materials acceptable for underground use include cast iron, heavy wall copper, ABS, PVC, extra-strength, vitrified clay pipe or concrete pipe.

Stormwater is usually conducted away from the drainage area at the same rate it collects. This flow rate is the basis for sizing the system. The discharge capacity is based on the size of the horizontal area to be drained, plus 50% of the adjacent wall areas and the design rate or rainfall. Local government departments, plumbing codes, and weather bureaus can usually supply climatological data on rainfall.

The accompanying chart will provide a guide for sizing. The chart is based on a maximum rate of rainfall of 4″ per hour. Where maximum rates are more or less than 4″ per hour, the figures for drainage area can be adjusted by multiplying by a factor of 4 and dividing by the local rate in inches per hour. The chart is applicable for round, square, or rectangular rainwater pipe, any of which may be used and considered equivalent when they can enclose a circle equivalent to the leader diameter.

Traps are not required for stormwater drains connected to a storm sewer. Leaders and drains connected to a combined sewer and floor drain connected to a storm drain need to be trapped. The size of the trap for individual conductors should be the same size as the horizontal drain to which they are connected.

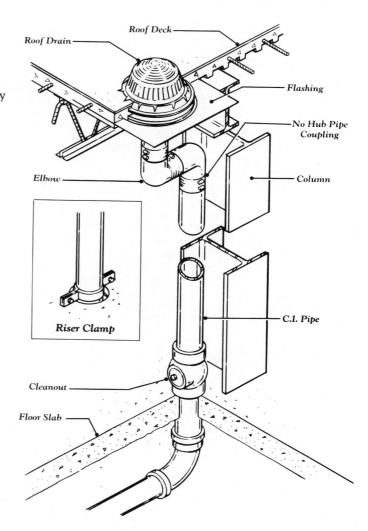

Man-hours

Description	m/hr						Units
	2″	3″	4″	5″	6″	8″	
Roof Drain	1.143	1.231	1.333	1.600	2.000	2.290	Ea.
Cleanout Tee	2.000	2.222	2.424	2.909	3.200	6.400	Ea.
C.I. Pipe (no hub)	.262	.276	.302	.324	.343	.533	lf
Elbow	1.230	1.334	1.454	1.714	2.000	3.368	Ea.
Tee	1.845	2.001	2.181	2.571	3.000	5.052	Ea.
Copper Tube	.348	.533	.696	.842	1.000	1.143	lf
Elbow	.727	1.455	1.778	2.667	2.667	3.000	Ea.
Tee	1.143	2.286	3.200	3.700	4.000	4.800	Ea.
Galvanized Steel Pipe	.250	.372	.444	.615	.774	.889	lf
Elbow	.889	1.600	2.667	3.200	3.429	4.000	Ea.
Tee	1.455	2.667	4.000	5.333	6.000	8.000	Ea.
DWV Pipe	.271	.302	.333	.370	.410	.500	lf
(ABS/PVC) Elbow	.571	.941	1.143	1.333	2.000	2.400	Ea.
Tee	.941	1.455	1.778	2.000	3.200	4.000	Ea.

Description

Pipe Diameter	Max. sf Roof Area	gal./min
2″	544	23
3″	1,610	67
4″	3,460	144
5″	6,280	261
6″	10,200	424
8″	22,000	913

8 MECHANICAL
PLUMBING FIXTURES AND ROUGHING-IN

The functions of a plumbing system are to supply potable water to the fixtures in a structure and to discharge used water and biological waste to a public sewer or private septic system. The supply water is drawn from wells or public water mains and piped under pressure to each fixture. The drainage piping originates at each fixture and discharges its contents by gravity into the public sewer or private septic system. These two piping networks must be kept separate to avoid contamination of the potable or public water supply. Strict adherence to local codes and the National Plumbing Code assures proper installation of these piping systems.

The water, waste, and vent piping, as well as the fixture carrier or support located in the immediate vicinity of the fixture and serving only that fixture, is called the "rough-in." The expense of the piping of the mains and branches serving all the fixtures must be added to the rough-in cost.

Properly installed traps and vents are important components of a building's drainage system because they prevent toxic and noxious sewer gas from backing up through drain fixtures and entering the structure. Sewer gas originates from both public sewer and private septic systems. Each drainage fixture is provided with a water

seal, or trap, built into the fixture or added to the drain pipe. This trap prevents sewer gas from exiting through the fixture as long as the water seal remains in the trap. When a large volume of drain water rushes through a trap, however, the water that makes up the seal tends to empty because of siphonage. To break the siphon and to prevent losing the seal, the traps are vented to the atmosphere. The vent, which is located on the house or discharge side of the trap, is piped through the roof of the structure and conducts the sewer gas safely to the open atmosphere. The vent also prevents a buildup of back pressure in the drainage piping by maintaining constant atmospheric pressure within the system.

Because the vent for each fixture does not have to be piped individually through the roof, the vents for separate fixtures may be interconnected, as allowed by codes and good plumbing practices. The interconnecting of vents minimizes the number of roof penetrations and reduces the amount of piping material required in the venting installation.

Carrier fittings for wall hung water closets, urinals and bidets may incorporate common vent and waste fittings for back to back or side by side fixtures.

Man-hours

Description	m/hr	Unit
For Setting Fixture and Trim		
Bath Tub	3.636	Ea.
Bidet	3.200	Ea.
Dental Fountain	2.000	Ea.
Drinking Fountain	2.500	Ea.
Lavatory		
Vanity Top	2.500	Ea.
Wall Hung	2.000	Ea.
Laundry Sinks	2.667	Ea.
Prison/Institution Fixtures		
Lavatory	2.000	Ea.
Service Sink	5.333	Ea.
Urinal	4.000	Ea.
Water Closet	2.759	Ea.
Combination Water Closet and Lavatory	3.200	Ea.
Shower Stall	8.000	Ea.
Sinks		
Corrosion Resistant	5.333	Ea.
Kitchen, Counter Top	3.330	Ea.
Kitchen, Raised Deck	7.270	Ea.
Service, Floor	3.640	Ea.
Service, Wall	4.000	Ea.
Urinals		
Wall Hung	5.333	Ea.
Stall Type	6.400	Ea.
Wash Fountain, Group	9.600	Ea.
Water Closets		
Tank Type, Wall Hung	3.019	Ea.
Floor Mount, One Piece	3.019	Ea.
Bowl Only, Wall Hung	2.759	Ea.
Bowl Only, Floor Mount	2.759	Ea.
Gang, Side by Side, First	2.759	Ea.
Each Additional	2.759	Ea.
Gang, Back to Back, First Pair	5.520	pair
Each Additional Pair	5.520	pair
Water Conserving Type	2.963	Ea.
Water Cooler	4.000	Ea.

Man-hours

Description	m/hr	Unit
For Roughing-In		
Bath Tub	7.730	Ea.
Bidet	8.990	Ea.
Dental Fountain	6.900	Ea.
Drinking Fountain	4.370	Ea.
Lavatory		
Vanity Top	6.960	Ea.
Wall Hung	9.640	Ea.
Laundry Sinks	7.480	Ea.
Prison/Institution Fixtures		
Lavatory	10.670	Ea.
Service Sink	17.978	Ea.
Urinal	10.738	Ea.
Water Closet	13.445	Ea.
Combination Water Closet and Lavatory	16.000	Ea.
Shower Stall	7.800	Ea.
Sinks		
Corrosion Resistant	7.920	Ea.
Kitchen, Counter Top	7.480	Ea.
Kitchen, Raised Deck	8.650	Ea.
Service, Floor	9.760	Ea.
Service, Wall	12.310	Ea.
Urinals		
Wall Hung	5.650	Ea.
Stall Type	8.040	Ea.
Wash Fountain, Group	8.790	Ea.
Water Closets		
Tank Type, Wall Hung	6.150	Ea.
Floor Mount, One Piece	8.250	Ea.
Bowl Only, Wall Hung	7.800	Ea.
Bowl Only, Floor Mount	8.890	Ea.
Gang, Side by Side, First	8.120	Ea.
Each Additional	7.480	Ea.
Gang, Back to Back, First Pair	9.090	pair
Each Additional Pair	8.840	pair
Water Conserving Type	8.250	Ea.
Water Cooler	3.620	Ea.

Drainage Requirements

Drainage lines must have a slope to maintain flow for proper operation. This slope should not be less than 1/4" per foot for 3" diameter or smaller pipe and not less than 1/8" per foot for 4" diameter or larger pipe. The capacity of building drainage systems is calculated on a basis of "drainage fixture units" (d.f.u.) as per the following chart.

Type of Fixture	d.f.u. Value	Type of Fixture	d.f.u. Value
Automatic Clothes Washer (2" Standpipe)	3	Service Sink (Trap Standard)	3
Bathroom Group (Water Closet, Lavatory and		Service Sink (P Trap)	2
Bathtub or Shower) Tank Type Closet	6	Urinal, Pedestal, Syphon Jet Blowout	6
Bathtub (with or without Overhead Shower)	2	Urinal, Wall Hung	4
Clinic Sink	6	Urinal, Stall Washout	4
Combination Sink and Tray with Food Disposal	4	Wash Sink (Circ. or Mult.) per Faucet Set	2
Dental Unit or Cuspidor	1	Water Closet, Tank Operated	4
Dental Lavatory	1	Water Closet, Valve Operated	6
Drinking Fountain	1/2	Fixtures not Listed Above	
Dishwasher, Domestic	2	Trap Size 1-1/4" or Smaller	1
Floor Drains with 2" Waste	3	Trap Size 1-1/2"	2
Kitchen Sink, Domestic with One 1-1/2" Trap	2	Trap Size 2"	3
Kitchen Sink, Domestic with Food Disposal	2	Trap Size 2-1/2"	4
Lavatory with 1-1/4" Waste	1	Trap Size 3"	5
Laundry Tray (1 or 2 Compartment)	2	Trap Size 4"	6
Shower Stall, Domestic	2		

Allowable Fixture Units (d.f.u.) for Branches and Stacks

For continuous or nearly continuous flow into the system from a pump, air conditioning equipment or other item, allow 2 fixture units for each gallon per minute of flow.

When the "drainage fixture units" (d.f.u.) for each horizontal branch or vertical stack is computed from the table above, the appropriate pipe size for each branch or stack is determined from the table below.

Pipe Diameter	Horiz. Branch (Not Incl. Drains)	Stack Size for 3 Stories or 3 Levels	Stack Size for Over 3 Levels	Maximum for 1 Story Building Stack
1-1/2"	3	4	8	2
2"	6	10	24	6
2-1/2"	12	20	42	9
3"	20*	48*	72*	20*
4"	160	240	500	90
5"	360	540	1100	200
6"	620	960	1900	350
8"	1400	2200	3600	600
10"	2500	3800	5600	1000
12"	3900	6000	8400	1500
15"	7000			

*Not more than two water closets or bathroom groups within each branch interval nor more than six water closets or bathroom groups on the stack.

Stacks sized for the total may be reduced as load decreases at each story to a minimum diameter of 1/2 the maximum diameter.

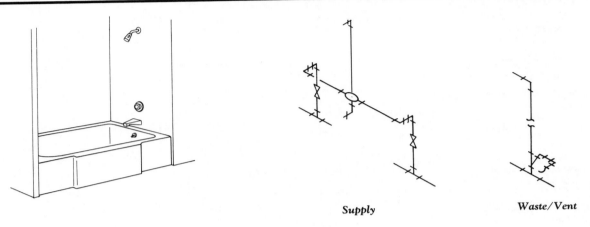

Supply   Waste/Vent

Recessed Bathtub

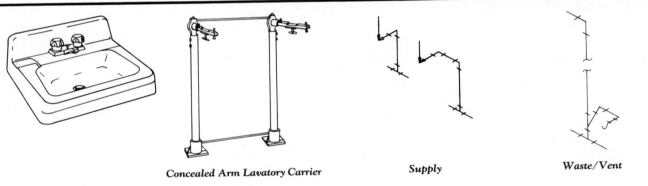

Concealed Arm Lavatory Carrier   Supply   Waste/Vent

Lavatory, Wall Hung

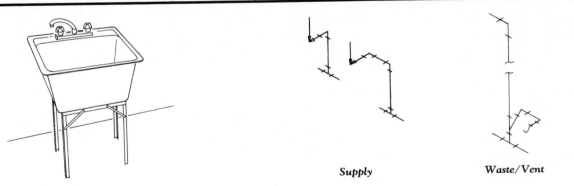

Supply   Waste/Vent

Single Compartment Laundry Sink

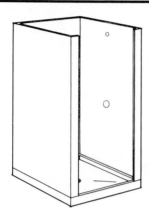

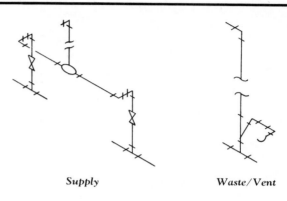

Supply   Waste/Vent

Shower Stall

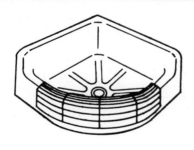

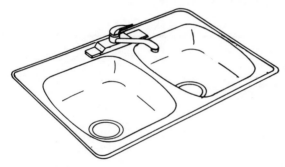

Supply   Waste/Vent

Double Bowl Sink

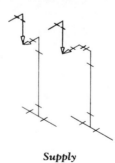

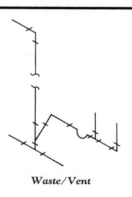

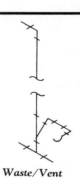

Supply   Waste/Vent

Service Sink, Corner Floor

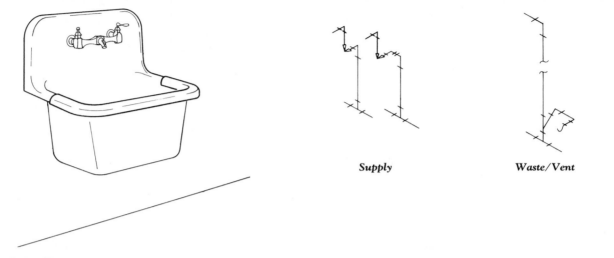

Service Sink, Wall Hung

Supply   Waste/Vent

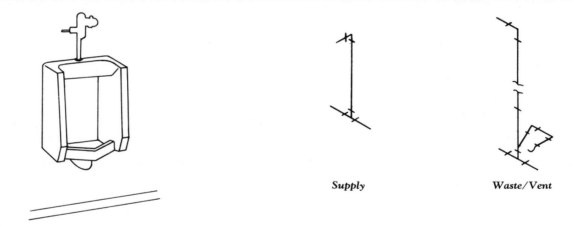

Wall Hung Urinal

Supply   Waste/Vent

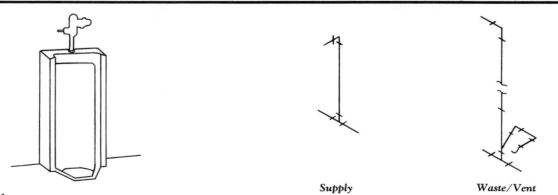

Stall Type Urinal

Supply   Waste/Vent

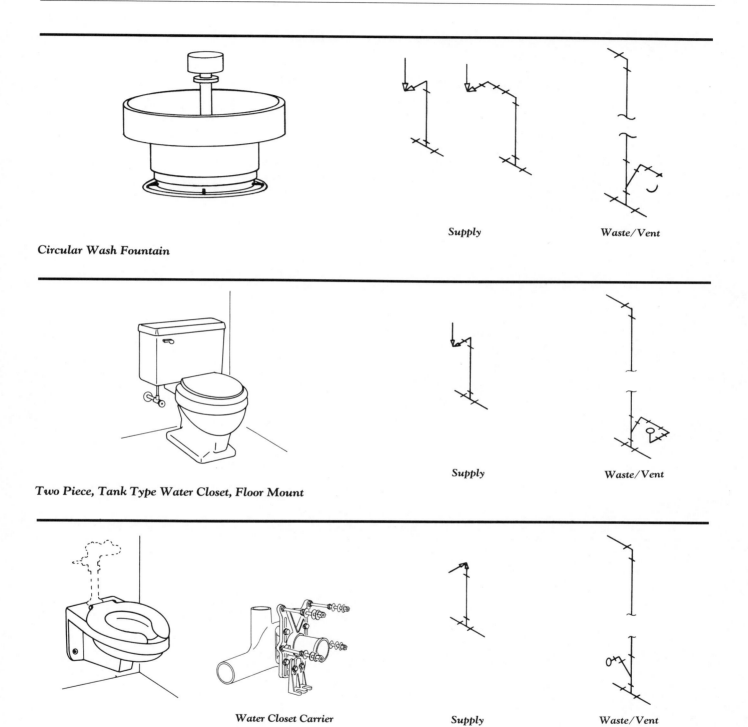

Circular Wash Fountain

Supply

Waste/Vent

Two Piece, Tank Type Water Closet, Floor Mount

Supply

Waste/Vent

Water Closet Carrier

Supply

Waste/Vent

Bowl Only Water Closet, Wall Hung

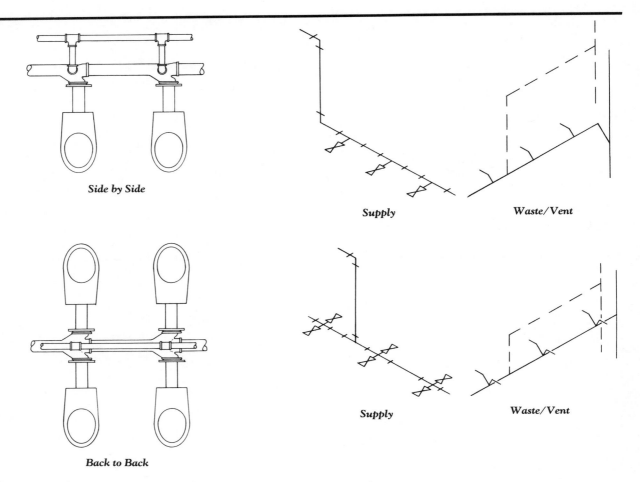

Side by Side

Supply　　　　　Waste/Vent

Back to Back

Supply　　　　　Waste/Vent

Water Closet Group

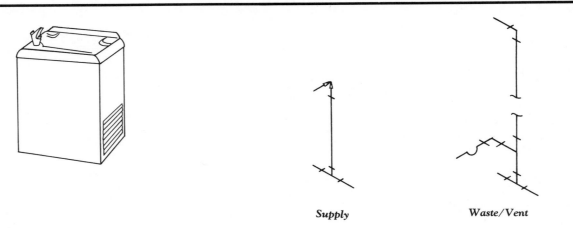

Water Cooler, Wall Hung

Supply　　　　　Waste/Vent

Minimum Plumbing Fixture Requirements

Type of Building/Use	Water Closets		Urinals		Lavatories		Bathtubs or Showers		Drinking Fountain	Other
	Persons	Fixtures	Persons	Fixtures	Persons	Fixtures	Persons	Fixtures	Fixtures	Fixtures
Assembly Halls Auditoriums Theater Public assembly	1-100 101-200 201-400	1 2 3	1-200 201-400 401-600	1 2 3	1-200 201-400 401-750	1 2 3			1 for each 1000 persons	1 service sink
	Over 400 add 1 fixt. for ea. 500 men: 1 fixt. for ea. 300 women		Over 600 add 1 fixture for each 300 men		Over 750 add 1 fixture for each 500 persons					
Assembly — Public Worship	300 men 150 women	1 1	300 men	1	men women	1 1			1	
Dormitories	Men: 1 for each 10 persons Women: 1 for each 8 persons		1 for each 25 men, over 150 add 1 fixture for each 50 men		1 for ea. 12 persons 1 separate dental lav. for each 50 persons recom.		1 for ea. 8 persons For women add 1 additional for each 30. Over 150 persons add 1 for each 20.		1 for each 75 persons	Laundry trays 1 for each 50 serv. sink 1 for ea. 100
Dwellings Apartments and homes	1 fixture for each unit				1 fixture for each unit		1 fixture for each unit			
Hospitals Room Ward Waiting room	Ea. 8 persons	1			Ea. 10 persons	1		1	1 for 100 patients	1 service sink per floor
Industrial Mfg. plants Warehouses	1-10 11-25 26-50 51-75 76-100	1 2 3 4 5	0-30 31-80 81-160 161-240	1 2 3 4	1-100 over 100	1 for ea. 10 1 for ea. 15	1 Shower for each 15 persons subject to excessive heat or occupational hazard		1 for each 75 persons	
	1 fixture for each additional 30 persons									
Public Buildings Businesses Offices	1-15 16-35 36-55 56-80 81-110 111-150	1 2 3 4 5 6	Urinals may be provided in place of water closets but may not replace more than 1/3 required number of men's water closets		1-15 16-35 36-60 61-90 91-125	1 2 3 4 5			1 for each 75 persons	1 service sink per floor
	1 fixture for ea. additional 40 persons				1 fixture for ea. additional 45 persons					
School Elementary	1 for ea. 30 boys 1 for ea. 25 girls		1 for ea. 25 boys		1 for ea. 35 boys 1 for ea. 35 girls		For gym or pool shower room 1/5 of a class		1 for each 40 pupils	
Schools Secondary	1 for ea. 40 boys 1 for ea. 30 girls		1 for ea. 25 boys		1 for ea. 40 boys 1 for ea. 40 girls		For gym or pool shower room 1/5 of a class		1 for each 50 pupils	

Drinking fountains and water coolers perform the same function. A refrigeration unit coupled with a drinking fountain changes it to a water cooler, but its function remains the same. These fixtures are manufactured in a wide variety of materials and finishes, including china, porcelain-enameled cast iron, aluminum, concrete, stone, fiberglass, stainless steel, and bronze. Architectural configurations are equally diversified. They may be wall-mounted, free-standing, recessed, deck-mounted, pedestal-mounted, dual height duplex, and multi-bubbler.

Available accessories include glass fillers, freeze-proof valves (for outdoor use), refrigerated compartments, pantry or laboratory faucets, aspirators for dental use, and hot-water capability. Special application models available in water coolers include explosionproof bottled water (requires no plumbing connections), restaurant/cafeteria type, and remote chiller type.

Remote water chillers can be piped to one or more drinking fountains. There are varied conditions that lend themselves to the remote chiller, satellite fountain combinations. Some of these conditions are economy of installation and operation, maintaining silence in a church or library, space saving in dental offices and laboratories.

Drinking Water Requirements (per Person)

Type of Service	Gal./hr per Person	
	Cup	Bubbler
Offices, Schools, Cafeterias, Hotels (per Room), Hospitals (per Bed and per Attendant)	0.033	0.083
Restaurants	0.040	0.100
Light Manufacturing	0.0573	0.143
Heavy Manufacturing	0.080	0.200
Hot, Heavy Manufacturing	0.100	0.250
Theaters per 100 Seats	0.4 gal./hr	1.0 gal./hr
Department Stores, Lobbies, Hotel and Office Buildings	1.6-2.0 gal./hr fountain	.4-5 gal./hr fountain

Components

Description	m/hr	Unit
Copper Tubing Type DWV, 50/50 Solder Joint, Hanger 10' On Center 1-1/4" Diameter	.267	lf
Wrought Copper DWV, Tee, Sanitary 2" Diameter	1.143	Ea.
P Trap, Copper Drainage, 1-1/4" Diameter	.444	Ea.
Copper Tubing Type L, 50/50 Solder Joint, Hanger 10' On Center 3/8" Diameter	.143	lf
Wrought Copper 90° Elbow for Solder Joints 3/8" Diameter	.364	Ea.
Wrought Copper Tee for Solder Joints, 3/8" Diameter	.571	Ea.
Stop and Waste, Straightway, Bronze, Solder, 3/8" Diameter	.333	Ea.

Man-hours

Description	m/hr	Unit
Drinking Fountain, One Bubbler, Wall Mounted		
Non-Recessed		
Bronze, No Back	2.000	Ea.
Cast Iron, Enameled, Low Back	2.000	Ea.
Fiberglass, 12" Back	2.000	Ea.
Stainless Steel, No Back	2.000	Ea.
Semi-Recessed, Polymarble	2.000	Ea.
Stainless Steel	2.000	Ea.
Vitreous China	2.000	Ea.
Full Recessed, Polymarble	2.000	Ea.
Stainless Steel	2.000	Ea.
Floor Mounted, Pedestal Type, Aluminum	4.000	Ea.
Bronze	4.000	Ea.
Stainless Steel	4.000	Ea.
Rough in, Supply Waste and Vent	6.452	Ea.

| No Back | Low Back | Semi-Recessed |

Drinking Fountains

Man-hours

Description	m/hr	Unit
Water Cooler, Electric, Wall Hung, 8.2 gal./hr	4.000	Ea.
Dual Height, 14.3 gal./hr	4.211	Ea.
Wheelchair Type, 7.5 gal./hr	4.000	Ea.
Semi Recessed, 8.1 gal./hr	4.000	Ea.
Full Recessed, 8 gal./hr	4.571	Ea.
Floor Mounted, 14.3 gal./hr	2.667	Ea.
Dual Height, 14.3 gal./hr	4.000	Ea.
Refrigerated Compartment Type, 1.5 gal./hr	2.667	Ea.
Cafeteria Type, Dual Glass Fillers, 27 gal./hr	6.400	Ea.
Rough in, Supply Waste and Vent	6.452	Ea.

Man-hours do not include electrical connections.

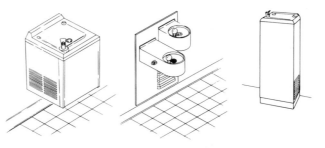

| Wall Hung | Wall Hung Wheelchair | Floor Mounted |

Water Coolers

8 MECHANICAL
SOLAR DOMESTIC HOT WATER

Solar heating systems are used for domestic hot water, space heating and swimming pool heating. The basic elements of a solar domestic hot-water system are the collectors, storage tank, controls, and the interconnected piping system. In most geographic locations an auxiliary or backup system must be provided for periods of little or no sun. The solar system itself may be the backup, or it may be used to preheat the domestic water to lessen the energy requirements of a conventional water heating system.

In areas subject to freezing temperatures, provision must be made to drain the exposed collectors and piping, or to employ a nonfreeze type of heat-transfer medium. When a nonfreeze heat-transfer fluid is used, a heat exchanger must be added to the system to transfer the heat from the antifreeze to the potable water. Therefore, there are two types of exchangers: the closed loop, immersed exchanger system, where the heat exchanger is submerged within the water storage tank; and, conversely, the external exchanger system, where the unit is remotely located from the storage tank.

There are two options for draining the exposed heat transfer fluid from the collector system. The first is the drainback, immersed exchanger system, where the heating medium drains back to a nonpressurized holding tank. The incoming pressurized, potable water flows through a heat exchanger (coil) immersed in the holding tank. The second is a direct collection drain-down system, where the potable water flows through the collectors and drains the collector loop when conditions are no longer suitable for heat collection. This loop is drained out of the system to waste by use of automatic valves and properly pitched piping. The previously heated water remains in the storage tank.

Solar collector panels are designed either to heat liquid or to heat air. Collectors may be singular or mounted in banks, referred to as an array. While there is a variety of styles, the basic components of all collectors are similar. They must have a casing, insulation, and a glazing absorber with fluid tubes or channels.

The flat plate collector is the most common. Collectors may be installed on the roof or on the ground in shade-free areas. They should face south for maximum efficiency and be angled up to 20 degrees plus the latitude to receive as much sunlight as possible. This angle varies according to geographic orientation.

Storage tanks and the piping systems should be adequately insulated to prevent heat loss during storage or transmission. Corrosion proof linings, as in conventional storage water heaters, are also necessary.

Controls, both operating and limit, must be installed and designed to operate the necessary circulating pumps, shut-off or diverting valves, and auxiliary heat, to initiate freeze protection, and to maintain desired temperatures. Space heating and pool heating follow the same basic principles as domestic hot water systems.

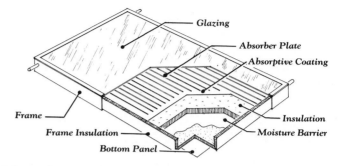

Solar Collector Panel - Liquid

Man-hours

	Description	m/hr	Unit
A, B	Controller	1.000	Ea.
A-1	Solenoid	.333	Ea.
B-1	Sensor	.800	Ea.
C	Thermometer	.250	Ea.
C-1	Heat Exchanger	5.333	Ea.
D, T	Fill-Drain Valve	.333	Ea.
D-1	Fan Control	1.000	Ea.
E, E-1	Air or Vacuum RLF	.250	Ea.
E-2	Thermostat	1.000	Ea.
F	Air Purger	.250	Ea.
F-1	Controller	1.000	Ea.
G	Expansion Tank	.941	Ea.
H	Strainer	.421	Ea.
I	Gate Valve	.400	Ea.
J	Vent Flashing	.333	Ea.
K, L	Circulator	1.333	Ea.
N	Relief Valve	.267	Ea.
N-1	Relief Valve	.267	Ea.
O	Pipe Covering	.077	lf
P	Pipe Covering	.070	lf
Q	Collectors	1.684	Ea.
	Roof Clamps	.229	Set
R, R-1	Check Valves	.400	Ea.
S	Gauge	.250	Ea.
U	Valve	.400	Ea.
V	Existing Tank		Ea.
W, W-1, W-2	Tanks	5.714	Ea.
X	Globe Valve	.400	Ea.
Y	Flow Control	.400	Ea.
Z, Z-1	Ball Valve	.400	Ea.
	Copper Tube L	.182	lf
	Copper Tube M	.174	lf
	Copper Fittings	.444	Ea.
	Sensor Wire	.800	CLF
	Ductwork	.094	lb
	Solar Fluid	.286	gal.
	PVC Piping	.222	lf
	PVC Fittings	.571	Ea.

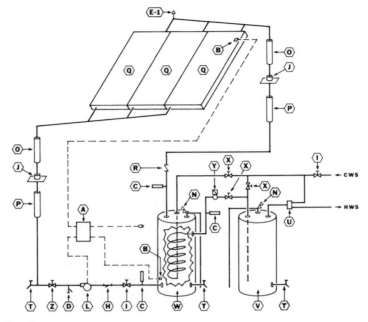

Drainback, Immersed Exchanger System

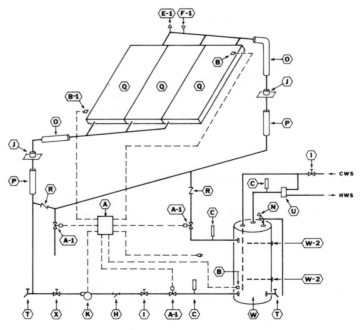

Direct Collection, Drain-down System

Refer to Man-hour Chart for Equipment Identifying Keys

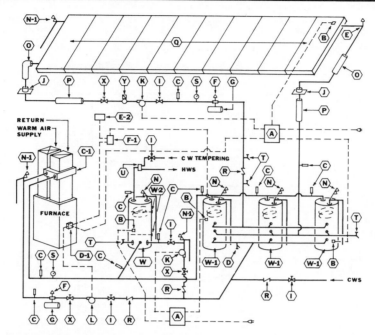

Closed Loop, Immersed Exchanger, Space Heating/Domestic Hot Water System

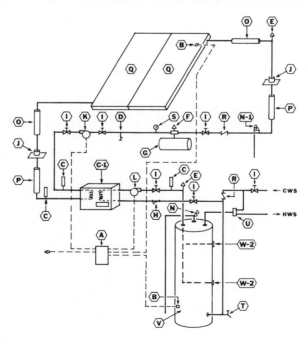

Closed Loop, External Exchanger System

Refer to Man-hour Chart for Equipment Identifying Keys

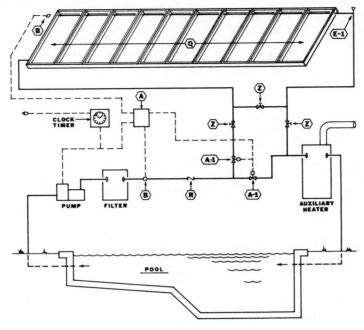

Swimming Pool Heating System

Refer to Man-hour Chart for Equipment Identifying Keys

8 MECHANICAL
AUTOMATIC SPRINKLER SYSTEMS

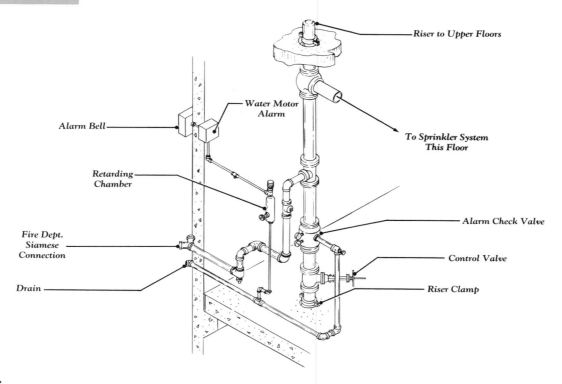

Riser to Upper Floors

Water Motor Alarm

Alarm Bell

To Sprinkler System This Floor

Retarding Chamber

Alarm Check Valve

Fire Dept. Siamese Connection

Control Valve

Riser Clamp

Drain

Wet Pipe

The wet pipe sprinkler system is the most popular type of automatic fire-suppression system in use today. The piping system is filled with water under pressure and connected to a municipal or private supply. This water is released immediately through one or more sprinkler heads that have been opened by fire or heat via a preset thermal element. Quick response and low first cost are the main reasons for the popularity of the wet pipe system.

In areas subject to freezing, a dry pipe sprinkler system is used. This system is quite similar to the wet pipe system except that the piping system is filled with compressed air maintained under a constant pressure, which must be higher than the available water pressure. These opposing pressures are contained and water is kept out of the system by means of the dry pipe valve. When the air pressure drops, water enters the piping system and is released through one or more sprinkler heads which have opened after sensing fire or heat. The dry pipe system does not have the quick response of the wet pipe system, and it costs more to install because of the compressed air requirement. Further, dry pendent heads and the piping must be installed without low points or pockets that cannot be drained. The dry pipe valve itself must be enclosed in an area not subject to freezing.

Preaction systems are used in areas subject to freezing or where accidental damage to sprinkler heads or piping and subsequent water leakage would be unacceptable. Because the preaction system is also more sensitive to fire than a sprinkler head, it provides a quicker response than the dry pipe system.

The preaction system is filled with air that does not have to be pressurized. The conventional closed type of sprinkler heads are used, but the initial detection of heat or fire occurs via heat-actuated devices that are more sensitive than the sprinkler heads. These devices detect a rapid temperature rise and open the preaction sprinkler valve, which fills the system with water and activates an alarm. With a further rise in temperature, one or more sprinkler heads in the fire area will open and begin the extinguishing process. This type of system has two important advantages: early warning, which allows occupants to evacuate, and quick notification of fire-fighting personnel. Like the dry pipe system, if the preaction system is installed in an area subject to freezing, pendent heads, if used, must be of the dry type and the other draining provisions must be adhered to. The preaction valve must be installed in a nonfreeze area.

A deluge system employs open type nozzles rather than heat-activated heads, piped to a water supply contained by a preaction deluge valve which is controlled by heat-activated devices. When this valve opens, it admits water to the entire system, or valved zone, and discharges water from all the nozzles in the protected area. This type of system, by using preaction fire-detection devices and by wetting down the entire area instantly, prevents the spread of the fire. *Note:* The deluge system is particularly suitable for storage or work areas involving flammable liquids.

Where water damage must be kept to a minimum in fire suppression, a firecycle system is employed. The firecycle system is a dry system with preaction fire-detection devices and a flow control valve. Electrical controls in this system have the capability to close the flow control valve after fire detectors sense that the fire is out. A time-delay feature built into the system allows water to flow through the opened sprinkler heads for a predetermined period before closing the flow valve. Should the fire reignite, the valve opens and the cycle begins again. Battery backup is a requirement for this type of system.

The firecycle system is always available for subsequent fires, since it does not have to be shut down for head replacement, etc. This system also eliminates the need for manual shut-off valves and position indicators.

Man-hours

Description	m/hr	Unit
Valve, Gate, Iron Body, Flanged OS and Y, 125 lb, 4″ Diameter	5.333	Ea.
Valve, Swing Check, w/Ball Drip, Flanged, 4″ Diameter	5.333	Ea.
Valve, Swing Check, Bronze, Thread End, 2-1/2″ Diameter	1.067	Ea.
Valve, Angle, Bronze, Thread End, 2″ Diameter	.727	Ea.
Valve, Gate, Bronze, Thread End, 1″ Diameter	.421	Ea.
Alarm Valve, 2-1/2″ Diameter	5.333	Ea.
Alarm, Water Motor, with Gong	2.000	Ea.
Fire Alarm Horn, Electric	.308	Ea.
Pipe, Black Steel, Threaded, Schedule 40,		
4″ Diameter	.444	lf
2-1/2″ Diameter	.320	lf
2″ Diameter	.250	lf
1-1/4″ Diameter	.180	lf
1″ Diameter	.151	lf
Pipe Tee, 150 lb Black Malleable		
4″ Diameter	4.000	Ea.
2-1/2″ Diameter	1.778	Ea.
2″ Diameter	1.455	Ea.
1-1/4″ Diameter	1.143	Ea.
1″ Diameter	1.000	Ea.
Pipe Elbow, 150 lb Black Malleable		
1″ Diameter	.615	Ea.
Sprinkler Head, 135° to 286°,		
1/2″ Diameter	.500	Ea.
Sprinkler Head, Dry Pendent		
1″ Diameter	.571	Ea.
Dry Pipe Valve, w/Trim and Gauges,		
4″ Diameter	16.000	Ea.
Deluge Valve, w/Trim and Gauges,		
4″ Diameter	16.000	Ea.
Deluge System Monitoring Panel 120 Volt	.444	Ea.
Thermostatic Release	.400	Ea.
Heat Detector	.500	Ea.
Firecycle Controls w/Panel, Batteries,		
Valves and Switches	32.000	Ea.
Firecycle Package, Check and Flow		
Control Valves, Trim, 4″ Diameter	16.000	Ea.
Air Compressor, Automatic, 200 gal.		
Sprinkler System 1/3 hp	6.154	Ea.

All of the above systems can be combined with fire standpipe systems.

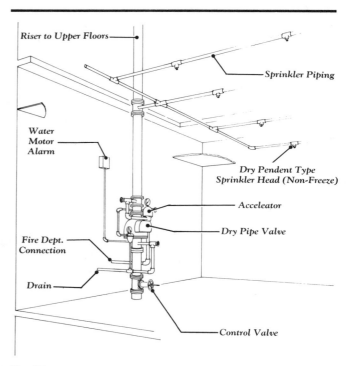

Dry Pipe

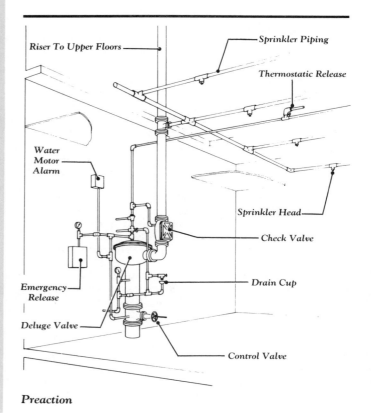

Preaction

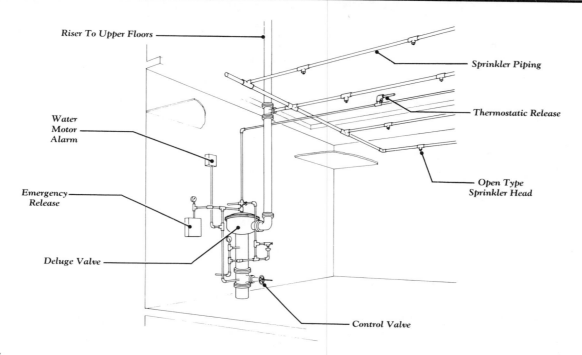

Riser To Upper Floors

Sprinkler Piping

Thermostatic Release

Water Motor Alarm

Open Type Sprinkler Head

Emergency Release

Deluge Valve

Control Valve

Deluge

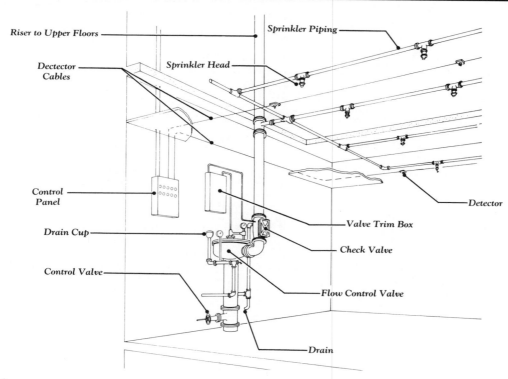

Riser to Upper Floors

Sprinkler Piping

Dectector Cables

Sprinkler Head

Control Panel

Drain Cup

Detector

Valve Trim Box

Control Valve

Check Valve

Flow Control Valve

Drain

Firecycle

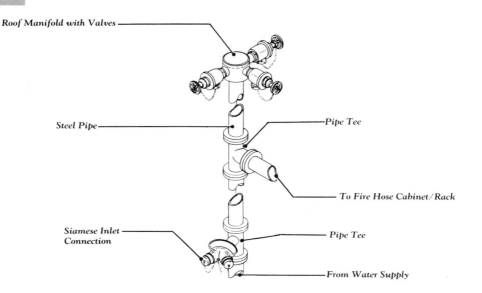

Roof Manifold with Valves

Steel Pipe

Pipe Tee

To Fire Hose Cabinet/Rack

Siamese Inlet Connection

Pipe Tee

From Water Supply

Standpipe System

The basis for standpipe system design is the National Fire Protection Association NFPA 14; however, the authority having jurisdiction should be consulted for special conditions, local requirements, and approval.

Standpipe systems, properly designed and maintained, are an effective and valuable time-saving aid for extinquishing fires, especially in the upper stories of tall buildings, the interior of large commercial or industrial malls, or other areas where construction features or access make the laying of temporary hose lines time consuming and/or hazardous. Adequate pressure is obtained by city water pressure, a reservoir at the roof or upper story of the building or by booster pumps. Standpipes are frequently installed with automatic sprinkler systems for maximum protection.

There are three general classes of service for standpipe systems:

Class I for use by fire departments and personnel with special training for heavy hose (2-1/2" hose connections).

Class II for use by building occupants until the arrival of the fire department (1-1/2" hose connector with hose).

Class III for use by either fire departments and trained personnel or by the building occupants (both 2-1/2" and 1-1/2" hose connections or one 2-1/2" hose valve with an easily removable 2-1/2" by 1-1/2" adapter).

Standpipe systems are also classified by the way water is supplied to the system. The four basic types are:

Type 1: Wet standpipe system having supply valve open and water pressure maintained at all times.

Type 2: Standpipe system so arranged through the use of approved devices to admit water to the system automatically when a hose valve is opened.

Type 3: Standpipe system arranged to admit water to the system through manual operation of approved remote-control devices located at the hose stations.

Type 4: Dry standpipe having no fixed water supply.

Standpipe systems are usually made of steel pipe, but other forms of ferrous or copper piping may be used where acceptable to the authority having jurisdiction. Joints may be flanged, threaded, welded, soldered, grooved, or whatever method is compatible with the approved piping materials.

Man-hours

Description	m/hr			Unit
	4"	6"	8"	
Black Steel Pipe	.444	.774	.889	lf
Pipe Tee	4.000	6.000	8.000	Ea.
Pipe Elbow	2.667	3.429	4.000	Ea.
Pipe Nipple, 2-1/2"	1.000	1.000	1.000	Ea.
Hose Valve, 2-1/2"	1.140	1.140	1.140	Ea.
Pressure Restricting Valve, 2-1/2"	1.140	1.140	1.140	Ea.
Check Valve with Ball Drip	5.333	8.000	10.667	Ea.
Siamese Inlet	3.200	3.478	3.478	Ea.
Roof Manifold with Valves	3.333	3.478	3.478	Ea.

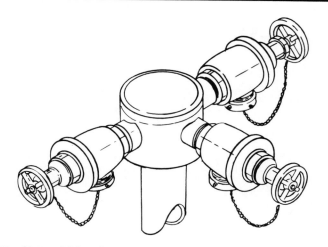

Roof Manifold with Valves

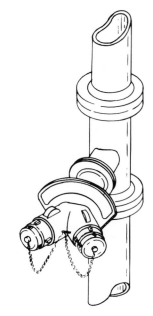

Siamese Inlet Connection

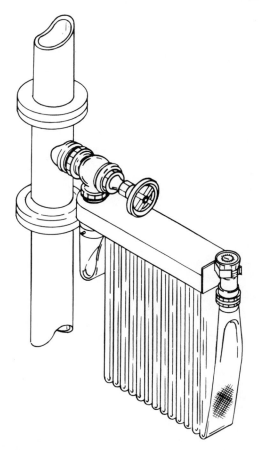

Fire Hose Rack

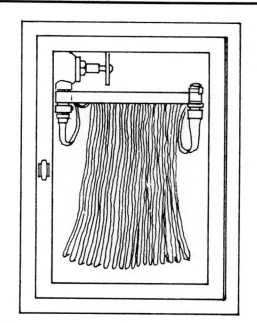

Fire Hose Cabinet

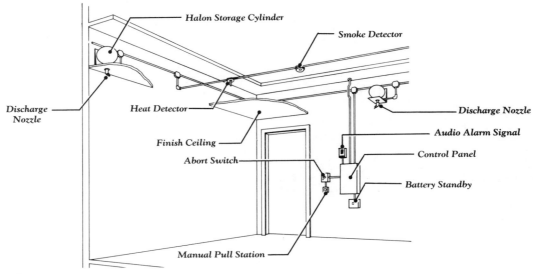

Clean Fire Suppression

Halon fire-suppression systems are used because they are fast, effective, and clean. These systems are called "clean" because they leave no residue that must be cleaned up or that could contaminate building contents, such as records or electronic equipment. Halon gas is a nonconductor of electricity; and because it is several times as heavy as air, it permeates the working area and penetrates cabinets or other electric or electronic enclosures where water or chemical powders cannot.

Depending on its concentration, halon gas ranges from nontoxic to low toxicity. Halon 1301 is normally used in a concentration of 5-7%. This concentration has no adverse effects on personnel in the area.

Seven to ten percent concentration requires evacuation within one minute of exposure; concentrations above 10% require evacuation prior to discharge. Halon 1301 is also colorless and has minimal visual impedance to hamper evacuation.

Halon may be applied to fires by portable extinguishers, local application of strategically located cylinders or prepackaged systems, or by a centrally located battery of storage cylinders connected to a piping distribution system and discharge nozzles. The local application or modular system is the most cost effective because it eliminates the piping installation costs encountered in a centrally located system.

Detection and actuation are critical requirements for fast extinguishing, to eliminate not only fire damage but also the accompanying risks of smoke, heat, carbon monoxide, and oxygen depletion. Fire and smoke detectors of the photoelectric or the ionization type are strategically located in the areas being protected. These detectors are wired to a control panel that activates the alarm systems, verifies or proves the existence of combustion, and releases the extinguishing agent, all in a matter of seconds.

The halon fire-suppression system is most effective in an enclosed area. The control system may also have the capability of closing doors and shutting off exhaust fans.

These systems are in popular use in the following places:

Aircraft (both cargo and passenger)
Libraries and museums
Bank and security vaults
Electronic data processing
Transformer and switchgear rooms
Tape and data storage vaults (rooms)
Telephone exchanges
Laboratories
Radio and television studios
Flammable liquid storage areas

Man-hours

Description	m/hr	Unit
Halon System, Filled,		
Including Mounting Bracket		
26 lb Cylinder	2.000	Ea.
44 lb Cylinder	2.286	Ea.
63 lb Cylinder	2.667	Ea.
101 lb Cylinder	3.200	Ea.
196 lb Cylinder	4.000	Ea.
Electro/Mechanical Release	4.000	Ea.
Manual Pull Station	1.333	Ea.
Pneumatic Damper Release	1.000	Ea.
Discharge Nozzle	.570	Ea.
Control Panel Single Zone	8.000	Ea.
Control Panel Multi-zone (4)	16.000	Ea.
Battery Standby Power	2.810	Ea.
Heat Detector	1.000	Ea.
Smoke Detector	1.290	Ea.
Audio Alarm	1.194	Ea.

Heating boilers are designed to produce steam or hot water. The water in the boilers is heated by coal, oil, gas, wood, electricity, or a combination of these fuels. Boiler materials include cast iron, steel, and copper. Several types of boilers are available to meet the hot water and heating needs of both commercial and residential buildings. Due to the high cost of fuels, efficiency of operation is a prime consideration when selecting boiler units, and because of this, innovations in the manufacturing field have provided the option to install efficient and compact boiler systems.

Cast iron sectional boilers may be assembled in place or shipped to the site as a completely assembled package. These boilers can be made larger on site by adding intermediate sections.

Steel boilers are usually shipped to the site completely assembled; large steel boilers may be shipped in segments for field assembly and testing. The components of a steel boiler consist of tubes within a shell, plus a combustion chamber. If the water being heated is inside the tubes, the unit is called a "water tube boiler." If the water is contained in the shell and the products of combustion pass through tubes surrounded by this water, the unit is called a "fire tube boiler." Water tube boilers are also manufactured with copper tubes or coils. Electric boilers have elements immersed in the water and do not fall into either category of tubular boilers.

Heating boilers are rated by their hourly output expressed in "British Thermal Units." The output available at the boiler supply nozzle is referred to as the gross output. The gross output in B.T.U. per hour divided by 33,475 will give the boiler horsepower rating.

The net rating of a boiler is the gross output less allowances for the piping tax and the pickup load. The net load should match the building heat load.

When selecting a boiler, efficiency is an important consideration. The Department of Energy has established test procedures to compare the "Annual Fuel Utilization Efficiency" (AFUE) of comparably sized boilers. Better insulation, heat extractors, intermittent ignition, induced draft, and automatic draft dampers contribute to the near 90% efficiencies claimed by manufacturers today.

In the search for higher efficiency, a new concept in gas-fired water boilers has recently been introduced. This innovation is the pulse- or condensing-type boiler, which relies on a sealed combustion system rather than on a conventional burner. The AFUE ratings for pulse-type boilers are in the low to mid 90% range. Pulse-type boilers cost more to purchase than conventional types, but savings in other areas help to offset this added cost. For example, because these units vent through a plastic pipe to a side wall, no chimney is required. Also, they take up less floor space, and their high efficiency saves on fuel costs.

Another innovation in the boiler field is the introduction from Europe of wall-hung, residential-size boilers. These gas-fired, compact, efficient (up to 84% AFUE) boilers may be directly vented through a wall or to a conventional

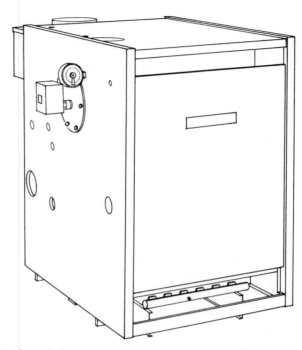

Packaged, Cast Iron Sectional, Gas Fired Boiler-Residential

flue. Combustion make-up air is directed to a sealed combustion chamber similar to that of the pulse-type boiler. Storage capacity is not needed in these boilers because the water is heated instantaneously as it flows from the boiler to the heating system. The boiler material consists mostly of steel. Heat-exchanger water tubes are made of copper or stainless steel, with some manufacturers using a cast iron heat exchanger.

Governing conditions in boiler selection include:

Accessibility. Both for initial installation and for future replacement, cast iron boiler sections can be delivered through standard door or window openings. Some steel water-tube boilers are made long and narrow to fit through standard door openings.

Economy of installation. Packaged boilers that have been factory fired require minimal piping, flue, and electrical field connections.

Economy of operation. In addition to the AFUE ratings, boiler output should be matched as closely as possible to the building heat loss. Installation of two or more boilers (modular), piped and controlled so as to step-fire to match varying load conditions, should be carefully evaluated. Only on maximum design load would all boilers be firing at once. This method of installation not only increases boiler life, but also provides for continued heating capacity in the event that one boiler should fail.

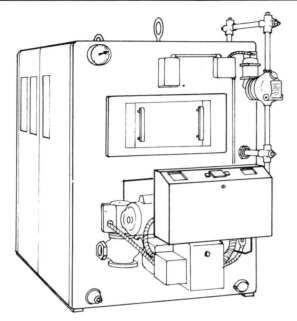

Packaged, Cast Iron Sectional,
Gas/Oil Fired Boiler-Commercial

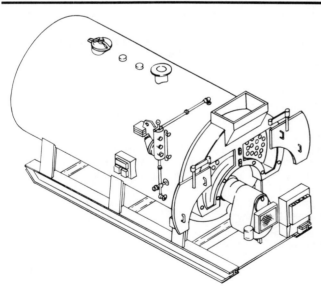

Packaged, Oil Fired,
Modified Scotch Marine Boiler-Commercial

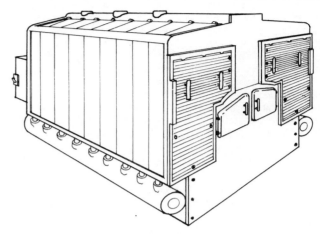

Cast Iron Sectional Boiler-Commercial

Man-hours

Description	m/hr	Unit
Electric Fired, Steel Output		
60 MBH	20.000	Ea.
500 MBH	36.293	Ea.
1000 MBH	60.000	Ea.
2000 MBH	94.118	Ea.
3000 MBH	114.286	Ea.
7000 MBH	177.778	Ea.
Gas Fired, Cast Iron Output		
80 MBH	21.918	Ea.
500 MBH	53.333	Ea.
1000 MBH	64.000	Ea.
2000 MBH	99.250	Ea.
3000 MBH	133.333	Ea.
7000 MBH	400.000	Ea.
Oil Fired, Cast Iron Output		
100 MBH	26.667	Ea.
500 MBH	64.000	Ea.
1000 MBH	76.000	Ea.
2000 MBH	114.286	Ea.
3000 MBH	139.130	Ea.
7000 MBH	400.000	Ea.
Scotch Marine Packaged Units,		
Gas or Oil Fired Output		
1300 MBH	80.000	Ea.
3350 MBH	80.000	Ea.
4200 MBH	85.000	Ea.
5000 MBH	90.000	Ea.
6700 MBH	90.000	Ea.
8370 MBH	110.000	Ea.
10,000 MBH	120.000	Ea.
20,000 MBH	130.000	Ea.
23,400 MBH	140.000	Ea.

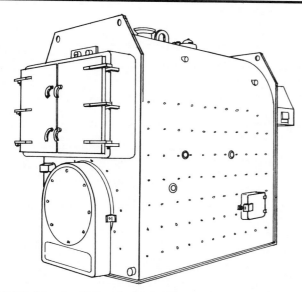

High Firebox, Steel, Fire Tube Boiler-Commercial

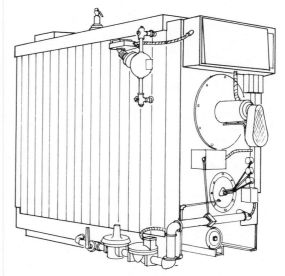

Packaged, Gas Fired, Steel, Watertube Boiler-Commercial

Electric Steel Boiler, Commercial

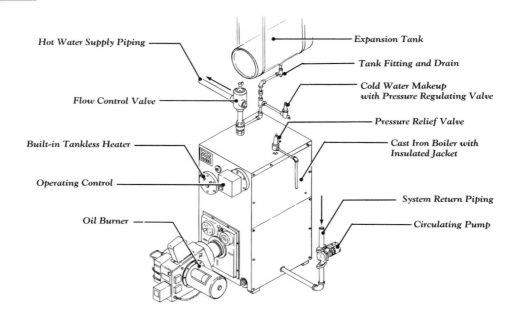

Oil Fired

Forced hot-water heating systems using direct radiation have replaced steam heating systems and cast iron radiators as the most common space heating systems in residential, commercial, and industrial buildings. The heart of these systems is the boiler, either cast iron or steel. The available energy source may be oil, gas, coal, wood, electric, or any combination thereof. In industrial and some commercial installations, waste heat or district heating is utilized in place of the boiler.

A forced hot-water system can range from the very simple mainless loop to the one-pipe and two-pipe direct-return system or the more complex, reverse-return system.

In the mainless loop system the heating medium flows from one radiator into the next without returning to a main pipe line. While this system is by far the most economical because of the minimal use of piping, it eliminates uniform temperature control in the heated spaces. The reverse-return system returns the heating medium from the first radiator and each succeeding radiator to a return main that is so designed that it returns the last radiator heated to the boiler first, and the first radiator last, etc. This type of system uses more piping than does a direct-return system, but it maintains a uniform water temperature in each radiator.

Forced hot-water systems utilize much smaller pipe sizes than a gravity-flow system does because of the positive circulation caused by the circulating pump. Higher water temperatures, 180 to 200 degrees Farenheit also permit smaller pipe and radiator sizes compared to the 150 degree temperature of the gravity systems. Higher water temperatures bring higher operating pressures, and may create a dangerous condition above 210 degrees, as any sudden drop in pressure (such as a leak or suddenly

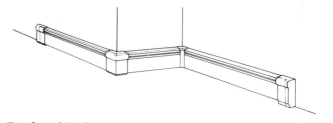

Baseboard Radiation

opened valve) can cause flash steam and, possibly a violent eruption.

High-temperature systems in the practical neighborhood of 450 degrees are used for very special applications and with the necessary precautions.

The terminal units used for forced hot-water systems are free standing or baseboard radiators, convectors, fin tube, coils, and fan coil units.

Man-hours

Description	m/hr	Unit
Hot Water Heating System, Area to 2400 sf		
Boiler Package, Oil Fired, 225 MBH	17.143	Ea.
Oil Piping System	4.584	Ea.
Oil Tank, 550 Gallon, with Black Steel		
Fill Pipe	4.000	Ea.
Supply Piping, 3/4" Copper Tubing	.182	lf
Supply Fittings, Copper	.421	lf
Baseboard Radiation	.667	lf

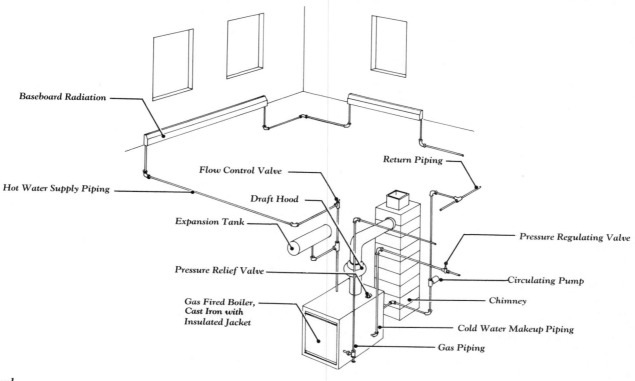

Baseboard Radiation

Return Piping

Hot Water Supply Piping

Flow Control Valve

Draft Hood

Expansion Tank

Pressure Regulating Valve

Pressure Relief Valve

Circulating Pump

Gas Fired Boiler,
Cast Iron with
Insulated Jacket

Chimney

Cold Water Makeup Piping

Gas Piping

Gas Fired

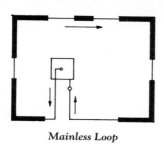

Mainless Loop

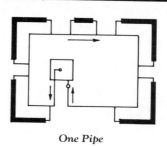

One Pipe

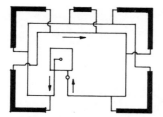

Two Pipe Reverse Return

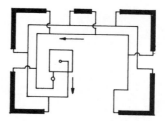

Two Pipe Direct Return

Diagramatic Piping Arrangements for Hot Water Heating Systems

8 MECHANICAL
DUCTWORK

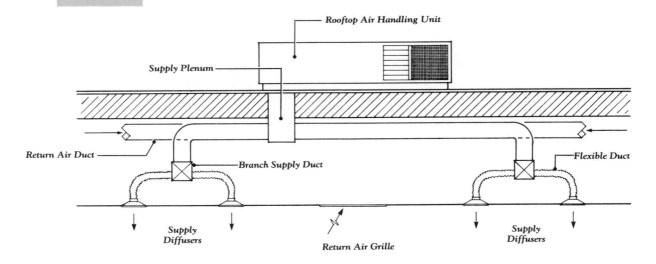

Ductwork is the conduit or pipe used in heating, ventilating, or air conditioning to transport air throughout the system.

Ductwork has traditionally been fabricated from sheets of galvanized steel, but it is also fabricated from aluminum sheets and fiberglass board. For use in a corrosive atmosphere, stainless steel or one of several plastics is the preferred material.

Ductwork may be rectangular or round in shape, depending on the requirements of size and application. Flexible or spiral-wound ductwork is also available for special installations.

There are fittings to accommodate changes in the direction, shape, and size of ductwork. Elbows, tees, transitions, reducers and increasers come under this category.

Industry standards regulate the thickness or gauge of metal duct material. The basis for these regulations is the size of the ductwork. The appropriate weight for ductwork material is determined using these same standards. Based on this weight, the fabrication, labor, and installation charge can be calculated. Fiberglass and flexible duct is estimated by the lineal foot.

Many accessories are needed to complete a ductwork system. Supply and return "faces" include diffusers, registers and grilles. Control devices include dampers and turning vanes.

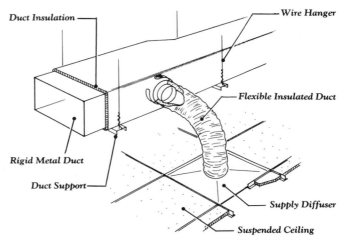

Ductwork System

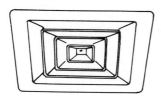

Supply Diffuser

Man-hours

Description	m/hr	Unit
Ductwork		
Fabricated Rectangular, Includes Fittings,		
Joints, Supports		
Allowance for Flexible Connections,		
No Insulation		
Aluminum , Alloy 3003-H14, Under 300 lb	.320	lb
300 to 500 lb	.300	lb
500 to 1000 lb	.253	lb
1000 to 2000 lb	.200	lb
2000 to 10,000 lb	.185	lb
Over 10,000 lb	.166	lb
Galvanized Steel, Under 400 lb	.102	lb
400 to 1000 lb	.094	lb
1000 to 2000 lb	.091	lb
2000 to 5000 lb	.087	lb
5000 to 10,000 lb	.084	lb
Over 10,000 lb	.080	lb
Stainless Steel, Type 304, Under 400 lb	.145	lb
400 to 1000 lb	.130	lb
1000 to 2000 lb	.120	lb
2000 to 10,000 lb	.107	lb
Over 10,000 lb	.102	lb
Flexible, Vinyl Coated Spring Steel or		
Aluminum, Pressure to 10″ (WG) UL-181		
Non-insulated, 3″ Diameter	.040	lf
4″ Diameter	.044	lf
5″ Diameter	.050	lf
Ductwork, Flexible, Non-insulated,		
6″ Diameter	.057	lf
7″ Diameter	.067	lf
8″ Diameter	.080	lf
9″ Diameter	.089	lf
10″ Diameter	.100	lf
12″ Diameter	.133	lf

Man-hours (cont.)

Description	m/hr	Unit
14″ Diameter	.200	lf
16″ Diameter	.267	lf
Insulated, 4″ Diameter	.047	lf
5″ Diameter	.053	lf
6″ Diameter	.062	lf
7″ Diameter	.073	lf
8″ Diameter	.089	lf
9″ Diameter	.100	lf
10″ Diameter	.114	lf
12″ Diameter	.160	lf
14″ Diameter	.200	lf
16″ Diameter	.267	lf
18″ Diameter	.356	lf
20″ Diameter	.369	lf
Fiberglass, Aluminized Jacket, 1-1/2″ Blanket		
4″ Diameter	.047	lf
5″ Diameter	.053	lf
6″ Diameter	.062	lf
7″ Diameter	.073	lf
8″ Diameter	.089	lf
9″ Diameter	.100	lf
10″ Diameter	.114	lf
12″ Diameter	.160	lf
14″ Diameter	.200	lf
16″ Diameter	.267	lf
18″ Diameter	.356	lf
Rigid Fiberglass, Round, .003″ Foil		
Scrim Jacket		
4″ Diameter	.052	lf
5″ Diameter	.058	lf
6″ Diameter	.067	lf
7″ Diameter	.073	lf
8″ Diameter	.089	lf
9″ Diameter	.100	lf
10″ Diameter	.114	lf
12″ Diameter	.160	lf
14″ Diameter	.200	lf
16″ Diameter	.267	lf
18″ Diameter	.356	lf
20″ Diameter	.369	lf
22″ Diameter	.400	lf
24″ Diameter	.436	lf
26″ Diameter	.480	lf
28″ Diameter	.533	lf
30″ Diameter	.600	lf
Rectangular, 1″ Thick, Aluminum Faced, No		
Additional Insulation Required	.069	sf surf

Sheet Metal Calculator (Weight in Lb/Ft of Length)

Gauge	26	24	22	20	18	16
Wt-Lb/SF	.906	1.156	1.406	1.656	2.156	2.656
SMACNA Max. Dimension - Long Side		30"	54"	84"	85" Up	
Sum-2 Sides						
2	.3	.40	.50	.60	.80	.90
3	.5	.65	.80	.90	1.1	1.4
4	.7	.85	1.0	1.2	1.5	1.8
5	.8	1.1	1.3	1.5	1.9	2.3
6	1.0	1.3	1.5	1.7	2.3	2.7
7	1.2	1.5	1.8	2.0	2.7	3.2
8	1.3	1.7	2.0	2.3	3.0	3.6
9	1.5	1.9	2.3	2.6	3.4	4.1
10	1.7	2.2	2.5	2.9	3.8	4.5
11	1.8	2.4	2.8	3.2	4.2	5.0
12	2.0	2.6	3.0	3.5	4.6	5.4
13	2.2	2.8	3.3	3.8	4.9	5.9
14	2.3	3.0	3.5	4.1	5.3	6.3
15	2.5	3.2	3.8	4.4	5.7	6.8
16	2.7	3.4	4.0	4.6	6.1	7.2
17	2.8	3.7	4.3	4.9	6.5	7.7
18	3.0	3.9	4.5	5.2	6.8	8.1
19	3.2	4.1	4.8	5.5	7.2	8.6
20	3.3	4.3	5.0	5.8	7.6	9.0
21	3.5	4.5	5.3	6.1	8.0	9.5
22	3.7	4.7	5.5	6.4	8.4	9.9
23	3.8	5.0	5.8	6.7	8.7	10.4
24	4.0	5.2	6.0	7.0	9.1	10.8
25	4.2	5.4	6.3	7.3	9.5	11.3
26	4.3	5.6	6.5	7.5	9.9	11.7
27	4.5	5.8	6.8	7.8	10.3	12.2
28	4.7	6.0	7.0	8.1	10.6	12.6
29	4.8	6.2	7.3	8.4	11.0	13.1
30	5.0	6.5	7.5	8.7	11.4	13.5
31	5.2	6.7	7.8	9.0	11.8	14.0
32	5.3	6.9	8.0	9.3	12.2	14.4
33	5.5	7.1	8.3	9.6	12.5	14.9
34	5.7	7.3	8.5	9.9	12.9	15.3
35	5.8	7.5	8.8	10.2	13.3	15.8
36	6.0	7.8	9.0	10.4	13.7	16.2
37	6.2	8.0	9.3	10.7	14.1	16.7
38	6.3	8.2	9.5	11.0	14.4	17.1
39	6.5	8.4	9.8	11.3	14.8	17.6
40	6.7	8.6	10.0	11.6	15.2	18.0
41	6.8	8.8	10.3	11.9	15.6	18.5
42	7.0	9.0	10.5	12.2	16.0	18.9
43	7.2	9.2	10.8	12.5	16.3	19.4
44	7.3	9.5	11.0	12.8	16.7	19.8
45	7.5	9.7	11.3	13.1	17.1	20.3
46	7.7	9.9	11.5	13.3	17.5	20.7
47	7.8	10.1	11.8	13.6	17.9	21.2
48	8.0	10.3	12.0	13.9	18.2	21.6
49	8.2	10.5	12.3	14.2	18.6	22.1
50	8.3	10.7	12.5	14.5	19.0	22.5
51	8.5	11.0	12.8	14.8	19.4	23.0
52	8.7	11.2	13.0	15.1	19.8	23.4
53	8.8	11.4	13.3	15.4	20.1	23.9
54	9.0	11.6	13.5	15.7	20.5	24.3
55	9.2	11.8	13.8	16.0	20.9	24.8

Gauge	26	24	22	20	18	16
Wt-Lb/SF	.906	1.156	1.406	1.656	2.156	2.656
SMACNA Max. Dimension - Long Side		30"	54"	84"	85" Up	
Sum-2 Sides						
56	9.3	12.0	14.0	16.2	21.3	25.2
57	9.5	12.3	14.3	16.5	21.7	25.7
58	9.7	12.5	14.5	16.8	22.0	26.1
59	9.8	12.7	14.8	17.1	22.4	26.6
60	10.0	12.9	15.0	17.4	22.8	27.0
61	10.2	13.1	15.3	17.7	23.2	27.5
62	10.3	13.3	15.5	18.0	23.6	27.9
63	10.5	13.5	15.8	18.3	24.0	28.4
64	10.7	13.7	16.0	18.6	24.3	28.8
65	10.8	13.9	16.3	18.9	24.7	29.3
66	11.0	14.1	16.5	19.1	25.1	29.7
67	11.2	14.3	16.8	19.4	25.5	30.2
68	11.3	14.6	17.0	19.7	25.8	30.6
69	11.5	14.8	17.3	20.0	26.2	31.1
70	11.7	15.0	17.5	20.3	26.6	31.5
71	11.8	15.2	17.8	20.6	27.0	32.0
72	12.0	15.4	18.0	20.9	27.4	32.4
73	12.2	15.6	18.3	21.2	27.7	32.9
74	12.3	15.8	18.5	21.5	28.1	33.3
75	12.5	16.1	18.8	21.8	28.5	33.8
76	12.7	16.3	19.0	22.0	28.9	34.2
77	12.8	16.5	19.3	22.3	29.3	34.7
78	13.0	16.7	19.5	22.6	29.6	35.1
79	13.2	16.9	19.8	22.9	30.0	35.6
80	13.3	17.1	20.0	23.2	30.4	36.0
81	13.5	17.3	20.3	23.5	30.8	36.5
82	13.7	17.5	20.5	23.8	31.2	36.9
83	13.8	17.8	20.8	24.1	31.5	37.4
84	14.0	18.0	21.0	24.4	31.9	37.8
85	14.2	18.2	21.3	24.7	32.3	38.3
86	14.3	18.4	21.5	24.9	32.7	38.7
87	14.5	18.6	21.8	25.2	33.1	39.2
88	14.7	18.8	22.0	25.5	33.4	39.6
89	14.8	19.0	22.3	25.8	33.8	40.1
90	15.0	19.3	22.5	26.1	34.2	40.5
91	15.2	19.5	22.8	26.4	34.6	41.0
92	15.3	19.7	23.0	26.7	35.0	41.4
93	15.5	19.9	23.3	27.0	35.3	41.9
94	15.7	20.1	23.5	27.3	35.7	42.3
95	15.8	20.3	23.8	27.6	36.1	42.8
96	16.0	20.5	24.0	27.8	36.5	43.2
97	16.2	20.8	24.3	28.1	36.9	43.7
98	16.3	21.0	24.5	28.4	37.2	44.1
99	16.5	21.2	24.8	28.7	37.6	44.6
100	16.7	21.4	25.0	29.0	38.0	45.0
101	16.8	21.6	25.3	29.3	38.4	45.5
102	17.0	21.8	25.5	29.6	38.8	45.9
103	17.2	22.0	25.8	29.9	39.1	46.4
104	17.3	22.3	26.0	30.2	39.5	46.8
105	17.5	22.5	26.3	30.5	39.9	47.3
106	17.7	22.7	26.5	30.7	40.3	47.7
107	17.8	22.9	26.8	31.0	40.7	48.2
108	18.0	23.1	27.0	31.3	41.0	48.6
109	18.2	23.3	27.3	31.6	41.4	49.1
110	18.3	23.5	27.5	31.9	41.8	49.5

Example: If duct is 34" x 20" x 15' long, 34" is greater than 30" maximum, for 24 ga. so must be 22 ga. 34" + 20" = 54" going across from 54" find 13.5 lb per foot. 13.5 x 15' = 202.5 lbs. For sf of surface area

202.5 ÷ 1.406 = 144 sf.

Note: Figures include an allowance for scrap.

284

8 MECHANICAL
FANS

Fans in the building construction industry are used to supply, circulate, or exhaust air for human comfort, safety, and health reasons. Fans may be exposed in the area being served, or they may be in a remote location and connected by ductwork to the served area. Fans may also be located outdoors, either on the building roof or on side walls. In general, a fan consists of an electric motor and drive, blades, a wheel or propeller. All of these may be contained in an enclosure. The drive assembly may operate via belts and pulleys or may be directly connected to the motor. While the direct-drive type fan is less expensive, objectionable noise can result as the size or speed of the fan increases. Belt drive affords greater flexibility in speed and performance. Proper fan selection is important, not only because of noise, but also to avoid the feeling of air movement or drafts due to excess velocity. Fans are sized according to the cubic feet of air they can handle in one minute.

Fans are classified in two general groups, centrifugal and axial-flow. Centrifugal fans are further classified by the position of the blades on the fan wheel, either forward-curved or backward-curved. Axial-flow fans, where the air flows around the axis of the blade and through the impeller, are classified as propeller, vane-axial, and tube-axial.

Self-contained air handling or air conditioning units usually depend on a centrifugal fan due to the fan's adaptability to duct configurations and its quiet and efficient operation. Air filters are used in air supply systems to protect the heating or cooling coils from dust or other particles picked up by the air flow.

The air handling capacity or volume delivered by a fan may be varied by a motor speed control, outlet damper control, inlet vane control, or a fan drive change. The most efficient method is a variable-speed motor, but this is also the most costly type of fan control.

Man-hours

Description	m/hr	Unit
Fans		
Belt Drive, in Line Centrifugal		
3800 CFM	5.882	Ea.
6400 CFM	7.143	Ea.
10,500 CFM	8.333	Ea.
15,600 CFM	12.500	Ea.
23,000 CFM	28.571	Ea.
28,000 CFM	50.000	Ea.
Direct Drive Ceiling Fan		
95 CFM	1.000	Ea.
210 CFM	1.053	Ea.
385 CFM	1.111	Ea.
885 CFM	1.250	Ea.
1650 CFM	1.538	Ea.
2960 CFM	1.818	Ea.
Direct Drive Paddle Blade Fan		
36", 4000 CFM	3.333	Ea.
52", 7000 CFM	5.000	Ea.
Direct Drive Roof Fan		
420 CFM	2.857	Ea.
675 CFM	3.333	Ea.
770 CFM	4.000	Ea.
1870 CFM	4.762	Ea.
2150 CFM	5.000	Ea.
Belt Drive Roof Fan		
1660 CFM	3.333	Ea.
2830 CFM	4.000	Ea.
4600 CFM	5.000	Ea.
8750 CFM	6.667	Ea.
12,500 CFM	10.000	Ea.
21,600 CFM	20.000	Ea.
Direct Drive Utility Set		
150 CFM	3.125	Ea.
485 CFM	3.448	Ea.
1950 CFM	4.167	Ea.
2410 CFM	4.545	Ea.
3328 CFM	6.667	Ea.

Man-hours (cont.)

Description	m/hr	Unit
Belt Drive Utility Set		
800 CFM	3.333	Ea.
1300 CFM	4.000	Ea.
2000 CFM	4.348	Ea.
2900 CFM	4.762	Ea.
3600 CFM	5.000	Ea.
4800 CFM	5.714	Ea.
6700 CFM	6.667	Ea.
11,000 CFM	10.000	Ea.
13,000 CFM	12.500	Ea.
15,000 CFM	20.000	Ea.
17,000 CFM	25.000	Ea.
20,000 CFM	25.000	Ea.
Belt Drive Propeller Fan		
12", 1000 CFM	.571	Ea.
14", 1500 CFM	.667	Ea.
16", 2000 CFM	.889	Ea.
30", 4800 CFM	1.143	Ea.
36", 7000 CFM	1.333	Ea.
42", 10,000 CFM	1.600	Ea.
48", 16,000 CFM	2.000	Ea.
Belt Drive Airfoil Centrifugal		
12,420 CFM	7.273	Ea.
18,620 CFM	8.000	Ea.
27,580 CFM	8.889	Ea.
40,980 CFM	10.667	Ea.
60,920 CFM	16.000	Ea.
74,520 CFM	20.000	Ea.
90,160 CFM	22.857	Ea.
110,300 CFM	32.000	Ea.
134,960 CFM	40.000	Ea.

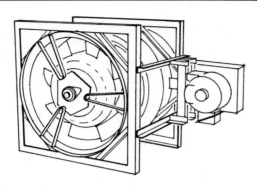

Axial Flow, Belt Drive, Centrifugal Fan

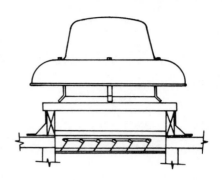

Centrifugal Roof Exhaust Fan

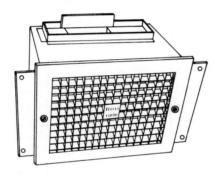

Ceiling Exhaust Fan

Belt Drive, Utility Set

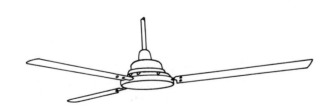

Paddle Blade Air Circulator

Belt Drive Propeller Fan with Shutter

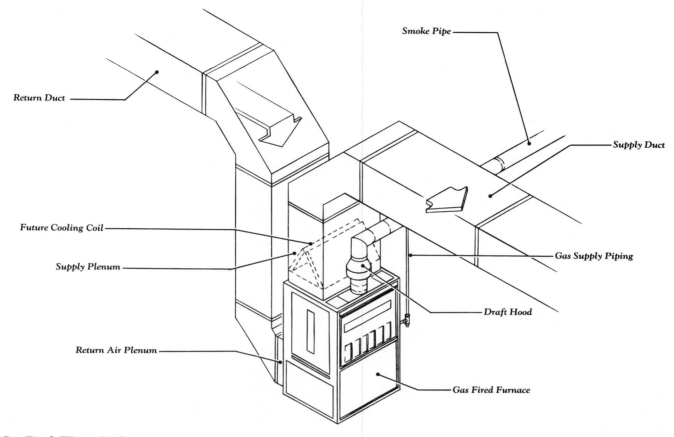

Return Duct

Smoke Pipe

Supply Duct

Future Cooling Coil

Supply Plenum

Gas Supply Piping

Return Air Plenum

Draft Hood

Gas Fired Furnace

Gas Fired, Warm Air System

Forced warm air heating systems are in use in many residential, commercial, and some industrial buildings. A built-in advantage of these systems is that in many cases the ductwork, with proper insulation, can be used for cooling also.

The heart of this type system is the warm air furnace, which is usually fired by gas or oil. The heated air is distributed by sheet metal or fiberglass ductwork to supply registers or diffusers, which direct the flow and amount of air to specific rooms or areas. In smaller systems, such as those used in individual residences, the cooler air is returned to the furnace from a centrally located, return air grille and a short run of duct back to the return plenum of the furnace or heating unit.

In some wood-frame residences, the return air duct consists of a piece of sheet metal or fiberglass board bridging two floor joists. The purpose is to achieve economy of cost. The floor joists become the sides of the duct, and the sub flooring acts as the top of the duct. In more sophisticated systems, a return air grille and branch duct is returned from each room or area to a main return duct, and then back to the heating unit. Dust filters, either cleanable or throw-away type, are built into the return air plenum. A desirable but more expensive option would be an electrostatic air cleaner. Another option would be a humidifier to maintain a constant level of humidity during the heating season. This option would add approximately 10% to the cost of the unit.

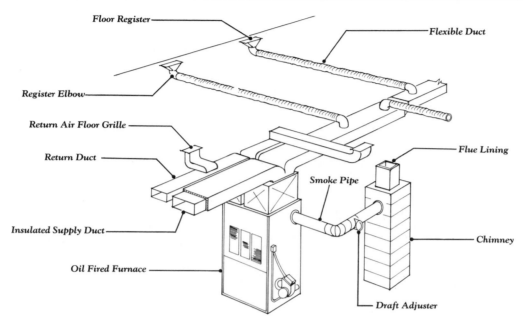

Oil Fired, Warm Air System

The same principles apply to large systems, except that rather than a furnace there would be a steam or hot-water source, such as a boiler, which would supply the heating medium through a network of pumps, piping, and controls to coils placed in the ductwork, or to fan coil unit in the ductwork system. A water chiller added to this arrangement, with the proper controls and appurtenances, is all that would be necessary for a complete air conditioning system (heating, cooling, cleaning, humidifying, or dehumidifying).

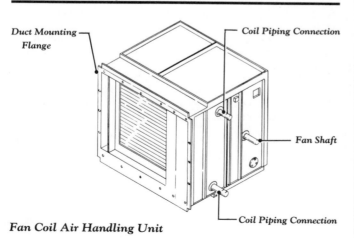

Fan Coil Air Handling Unit

Man-hours

Description	m/hr	Unit
Heating Only, Gas Fired Hot Air,		
One Zone, 1200 sf Building		
Furnace, Gas, Up Flow	4.710	Ea.
Intermittent Pilot	4.710	Ea.
Supply Duct, Rigid Fiberglass	.007	lf
Return Duct, Sheet Metal, Galvanized	.102	lb
Lateral Ducts, 6″ Flexible Fiberglass	.062	lf
Register, Elbows	.267	Ea.
Floor Registers, Enameled Steel	.250	Ea.
Floor Grille, Return Air	.364	Ea.
Thermostat	1.000	Ea.
Plenum	1.000	Ea.
Ductwork		
Fabricated Rectangular, Includes		
Fittings, Joists, Supports,		
Allowance for Flexible Connections,		
No Insulation		
Aluminum, Alloy 3003-H14,		
Under 300 lb	.320	lb
300 to 500 lb	.300	lb
500 to 1000 lb	.253	lb
1000 to 2000 lb	.200	lb
2000 to 10,000 lb	.185	lb
Over 10,000 lb	.166	lb
Galvanized Steel, Under 400 lb		
400 to 1000 lb	.094	lb
1000 to 2000 lb	.091	lb
2000 to 5000 lb	.087	lb
5000 to 10,000 lb	.084	lb
Over 10,000 lb	.080	lb

8 MECHANICAL
AIR HANDLING UNITS

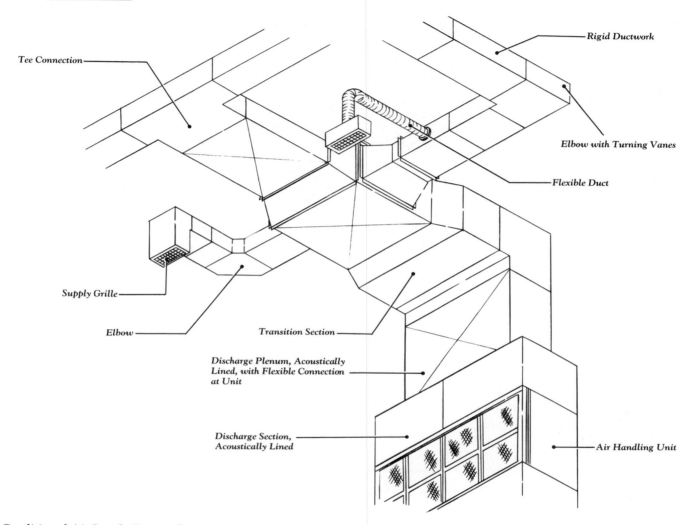

Tee Connection

Rigid Ductwork

Elbow with Turning Vanes

Flexible Duct

Supply Grille

Elbow

Transition Section

Discharge Plenum, Acoustically Lined, with Flexible Connection at Unit

Discharge Section, Acoustically Lined

Air Handling Unit

Conditioned Air Supply Ductwork

Air handling units, which consist basically of a fan and a coil within the same enclosure, are used to distribute clean, cooled, or heated air to a building. These units are available in a wide range of capacities, from 200 cubic feet per minute to tens of thousands of cubic feet per minute. The units also vary in complexity of design and versatility of operation. Small units tend to be relatively simple in their coil and filter arrangements, and modest in the size of their fan motors. Large, sophisticated units usually require remote placement. Because of the need to overcome losses caused by intake and supply ductwork, and by complex coil and filter configurations, fan-motor horsepower must be dramatically increased.

Determining the proper size, number, capacity, type, and configuration of coils in the unit is a prime consideration when selecting and/or designing an air handling unit. As a general rule, the amount of air (in cubic feet per minute) to be handled by the unit determines the size and number of the various coils. Electric or hydronic coils are used for heating; chilled water or direct expansion coils are used for cooling. As the units increase in size and complexity, the coil configurations and arrangements become limitless. A simple heating and cooling unit, for example, uses the same coil for either hot or chilled water; a large unit usually demands different types of coils to perform many separate functions. In humid conditions, the air temperature may be intentionally lowered to remove moisture, so a reheat coil is added to return the temperature to its desired level. Conversely, in dry conditions, a humidifier component is built into the unit. If outside air is introduced to the unit at subfreezing temperatures, then a preheat coil is placed in the outside air intake duct.

289

Certain precautions should be taken to prevent damage to and to assure the efficiency of the unit's coils and other components. To protect the coil surfaces from accumulating dust and other airborne impurities, a filter section is a necessary addition to the unit. If a unit is designed to cool air, then a drain pan must be included beneath the cooling coil section. This pan is then piped to an indirect drain to draw off the unwanted condensation. Another important consideration for the protection of the unit is to insulate the fan coil casing internally. If this precaution is not taken to protect the cooling coil section and all other sections "down-stream", corrosive or rust-causing condensation will damage the casing and discharge ductwork. Insulating this casing also helps to deaden the noise of the fan. Noise may also be controlled by installing flexible connections between the unit and its ductwork and by mounting the unit on vibration-absorbing bases, if it is located on the floor, or by suspending the unit from vibration-absorbing hangers, if it is secured to the ceiling or wall.

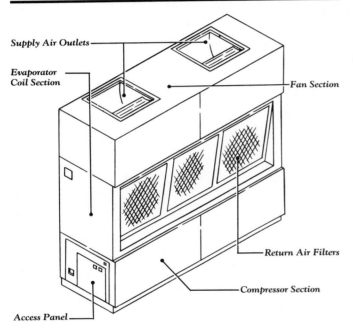

Packaged Vertical Fan Coil Air Handling Unit

Small air handling units may be located and mounted in a variety of settings and by different methods. They may be mounted on the floor or hung from walls or ceilings with no discharge ductwork required in the room they service. As the units increase in size, space restrictions and noise considerations mandate remote locations in corridors or equipment rooms. Sheet metal or fiberglass ductwork is then used to distribute the conditioned air to the desired locations. Several separate areas or zones may be serviced by a single unit, as a series of dampers and controls conduct the desired amount of conditioned air to any or all of the individual zones. Units that perform this function are called multi-zone units.

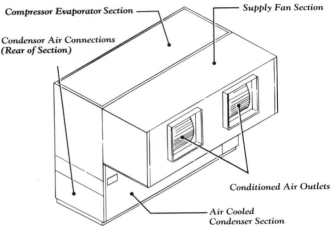

Fan Coil Direct Expansion Cooling Unit

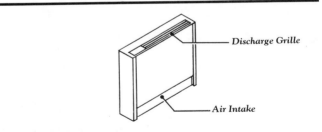

Room Size Fan Coil Unit

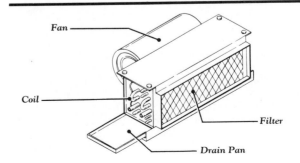

Concealed Fan Coil Element

Economical and efficient operation of air handling units can be assured if the following suggestions are applied: Plan the system so as to limit within the requirements of building codes, the amount of outside air drawn into the system, especially during very warm (for cooling units) and very cold (for heating units) times of year. Locate the unit as close as possible to the areas being served, in order to reduce the amount of ductwork.

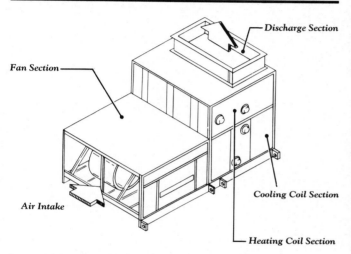

Central Station Air Handling Unit

Man-hours		
Description	**m/hr**	**Unit**
Fan Coil Unit, Free Standing		
Finished Cabinet, 3 Row Cooling		
or Heating Coil, Filter		
200 CFM	2.000	Ea.
400 CFM	2.667	Ea.
600 CFM	2.909	Ea.
1000 CFM	3.200	Ea.
1200 CFM	4.000	Ea.
4000 CFM	8.571	Ea.
6000 CFM	16.000	Ea.
8000 CFM	30.000	Ea.
12,000 CFM	40.000	Ea.
Direct Expansion Cooling		
Coil, Filter		
2000 CFM	5.333	Ea.
3000 CFM	5.333	Ea.
4000 CFM	9.231	Ea.
8000 CFM	34.286	Ea.
12,000 CFM	40.000	Ea.
16,000 CFM	53.333	Ea.
20,000 CFM	63.158	Ea.
Central Station Unit, Factory		
Assembled, Modular, 4, 6 or		
8 Row Coils, Filter and Mixing Box		
1500 CFM	13.333	Ea.
2200 CFM	14.545	Ea.
3800 CFM	20.000	Ea.
5400 CFM	30.000	Ea.
8000 CFM	40.000	Ea.
12,100 CFM	52.174	Ea.
18,400 CFM	72.727	Ea.
22,300 CFM	82.759	Ea.
33,700 CFM	126.316	Ea.
52,500 CFM	200.000	Ea.
63,000 CFM	246.154	Ea.

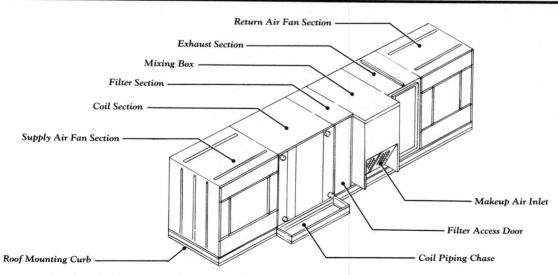

Central Station Air Handling Unit for Rooftop Location

8 MECHANICAL
ROOF VENTILATORS

Roof-mounted ventilators are designed to remove air from a building without the use of motor-driven fans. In some cases, this process is achieved by a rising of warm air and its displacement by denser or heavier cold air. Some ventilators, however, use the action of the wind to syphon air through the ventilator. Relief hoods may be used for exhaust air, as well as for makeup air intakes, through the use of dampers or self-acting shutters. Hoods and ventilators are usually constructed of galvanized steel or aluminum.

Gravity roof ventilators, not being too efficient, are used in situations where rapid removal of stale air is not a factor, motor-driven roof fans cost two to three times as much as a gravity ventilator.

Man-hours

Description	m/hr	Unit
Rotary Syphons		
6" Diameter 185 CFM	1.000	Ea.
8" Diameter 215 CFM	1.143	Ea.
10" Diameter 260 CFM	1.333	Ea.
12" Diameter 310 CFM	1.600	Ea.
14" Diameter 500 CFM	1.600	Ea.
16" Diameter 635 CFM	1.778	Ea.
18" Diameter 835 CFM	1.778	Ea.
20" Diameter 1080 CFM	2.000	Ea.
24" Diameter 1530 CFM	2.000	Ea.
30" Diameter 2500 CFM	2.286	Ea.
36" Diameter 3800 CFM	2.667	Ea.
42" Diameter 4500 CFM	4.000	Ea.
Spinner Ventilators		
4" Diameter 180 CFM	.800	Ea.
5" Diameter 210 CFM	.889	Ea.
6" Diameter 250 CFM	1.000	Ea.
8" Diameter 360 CFM	1.143	Ea.
10" Diameter 540 CFM	1.333	Ea.
12" Diameter 770 CFM	1.600	Ea.
14" Diameter 830 CFM	1.600	Ea.
16" Diameter 1200 CFM	1.778	Ea.
18" Diameter 1700 CFM	1.778	Ea.
20" Diameter 2100 CFM	2.000	Ea.
24" Diameter 3100 CFM	2.000	Ea.
30" Diameter 4500 CFM	2.286	Ea.
36" Diameter 5500 CFM	2.667	Ea.
Stationary Gravity Syphons		
3" Diameter 40 CFM	.667	Ea.
4" Diameter 50 CFM	.800	Ea.
5" Diameter 58 CFM	.889	Ea.
6" Diameter 66 CFM	1.000	Ea.
7" Diameter 86 CFM	1.067	Ea.
8" Diameter 110 CFM	1.143	Ea.
10" Diameter 140 CFM	1.333	Ea.
12" Diameter 160 CFM	1.600	Ea.
14" Diameter 250 CFM	1.600	Ea.
16" Diameter 380 CFM	1.778	Ea.
18" Diameter 500 CFM	1.778	Ea.
20" Diameter 625 CFM	2.000	Ea.
24" Diameter 900 CFM	2.000	Ea.
30" Diameter 1375 CFM	2.286	Ea.
36" Diameter 2000 CFM	2.667	Ea.
42" Diameter 3000 CFM	4.000	Ea.

Rotary Syphon

Spinner Ventilator

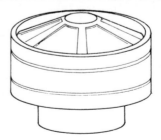

Stationary Gravity Syphon

Man-hours

Description	m/hr	Unit
Rotating Chimney Caps		
4″ Diameter	.800	Ea.
5″ Diameter	.889	Ea.
6″ Diameter	1.000	Ea.
7″ Diameter	1.067	Ea.
8″ Diameter	1.143	Ea.
10″ Diameter	1.333	Ea.
Relief Hoods, Intake/Exhaust		
500 CFM 12″ x 16″	2.000	Ea.
750 CFM 12″ x 20″	2.222	Ea.
1000 CFM 12″ x 24″	2.424	Ea.
1500 CFM 12″ x 36″	2.759	Ea.
3000 CFM 20″ x 42″	4.000	Ea.
6000 CFM 20″ x 84″	6.154	Ea.
8000 CFM 24″ x 96″	6.957	Ea.
10,000 CFM 48″ x 60″	8.889	Ea.
12,500 CFM 48″ x 72″	10.000	Ea.
15,000 CFM 48″ x 96″	12.308	Ea.
20,000 CFM 48″ x 120″	13.333	Ea.
25,000 CFM 60″ x 120″	17.778	Ea.
30,000 CFM 72″ x 120″	22.857	Ea.
40,000 CFM 96″ x 120″	26.667	Ea.
50,000 CFM 96″ x 144″	32.000	Ea.

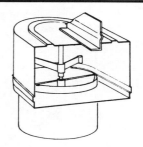

Rotating Chimney Cap

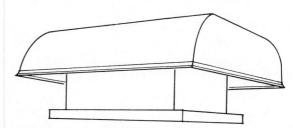

Relief Hood, Intake/Exhaust

8 MECHANICAL
PACKAGED ROOFTOP AIR CONDITIONER

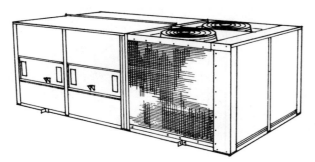

Packaged Rooftop Air Conditioner

Man-hours

Description	m/hr	Unit
Single Zone, Electric Cool, Gas Heat		
2 Ton Cooling, 60 M Btu/hr Heating	10.667	Ea.
4 Ton Cooling, 95 M Btu/hr Heating	14.545	Ea.
5 Ton Cooling, 112 M Btu/hr Heating	28.571	Ea.
10 Ton Cooling, 200 M Btu/hr Heating	52.174	Ea.
15 Ton Cooling, 270 M Btu/hr Heating	77.419	Ea.
20 Ton Cooling, 360 M Btu/hr Heating	100.000	Ea.
30 Ton Cooling, 540 M Btu/hr Heating	145.455	Ea.
40 Ton Cooling, 675 M Btu/hr Heating	200.000	Ea.
50 Ton Cooling, 810 M Btu/hr Heating	246.154	Ea.
Multi-Zone, Electric Cool, Gas Heat, Economizer		
15 Ton Cooling, 360 M Btu/hr Heating	145.455	Ea.
20 Ton Cooling, 360 M Btu/hr Heating	152.381	Ea.
25 Ton Cooling, 450 M Btu/hr Heating	177.778	Ea.
28 Ton Cooling, 450 M Btu/hr Heating	200.000	Ea.
30 Ton Cooling, 540 M Btu/hr Heating	213.333	Ea.
37 Ton Cooling, 540 M Btu/hr Heating	246.154	Ea.
70 Ton Cooling, 1500 M Btu/hr Heating	355.556	Ea.
80 Ton Cooling, 1500 M Btu/hr Heating	400.000	Ea.
90 Ton Cooling, 1500 M Btu/hr Heating	457.143	Ea.
105 Ton Cooling, 1500 M Btu/hr Heating	533.333	Ea.

A packaged rooftop air conditioner is a self-contained air handling unit shipped to the job site completely assembled. The unit has been assembled, wired, piped, charged with refrigerant, tested, and operated before leaving the factory.

The only field connections required are electrical hook-ups for power and control (sometimes for resistance heating), piping for gas heat and condensate drain, and ductwork for supply and return air.

The most common of these units use electricity to cool and gas to heat. Less popular systems use electric heat and, rarely, oil heat.

Single zone units can use two methods of air distribution. One is conventional supply and return ductwork to diffusers in the conditioned space; the other is a combination supply and return diffuser mounted directly under the unit. Strategic location of these units gives the cost advantage of minimum ductwork and insulation. An economizer cycle, which permits cooling via outside air with no mechanical cooling required if the outside air temperature is lower than the building temperature, is available as an optional accessory.

A factory-prefabricated roof-mounting curb or frame is available from most unit manufacturers. This curb should be shipped to the site separately in time to be built into the roof structure. The curb not only ensures a weathertight fit and structural integrity for the chosen unit, but also allows completion of duct installation before the unit is set in place. Units are set directly in place either by crane or, in some cases, by helicopter; either method involves considerable expense.

Low installation and operating costs, ease of maintenance, and the unit's flexibility make the packaged rooftop air conditioner a desirable and economical method of heating and cooling low-rise buildings.

These units range in size from less than 1 ton to over 100 tons cooling with a similar range for heating. Multizone rooftop units are also available for more complex and confined areas.

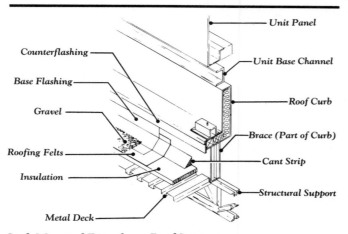

Curb Mounted Directly on Roof Supports

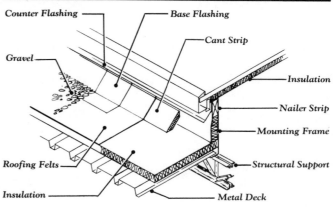

Flashing for Roof Mounting Frame

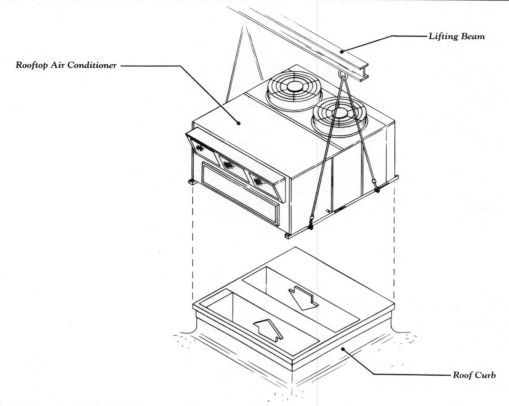

Lifting Beam

Rooftop Air Conditioner

Roof Curb

Installation of Rooftop Air Conditioner

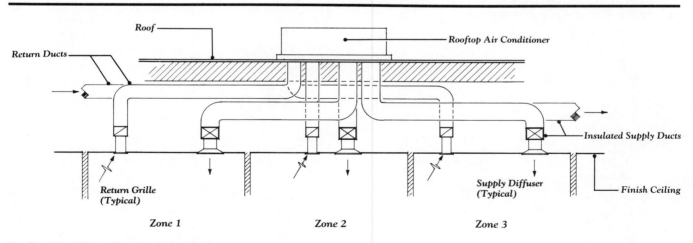

Roof

Return Ducts

Rooftop Air Conditioner

Insulated Supply Ducts

Return Grille (Typical)

Supply Diffuser (Typical)

Finish Ceiling

Zone 1   Zone 2   Zone 3

Rooftop Multizone Air Conditioning System

The central component of any chilled-water air conditioning system is the water chiller. Packaged water chillers are available in three basic designs: the reciprocating compressor, direct-expansion type; the centrifugal compressor, direct-expansion type; and the absorption type. Chillers and other air conditioning equipment are sized by the ton; that is, a ton of cooling equals the melting rate of one ton of ice in a 24-hour period, or 12,000 (British Thermal Units per Hour). The three types of chillers vary significantly in their cooling power, as well as in their operations. The reciprocating compressor chiller, which generates cooling capacities in the range of 10 to 200 tons, is usually powered by an electric motor. The centrifugal compressor type, which generates cooling capacities ranging from 100 to several thousand tons, is also commonly powered by an electric motor, but it may be designed for a steam-turbine drive as well. In some instances both of these types of chillers may be powered by internal combustion engines.

The absorption-type chiller provides cooling capacities ranging from 3 to 1600 tons. Because it uses water as a refrigerant and lithium bromide or other salts as an absorbant, this system consumes about 10% of the electrical power required to operate the conventional reciprocating and centrifugal direct-expansion chillers. This advantage is particularly desirable in buildings where an electrical power failure triggers an emergency backup system, such as in a hospital, a data processing center, electronic switching systems location, and other buildings that must continue to function on auxiliary power. Absorption-type chillers are also economically advantageous in areas where electric power is scarce or costly, where gas rates are low, or where waste or process steam or hot water is available during the cooling season. Solar power may also be used in some areas to generate the heat required for the absorption process.

The chillers themselves may be air or water cooled. Very small chillers are available with an air-cooled condenser built into the package. For larger systems and for systems that may create too much noise for their location, the air-cooled condenser is installed at a distance from the chiller, and the two units are connected with refrigerant piping. The condenser for water cooled chillers is piped into a remote water source, such as a cooling tower, pond, or river, which is called the "condenser water system". Completely packaged systems include built-in chilled water pumps, condenser water pumps, and all piping, wiring, and controls. All of the components of the completely packaged unit are factory installed and tested prior to shipping to the installation site.

Man-hours

Description	m/hr	Unit
Packaged Water Chillers		
Centrifugal, Water Cooled, Hermetic		
200 Ton	80.000	Ea.
400 Ton	96.000	Ea.
1000 Ton	174.000	Ea.
1300 Ton	294.000	Ea.
1500 Ton	324.000	Ea.
Reciprocating, Air Cooled		
20 Ton	91.429	Ea.
40 Ton	133.333	Ea.
65 Ton	228.571	Ea.
100 Ton	320.000	Ea.
110 Ton	355.556	Ea.
125 Ton	400.000	Ea.
Water Cooled Multiple Compressor		
Semi Hermetic		
15 Ton	80.000	Ea.
20 Ton	88.889	Ea.
25 Ton	100.000	Ea.
30 Ton	118.519	Ea.
40 Ton	133.333	Ea.
50 Ton	152.381	Ea.
60 Ton	168.421	Ea.
80 Ton	213.333	Ea.
100 Ton	266.667	Ea.
120 Ton	320.000	Ea.
140 Ton	355.556	Ea.
Air Cooled for Remote Condenser		
40 Ton	118.519	Ea.
60 Ton	152.381	Ea.
80 Ton	200.000	Ea.
100 Ton	290.909	Ea.
120 Ton	320.000	Ea.
160 Ton	355.556	Ea.
Absorption		
Gas Fired, Air Cooled		
5 Ton	26.667	Ea.
10 Ton	40.000	Ea.
Steam of Hot Water, Water Cooled		
100 Ton	108.000	Ea.
600 Ton	126.000	Ea.
1100 Ton	165.000	Ea.
1500 Ton	210.000	Ea.

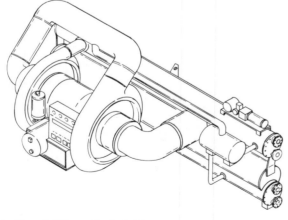

Centrifugal, Water Cooled, Hermetic

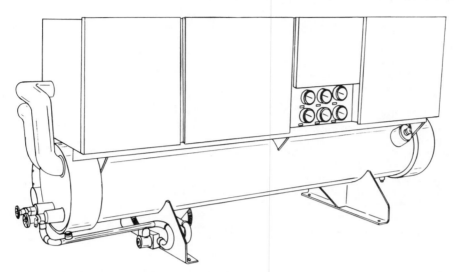

Reciprocating, Water Cooled Multiple Compressor, Semi Hermetic

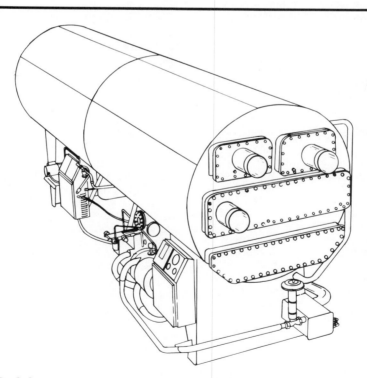

Absorption, Gas Fired, Air Cooled

8 MECHANICAL
COOLING TOWERS

For air conditioning systems that use refrigerants as the cooling medium, the basic process is to cool and condense back to liquid form the refrigerant gas that has become heat ladened during the evaporation stage of the cycle. This condensing process is achieved by cooling the gas with air, water, or a combination of both. Air-cooled condensers cool the refrigerant by blowing air directly across the refrigerant coil; evaporative condensers use the same method of cooling, with the addition of a spray of water over the coil to expedite the process.

When water is used as the cooling medium and is abundant enough not to require recycling, it may be piped to a drain after performing its cooling function and returned to its source, such as a river, pond, or the ocean. Where the supply of water is limited, expensive, or regulated by environmental restrictions, a water conserving or recycling system must be employed. Several types of systems may be installed to alleviate these limitations. For example, a water regulating valve, a spray pond, a natural draft cooling tower, or a mechanical draft cooling tower may be used.

In very small cooling systems a temperature-controlled, water-regulating valve may be used, provided that such a system is permitted by local environmental and/or building codes. The regulating valve system functions by allowing cooling water to flow when the condenser temperature rises, and, conversely, by stopping the flow as the temperature falls. The problem with this system, however, is that the heated condenser water cannot be recycled, and is therefore wasted during the flow cycle.

The spray pond and natural draft cooling tower systems, although they are viable and available methods of cooling, are not commonly used for building air conditioning. Because of water loss caused by excessive drift and the large amount of space required for their installation and operation, they are less desirable than the mechanical draft cooling tower method.

Mechanical draft cooling tower systems are classified in two basic designs: induced draft and forced draft. In an induced draft tower, a fan which is positioned at the top of the structure draws air upwards through the tower as the warm condenser water spills down. A cross-flow induced draft tower operates on the same principle, except that the air is drawn horizontally through the spill area from one side and then is discharged through a fan located on the opposite side. In a forced draft tower, the fan is located at the bottom or the side of the structure. Air is forced by the fan into the water spill area, through the water, and then discharged at the top. All designs of mechanical draft towers are rated in tons of refrigeration, usually 3 gallons of condenser water per minute per ton is an approximate tower sizing method.

After the water has been cooled in the tower, it passes through a heat exchanger, or condenser, in the refrigeration unit. Here, it again picks up heat and is pumped back to the cooling tower. The piping system is called the condenser water system.

The actual process of cooling within the mechanical draft cooling tower takes place when air is moved across or counter to a falling stream of water that falls through a

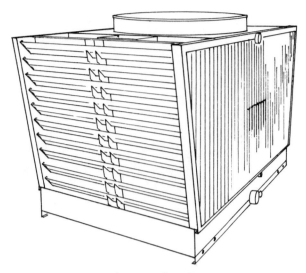

Induced Air, Double Flow, Cooling Tower

system of baffles, or "fill," to the tower basin. After the cooled water reaches the basin, it is piped back to the condenser. Some of the droplets created by the fill are carried away by the moving air as "drift," and some of the droplets evaporate. This limited loss of water is to be expected as part of the operation of the tower system. Because of the loss of water by drift, evaporation, and bleed off, replenishment water must be added to the tower basin to maintain a predetermined level in order to assure continuous operation of the system. To prevent scale buildup, algae, bacterial growth, or corrosion, tower water should be treated with chemicals or ozone applications.

The materials used in constructing mechanical draft cooling towers include redwood, which is the most commonly employed material, other treated woods, asbestos, various metals, plastics, concrete, or ceramic materials. The fill, which is the most important element in the tower's operation, may be manufactured from the same wide variety of materials used in the tower structure. Factory assembled, prepackaged towers are available and are usually preferable to built-in-place units, with multiple tower installations now being used for large systems.

The location of cooling towers is an important consideration for both practical and aesthetic reasons. They may be located outside of the building, on its roof, or on the ground. If space permits, they may be installed indoors by substituting centrifugal fans for the conventional noisy propeller type and by adding air intake and exhaust ductwork. Care should be taken to avoid directing the tower discharge into the prevailing wind, or into doors, windows, and building air intakes. In general, common sense should be used when determining tower placement so that the noise, heat, and humidity the system creates do not interfere with the building's operation and comfort. The manufacturer's guidelines for installation should be strictly followed, especially those sections that address clearances for maximum air flow, maintenance, and future unit replacement.

Economical operation of a cooling tower system may be achieved through effective control and management of several critical aspects of its operation, including: careful monitoring of water treatment, selecting and maintaining the most efficient condensing temperature, and controlling water temperature with fan cycling. A recent important development in water system operation allows the tower to substitute for the water chiller under certain favorable climatic conditions. This new method cannot be implemented in all cases. During periods when it can be employed, substantial savings result in the reduced cost of chiller operation.

Man-hours

Description	m/hr	Unit
Labor to Set in Place - Rigging not Included		
60 Ton Single Flow	8	Ea.
90 Ton Single Flow	16	Ea.
100 Ton Single Flow	16	Ea.
125 Ton Double Flow	24	Ea.
150 Ton Double Flow	30	Ea.
300 Ton Double Flow	40	Ea.
600 Ton Double Flow	80	Ea.
840 Ton Double Flow	136	Ea.
1000 Ton Double Flow	152	Ea.

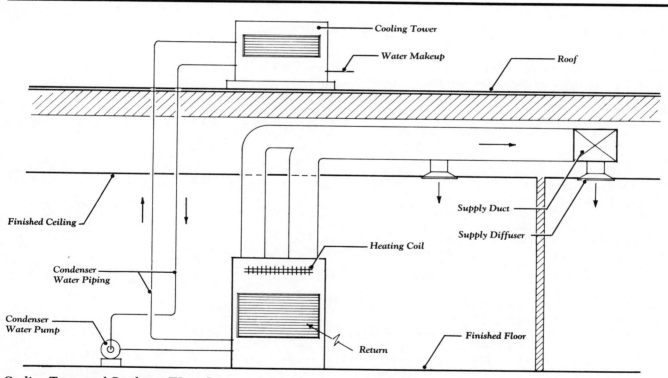

Cooling Tower and Condenser Water System

8 MECHANICAL
COMPUTER ROOM COOLING

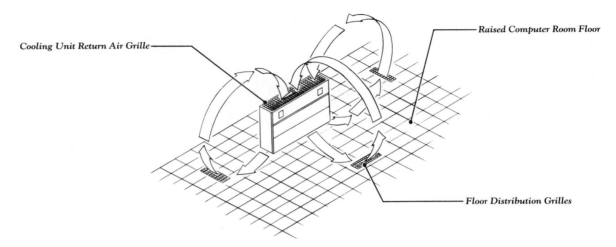

Cooling Unit Return Air Grille

Raised Computer Room Floor

Floor Distribution Grilles

Self-Contained Computer Room Cooling Unit - Under Floor Distribution

Computer rooms impose special needs on air conditioning systems. A prime requirement is reliability, because of the potential monetary loss that could be incurred by a system failure. A second basic requirement is the tolerance of control with which temperature and humidity are regulated, and dust eliminated. As the air conditioning system reliability is so vital, the additional cost of reserve capacity and redundant components is often justified.

Computer areas may be environmentally controlled by one of three methods, as follows:

1. **Self-Contained Units**
   These are units built to higher standards of performance and reliability. They usually contain alarms and controls to indicate component operation failure, filter change, etc. It should be remembered that these units in the room will occupy space that is relatively expensive to build, and that all alterations and service of the equipment will also have to be accomplished within the computer area.

2. **Decentralized Air Handling Units**
   In operation these are similar to the self-contained units except that their cooling capability comes from remotely located refrigeration equipment as refrigerant or chilled water. As no compressors or refrigerating equipment are required in the air units, they are smaller and require less service than self-contained units. An added plus for this type of system occurs if some of the computer components themselves also require chilled water for cooling.

3. **Central System Supply**
   Cooling is obtained from a central source which, since it is not located within the computer room, may have excess capacity and permit greater flexibility without interfering with the computer components. System performance criteria must still be met.

Note: The man-hours shown do not include ductwork or piping

Man-hours

Description	m/hr	Unit
Computer Room Unit, Air Cooled, Includes Remote Condenser		
3 Ton	32.000	Ea.
5 Ton	35.556	Ea.
8 Ton	59.259	Ea.
10 Ton	64.000	Ea.
15 Ton	72.727	Ea.
20 Ton	82.759	Ea.
23 Ton	85.714	Ea.
Chilled Water, for Connection to Existing Chiller System		
5 Ton	21.622	Ea.
8 Ton	32.000	Ea.
10 Ton	32.653	Ea.
15 Ton	33.333	Ea.
20 Ton	34.783	Ea.
23 Ton	38.095	Ea.
Glycol System, Complete Except for Interconnecting Tubing		
3 Ton	40.000	Ea.
5 Ton	42.105	Ea.
8 Ton	69.565	Ea.
10 Ton	76.190	Ea.
15 Ton	92.308	Ea.
20 Ton	100.000	Ea.
23 Ton	109.091	Ea.
Water Cooled, Not Including Condenser Water Supply or Cooling Tower		
3 Ton	25.806	Ea.
5 Ton	29.630	Ea.
8 Ton	44.485	Ea.
15 Ton	59.259	Ea.
20 Ton	63.158	Ea.
23 Ton	70.588	Ea.

DIVISION 9

ELECTRICAL

9 ELECTRICAL
POLE LINE DISTRIBUTION

The most common method of distributing electrical power is the pole line system. Generally, pole line systems are comprised of one of three types of cable, each of which is supported differently. The first is bare or weatherproof insulated cable supported by insulators which are attached to crossarms at the highest level of sectioned poles. The second is weatherproof insulated wire mounted on racks located on the sides of poles. The third is insulated aerial cable supported either by means of clamps bolted to the sides of poles or crossarms or by a messenger cable. Pole line distrbution systems consist of three fundamental components: the pole, an arrangement of cable supports and insulators, and the line cable.

Four types of poles are commonly used, depending on the amount of weight and size of the line wire to be supported. Wood poles, ranging from 16' to 90' in length, are made from northern white cedar, western red cedar, creosoted southern pine, and chestnut and are used for light cable applications. The American Standards Institute has standardized specifications and sizes of wood poles, which are classified according to circumference at the top of the pole and at the 6' point. Steel poles are used for medium voltage transmission cables which span from 250' to 350' between poles. Steel poles are classified in three types: latticed, expanded truss, and tubular. Fabricated steel towers are used to transmit high voltage lines which span over 350'. Reinforced concrete poles are the most durable type; but because they are expensive to manufacture and install, cost limitations prevent extensive use in routine transmission situations.

The size and spacing of poles depend on local conditions and municipal ordinances. In urban areas, 35' poles are commonly used when one or two crossarms must be supported; and 40' poles, when four or more crossarms are involved. In suburban areas, 30' poles are commonly employed for all transmission needs. Poles are normally spaced at intervals of 125' to 150' in urban and suburban locations, and spans of 200' to 250' are often acceptable in rural areas. Spans of up to 500' are not uncommon for high voltage lines supported by steel poles and steel towers. The length of the span decreases when curves and changes in direction of the transmission line occur.

The type of cable supports and insulators which are attached to the pole vary according to the type of pole and the size of the cable. Wood crossarms, made from yellow pine, Norway pine, or Douglas fir, are generally installed to support light cable on wood poles. Galvanized steel crossarms cost more than wood crossarms, but can carry much heavier loads and higher voltage lines. Galvanized steel crossarms range from 28" to 116" in length for power transmission cables and 20" to 100" in length for communication lines. Several types of insulators are used to attach transmission and guy wires from the pole structure. Types of insulators include pin-type, suspension-type, spool, strain, and cable-holder models.

The cable itself varies in size and type, according to the amount of power transmitted and other variables. The smallest size used for line wire is #6 AWG. Solid wire is used for sizes up to #2/0, and stranded wire for larger conductors. Triple-braid weatherproof covering is the standard insulation for aerial cable. Annealed, or soft-drawn, wire is normally used in all installations because of its ease of handling and high tensile strength.

Two methods are used to string the wire for pole line distribution systems. One method is to draw the wire over the crossarms by means of a rope attached to the lead end of the wire which is then payed off a stationary reel located at one end of the line. Up to 2000' of wire can be strung in one run with this method. Another method is to place the reel of wire on a cart and hoist the wire onto the crossarm as each span is completed. Guy wires must be used to stabilize poles which are pulled from their normal position by the tension of transmission wires. A guy assembly normally consists of the guy wire itself and one or two strain insulators. The assembly is secured to the pole at one end and to the lower part of another pole or to a ground anchor and anchor rod at the other end.

Man-hours

Description	m/hr	Unit
Poles, Wood, Creosoted 20' High	7.800	Ea.
25' High	8.300	Ea.
30' High	9.300	Ea.
35' High	10.000	Ea.
40' High	10.500	Ea.
45' High	14.100	Ea.
Crossarms with 4' Long	3.200	Ea.
5' Long	3.330	Ea.
6' Long	3.640	Ea.
Steel, Galvanized 20' High	9.200	Ea.
30' High	10.500	Ea.
35' High	10.900	Ea.
40' High	14.100	Ea.

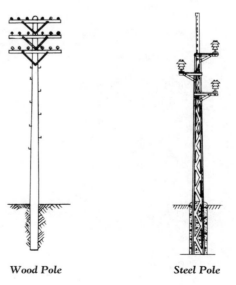

Wood Pole   *Steel Pole*

Pole Types

A manhole is a subterranean vault usually constructed of concrete, brick, or a combination of these materials, and used for the support, splicing, and general manipulation of cables in underground transmission lines. Brackets located on the sides of the vault support the cables which enter the manhole on one side and exit on the other.

Manholes are shaped and sized according to design needs and local conditions, but generally they take the form of an ellipse to provide a curved sidewall contour. This shape allows for easier manipulation of the cables and eliminates sharp bends within the line. If a rectangular-shaped manhole is called for, extreme care should be taken when forming the bends in the line to assure protection of the cable's covering. The size of a manhole varies according to the number of cables to be accommodated and the work area needed to manipulate them. The smallest manholes measure 3' by 4', and 5' by 7' units are large enough to accommodate most cable splicing and redirectioning operations and support systems. When transformers are to be located within the manhole, the volume of the vault should be increased over its normal size at a rate of 3 cubic feet per kilovolt-ampere of transformer rating.

Concrete manholes consist of a poured concrete floor and formed sidewalls of approximately 8" thickness. The exterior of the sidewall may be formed by the perimeter of the excavation if the soil is hard enough. Steel reinforcing is placed in the concrete top of large manholes, but it may not be required in smaller ones. The concrete is placed by direct chute for both the sidewalls and top. Precast manhole units are also available and have grown in popularity in recent years. Brick manholes are constructed of 12" thick brick sidewalls erected on a poured concrete

floor pad. The roof of large brick manholes is constructed of steel-reinforced concrete. For small brick manholes, the cast steel lid usually doubles as both roof and cover.

Both concrete and brick manholes normally include heads made of cast steel or cast iron and a cover of cast steel. The cover should be ventilated to provide means of escape of any gases which may accumulate in the vault. Because it is impractical to make manholes watertight, provisions for drainage of surface water and condensation should be made. A sewer connection at the base of the vault is the best way of assuring proper drainage. When a sewer connection is impossible or impractical, a hole should be made in the concrete floor and a bed of crushed stone or gravel placed to provide drainage.

The methods of installation and securing of the cables within the manhole vary according to distance between manholes, cable size, and ductbank arrangement. In general, cables are not pulled in lengths that exceed 700' because the mechanical strain on the conductors and splices becomes too great during the cable pulling operation. The recommended distance between manholes is 500', but this measurement is reduced when curves and other changes of direction occur within the distribution line. Within the manhole, the cables are supported by metal cable racks which are available in many sizes and configurations. Eyebolts should be set in the sidewalls to provide means of attachment during the cable pulling process. Incoming ductbank arrangements should be carefully planned to minimize cable crossing and recrossing. Whenever possible, cables should enter and exit the vault at the same relative position.

Concrete Manhole

Manhole Cover — Manhole Head — Conduit Entrance — Cable Rack — Pulling Eye — To Drainage System

Man-hours

Description	m/hr	Unit
Manholes, Brick 4' Inside Diameter		
4' Deep	16.000	Ea.
6' Deep	23.000	Ea.
8' Deep	32.000	Ea.
For Depths Over 8', Add per Foot	4.000	vlf
Concrete, Cast in Place 8" Thick		
4' Deep	12.000	Ea.
6' Deep	16.000	Ea.
8' Deep	24.000	Ea.
For Depths Over 8', Add per Foot	3.000	vlf
Precast, 4' Inside Diameter		
4' Deep	6.000	Ea.
6' Deep	8.000	Ea.
8' Deep	12.000	Ea.
Slab Tops Precast, 8" Thick		
4' Diameter Manhole	3.000	Ea.
5' Diameter Manhole	3.200	Ea.
6' Diameter Manhole	3.400	Ea.
Frames and Covers	3.000	Ea.
Steps, Cast Iron	.200	Ea.
Galvanized Steel	.200	Ea.

9 ELECTRICAL
RESIDENTIAL SERVICE — UNDERGROUND AND OVERHEAD

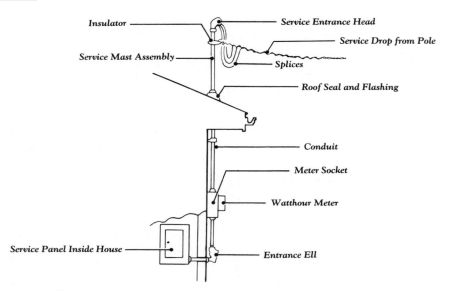

Insulator

Service Entrance Head

Service Drop from Pole

Service Mast Assembly

Splices

Roof Seal and Flashing

Conduit

Meter Socket

Watthour Meter

Service Panel Inside House

Entrance Ell

Through Roof Service Installation

A residential service includes all of the materials and equipment necessary to deliver electrical power from the utility supply lines to the distribution system within the residence. The electrical service may be brought to the residence by overhead or underground supply lines. Residential service equipment consists of four basic components: a service drop conductor (for overhead service) or a service lateral conductor (for underground service); metering equipment; overcurrent protection; and a distribution panel.

The service entrance, which connects the residence directly to the utility supply lines, is run either overhead or underground. In overhead services, the connecting cable runs from the utility pole to a bracket installed on the building. This aerial cable, called the service drop, is installed and maintained by the utility company. In underground services, the connection between the utility pole and residence may be run from either a utility pole or a utility-owned underground supply line. Although overhead service is the most commonly used system for residential installations, underground service, or "Underground Residential Distribution," is steadily increasing in popularity.

Residential service layouts vary widely, depending on voltage and amperage rating and the type of residence being served. In most cases, residential service is 100A, but 200A systems are being installed with increasing frequency to accommodate the many convenience appliances in today's homes. The increase in installation cost of 200A service over that of 100A service is about 30%. A typical overhead service is comprised of the following components and equipment: service head; service lateral, which consists of type SE cable or PVC, EMT, or rigid steel pipe with individual conductors; meter box (sometimes supplied by the utility); LB fitting, which is used to gain entrance from

outside the dwelling to the distribution panel inside; distribution panel, with a 100A, 150A, or 200A main breaker and space for 20, 30, or 42 circuits 110/208V single-phase; ground rod and bare copper cable for the grounding system; and branch circuit breakers of 15A to 30A single-pole or 15A to 60A double-pole.

The layout for a typical underground system consists of service lateral conductors made of URD direct burial cable, PVC pipe with individual cables, or galvanized steel pipe with individual cables. In cases where URD cable is used, an approved protective sleeve must be installed where the cable is exposed at each end of the service lateral conductor (between the ground and connection to the supply line on the utility pole and between finish grade and the meter socket on the dwelling). The layout for a typical underground system is the same as that for an overhead system in all other respects.

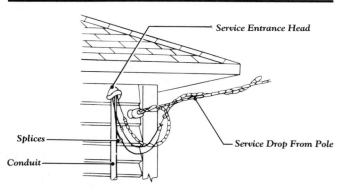

Service Entrance Head

Splices

Service Drop From Pole

Conduit

Under Eave Service Installation

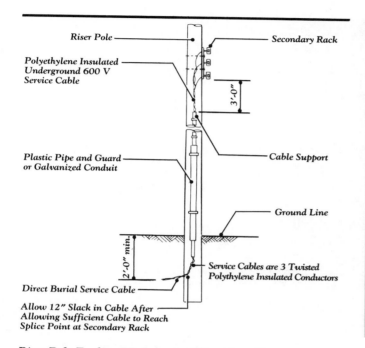

Riser Pole — Riser Pole

Polyethylene Insulated
Underground 600 V
Service Cable

Secondary Rack

3'-0"

Plastic Pipe and Guard
or Galvanized Conduit

Cable Support

Ground Line

2'-0" min.

Direct Burial Service Cable

Service Cables are 3 Twisted
Polythylene Insulated Conductors

Allow 12" Slack in Cable After
Allowing Sufficient Cable to Reach
Splice Point at Secondary Rack

Riser Pole Feeding Underground Secondary Service

Man-hours

Description	m/hr	Unit
Service Cap 100 Amps	.660	Ea.
200 Amps	1.000	Ea.
SE Cable #2	7.270	CLF
#4/0	10.000	CLF
URD Cable #2	1.800	CLF
#4/0	2.600	CLF
PVC 1-1/4"	0.070	lf
2"	0.080	lf
EMT 1-1/4"	0.080	lf
2"	0.100	lf
Galv. Steel 1-1/4"	0.123	lf
2"	0.178	lf
#3 THW Copper	1.600	CLF
#3/0 THW Copper	3.200	CLF
#6 Bare Copper	0.800	CLF
8' Ground Rod, 1/2" Diameter	1.600	CLF
1-1/4" LB Fitting	1.000	Ea.
2" LB Fitting Terminal	1.600	Ea.
Meter Socket 4 Terminal 100 Amps	2.500	Ea.
4 Terminal Amps	4.210	Ea.
Panel 120/240V IP 100 Amps		
W/M.B. 18 Circuits	10.00	Ea.
Panel 120/240V 1P 200 Amps		
W/M.B. 40 Circuits	17.78	Ea.

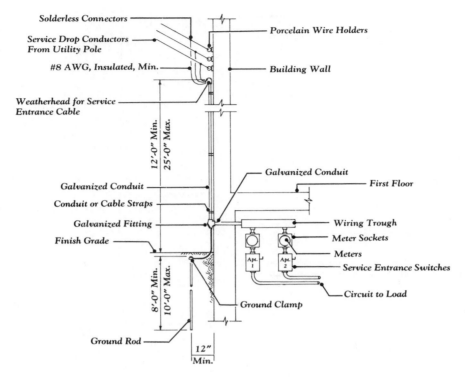

Service entrances for small apartment and commercial buildings consist of service conductors and service equipment. Service conductors, which may be run overhead or underground, are the supply lines that connect the service equipment of the building to the utility distribution main or transformer located on the street. Generally, the service equipment includes the disconnect, overcurrent protection, and metering devices within the building.

In an overhead system, the service conductors and service entrance conductors consist of two components: service drop conductors and service entrance conductors. Service drop conductors are the lines which run from the utility connection on the street to a point on the building where they are connected to the service entrance conductors. Service entrance conductors run from that point of connection to the service equipment within the building.

The service equipment for both small apartment and commercial buildings is comprised of a required externally operated disconnecting device, overcurrent protection devices, and metering system. The disconnecting device must be able to disconnect service conductors from the source of supply. It may be a single switch or circuit breaker, or not more than six switches or circuit breakers. If there are less than six tenants using the service, a master switch is not required. However, each tenant must be able to disconnect his circuit without disturbing power supply to the rest of the building.

The metering system within the service equipment includes a kilowatt-hour meter for each tenant circuit within the apartment or small commercial building. The branch circuit distribution equipment is usually grouped with the other components of the service equipment. Usually, the disconnect circuit breaker is located in the same compartment as the meter socket and main service disconnect. This arrangement is called a meter bank.

In commercial buildings, 277/480V three-phase service is often required. In this case, a step-down dry transformer is installed to drop the voltage to 120/208V. The size of the transformer is determined by the electrical needs of the tenant and is located between the primary disconnect and the distribution panel. The distribution panels can be either three-phase or single-phase, with branch circuit capacity of 20 to 42 circuits of 100A, 150A, or 200A.

Man-hours

Description	m/hr	Unit
Main Circuit Breaker 3P 4 Wire		
400A	10.000	Ea.
Meter Bank 6 Meter		
w/Breakers	26.670	Ea.
Panel 100A 20 Circuit		
120/240V	12.310	Ea.
4" Galvanized Steel Pipe	.400	lf
4" PVC	.170	lf
4" EMT	.200	lf
#500 MCM Cable	5.000	CLF
SER Cable 3#2 & 1#4	7.300	CLF
#6 Bare Copper	.800	CLF
8' Ground Rod	1.600	CLF
Service Head	2.000	Ea.
4" lb	2.000	Ea.

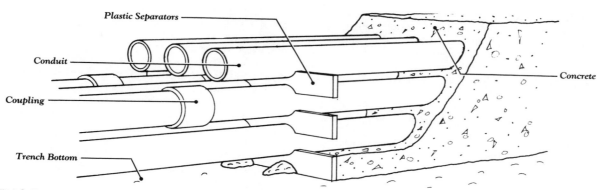

Duct Bank System

A duct bank consists of two or more conduits installed together and, depending on the type of installation required, with or without a concrete encasement. Duct lines are used for general underground distribution when more than one cable is used in a run, or if future needs will require replacement with larger or additional cables. Duct lines are installed in trenches with the top run at least 2' to 3' underground and terminate in underground vaults called manholes. Cables are drawn through the ducts from one manhole to the next to form an underground distribution system.

In most situations conduit duct is made of plastic PVC, but galvanized steel duct is occassionally used when maximum cable protection is called for. PVC is the preferred material because its material and installation costs are less than that of steel and because it is noncorrosive. Because PVC is available in lengths of up to 30', installation time is also reduced considerably. Both PVC and galvanized steel duct are available in standard diametrical sizes of 2", 3", 4", 5", and 6". Nonmetallic conduit, usually PVC, is frequently placed without concrete encasement for low voltage power and signal wiring. However, some plastic conduits require concrete encasement.

The installation of the duct system requires first the excavation of the trench and the placing of forms for the concrete encasement. If the soil is hard-packed and no outside form is necessary, the width of the trench should allow 3" on each side of the conduit. If the soil conditions require that outside forms be placed, then an additional 3" of width on each side is necessary.

After the trench has been excavated to adequate depth and width, the conduit is placed by one of two methods: the monolithic method or the tier-by-tier method. With the monolithic method, plastic separators are first placed between the tiers of conduit and then concrete is poured into the trench and around the conduit. With the tier-by-tier method, a layer of concrete is placed on top of each tier of conduit after it is installed. The alternating layers of concrete and conduit are built up until the desired number of tiers of conduit have been placed. Although the monolithic method is generally faster, careless spreading of the concrete can cause voids and air pockets. The tier-by-tier method is slower, but it ensures a solid concrete encasement free from voids. Both methods are usually completed in 100' sections to allow for the staggering of conduit and concrete placement operations.

Man-hours

Description	m/hr	Unit
Underground Duct, PVC Type EB		
2 @ 2" Diameter	.060	lf
4 @ 2" Diameter	.130	lf
2 @ 4" Diameter	.100	lf
4 @ 4" Diameter	.200	lf
6 @ 4" Diameter	.290	lf
4 @ 6" Diameter	.320	lf
6 @ 6" Diameter	.530	lf
Galvanized Steel 2 @ 2" Diameter	.080	lf
4 @ 2" Diameter	.170	lf
2 @ 4" Diameter	.220	lf
4 @ 4" Diameter	.470	lf
6 @ 4" Diameter	.720	lf
4 @ 6" Diameter	.800	lf
6 @ 6" Diameter	1.140	lf

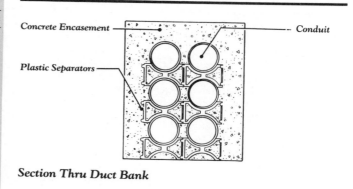

Section Thru Duct Bank

9 ELECTRICAL
CONDUIT SYSTEMS

Electrical conduits protect, support, and distribute wires and cables which are enclosed in them. Conduits shield the wiring from physical damage and corrosion and protect personnel against shock hazards. Conduits may also serve as a continuous ground path. They help protect surrounding areas from electrical fires (which may occur because of arcing or overheating of electrical conductors). Six types of materials, each designed for specific installation conditions, are commonly used for conduit systems: heavy wall galvanized steel conduit, also referred to as "rigid steel conduit"; aluminum conduit; intermediate metal conduit, usually called "IMC"; electrical metal tubing, normally called "EMT" or "thin wall" conduit; PVC-coated steel conduit; and "PVC" nonmetallic plastic conduit.

Because rigid steel conduit is the heaviest conduit material, it is employed for installations which require maximum protection for enclosed conductors. It is manufactured in standard lengths of 10' and in nominal trade sizes of 1/2", 3/4", 1", 1-1/4", 1-1/2", 2", 2-1/2", 3", 3-1/2", 4", 5", and 6". It features factory threaded ends which are joined by threaded couplings in the field. When rigid steel conduit terminates at junction boxes or is joined at other places along the run, which are located at less than the standard 10' intervals, it must be field-cut and threaded. Care must be exercised to ream smooth the field-cut ends so that the wires are protected from cutting and chafing. For further protection bushings must be installed to all conduit ends.

Aluminum conduit is used as an alternate to rigid steel conduit in applications where corrosion or weight is a consideration. Labor and material costs for aluminum run lower than rigid steel. Aluminum is not suitable for embeddment in concrete.

IMC is similar to rigid steel conduit in its application, but weighs less because it is manufactured with thinner walls. IMC is available in the same standard lengths as rigid steel conduit and in nominal trade sizes of 1/2", 3/4", 1", 1-1/4", 1-1/2", 2", 2-1/2", 3", 3-1/2', and 4". IMC installations also use the same fittings and threading format as rigid steel conduit. Both IMC and rigid steel are approved for installation situations which require conduits to be embedded in concrete.

EMT features even thinner walls than IMC to yield a larger inside diameter for easier wire pulling. In addition to this advantage, the light weight of EMT reduces the cost of material and provides labor savings, especially in installations which require substantial field bending and handling. Because of its thin wall construction, EMT cannot be threaded; consequently, set screw or pressure-type fittings must be used for joints and connections. EMT is manufactured in standard 10' lengths and in nominal trade sizes to 4". EMT cannot be installed in locations which would subject it to severe physical damage or exposure to corrosive materials. If the fittings that are used to join or terminate the conduit are classified as rain-tight or concrete-tight, EMT may be installed outdoors or buried in concrete.

As an alternative to standard rigid steel and IMC, PVC-coated conduit may be installed in situations which require

protection for the exterior of the conduit because of its exposure to oils, acids, alkalies, or excessive moisture. This type of conduit is manufactured from standard galvanized steel conduit with a seamless, flame retardant covering of polyvinyl chloride plastic. Like rigid steel conduit, PVC-coated conduit is available in standard lengths and in nominal trade sizes in standard increments up to 6". After joints and/or conduit terminations are completed with PVC-coated fittings, a PVC liquid coating should be applied to seal the connections and any exposed metal.

PVC rigid nonmetallic conduit, under certain circumstances, may be used in place of metal conduit materials; but, generally, it lacks the strength and durability of metallic piping. For example, it may serve effectively as a conduit material in locations which are wet, which are exposed to corrosive materials, or which require direct burial in earth. It may also be installed in walls, floors, and ceilings, but only in situations where it is not exposed to physical damage or where ambient temperatures do not exceed the conduit's rated limits. Because PVC lacks strength, it requires twice as many supports as metallic conduit to carry the same weight. PVC is manufactured in standard 10' lengths and in standard trade sizes up to 6" in diameter. Nonthreaded couplings, which are cemented directly to the conduit, are used to join and terminate sections. Bushings are not required, since PVC can be cleanly cut with fine-tooth hacksaw blade and reamed smooth with a pen knife. Turns and bends along the run of a PVC conduit system are accomplished with manufactured bends or by field heating and bending with a "hot box" or heat wrap blanket.

Conduit may be installed in exposed locations or, with approved conduit materials, in earthen trenches and concrete. When the conduit is installed in exposed locations, various support systems are employed, including devices which range from simple one-hole clips to complex network racking systems that support numerous conduits on the same hanger arrangement. When approved conduit is buried in the earth, it should be placed in a specially prepared trench which is first graded true and cleared of stones and soft spots. After a gravel bedding has been installed, the conduit is placed and covered with a selected backfill. A sand backfill then covers the selected backfill to indicate the conduit's location for future excavation. The sand backfill is covered by ordinary backfill which is tamped around the sides of the trench wall and leveled at grade.

When conduit is placed in concrete, two methods of installation are commonly employed. One method is to form duct banks by stacking the conduit in a trench with plastic spacers separating the layers of piping. When the specified number of conduit ducts are built up, the concrete is placed in the trench and around the sections of conduit. A second method of placing conduit in concrete is to form the duct banks in a tier-by-tier arrangement in which a row of conduit is installed and then covered with a layer of concrete. This layering process is repeated until the specified number of conduit ducts has been installed.

Man-hours

Description	m/hr	Unit
Rigid Galvanized Steel 1/2″ Diameter	.089	lf
1-1/2″ Diameter	.145	lf
3″ Diameter	.320	lf
6″ Diameter	.800	lf
Aluminum 1/2″ Diameter	.080	lf
1-1/2″ Diameter	.123	lf
3″ Diameter	.178	lf
6″ Diameter	.400	lf
IMC 1/2″ Diameter	.080	lf
1-1/2″ Diameter	.133	lf
3″ Diameter	.267	lf
4″ Diameter	.320	lf
Plastic Coated Rigid Steel 1/2″ Diameter	.100	lf
1-1/2″ Diameter	.178	lf
3″ Diameter	.364	lf
6″ Diameter	.800	lf
EMT 1/2″ Diameter	.047	lf
1-1/2″ Diameter	.089	lf
3″ Diameter	.160	lf
4″ Diameter	.200	lf
PVC Nonmetallic 1/2″ Diameter	.042	lf
1-1/2″ Diameter	.080	lf
3″ Diameter	.145	lf
6″ Diameter	.267	lf

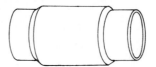

Rigid Steel, Plastic Coated Coupling

PVC Conduit

PVC Elbow

EMT Set Screw Connector

EMT Connector

EMT to Conduit Adapter

EMT to Greenfield Adapter

Aluminum Conduit

Aluminum Elbow

Rigid Steel, Plastic Coated Conduit

Rigid Steel, Plastic Coated Elbow

EMT Compression Coupling

Standard Locknut

Plastic Bushing

Grounding Bushing, Insulated

LB Fitting with Cover

T Fitting with Cover

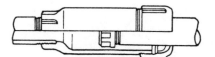

Expansion Coupling

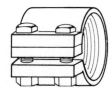

Split Coupling

Union

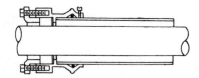

Through - Wall Seal

One Hole Clip

Clamp Back Spacer

Close Nipple with Locknut and Bushing

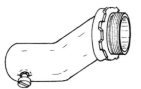

Offset Connector

Conduit Beam Clamp

310

Parallel Type, Conduit Beam Clamp

Right Angle, Conduit Beam Clamp

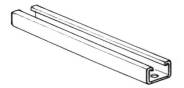

Channel

Channel Strap

Conduit Hanger

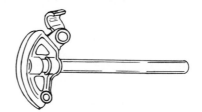

Hand Bender

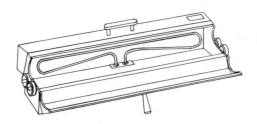

"Hot Box" PVC Bender

Conduit Benders

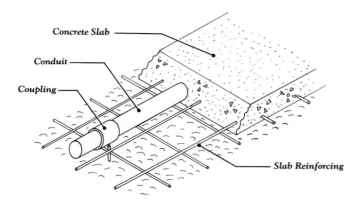

Conduit with Coupling

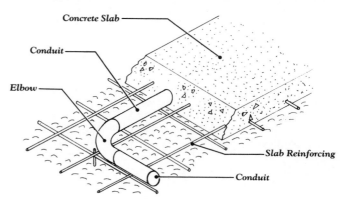

Conduit with Elbow

PVC Conduit in Concrete Slab

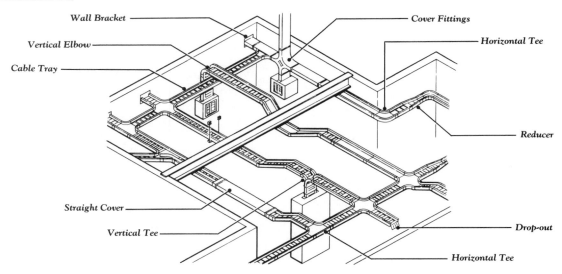

Wall Bracket

Vertical Elbow

Cable Tray

Cover Fittings

Horizontal Tee

Reducer

Straight Cover

Vertical Tee

Drop-out

Horizontal Tee

Cable Tray System

A cable tray system is an assembly of components, associated fittings, and accessories designed to form a rigid, uninterrupted support system for wires and cables. Two materials commonly used for fabricating the components are aluminum and electro-galvanized steel. Hot-dipped galvanized steel, stainless steel, PVC-coated steel, and fiberglass may also be used to suit special applications and environments. Using a cable tray system in lieu of conduit offers the following advantages: free air ampacity ratings for power cables, flexibility for design changes, rapid installation, and relatively low cable installation costs.

There are three basic types of tray systems: the ladder type, ventilated trough tray, and the solid bottom tray. The ladder tray, consists of I-beam side rails with cross-rungs spaced to 6", 9", 12", or 18" intervals; this system is recommended for heavy power cables. The ventilated trough tray is used primarily for control, communication, and low voltage power cables. It is formed of solid sheet metal with corrugated bottom slots for air circulation. Solid bottom trays, are recommended for instrumentation, low voltage signal, and control cables, since maximum shielding is provided. The widths for all three styles range from 6" to 36"; common rail depths are 3", 6", and 7". Standard straight section lengths are 12' and 24'.

There are many fittings available for the purpose of changing direction (either horizontally or vertically) or tray size, and for forming Tees, Wyes and X's. Cable tray fittings also provide for cable entry and exit from the system via bushings and drop-outs, either directly into the equipment or via short branch conduits to remote equipment, valves, motors, etc.

Barrier strips may be installed in trays in order to separate wiring of different voltage classes (with varying insulation ratings). Trays may also be fitted with covers to protect cables from falling debris, dirt and dust. Support hardware for cable trays includes wall brackets, trapeze hangers, hold-down clamps, and beam clamps.

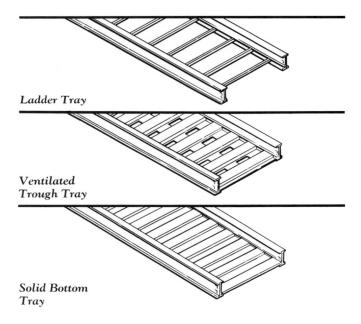

Ladder Tray

Ventilated Trough Tray

Solid Bottom Tray

Man-hours

Description	m/hr	Unit
Cable Tray		
Ladder Type 36" Wide	.267	lf
Elbows Vertical 36"	3.810	Ea.
Elbows Horizontal 36"	3.810	Ea.
Tee Horizontal 36"	5.330	Ea.
Tee Vertical 36"	4.440	Ea.
Drop-Out 36"	1.000	Ea.
Reducer 36" to 12"	2.290	Ea.
Wall Bracket 12"	.364	Ea.
Cover Straight 36"	.100	lf
Cover Elbow 36"	.320	Ea.

Horizontal Elbow

Vertical Elbow - Inside

Vertical Elbow - Outside

Horizontal Tee

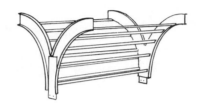

Vertical Tee

Straight Reducer

Straight Cover

Vertical Inside Elbow Cover

Vertical Outside Elbow Cover

Horizontal Elbow Cover

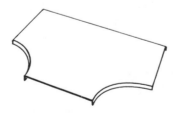

Horizontal Tee Cover

Straight Reducer Cover

Drop-out

Wall Bracket

Cable Tray Fittings

9 ELECTRICAL
UNDERFLOOR RACEWAY SYSTEMS

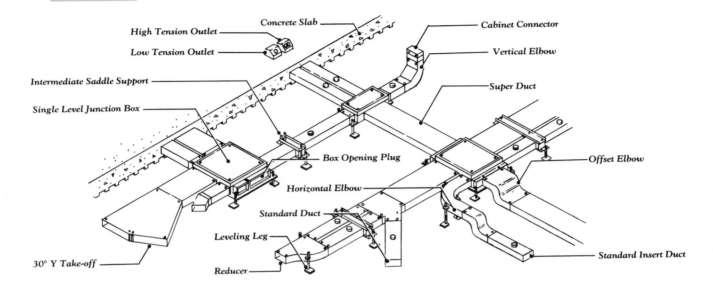

Underfloor raceway systems are used extensively in structures that house offices, as they provide an accessible, flexible system for the ever-changing power and relocation needs of an office environment. Raceways may be used for conducting electrical power cables, as well a telephone or signal systems wires. They are designed for two types of distribution systems: the two-level system, which is used when electrical or communications growth is anticipated, and the single-level system, which is used when such growth is not anticipated and when installation cost is determined by present needs.

Underfloor systems consist of steel ducts, manufactured in standard lengths of 10', that are supported by adjustable saddle supports and joined by junction boxes or steel couplings. The two most commonly used cross-section sizes of duct are 1-3/8" deep by 3-1/8" wide, and 1-3/8" deep by 7-1/4" wide.

Raceway ducts are manufactured in two types: blank or insert. Blank ducts are used when access to the raceway is not required; for example, to span the distance between the power panel and the first junction box in an office area. Because this section of raceway serves as a feeder duct, it requires no access between the two points. Insert ducts are manufactured with threaded openings every 2'. When the duct is installed, these openings are capped with a special plug that adjusts the access height of the opening to a point just below the finished surface of the concrete floor.

Underfloor raceway systems are installed prior to the placing of the concrete floors in which they travel. In general, at least 3/4" of concrete should cover the ducts. The ducts should form a straight line from junction box to junction box, so that access locations for inserts can be easily measured and located. Nonmetallic materials should not be used to join components, as the metal duct raceway must form a continuous grounded system.

Wire with ratings of up to 600 volts can be run in underfloor raceway ducts, but power cables cannot be run in the same ducts as telephone and signal wires. For this

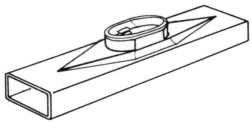

'Standard' Duct with Insert

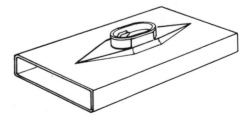

'Super' Duct with Insert

Underfloor Ducts

reason, separate ducts are run parallel to each other and are classified as high-tension (power) and low-tension (telephone and signal) ducts. When high- and low-tension ducts terminate at a junction box, they must be isolated from each other by partitioning the junction box or by installing separate junction boxes for each duct.

The junction boxes for underfloor raceways are available in single-compartment types or combination boxes; the latter are used to accommodate multiple ducts of various sizes and capacities. Some of the more common junction-box configurations include: single-level single duct, single-level double duct, single-level triple duct, two-level single duct, two-level double duct, and two-level triple duct. Junction boxes that can accept various combinations of duct widths are available by special order.

Fittings for underfloor duct systems are used to support, connect, or change direction of ducts. They can accommodate both single- and dual-level systems. They include such items as duct supports for single, double, and triple cells, as well as offsets, panel connectors, vertical and horizontal elbows, and a variety of other adapters to meet special needs.

The pedestal-type fittings for underfloor raceway systems are classified as either high or low tension. When the outlet is installed, a small amount of concrete is removed from the depression of the outlet plug. After the plug is removed, a service fitting, or nipple, is screwed into place. The nipple then secures the pedestal outlet box with locknuts. Special tools may be used to install an outlet that must be located in a section of duct that does not contain an insert. These tools are available from the manufacturer of the duct.

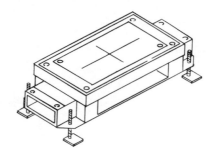

Single Level

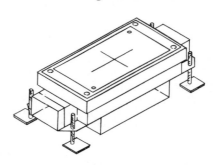

Two Level

Junction Boxes

Man-hours

Description	m/hr	Unit
Blank Duct	.100	lf
Insert Duct	.110	lf
Elbow (Vertical)	.800	Ea.
Elbow (Horizontal)	.300	Ea.
Panel Connector	.250	Ea.
Junction Box, Single Duct	2.000	Ea.
Junction Box, Double Duct	2.500	Ea.
Junction Box, Triple Duct	2.950	Ea.
Saddle Support, Single Duct	.290	Ea.
Saddle Support, Double Duct	.500	Ea.
Saddle Support, Triple Duct	.720	Ea.
Insert to Conduit Adapter	.250	Ea.
Outlet, Low Tension (Telephone and Signal)	1.000	Ea.
Outlet, High Tension (Power)	1.000	Ea.
Offset Duct Type	.300	Ea.

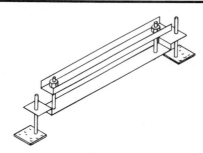

Intermediate Support

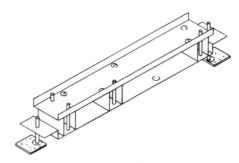

Butt End Support Coupler

Adjustable Saddle Supports

9 ELECTRICAL
TRENCH DUCT

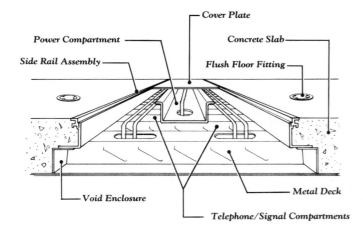

Cover Plate

Power Compartment

Concrete Slab

Side Rail Assembly

Flush Floor Fitting

Void Enclosure

Metal Deck

Telephone/Signal Compartments

section of cover plate. This arrangement allows the elbow to be mounted conveniently to the siderail assembly in the same way as a standard cover plate. Risers are designed to slip fit on the vertical section of the elbow and contain an access plate for cable pulling and maintenance. Cabinet connectors are designed to slip fit on the riser and attach conveniently to the bottom of standard-depth power panel enclosures.

Trench duct is the most commonly used raceway system for conducting power and communication cables to a cellular steel floor which utilizes its cells as distribution ducts. The depth of the trench duct in these cases is determined by the depth of concrete fill to be placed over the cellular steel subfloor. The width of the trench duct system is determined by the area of raceway required to feed the floor area.

Trench duct is a flush electrical raceway which is used as a feeder for cells of a cellular steel floor system or as a distribution duct for an underfloor duct system. Trench duct is also utilized as a self-contained raceway system for computer rooms, research laboratories, hospitals, and other locations where underfloor power and data cables are called for.

Trench duct is manufactured in three basic types: assembled bottom type, unassembled bottom type, and unassembled bottomless type. Basically, all three types consist of the same components, and they differ only in method of assembly. The basic components of all systems include a siderail assembly, cover plates, bottom plates, dividers, and an assortment of elbow, riser, and cabinet connectors. All three systems are manufactured in duct depths of 2-1/2" and 3-1/4" with a 1" adjustment range.

The siderail assembly, is comprised of standard 10' lengths which are installed parallel to one another and then coupled together with spacing bars. The spacing bars are adjusted to the desired width of the duct. A longitudinal slot on the top of the siderail assembly serves as the seat for the cover plate hold-down screws. The bottom flange of the assembly forms a continuous weld tab along the entire length of the trench.

The cover plates are manufactured from 1/4" steel with a baked enamel finish in widths of 9", 12", 18", 24", 27", 30", and 36". Predrilled screw holes in the plate can accept 1/8" high raised panhead screws for tile floor finishes or flathead screws for flush carpet installations. The cover plates are placed or removed by means of a lifting bolt threaded into 3/8" holes tapped in opposite corners of the plate or by a special suction cup tool available from the manufacturer. Bottom plates are manufactured from 14 gauge galvanized steel in standard 5' lengths. To allow for the attachment of the siderail assembly, the width of the bottom plate must be 3" wider than that of the cover plate.

In addition to the basic components, various accessories are available for trench duct systems. Dividers are used in duct runs containing two or more types of services which require separate ducts. In a bottomless trench duct installation, a U-trough is used as the power compartment. A vertical elbow consists of a short riser welded to a

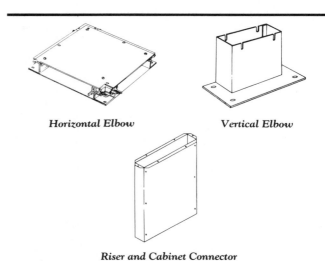

Horizontal Elbow   Vertical Elbow

Riser and Cabinet Connector

Trench Duct Fittings

Man-hours

Description	m/hr	Unit
Trench Duct, Straight Section, Single Compartment		
12" Wide	.500	lf
24" Wide	.720	lf
36" Wide	1.000	lf
Two Compartment 12" Wide	.530	lf
24" Wide	.800	lf
36" Wide	1.140	lf
Three Compartment 12" Wide	.570	lf
24" Wide	.880	lf
36" Wide	1.330	lf
Vertical Elbow 12" Wide	3.480	Ea.
24" Wide	5.000	Ea.
36" Wide	6.670	Ea.
Riser and Cabinet Connector 12" Wide	3.480	Ea.
24" Wide	5.000	Ea.
36" Wide	8.000	Ea.

9 ELECTRICAL
CELLULAR CONCRETE FLOOR RACEWAY SYSTEM

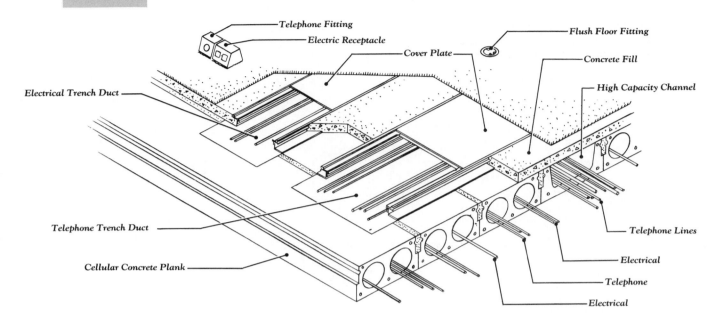

Telephone Fitting
Electric Receptacle
Cover Plate
Flush Floor Fitting
Concrete Fill
High Capacity Channel
Electrical Trench Duct
Telephone Trench Duct
Cellular Concrete Plank
Telephone Lines
Electrical
Telephone
Electrical

Cellular concrete floor raceways are used to provide a safe and effective underfloor electrical distribution system. The floor consists of precast reinforced concrete members that contain smooth, round longitudinal cells which form the raceways for the power and signal wiring systems. Standard precast concrete floor raceway slabs are manufactured in two sizes, 4-1/8″ diameter and 6-1/8″ diameter cells; however, high-stress cellular decking is also available in cell cross-sectional area sizes of 13.9, 14.1, 27.1, 50.2, and 67 square inches. Concrete floor raceway systems also consist of metal underfloor header ducts which carry the electrical supply to the cellular raceways. Standard underfloor header duct is available in 3-1/8″ and 7-1/4″ widths, but high capacity trench duct header raceways are also manufactured in 9″, 12″, 18″, 24″, 27″, 30″, and 36″ widths.

The installation of the wiring to be run in a cellular concrete floor raceway system involves the placing of supply wires across the slabs and tying them into the various branches of the electrical distribution system within the cells. In most cases, the supply wiring is run from a distribution panel via metal header ducts across the line of the cells. The header ducts are then covered with concrete fill which forms the finished floor. Supply access from the headers to the cells is accomplished by installing handhole metal junction boxes at appropriate places along the header run. Outlet access locations, which may be placed at any point along the raceway, are installed by drilling through the floor with a 1-7/8″ core bit. An outlet drive pin plug with an interior thread is inserted into the hole to provide a secure receptacle for the nipple of the outlet device. The supply wiring is then run to the outlet from a handhole junction box. The various fittings used for header duct installations are described in detail in the section ''Underfloor Raceway Systems.''

Although cellular concrete floor raceway systems provide a safe and effective means of electrical and signal wiring distribution, certain limitations of their use should be noted. Cellular concrete raceways may **not** be employed in the following situations: in areas containing corrosive vapors; in all hazardous locations with exception of Class 1, Division 2 locations; in commercial garages. In appropriate installation situations, electrical wiring may not be run in cells which also contain piping for steam, water, air, gas, or wiring services other than electrical. Because a cell should never contain both branch electrical power and telephone wiring, common wiring practice is to stagger or alternate power wiring cells and telephone wiring cells. A final limitation is that the maximum size conductor allowed for concrete cellular raceways is 1/0 AWG. The maximum allowable cross-sectional area of all conductors within a given cell must not exceed 40% of the cell's interior cross-sectional area.

Man-hours

Description	m/hr	Unit
Underfloor Header Duct 3-1/8″ Wide	.100	lf
Underfloor Header Duct 7-1/4″ Wide	.133	lf
Header Duct, Single Compartment 9″ Wide	.400	lf
Header Duct, Single Compartment 24″ Wide	.727	lf
Header Duct, Double Compartment 24″ Wide	.800	lf
Header Duct, Three Compartment 36″ Wide	1.330	lf
Outlet Drive Plug	.670	Ea.
Location Market Plug	.250	Ea.

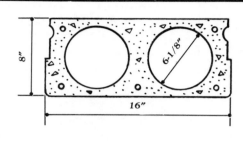

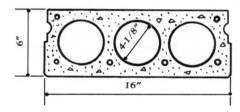

Cellular Concrete Floor Slabs

Deck Size		Cell Area (sq in.)

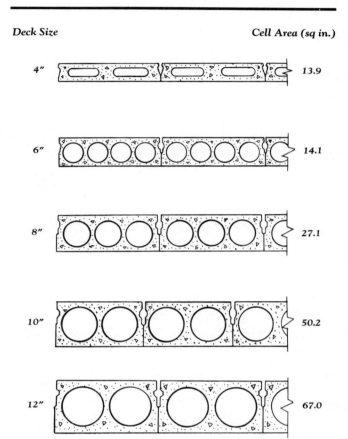

4″		13.9
6″		14.1
8″		27.1
10″		50.2
12″		67.0

Hi-stress Deck Configurations

9 ELECTRICAL
ELECTRICAL CONDUCTORS: WIRE/CABLE

Electrical conductors are the means by which current flows in an electrical distribution system. The primary elements of the distribution system are wires and cables, which are classified and rated by such variables their diametric size, type of insulated covering, and current-carrying capacity. Basically, wires and cables are comprised of the conductor itself, either copper or aluminum, and usually some type of insulated covering. The most comprehensive source of detailed information on wire and cable material is the National Electrical Code, which provides guidelines and data on all aspects of wire and cable, including their characteristics, allowable operating temperatures, ampacities, and application provisions.

The only difference between wire and cable is that of size, because their function and material composition are the same. When the circular mils area of a round cross-sectional conductor is #6 AWG or larger, it is referred to as cable; when the cross-sectional area is #8 AWG or smaller, it is called wire. Confusion often arises in the size designations of small round cross-sectional conductors as established by the American Wire Gauge system (AWG), because the AWG numbers increase as the wire size decreases, and, conversely, the AWG numbers decrease as the wire size increases. For example, #6 AWG conductor is larger than #12 AWG conductor and smaller than #1 AWG conductor. This AWG numbering system is maintained to #0 AWG, more commony designated #1/0 AWG, when a different numeric designation is employed for conductors. Beginning with #1/0 AWG, the designation numbers increase with the increase in the size of the cable to the largest AWG rating of #0000, or #4/0. For example, a cable sized at #3/0 is larger than a cable sized at #2/0. For very heavy cables larger than #4/0 AWG, the designation changes again to MCM, or thousand circular mils. With this designation system, the MCM number increases with the increase in cable size; for example, 500 MCM cable is larger or heavier than cable sized at 300 MCM. The reason for the corresponding increase of designation number and cable size is that the circular mil is an artificial area measurement which represents the square of the cable diameter measured in mils, or thousandths of an inch. For example, a solid conductor of 1/2", or 500 mils, in diameter is sized at 250,000 circular mils in area, or 250 MCM.

Most current-carrying conductors are covered with some type of insulated covering which protects them from physical damage and provides an obvious safety function. Insulation also serves as a shield against heat and moisture and prevents arcing and short circuits. Wire and cable insulation is rated by voltage, with the most commonly used ratings of 250, 600, 1,000, 3,000, 5,000, and 15,000 volts. If wire or cable is required to conduct voltages higher than its insulation's rating, the insulation eventually breaks down and short circuits may result. If the insulation deteriorates extensively because of the excessive voltage, arcing and, consequently, a fire hazard are possible.

The conductor or current-carrying component of a wire or cable is manufactured from copper or aluminum. These two materials are used because they possess the low resistance required for efficient electrical conduction. Generally, larger aluminum conductors are necessary to meet the same ampacity needs of copper conductors. Copper, because it is less resistant than aluminum and possesses other advantageous properties, is the more commonly employed material of the two, especially for smaller wires and cables. Because aluminum costs and weighs less than copper, it is most often used in larger size cables. Difficulties may arise when aluminum is employed, however, as aluminum cold flow characteristics when under pressure may cause joints to loosen. Also, aluminum conductor is prone to rapid oxidation if exposed to the air. Because the oxide which forms on the surface of the wire or cable is not conductive, it creates high resistance to the electrical flow and must be removed and prevented from reforming. Experience and skilled craftsmanship have reduced the risk of aluminum conductor oxidation, but many states have banned the use of aluminum conductors in branch wiring to prevent unskilled homeowners from installing them. The National Electrical Code provides charts and other sources of information on the ampacities of aluminum and the corresponding ampacities for copper conductors.

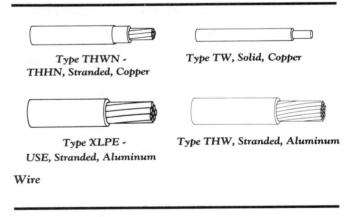

Type THWN - THHN, Stranded, Copper

Type TW, Solid, Copper

Type XLPE - USE, Stranded, Aluminum

Type THW, Stranded, Aluminum

Wire

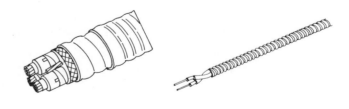

15 KV, Copper, 3 Conductor with PVC Jacket, Non-Shielded, in Cable Tray

BX, 2 Wire, Copper

Armored Cable

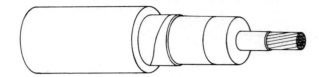

Shielded Cable, Ungrounded Neutral, Copper

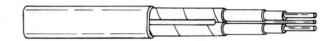

Non-metallic Sheathed Cable with Ground Wire, Copper

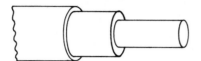

Mineral Insulated Cable, 1 Conductor

2 Conductor   **3 Conductor**   **4 Conductor**   **7 Conductor**

Mineral Insulated Cable Terminator, 2 Conductor

Mineral Insulated Cable

PVC Jacket Connector   **SER, Insulated, Aluminum**

600 Volt, Armored   **5 KV Armored**

Cable Connectors

Compression Adapter for   **Split Bolt Connector, Tapped**
Aluminum Wire

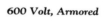

Crimp, 2-Way Connector   **Crimp, 1-Hole Lug**

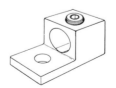

Terminal Lug, Solderless   **Compression Hand Tool**

Cable Terminations, Solderless

Cable Support   **Service Entrance Cap**

Man-hours

Description	m/hr	Unit
600V Copper #14 AWG	.610	CLF
#12 AWG	.720	CLF
#10 AWG	.800	CLF
#8 AWG	1.000	CLF
#6 AWG	1.230	CLF
#4 AWG	1.510	CLF
#3 AWG	1.600	CLF
#2 AWG	1.780	CLF
#1 AWG	2.000	CLF
#1/0	2.420	CLF
#2/0	2.760	CLF
#3/0	3.200	CLF
#4/0	3.640	CLF
250 MCM	4.000	CLF
500 MCM	5.000	CLF
1000 MCM	9.000	CLF

9 ELECTRICAL
BUS DUCT

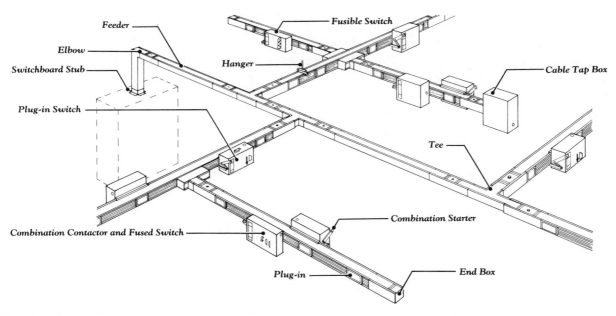

Feeder

Elbow

Switchboard Stub

Plug-in Switch

Combination Contactor and Fused Switch

Hanger

Fusible Switch

Cable Tap Box

Tee

Combination Starter

Plug-in

End Box

Bus Duct System

Bus duct consists of either copper or aluminum bus bars mounted within a sheet metal enclosure. Used primarily in commercial, industrial, and institutional buildings, bus duct provides a flexible, high-capacity, electrical power feeder system throughout the structure. The two most commonly used types of bus duct are low-voltage-drop feeder bus duct **without** plug-in provision, and low-voltage-drop feeder bus duct **with** plug-in provision.

Low-voltage-drop feeder bus duct without plug-in provision, commonly called "feeder duct," is manufactured in ratings of 600, 800, 1,000, 1,200, 1,350, 1,600, 2,000, 2,500, 3,000, 3,500, 4,000, and 5,000 amps. Because it contains no power tap-offs, it is commonly used in industrial buildings for power connections between the service entrance and the main switchboard, or high-amperage centers. Feeder duct can also function as risers in office and high-rise apartment buildings and as direct feeders to high-capacity welders and high-frequency equipment. The standard length of feeder duct sections is 10′, but any length up to 10′ is available, although the linear cost is greatly increased for nonstandard lengths.

Low-voltage-drop feeder bus duct with plug-in provision, also called "plug-in duct," has the same high-capacity and low-voltage-drop characteristics as feeder duct, but it also provides power tap-off flexibility. Plug-in duct is manufactured in ratings of 225, 400, 600, 800, 1,000, 1,200, 1,350, 1,600, 2,000, 2,500, 3,000, and 4,000 amps for applications in systems of up to 600 volts. Three-pole tap-off systems are available in 1′, 2′, 3′, 5′, and 10′ lengths and contain from ten to twelve tap-off provisions per 10 linear feet. The access provisions, located half on one side of the duct and half on the other, are equipped with

sliding, accidental-exposure covers that are closed when the tap-off is not being used.

Plug-in devices used to tap power from bus duct include the following configurations: nonfusible plug-in type, with ratings of 30, 60, 100, and 200 amps; fusible plug-in type, with ratings of 30, 60, 100, 200, and up to 1,600 amps; safety switch plug-in type, both fused and nonfused, with ratings of 30 to 600 amps; circuit breaker plug-in type, with ratings of 15 to 1,600 amps; transformer plug-in type of up to 10,000 volt - amperes single-phase loads; and combination starters for the localization of motor controls.

Fittings for both types of bus duct are available in the same voltage and amperage variations as the straight lengths of duct and provide a means of supplying power to and from the duct. Standard access fittings include: end tap boxes, for attaching conduit at either end of a bus duct; tap boxes, for feeding power into or tapping power from any point of a bus duct run; and plug-in tap boxes, for tapping power from access locations along plug-in duct. Fittings used to change the direction of bus duct include elbows, offsets, and tees. Connectors for entering or leaving switchboards and power panels are also available.

The process of installing both feeder and plug-in bus duct is the same. The bus duct is hoisted by sections into position on hangers or brackets that have been previously installed at intervals of 5′. The lengths of duct are then butted together, leveled, and fastened by means of a splicing plate attached to one end of the section of duct. After the splicing plate has been installed, the bus bars are connected with carriage bolts and tightened by torque wrench to the manufacturer's specification.

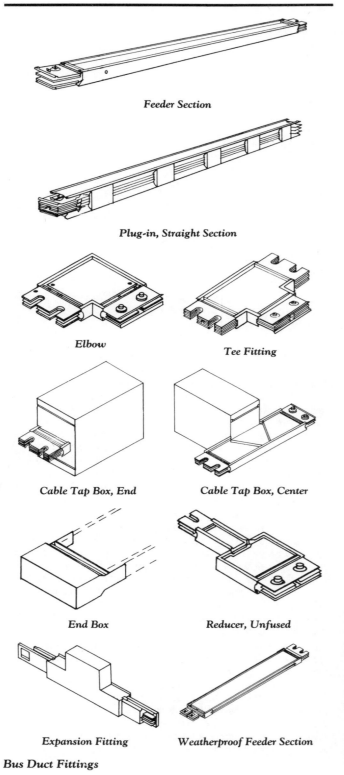

Feeder Section

Plug-in, Straight Section

Elbow

Tee Fitting

Cable Tap Box, End

Cable Tap Box, Center

End Box

Reducer, Unfused

Expansion Fitting

Weatherproof Feeder Section

Bus Duct Fittings

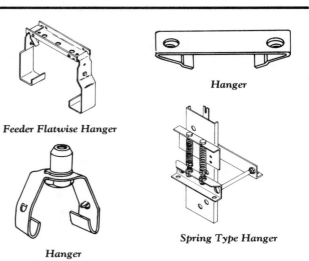

Hanger

Feeder Flatwise Hanger

Hanger

Spring Type Hanger

Bus Duct Hangers

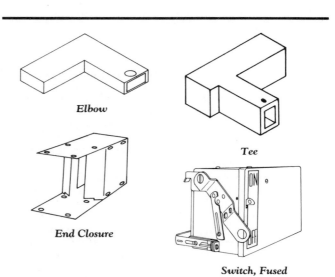

Elbow

Tee

End Closure

Switch, Fused

Bus Duct, 100 Amp and Less, Plug-in

Man-hours

Description	m/hr	Unit
Bus Duct 1000A 3-Pole 600V Copper	.800	lf
Elbows 1000A	5.350	Ea.
Tee Fitting 1000A	10.000	Ea.
Offset Fitting 1000A	6.650	Ea.
Expansion Fitting 1000A	4.450	Ea.
Cable Tap Box (Center) 1000A	8.900	Ea.
Cable Tap Box (End) 1000A	8.900	Ea.
Plug-in Fusible 200A 600V	5.000	Ea.
Plug-in Fusible with Starter 200A	5.700	Ea.
Plug-in Circuit Breaker 225A	5.000	Ea.
Plug-in Transformer 10KVA	9.000	Ea.

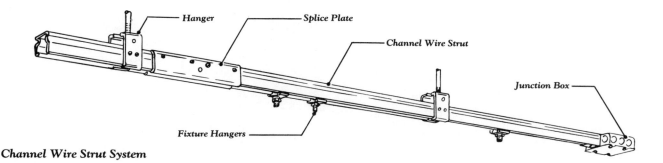

Channel Wire Strut System

A flat cable system, or Type F. C. system, is an assembly of three or four parallel #10 AWG stranded copper conductors that are enclosed in a special plastic housing and are designed to be inserted into a matching U-channel raceway. Tap off devices can be inserted into the open side of the raceway at any point along its run to provide a convenient and rapid means of access to the electrical supply. The tap off devices connect to the flat cable by means of pin-type contacts which penetrate the insulation of the flat cable assembly when they are fastened into place. The penetrating pins come into contact with the matching stranded copper conductor in the following phase sequence: phase 1 and neutral, phase 2 and neutral, and phase 3 and neutral. For safety precautions, covers are required for the open side of the U-channel if the installation height is less than 8' above the finished floor. Because flat cable assemblies are manufactured with only #10 AWG wire, their maximum branch circuit rating is 30 amps at 120, 230, 277, or 480 volts. The cable is available in bulk on 1,000' reels, or it may be cut to shorter lengths to meet specific installation needs.

Because flat cable systems are not designed for heavy power loads, they are most commonly used to supply branch circuit power to lighting fixtures. A typical arrangement for this application of flat cable employs a tap off device which is spliced to cord wires in a junction box. A fixture hanger, which is designed to grip the lips or overlap of the U-channel, is then installed to support the lighting fixture. The cord from the lighting fixture is connected to the junction box with a strain relief connector, and the connections between the supply wires and the fixture cord are made with wire nuts within the junction box. Many switching arrangements and combinations are possible when the alternate pin method is applied.

After the U-channel raceway has been placed and fastened, the flat cable is then installed in a series of fairly simple steps. First, the center conductors of the cable are stripped and guide wires are cut back at a 45° angle to reduce friction and snagging of the lead end of the cable assembly while it is being pulled through the U-channel. Next, the two center conductors are attached to a fish tape and the cable is fed into and carefully pulled through the U-channel. A pull-in guide and guide rollers assist in maintaining cable alignment during this step of the installation process. After the flat cable has been pulled through the U-channel, 4" to 6" of the lead end of the

assembly should extend beyond the end of the channel to allow enough material for splicing. The final step is to install a splice box at the end of the U-channel, which serves as the housing for the connections and to splice the flat cable and the feeder wiring. Conventional raceways may be used to run the feeder wires to the splice box, including EMT, rigid steel conduit, and IMC systems.

Man-hours

Description	m/hr	Unit
Cable, 30 Amp 3 Phase	.040	lf
Junction Box	1.000	Ea.
Insulating End Cap	.200	Ea.
U-Channel Splice Plate	.200	Ea.
Tap	.200	Ea.
Fixture Hanger	.130	Ea.
U-Channel 1-5/8" x 1-5/8"	.110	lf

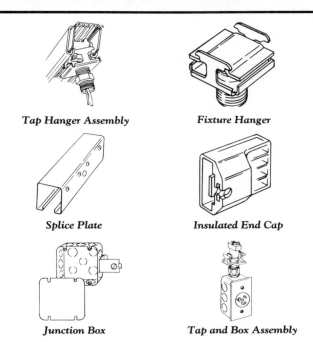

Tap Hanger Assembly *Fixture Hanger*

Splice Plate *Insulated End Cap*

Junction Box *Tap and Box Assembly*

Fittings

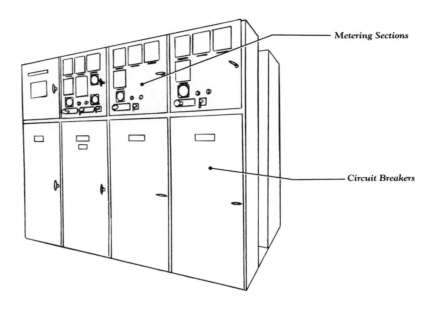

Metering Sections

Circuit Breakers

Metal-clad switchgear units provide safe and convenient means of centralizing the control of high voltage electrical power systems. The switchgear units consist of heavy-duty electrical switching equipment and the specialized housing within which the gear is enclosed. Metal-clad switchgear units are available in many modular configurations, electrical ratings, and enclosure sizes, for indoor and outdoor locations. Indoor installations should be given first priority when determining the unit's location, as the outdoor enclosures are much more expensive.

Indoor metal-clad switchgear units consist of heavy-duty circuit breakers which are enclosed in a unique housing with various compartments to separate the various electrical components within the structure. These units differ from low voltage switchgear enclosures in several ways. As in low voltage models, the breakers used in metal-clad systems are the removable type and barriered from other components; but the buses, potential transformers, control power transformers, and cable terminations within metal-clad units must also be enclosed in separate metal compartments. Further, all metal barriers within metal-clad switchgear structures must be properly grounded. Shutters, which close automatically when a breaker is withdrawn, are also provided in metal-clad switchgear systems to prevent contact with the main bus, which may be energized at the time of the breaker's withdrawal. Instruments and relays may be mounted on the door through which the breaker is inserted into its cubicle, but a barrier must be provided between the instruments and the breaker. The circuit breakers of most manufacturers of metal-clad switchgear units are installed or removed by the horizontal drawout method, but the vertical lift method for installation or removal is also available.

The electrical components and their operation are the same for outdoor and indoor switchgear units, but the special weatherproof housing for the outdoor units makes them larger and more costly. This weatherproof enclosure is constructed with an aisle that is large enough to allow workmen to withdraw the breakers from their cubicles. The enclosures are designed in two models, those with standard aisles and those with common aisles. Models which are designed with common aisles are used when two lines of breaker and auxiliary units face each other. Lights and convenience outlets are also provided with outdoor switchgear enclosures, as well as heaters to prevent a build-up of condensation within the structure.

Among the important electrical components within the metal-clad switchgear enclosures are bus bars, which may be manufactured from copper or aluminum. Although copper is the preferred bus bar material because of its efficient conductive properties, its high cost in recent years had made aluminum the standard and copper the premium material. Aluminum bus bars are installed with welded or bolted joints. If the joints of the bus are bolted, they must be silver plated or tinned at the connection point; and Belleville or similar washers should be employed to provide a large compression area. The use of these large washers minimizes cold flow to prevent the loosening of the bus bar joints. When copper bus bars are used, the joints are always bolted, and the copper is plated with silver at the connections. Belleville-type washers are not required for the joints of copper bus bars.

For safety and operational reasons, all of the bus bars and bus connections must be covered with an insulation material, and the switchgear unit must be covered with an insulation material, and the switchgear unit must be protected with a ground bus. Usually, the insulation for the bus bar itself is comprised of a specialized tubing which is installed over the bus before it is bolted into place. The bus bar joints are covered with a "boot" after the bolting procedure has been completed. The insulated bus assembly is supported on glass polyester or porcelain insulators which are flame retardant and track resistant. A ground bus, which is installed in addition to the main bus, extends the entire length of the switchgear and is bolted to each switchgear component. After the ground bus has been installed, it is connected to the station ground.

The circuit breakers which are used in 13.8KV switchgear systems are rated in three separate areas: voltage rating, interrupt rating, and continuous current rating. Standard voltage ratings for metal-clad switchgear are 4.16, 7.2, 13.8, and 34.5 kilovolts. Because preferred interrupting and continuous current ratings are determined by the load carried through the breaker, they should be carefully engineered or recommended by the manufacturer. The continuous current ratings for most types of breakers are 1,200, 2,000, and 3,000 amps. Switchgear that is rated at 34.5KV capacity and above, require oil type circuit breakers.

The selection of the various relays and meters which service each breaker is determined by the specific use of the breaker. Some of the available options include changeover switches, control switches, governor control switches, voltage regulators, ammeters, voltmeters, frequency meters, watt meters, and watt hour meters. These and other relay and monitoring devices should be installed on the recommendation of the switchgear's manufacturer.

Layouts and sizes of metal-clad switchgear units vary with the load and location of the system. In 13.8KV switchgear units, unlike stacked low voltage gear, contain breakers which are individually installed in each breaker unit. Indoor 5KV switchgear units for 1,200 and 2,000 amp breakers measure 26″ in width by 72″ to 92″ in height by 56″ to 80″ in depth, with aisle space of 28″ to 50″ in width. The height, depth, and aisle width measurements for 5KV switchgear units for 1,200 and 2,000 amp breakers vary with the manufacturer. Indoor 5KV switchgear units for 3,000 amp breakers measure 36″ in width by 64″ to 94″ in depth by 90″ in height. The measurements for 15KV metal-clad switchgear units for breakers of all ratings are 36″ in width by 81″ to 105″ in depth by 90″ in height.

Man-hours

Description	m/hr	Unit
Metal-clad Structures 1200 Amp	17.500	Ea.
Metal-clad Structures 2000 Amp	19.500	Ea.
Metal-clad Structures 3000 Amp	24.000	Ea.
Breakers 1200 Amp	10.000	Ea.
Breakers 2000 Amp	13.000	Ea.
Breakers 3000 Amp	18.000	Ea.
Instrument Wiring	3.500	Ea.
Bus Bar Connections per Structure 1200 Amp	10.000	Ea.
Bus Bar Connections per Structure 2000 Amp	15.000	Ea.
Bus Bar Connections per Structure 3000 Amp	19.000	Ea.
Ground Bus Connection per Structure	6.000	Ea.

9 ELECTRICAL
LOW VOLTAGE METAL-ENCLOSED SWITCHGEAR

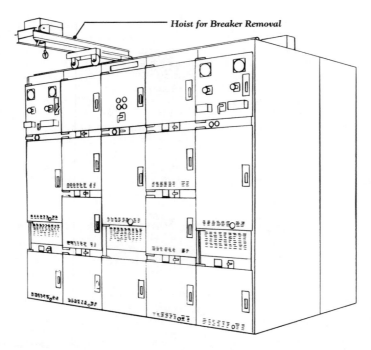

Hoist for Breaker Removal

Low voltage metal-enclosed switchgear units provide a safe and convenient means of centralizing the control of low voltage electrical power systems. The units are manufactured for both indoor and outdoor installations in a variety of configurations, capacities, and enclosure sizes. Wherever possible, indoor switchgear units should be installed, as the larger weatherproof outdoor enclosure adds considerably to the unit's cost. Low voltage metal-enclosed switchgear units consist of six basic components and assemblies: the structure, bus bars, low voltage circuit breakers, protective devices, meters, and secondary wiring.

Two basic types of low voltage metal-enclosed switchgear structures are available: the indoor and outdoor type. In most new installations, the switchgear unit is located indoors if the "end use" of the power load is intended for inside applications; however, if the "end use" of the load is intended for outside operations, the more expensive outdoor type of switchgear unit may have to be installed. Indoor low voltage metal-enclosed switchgear consists of three sections: a front section, which contains the circuit breakers, meters, relays, and controls; a bus section; and a cable entrance section. Each circuit breaker is contained in its own compartment and isolated from all other equipment. Each circuit breaker compartment is also standardly equipped with vents to cool the breaker assembly and to allow means of escape for ionized gasses which are generated when the circuit breaker opens to interrupt fault currents. Vents are also included in other sections of the switchgear structure.

Outdoor low voltage metal-enclosed switchgear units are essentially the same as indoor units, with the exception of the weatherproof enclosure which is designed to protect the electrical equipment from the elements. Outdoor switchgear structures include an enclosed walk-in aisle in front of the circuit breaker section to provide safe and convenient weather protection for the equipment and workers during maintenance. Outdoor switchgear enclosures also provide lighting in the aisle space, a 120V receptacle for power tools, and heating units to prevent condensation within the enclosure. Outdoor switchgear units are painted dark gray ANS 1 24; indoor switchgear units, light gray ANS 1 61.

The two materials used in the manufacture of bus bars for low voltage metal-enclosed switchgear are copper and aluminum. Copper is the preferred material; but aluminum, because of the premium cost of copper in recent years, is used by most switchgear manufacturers today. Generally, the details of bus design should be left to the manufacturer of the switchgear unit. Because each vertical section of low voltage switchgear contains from one to four circuit breakers which are part of the system of branches and taps off the main bus, the bus bars must be properly secured with insulated supports. These glass polyester insulators are flame retardant, track resistant, and nonhygroscopic.

The circuit breakers which are used in low voltage metal-enclosed switchgear units are either fixed or drawout models. The drawout type circuit breakers cost more, but they are more commonly installed than the fixed type because they can be removed for maintenance or replacement without de-energizing the main bus. Circuit breakers for low voltage metal-enclosed switchgear are manufactured in frame sizes ranging from 225 to 4,000 amps in voltage ratings of 208, 240, 480, and 600 volts. The most commonly installed breakers are the 600 and 1,600 amp sizes. The frame size and ratings for the main and feeder breakers vary with size of the main transformer.

The protective devices which are used in the circuit breakers for low voltage switchgear systems include magnetic or static trip devices in place of nonautomatic overcurrent trip devices. Feeder circuit breakers may also be equipped with ground fault trip components to protect against arcing faults within motor control centers and panelboards and in the cables on the load side of the breaker.

The metering and secondary wiring circuits employed within the low voltage metal-enclosed switchgear structures are designed to monitor the various aspects of the electrical flow through the breakers. Usually, the minimum metering requirement includes a single-phase voltmeter on the line power source and a single-phase ammeter to indicate total current. If the voltage is greater than 240V, a potential transformer is required, since trade standards do not permit voltages on the panels to exceed 250V to ground. Current transformers are required for all ammeters. The wiring systems which supply the various meters and control circuits must be composed of #14 AWG wire or larger and rated at 90° C or more. The secondary wiring is installed with crimp-type terminals and, at extra cost, with optional wire markers for circuit identification.

The layouts and sizes of indoor and outdoor low voltage switchgear components and sections vary with each manufacturer. Breakers with 225 or 600 amp frame sizes can be stacked four high and range from 18″ to 20″ in width; those with frame sizes of 1,600 amps can be stacked three or four high with widths ranging from 24″ to 27″. Only one breaker rated at 4,000 amps can be included within each low voltage switchgear unit. The width of this breaker measures from 30″ to 38″, depending on its manufacturer. Indoor low voltage switchgear sections range from 54″ to 60″ in depth and 90″ in height; outdoor sections measure 94″ in depth and approximately 112″ in height.

Man-hours

Description	m/hr	Unit
Structures 1600 Amps	17.500	Ea.
Structures 4000 Amps	26.000	Ea.
Breakers Draw-out 225 Amps	4.000	Ea.
600 Amps	8.000	Ea.
1600 Amps	11.000	Ea.
4000 Amps	23.000	Ea.
Bus Bar Connections 1600 Amps Per Structure	12.000	Ea.
4000 Amps Per Structure	21.000	Ea.

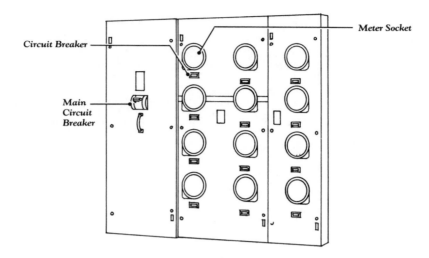

Circuit Breaker

Meter Socket

Main Circuit Breaker

Metering switchboards provide an efficient means of grouping individual metered services into one multi-meter service center. They are designed to accommodate services of up to 2000A main capacity for single-phase, three-wire 120/240V and for three-phase, four-wire 120/240V and 120/208V. They are most commonly installed in apartment buildings, commercial office buildings, malls and other locations which require separate metering of electrical power.

Many types of metering switchboard arrangements are available, depending on the electrical needs of the facility in which it is installed. Individually metered single- or three-phase subservice with meter sockets and circuit breakers rated up to 200A are presently manufactured. Meter units are stacked and can be combined with circuit breakers, fusible switches, and other components to suit the specifications and design of the installation.

Metering switchboards are manufactured in sections which measure 14″ in depth and 90″ in height. They are constructed in single sections or multi-section units, which are provided with all of the necessary crossover busing.

The main bus in each section or sections is manufactured in capacities of 400A, 600A, 800A, 1000A, 1200A, 1600A, and 2000A. Main lugs, molded case circuit breakers, and bolted pressure contact switches are rated up to 2000A. Fusible switches are available in capacities up to 1200A. Line side service cable can enter a section at the top or bottom of the switchboard. Bus duct terminations can enter at the top or bottom, but in some cases they may require an auxiliary section or pull box.

The basic component of a multi-metering switchboard is the single or twin metering unit which is preassembled for mounting in the switchboard frame. The meter sockets and breakers are mounted on a formed steel mounting pan that is bolted to the front channels in the switchboard frame. The breaker compartments form a barriered load wireway when the stack meter units are installed in the switchboard frame.

Stack meter units are available with 100A meter sockets with breaker ratings of 15A to 100A, and with 200A meter sockets with breaker ratings of 125A to 200A. All stack meter units are rated at 240V AC maximum. For single-phase, three-wire 120/240V systems, the meter units are equipped with four jaw meter sockets and two-pole breakers. For three-phase four-wire systems of either 120/240V, the units are supplied with seven jaw meter sockets and three-pole breakers.

Meter Socket

Man-hours

Description	m/hr	Unit
Main Breaker Section 800 Amps	17.800	Ea.
1200 Amps	21.000	Ea.
1600 Amps	23.500	Ea.
Main Section 100 Amps 6 Meters	26.700	Ea.
8 Meters	30.800	Ea.
10 Meters	33.300	Ea.

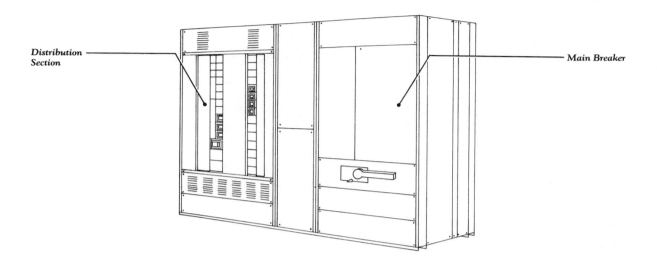

Distribution Section

Main Breaker

Multi-section switchboards function as a complete service entrance or as a secondary distribution center within a substation (See "Secondary Unit Substations"). They are most commonly used in commercial, industrial, and institutional applications. Multi-section switchboards consist of three major components: the auxiliary section; the main breaker, or fusible switch, section; and the distribution section.

The auxiliary section is used for cable transition of underground feeders to the top lugs of the main breaker, or fusible switch, section. It provides additional space to form cables and can be used to house main bus lug kits. A main bus lug kit provides load side cable terminations from a main section or line side cable terminations to a distribution section. Main bus kits are available in 1200A and 2000A sizes. They are supplied with four lugs per phase in the 1200A kit and six lugs per phase in the 2000A kit. Each lug accepts either copper or aluminum cable in sizes ranging from #3/0 to 750 MCM. Main bus lug kits are designed to feed from left or right and include three bus splice plates. Auxiliary sections measure approximately 91.5" in height, 24" in width, and 24" in depth, and weigh about 400 lbs. These measurements may vary slightly from manufacturer to manufacturer.

The main breaker section contains a 75,000 RMS amp, high interrupting capacity 600V molded case circuit breaker. These breakers are available in 1200A, 1600A, and 2000A ratings. Line side lugs are located at the top of the main breaker. Load side bus connectors connect the breaker to a through bus. The main breaker is furnished with a ground fault protection system, consisting of a ground fault sensor, ground relay, and monitor panel. The main breaker is also equipped with a ground trip which is powered by the output from the ground fault protection system. The main breaker section is furnished with one set of side plates which can be mounted on either side of the section. The front of the section is completely front-accessible so that units can be mounted snugly against a

wall. The weight of 1200A sections is approximately 700 lbs.; the weight of 1600A and 2000A sections is about 725 lbs. Main breaker sections measure 91.5" in height, 30" in width, and 24" in depth.

Main switch sections are available for use in systems with up to 200,000 RMS symmetrical amperes fault current at 480V AC. The switch has provisions for mounting Class L current limiting fuses. The 480V main switch can also be used for 120/208 and 240V applications.

The distribution section contains a branch circuit breaker distribution panel of double-row construction and provides 117" of breaker mounting space. The panel is connected to the through bus by formed bus bars. The branch breakers that mount in the left vertical row are available in sizes up to 800A in the following types: FY, FA, FH, IF, Q2, Q2-H, KA, Q4, LA, KH, LH, MA, and MH. The right vertical row accepts FY, FA, FH, Q2, and Q2-H breaker arrangements. The distribution section measures 91.5" in height, 48" in width, and 24" in depth. The 2000A model weighs approximately 1125 lbs.

Man-hours

Description	m/hr	Unit
Main Switchboard Section 1200 Amps	18.000	Ea.
1600 Amps	19.000	Ea.
2000 Amps	20.000	Ea.
Main Ground Fault Protector 1200-2000 Amps	2.960	Ea.
Bus Way Connections 1200 Amps	6.150	Ea.
1600 Amps	6.670	Ea.
2000 Amps	8.000	Ea.
Auxiliary Pull Section	8.000	Ea.
Distribution Section 1200 Amps	22.220	Ea.
1600 Amps	24.240	Ea.
2000 Amps	25.810	Ea.
Breakers, 1 Pole 60 Amps	1.000	Ea.
2 Pole 60 Amps	1.150	Ea.
3 Pole 60 Amps	1.500	Ea.

9 ELECTRICAL
SECONDARY UNIT SUBSTATIONS

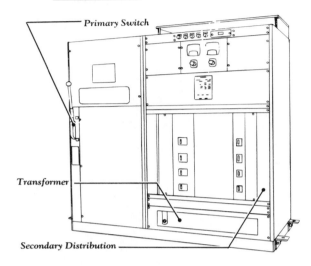

Primary Switch

Transformer

Secondary Distribution

A secondary unit substation serves as the center of all industrial electrical distribution systems. The basic function of the substation is to step down the primary incoming voltage to the utilization voltage level of load centers throughout the facility. In most cases, the primary voltage is less than 15 KV; and the utilization voltage, 480V or less.

As a general rule, secondary unit substations with operating voltages in over 15 KV are located outdoors due to their bulk. With operating voltages of 15 KV and less, the substation can be located indoors, outdoors, or a combination of both. Indoor equipment is less expensive and easier to maintain than outdoor equipment; but because outdoor installations cost more, these advantages may be outweighed.

A secondary unit substation is comprised of three basic components: an incoming line section, a transformer section, and an outgoing line section. Descriptions of the various possible arrangements for these components follow:

The INCOMING LINE SECTION can consist of any of the following combinations:

A. Oil-filled switches in combination with a terminal chamber to provide termination of primary cables, which generally enter from below, and to allow a primary means for disconnecting the transformer without interrupting service to other primary lines. Oil switches are available for primary voltages up to and including 69 KV.

B. Air-filled load interrupter switches to provide short circuit protection and disconnecting means. They are available for primary voltages up to 34.5 KV, with or without current limiting power fuses.

C. Metal-clad switchgear arrangement with drawout circuit breaker to provide short circuit protection and disconnecting means. By far the most expensive option, this type of arrangement is available for primary voltages up to 34.5 KV.

D. Oil-filled cutouts mounted in an air-filled terminal chamber for terminating 3-phase feeders. They are available in fused or unfused variations. Oil-filled cutouts can be used only with maximum transformer ratings of 300 KVA at 2400V and 500 KVA at 4160V with fuses, and 750 KVA at 2400V and 1000 KVA at 4160V without fuses.

The TRANSFORMER SECTION for a secondary unit substation is usually arranged in one of these four basic designs:

A. Oil-immersed transformers for both indoor and outdoor installations. They are available in voltage ratings ranging from 112.5 KVA to 1000 KVA at 208y/120, 240V and from 112.5 KVA to 2500 KVA at 480y/277, 480V.

B. Askarel (nonflammable)-immersed transformers for indoor and outdoor installations. They are available in the same voltage ratings as oil-immersed transformers.

C. Open-ventilated dry-type transformers for indoor installations only. They are available in the same voltage ratings as oil- and askarel- immersed transformers.

D. Sealed dry (gas-filled) transformers for indoor and outdoor installations. They are available in voltage ratings ranging from 750 KVA to 1000 KVA at 208y/120, 240V and from 750 KVA to 2500 KVA at 480y/120, 480V.

The OUTGOING LINE SECTION of a secondary unit substation usually consists of one of the following combinations:

A. Metal enclosure switchgear with stationary or drawout low voltage power circuit breakers (See "Metal-clad Low Voltage Switchgear").

B. Metal-enclosed switchgear or switchboard with molded case circuit breakers.

C. Metal-enclosed switchboard with switch and fuse combinations.

D. Motor control center assemblies (See "Motor Control Centers").

E. Combinations of the above.

Man-hours

Description	m/hr	Unit
Load Interrupter Switch, 300 KVA and below	60.000	Ea.
400 KVA and above	63.000	Ea.
Transformer Section 112 KVA	37.000	Ea.
300 KVA	59.000	Ea.
500 KVA	67.000	Ea.
750 KVA	83.000	Ea.
Low Voltage Breakers, 2-Pole, 15 to 60 Amps,		
Type FA	1.430	Ea.
3-Pole, 15 to 60 Amps, Type FA	1.510	Ea.
2-Pole, 125 to 225 Amps, Type KA	2.350	Ea.
3-Pole, 125 to 225 Amps, Type KA	2.500	Ea.
2-Pole, 700 and 800 Amps, Type MA	5.330	Ea.
3-Pole, 700 and 800 Amps, Type MA	6.150	Ea.

9 ELECTRICAL
PANELBOARDS

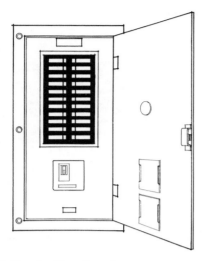

Main Circuit Breaker Panelboard

Panelboards provide a safe means of distributing power to the various circuits in an electrical system. Basically, they consist of copper or aluminum bus bars, or **mains**, and overcurrent protective devices for the various circuits within a metal enclosure. For flexibility, most panelboards are designed to accommodate fusible or circuit breaker assemblies which are plugged into or bolted onto the busbars and which provide safe, systematic access to the mains. The overcurrent devices which may be installed at the various tap off locations are available in single-, two-, and three-pole units. The enclosures which house the bus bars and overcurrent devices are generally constructed from sheet steel of #16 manufacturer's standard gauge or thicker. The steel must be galvanized to prevent rust and corrosion. The space between the panelboard's main lugs and outer edge of the enclosure is called the gutter and is used as a channel for the access wiring. All types of panelboards are rated by amperage and voltage capacities, with the amperage rating determined by the ampacity of the bus bar to which the overcurrent devices are connected. Panelboards are divided into four classifications: lighting and appliance branch circuit panelboards, service equipment panelboards, feeder distribution panelboards, and distribution panelboards.

Of these four classifications, the lighting and appliance branch circuit panelboard is the most commonly installed unit. The National Electrical Code defines this type of panelboard as one having more than 10% of its overcurrent devices rated at 30 amps or less for which neutral connections are provided. Lighting and appliance panelboards may contain up to forty-two overcurrent devices. Single-pole circuit breakers count as one overcurrent device; two-pole, two overcurrent devices; and three-pole, three overcurrent devices. Lighting and appliance panelboards are manufactured in three- or four-wire units rated at 120/240V, 120/208V, and 277/480V. They are available with main lugs or main circuit breakers

of up to 600 amp ratings, with standard ampere increments for main lug or circuit breakers of 100, 225, 400, and 600 amps. The branch capacities for the various lighting and appliance panelboards are 10, 12, 14, 18, 24, 30, 34, 36, and 42 spaces. The enclosure for this type of panelboard may be surface- or recess-mounted with appropriate panel covers available for either mounting situation.

The remaining three classifications of panelboards are more specialized in their function and normally designed for heavier power distribution loads. Service equipment panelboards are used for tap off loads of up to 800 amps which contain six or fewer main fused switches, fused pull-outs, or circuit breakers. Feeder distribution panelboards contain circuit overcurrent devices which are rated at more than 30 amps to protect sub-feeders that supply smaller branch circuit panels. Power distribution panelboards contain bus bars which are rated up to 1,200 amps at 600 volts or less and feature control and overcurrent devices to match connected motor or other power circuit loads. In most cases, these control and overcurrent devices are three-phase units.

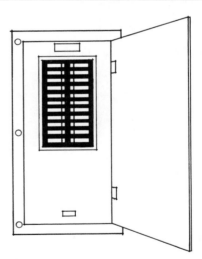

Main Lugs Only Panelboard

Man-hours

Description	m/hr	Unit
Panelboard 3-Wire 225 Amps Main Lugs		
38 Circuit	22.220	Ea.
4-Wire 225 Amps Main Lugs 42 Circuit	23.530	Ea.
3-Wire 400 Amps Main Circuit Breaker		
42 Circuit	32.000	Ea.
4-Wire 400 Amps Main Circuit Breaker		
42 Circuit	33.330	Ea.
3-Wire 100 Amps Main Lugs 20 Circuit	12.310	Ea.
4-Wire 100 Amps Main Circuit Breaker		
24 Circuits	17.020	Ea.

1 Pole

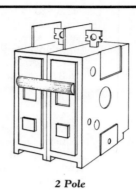

2 Pole

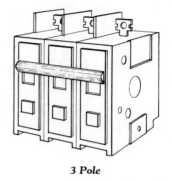

3 Pole

Bolt-on Circuit Breakers

Plug-In Circuit Breaker, 1 Pole

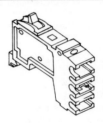

Circuit Breaker, 1 Pole

9 ELECTRICAL
TRANSFORMERS

The function of a transformer is to step up or step down the voltage of alternating current for appropriate power distribution. For example, a transformer may be used to step down an incoming 4,160 volts service to 480 volts to distribute power within a building; another transformer may step down the voltage again from 480 volts to 120 volts for lighting systems. Transformers usually contain, two sets of terminals: the primary terminals, for incoming wires and the secondary terminals, for outgoing wires. Because transformers can step up or step down voltage, the primary terminals are not always high voltage, nor are the secondary terminals always the low voltage side. This distinction is an important one, as the reverse function of transformers is often misunderstood. Another important fact that is often overlooked is that transformers are employed for alternating current only, never for direct current systems.

Transformers are manufactured in single-phase or three-phase construction. Their power capacity is rated in VA or KVA (kilovolt-amperes). The voltages differ at the primary and secondary terminals while the KVA rating is constant. Therefore, the amount of primary and secondary current differs proportionately. For example, a 200 KVA 2,400/120V transformer will carry full load current at each terminal as computed below:

$$\text{Primary current} \quad \frac{200,000 \text{ VA}}{2,400 \text{ V}} = 83.2 \text{ amperes}$$

$$\text{Secondary current} \quad \frac{200,000 \text{ VA}}{120 \text{ V}} = 1,664 \text{ amperes}$$

Transformers are classified according to the type of material used to cool their coils and the function of the unit. The coils may be oil filled, askeral filled, or dry (air) in their cooling method. Some of the common transformer types include:

1. General purpose dry type. This type of transformer uses air for cooling and insulation. Its primary voltages are 120, 208, 240, 480, and 600; and its secondary voltages are 120, 208, 240, 480, and 600, with a maximum three-phase capacity of 750 KVA. This type of transformer is used for lighting and power circuits.

2. Load center type. This type of transformer uses air, silicone, or oil for cooling and insulation. Its primary voltages are 2,400, 4,160, 7,200, 12,470, 11,000, 13,200, and 13,800 volts; and its secondary voltages are 120, 208, 240, 480, and 600, with a maximum three-phase capacity of 2,000 KVA. This type of transformer is used for building services, unit substations, and indoor and outdoor load centers.

3. Line distribution type. This type of transformer uses silicone or oil for cooling and insulation. Its primary voltages are 2,400, 4,160, 7,200, 13,200, and 13,800; and its secondary voltages are 120, 208, 240, 480, and 600, with a maximum three-phase capacity of 750 KVA. This type of transformer is installed on a pole, a pole platform, or exterior concrete pad.

4. Substation type. This type of transformer uses oil for cooling and insulation. Its primary voltages are 2,400, 4,160, 7,200, 12,470, 13,200, 22,000, and 34,000; and its secondary voltages are 480, 600, 2,400, and 4,160, with a three-phase capacity of over 500 KVA. This type of transformer is used outside for groups of buildings or large single structures.

The most commonly installed transformers are the single- and three-phase dry type. The most popular single-phase sizes are 3, 5, 10, 15, 25, 37.5, 50, 75, 100, 167, and 250 KVA. The most common three-phase sizes are 45, 75, 112.5, 150, 225, 300, and 500 KVA. The size and weight of dry indoor transformers increase with their capacities. For example, a 3 KVA single-phase unit, which measures 15-1/4" high by 8-5/8" wide by 7-3/4" deep and weighs 97 lbs., can be mounted directly to a wall; a 500 KVA three-phase transformer, which measures 70-1/2" high by 69" wide by 39-1/4" deep and weighs 4,100 lbs., must be mounted on the floor.

Oil Filled Transformer   Dry Type Transformer, Single Phase

 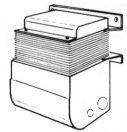

Dry Type Transformer, 3 Phase   Buck - Boost Transformer

Transformer Types

Man-hours

Description	m/hr	Unit
Oil Filled 5 KV Primary 277/480 Volt Secondary		
3 Phase 150 KVA	30.770	Ea.
1000 KVA	76.920	Ea.
3750 KVA	125.000	Ea.
Silicon Filled 5 KV Primary 277/480 Volt		
Secondary 3 Phase 225 KVA	36.360	Ea.
1000 KVA	76.920	
2500 KVA	105.000	Ea.
Dry 480 Volt Primary 120/208 Volt Secondary		
3 Phase 15 KVA	14.550	Ea.
112 KVA	23.530	Ea.
500 KVA	44.440	Ea.

A fluorescent lamp consists of a hot cathode in a phosphor-coated tube which contains inert gas and mercury vapor. When energized, the cathode causes a mercury arc to produce ultraviolet light and fluorescence on the phosphor coating of the tube. The color of the light varies according to the type of phosphor used in the coating. Fluorescent lamps are high in efficiency, and with limited switching on and off, they have a life in excess of 20,000 hours. A ballast is required in the lamp circuit to limit the current. Ballasts are available in various watt-saving types and can be matched with special energy-saving lamps. Special ballasts are required for dimming.

Fluorescent tubes are produced in many different wattages, sizes, and types. One manufacturer lists lamps of 4 watts to 215 watts with lengths of 6″ to 96″. Three basic types of fluorescent lamps are presently manufactured: preheat, instant start, and rapid start. The preheat lamp, which is the oldest type, requires a starter. The instant start lamp, or slimline, was developed after the preheat type. The rapid start lamp, which is the most commonly used fluorescent lamp today, operates at 425 mA. High output lamps operate at 800 mA.; very high output, at 1500 mA. Because the ballasts used in high output and very high output lamps tend to be noisy, these types of fluorescent lamps are not recommended for use in quiet areas.

The most commonly used fluorescent lamp for general lighting applications is the rapid start 48″ 40 watt size. Other popular lamps are 24″ 20 watt, 36″ 30 watt, 72″ 55 watt, and 96″ 75 watt sizes. For office lighting, the most commonly used fixture is the 2′ by 4′ troffer with four 48″ lamps and an acrylic lens. The troffer is also available with two lamps for corridor lighting and for other locations requiring only limited lighting. The troffers are also available in air-handling models. Many different types of lenses are also manufactured for troffers.

Fluorescent lamps are also available for specialized uses in many different shapes and configurations. In addition to straight tubes, U-shaped and circline (circular) models are manufactured. Parabolic reflectors provide efficient lighting while allowing greater spacing between fixtures. Surface-mounted fixtures are available in one-, two-, three-, and four-lamp models, in lengths of 4′ and 8′. These fixtures can be ordered with lenses, in wrap-around configurations, or as plain strips with or without reflectors. Basic industrial fixtures for use in shops, supermarkets, and manufacturing areas usually do not require lenses. Fixtures of this type normally consist of a strip of lamps and a reflector with an acrylic or porcelain finish.

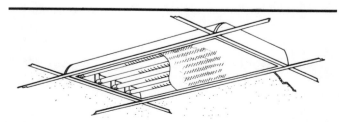

Troffer Mounted Fixture with Acrylic Lens, 4 Tube

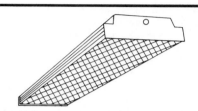

Surface or Pendant Mounted Fixture with Wrap Around Acrylic Lens, 4 Tube

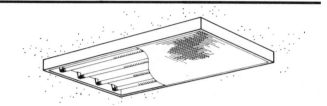

Surface Mounted Fixture with Acrylic Lens, 4 Tube

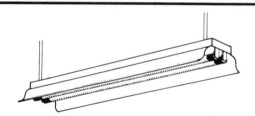

Pendant Mounted Industrial Fixture, 2 Tube

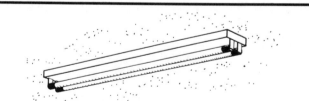

Surface Mounted Strip Fixture, 2 Tube

Man-hours

Description	m/hr	Unit
Troffer with Acrylic Lens 4-40W RS		
2′ x 4′	1.700	Ea.
2-40W URS 2′ x 2′	1.400	Ea.
Surface Mounted Acrylic Wrap-around Lens		
4-40W RS 16″ x 48″	1.500	Ea.
Industrial Pendant Mounted		
4′ Long, 2-40W RS	1.400	Ea.
8′ Long, 2-75W SL	1.820	Ea.
8′ Long, 2-110W HO	2.000	Ea.
8′ Long, 2-215W VHO	2.110	Ea.
Surface Mounted Strip, 4′ Long, 1-40W RS	.940	Ea.
8′ Long, 1-75W SL	1.190	Ea.

ELECTRICAL
INCANDESCENT LIGHTING

An incandescent lamp is a glass bulb which contains a tungsten filament in a mixture of argon and nitrogen gas. The base of the bulb is usually capped with a screw base made of brass or aluminum. Incandescent lamps are versatile sources of light, as they are manufactured in many different sizes, shapes, wattages, and base configurations. Some of these variations include bulbs which feature clear, frosted, and hard glass (for weatherproof applications); aluminized reflectors, wide, narrow, and spot beam prefocused; and three-way wattage switching. For general applications, incandescent lamps are rated from 2 watts to 1500 watts, but some street lighting lamps may be rated as high as 15,000 watts.

Along with the advantage of variety of lamp sizes and special features, the relatively small size of incandescents allows them to be fit easily into the design of the fixtures which hold them. They are low in cost, soft in color, and easy to use with dimmers. The disadvantages of incandescent lamps include relatively short life, usually less than 1000 hours, and higher energy consumption than fluorescent and mercury vapor lamps. If the incandescent lamp is used in a system with a higher voltage than that recommended by the manufacturer, the life of the lamp decreases significantly. An excess of just 10 volts above the recommended voltage can reduce the lamp's life considerably.

The quartz lamp, or tungsten halogen lamp, is a special type of incandescent lamp which consists of a quartz tube with various configurations. Some quartz lamps are simple double-ended tubes, while others include a screw base or a mounted inside an R- or Par-shaped bulb. A quartz lamp maintains maximum light output throughout its life, which varies according to its type and size from 2000 to 4000 hours. Generally, quartz lamps are more energy efficient than regular incandescents, but their purchase price is higher. Quartz lamps are available in sizes from 36 watts to 1500 watts.

Fixtures for regular incandescent lamps vary widely, depending on their function, location, and desired appearance. Simple lampholders, decorative multi-lamp chandeliers, down lights, spotlights, accent wall lights, and track lights are just a few of the many types of fixtures. Some of the fixtures used for quartz lamps include exterior flood, track, accent, and emergency lights, to name a few.

Track Lighting Spotlight

Exterior Fixture, Wall Mounted, Quartz

Fixtures

Round Ceiling Fixture, Recessed, with Alzak Reflector

Round Ceiling Fixture with Concentric Louver

Round Ceiling Fixture with Reflector, No Lens

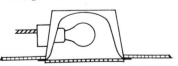

Square Ceiling Fixture, Recessed, with Glass Lens, Metal Trim

Fixtures

Man-hours

Description	m/hr	Unit
Ceiling, Recess Mounted Alzak Reflector		
150W	1.000	Ea.
300W	1.190	Ea.
Surface Mounted Metal Cylinder		
150W	.800	Ea.
300W	1.000	Ea.
Opal Glass Drum 10″ 2-60W	1.000	Ea.
Pendant Mounted Globe 150W	1.000	Ea.
Vaportight 200W	1.290	Ea.
Chandelier 24″ Diameter x 42″ High		
6 Candle	1.330	Ea.
Track Light Spotlight 150W PAR	.500	Ea.
Wall Washer Quartz 250W	.500	Ea.
Exterior Wall Mounted Quartz 500W	1.510	Ea.
1500W	1.900	Ea.
Ceiling, Surface Mounted Vaportight		
100W	2.650	Ea.
150W	2.950	Ea.
175W	2.950	Ea.
250W	2.950	Ea.
400W	3.350	Ea.
1000W	4.450	Ea.

High intensity discharge lighting (HID) sources include mercury vapor, metal halide, high pressure sodium, and low pressure sodium lamps. HID lamps are usually installed to light large indoor and outdoor areas, such as factories, gymnasiums, sports complexes, parking lots, building perimeters, streets, and highways.

A mercury vapor lamp is a glass bulb which contains high pressure mercury vapor. It works on the same principle as a fluorescent lamp, except that the pressure of the mercury vapor within the bulb is much higher than that of a standard fluorescent tube. Like the incandescent lamp, the mercury vapor lamp usually contains a metal screw base and is manufactured in a variety of shapes. Some models are also available with a side-prong base. The size of mercury vapor lamps ranges from 40 watts to 1000 watts, with clear or coated surfaces. Clear mercury vapor lamps, which produce a blue-green color, are not recommended for indoor use; but coated models are comparable in color to fluorescent lamps and may be used indoors. Mercury lamps are small enough to be used in many types of indoor and outdoor fixtures. Like fluorescent lamps, they require a ballast, which may be mounted in the fixture or at a remote location. Discretion should be used in locating the ballasts, as they are noisy.

Like other lighting sources, mercury vapor lamps have their advantages and disadvantages. Among the advantages is a rated life of 24,000+ hours and greater energy efficiency than incandescents when installed in regular mercury lamp and ballast combinations. They are less efficient when installed with adapter kits into standard incandescent sockets. Self-ballasted lamps are also less efficient. They are also less efficient than metal halide and sodium lamps. Another disadvantage is that mercury vapor lamps require a warm-up period of several minutes before reaching full output, as well as a cool-down period before restarting. Quartz lamps are usually included in circuit for temporary lighting during these periods.

A metal halide lamp is basically a mercury vapor lamp with modifications in its arc tube arrangement. Generally, the color, as well as the efficiency, is better in this type of lamp when compared to conventional mercury vapor models. The rated life of metal halide lamps, however, falls below that of mercury lamps. The size of these lamps ranges from 175 watts to 1500 watts.

High pressure sodium lamps differ from mercury lamps primarily in the type of vapor contained in the arc tube. These lamps use a mixture of sodium, mercury, and xenon to produce a slightly yellow color. Like mercury lamps, sodium lamps are efficient in energy consumption and high in rated lamp life. They have an advantage over mercury lamps in that their start and restart times are faster. High pressure sodium lamps range in size from 35 watts to 1000 watts. They are also available in medium base sizes up to and including 150 watts. Some models can be used as a retrofit with certain existing mercury ballasts. High pressure sodium lamps are gaining in popularity for use in office lighting because they are small, convenient, and more efficient than fluorescent lamps. However, their color is sometimes found objectionable for office use.

The low pressure sodium lamp is more efficient than the high pressure sodium type, but its intense yellow color makes it unsuitable for indoor use. This type of lamp is used primarily for lighting large outdoor areas, such as roadways, parking lots, and places that require security lighting. Low pressure sodium lamps range in size from 18 watts to 180 watts in lengths of 8-1/2" to 44-1/8". The rated life of a low pressure sodium lamp is 18,000 hours.

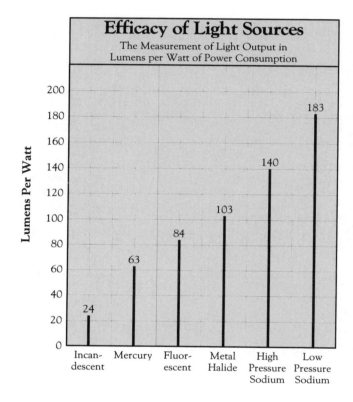

Efficacy of Light Sources
The Measurement of Light Output in Lumens per Watt of Power Consumption

Source	Lumens Per Watt
Incandescent	24
Mercury	63
Fluorescent	84
Metal Halide	103
High Pressure Sodium	140
Low Pressure Sodium	183

Man-hours

Description	m/hr	Unit
Ceiling Recessed Mounted Prismatic Lens		
Integral Ballast 2' x 2' HID 150W	2.500	Ea.
250W	2.500	Ea.
400W	2.760	Ea.
Surface Mounted 250W	2.960	Ea.
400W	3.330	Ea.
High Bay Aluminum Reflector 400W	3.480	Ea.
1000W	4.000	Ea.

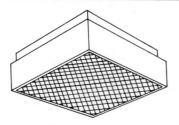

Mercury Vapor Ceiling Fixture, Recessed, Integral Ballast

Mercury Vapor Fixture, Surface Mounted

Mercury Vapor Fixture, Round, Pendent

Mercury Vapor Fixture, Square, Pendent Mounted

Mercury Vapor Fixture, Square, Wall Mounted

High Pressure Sodium Fixture, Round, Surface

High Pressure Sodium, Round, Wall Mounted

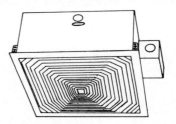

Vaporproof, High Pressure Sodium Fixture, Recessed

Vaporproof, High Pressure Sodium Fixture, Wall Mounted

Metal Halide Fixture, Square, Recessed

337

Vaporproof, Metal Halide Fixture, Surface Mounted

Bollard Light Fixture with Polycarbonate Lens

Vaporproof, Metal Halide Fixture, Pendent Mounted

9 ELECTRICAL
HAZARDOUS AREA WIRING

All hazardous locations require special material and wiring methods as determined by the classifications defined in the National Electrical Code. The descriptions of hazardous locations listed below have been abbreviated to provide examples of some of the common classifications. The National Electrical Code should be consulted for complete descriptions.

CLASS 1 LOCATIONS are those in which flammable gases or vapors are or may be present in the air in quantities sufficient to produce explosive or ignitable mixtures.

Class 1, Division 1: This classification includes locations where hazardous atmosphere is expected during normal operations; locations where a breakdown in operation of processing equipment results in release of hazardous vapors and the simultaneous failure of electrical equipment.

Class 1, Division 2: This classification includes locations in which volatile flammable liquids or gases are handled, processed, or used, but will normally be confined to closed containers or systems from which they can escape only by accidental rupture or breakdown; locations where hazardous conditions will occur only under abnormal circumstances.

CLASS 2 LOCATIONS are those in which hazardous conditions exist because of the presence of combustible dust.

Class 2, Division 1: This classification includes locations where combustible dust may be suspended in the air under normal conditions in sufficient quantities to produce explosive or ignitable mixtures; locations where a breakdown in operation of machinery or equipment might cause a hazardous condition to exist while creating a source of ignition with the simultaneous failure of electrical equipment; locations in which combustible dust of an electrically conductive nature may be present.

Class 2, Division 2: This classification included locations where air-suspended combustible dust is not at hazardous levels, but where an accumulation of dust may interfere with the safe dissipation of heat from electrical equipment, or may be ignited by arcs, sparks, or burning material located near electrical equipment.

CLASS 3 LOCATIONS are those in which the presence of easily flammable fibers or flyings are present in the air, but not in sufficient quantities to produce ignitable mixtures under normal conditions.

Class 3, Division 1: This classification includes locations where easily ignitable fibers or materials producing combustible flyings are handled, manufactured, or used.

Class 3, Division 2: This classification includes locations where easily ignitable fibers are stored or handled.

As noted earlier, special material and wiring methods are required for hazardous area wiring installations. Class 1 and Class 2, Division 1 installations must include rigid conduit or steel IMC with at least five full tapered threads tightly engaged in the enclosure. MI cable is also allowed in most cases; and MC, SNM, MV, TC, an PLTC cables are acceptable in some instances. All conduit boxes and enclosures must be of threaded or flat-jointed bolted construction. The matching surfaces of both the cover an enclosure should be lubricated and free from nicks and other imperfections.

Sealing fittings are also required for several reasons. Sealing fittings restrict the passage of gases, vapors, and flames form one part of the electrical installation to another at atmospheric pressure and normal ambient temperatures. They also help to contain explosions that may occur in electrical enclosures and prevent precompression, or "pressure piling," in conduit systems. The sealing fittings must be installed with both an acceptable fiber, to create a dam, and sealing compound. The fittings are available in several configurations.

Sealing fittings must be installed where conduit enters any enclosure that houses arcing or high temperature equipment. Sealing fittings are also required where conduit leaves a Division 1 or Division 2 location and enters a nonhazardous area and where 2″ or larger conduit enters enclosures that house terminals, splices, or taps. Sealing fittings are not required for enclosures or devices which are factory sealed.

Explosionproof Sealing Fitting

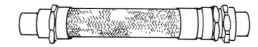

Explosionproof Flexible Coupling

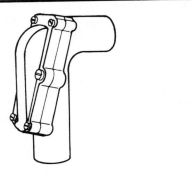

Explosionproof Pulling Elbow

Explosionproof Round Box with Cover, 3 Threaded Hubs

Explosionproof NEMA 7, Surface Mounted, Pull Box

Man-hours

Description	m/hr	Unit
Sealing Fitting 1/2" Diameter	.660	Ea.
2" Diameter	1.600	Ea.
3" Diameter	2.000	Ea.
4" Diameter	2.670	Ea.
Flexible Coupling 3/4" Diameter x 12" Long	.800	Ea.
2" Diameter x 12" Long	1.740	Ea.
3" Diameter x 12" Long	2.670	Ea.
4" Diameter x 12" Long	3.330	Ea.
Pulling Elbow 3/4" Diameter	1.000	Ea.
2" Diameter	2.000	Ea.
3" Diameter	2.670	Ea.
Conduit LB 3/4" Diameter	1.000	Ea.
T 3/4" Diameter	1.330	Ea.
Cast Box NEMA 7		
6" L x 6" W x 6"D	4.000	Ea.
12"L x 12"W x 6"D	8.000	Ea.
18"L x 18"W x 8"D	16.000	Ea.

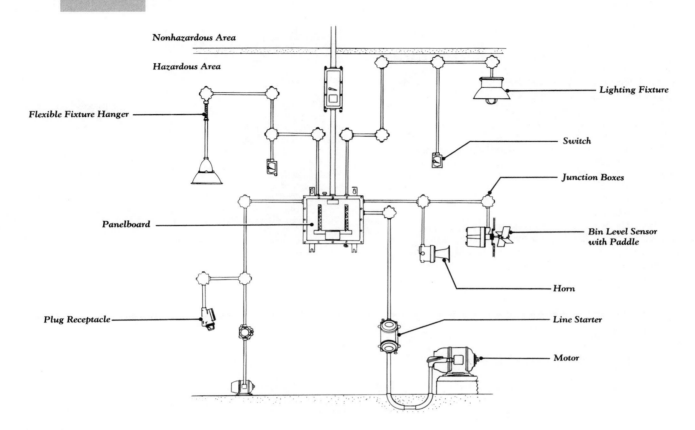

Nonhazardous Area

Hazardous Area

Flexible Fixture Hanger

Lighting Fixture

Switch

Junction Boxes

Panelboard

Bin Level Sensor with Paddle

Horn

Plug Receptacle

Line Starter

Motor

Class II Lighting Installation

Hazardous area lighting fixtures are available in fluorescent, incandescent, and high intensity discharge (HID) lamp designs. Complete descriptions of fittings for the various types of hazardous area lighting fixtures can be found in the National Electrical Code.

Hazardous area lighting fixtures are manufactured in many styles, wattages, and mounting arrangements. Fluorescent fixtures are available for 40 watt lamps in 2-, 3-, and 4-lamp configurations for surface or pendant mounting. Incandescent fixtures range from 100 watt to 500 watt ratings in several designs and lamp configurations. Reflectors and guards for standard fixtures, exit signs, and 100 watt portable handlamps are also available for

incandescent lighting fixtures. HID hazardous lighting fixtures are also manufactured in several types, some with factory-sealed integral ballast. Reflectors, guards, and stanchion-mounted models are also available. The wattage ratings for the various HID lamps used in hazardous area lighting fixtures are as follows: mercury vapor, 100 watt to 400 watt; high pressure sodium, 50 watt to 400 watt; metal halide sodium, 175 watt to 400 watt.

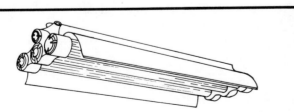

Explosionproof Fluorescent Fixture,
Pendent Mounted, 3 Tube

Man-hours

Description	m/hr	Unit
Fixture, Pendant Mounted, Fluorescent 4' Long		
2-40W RS	3.480	Ea.
4-40W RS	4.710	Ea.
Incandescent 200W	2.290	Ea.
Ceiling Mounted, Incandescent 200W	2.000	Ea.
Ceiling, HID, Surface Mounted 100W	2.670	Ea.
150W	2.960	Ea.
250W	2.960	Ea.
400W	3.330	Ea.
Pendent Mounted 100W	2.960	Ea.
150W	3.330	Ea.
250W	3.330	Ea.
400W	3.810	Ea.

9 ELECTRICAL
DEVICES: STARTERS HAZARDOUS AREAS

Because of the large assortment of electrical materials and devices available for the many hazardous area classifications, the list of equipment which follows is not complete. The National Electrical Code and various manufacturers' catalogs should be consulted for a complete listing of materials and methods of installation.

Plugs and receptacles are manufactured in several types with a range of 7A to 200A and up to 600V. Some receptacles are available in factory-sealed models to eliminate the need for field seal installations. Some receptacles are also manufactured in interlocking design with safety switches or circuit breakers. Combination motor/starters of up to 200 horsepower, 600V are also available. Circuit breakers for hazardous area starters range up to 600A, 600V. Many styles, models, and configurations of the following materials and devices are also manufactured for hazardous area starters: push buttons, pilot lights and selector switches, panelboards, wall switches, safety switches, manual starters, signal bells, horns, sirens, thermostats, and telephones.

Man-hours

Description	m/hr	Unit
Circuit Breaker NEMA 7 600 Volts 3 Pole		
50 Amps	3.480	Ea.
150 Amps	8.000	Ea.
400 Amps	13.330	Ea.
Control Station Stop/Start	1.330	Ea.
Stop/Start Pilot Light	2.000	Ea.
Magnetic Starter FVNR 480 Volts 5 HP Size O	5.000	Ea.
25 HP Size 2	8.890	Ea.
Combination 10 HP Size 1	8.000	Ea.
50 HP Size 3	20.000	Ea.
Panelboard 225 Amps M.L.O. 120/208 Volts		
24 Circuit	40.000	Ea.
Main Breaker	53.330	Ea.
Wall Switch, Single Pole 15 Amps	1.510	Ea.
Receptacle 15 Amps	1.510	Ea.

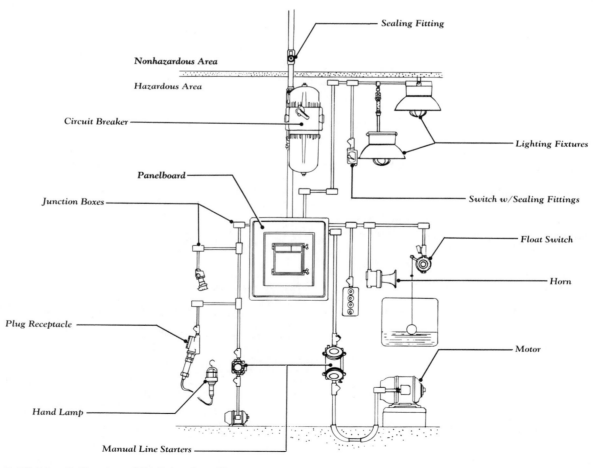

Class I, Division 2, Power and Lighting Installation

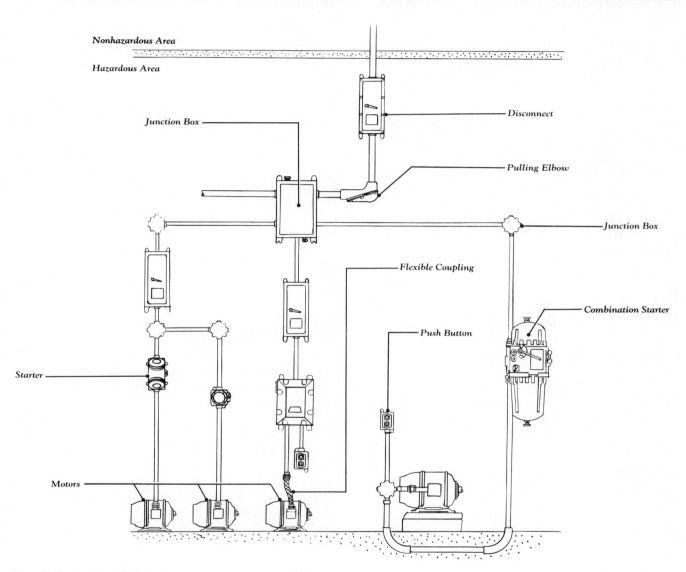

Nonhazardous Area

Hazardous Area

Disconnect

Junction Box

Pulling Elbow

Junction Box

Flexible Coupling

Combination Starter

Push Button

Starter

Motors

Class II Power Installation

At one time the aesthetic impact of parking area lighting systems was far outweighed by practical considerations. Today, however, manufacturers and designers of exterior area lighting systems offer a wide variety of styles, materials, and sizes to enhance most architectural designs. Generally, all parking area lighting systems still consist of four basic components: the luminaire, or light fixture; the slip fit and mounting bracket for the fixture; the pole; and the pole base.

In most installations, the luminaire is specifically designed to accommodate one of the three types of HID lamps. The type of lamp and popular sizes are as follows: mercury vapor, 175/250/400/1000 watt; metal halide, 175/400/1000 watt; high pressure sodium, 100/150/250/400/1000 watt. The ballast voltage for the HID lamp must be specified as 120V, 208V, 240V, 277V, or 480V. In most cases, the outside of the fixture includes a lightweight cast aluminum doorframe and a door, usually made of alzak aluminum, which serves as both the lamp enclosure and primary reflector. Within the fixture is a lamp holder, prewired ballast pack, secondary reflector, and the lamp. The fixture lens is normally heat tempered and shock resistant. The reflector arrangement produces a highly controlled rectangular pattern which distributes the lighting efficiently in conventionally shaped parking areas. The units can be mounted in configurations of one, two, three, or four fixtures per pole.

Because each installation is different, the mounting brackets used in each system must be carefully selected according to the type of fixture and the pole on which it will be mounted. Generally, two methods are used to attach the bracket to the pole. The first is the slip fit method in which the bracket is slipped onto a 3″ top tenon of the pole and then secured by set screws and thru bolt. In the second method, the bracket is bolted directly to pretapped holes in the pole. This method usually provides for the mounting of four fixture brackets per pole. If fewer than four brackets are to be mounted on a pole, the unused pretapped holes are covered with an aluminum shield.

The poles are used to support the lighting fixtures and brackets are usually manufactured from aluminum, steel, or fiberglass. Wood is also used, but not as often as these materials. Aluminum poles are available in round, square, and tapered models. They range in length from 20′ (weight of about 80 lbs.) to 40′ (weight of about 360 lbs.). Steel poles are also available in round, square, and tapered designs. They range in length from 20′ (weight of about 152 lbs.) to 50′ (weight of about 721 lbs.). Fiberglass poles are manufactured in 20′ lengths (weight of about 70 lbs.) and 40′ (weight of about 162 lbs.). In most cases, the bases for all aluminum, steel and fiberglass poles are molded into the pole's construction. The base is usually secured by four anchor bolts preset in a concrete pad.

If the electrical design for the lighting system requires that the fixture ballast be located near ground level, a steel transformer base is usually installed. This compartment may be up to 24″ high with an 18″ bolting base. The transformer base features two removable side panels which allow easy and convenient access to the ballast without the need of specialized equipment. This type of base ranges from 90 lbs. to 150 lbs. in weight and can accommodate up to four fixture ballasts.

Man-hours

Description	m/hr	Unit
Luminaire HID 100 Watt	2.960	Ea.
150W	2.960	Ea.
175W	2.960	Ea.
250W	3.330	Ea.
400W	3.640	Ea.
1000W	4.000	Ea.
Bracket Arm 1 Arm	1.000	Ea.
2 Arm	1.000	Ea.
3 Arm	1.510	Ea.
4 Arm	1.510	Ea.
Aluminum Pole 20′ High	8.300	Ea.
30′ High	9.250	Ea.
40′ High	12.000	Ea.
Steel Pole 20′ High	9.250	Ea.
30′ High	10.500	Ea.
40′ High	14.200	Ea.
Fiberglass Pole 20′ High	6.000	Ea.
30′ High	6.700	Ea.
40′ High	8.600	Ea.
Transformer Base	2.670	Ea.

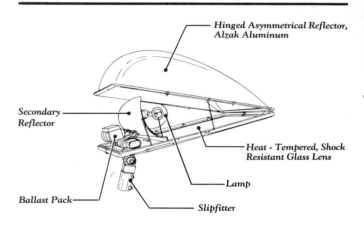

Hinged Asymmetrical Reflector, Alzak Aluminum

Secondary Reflector

Heat - Tempered, Shock Resistant Glass Lens

Lamp

Ballast Pack

Slipfitter

Luminaire Construction Features

Large Luminaire Light Fixture, 1000 Watt

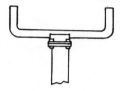

Top Bracket, 2 in Line Tenons

Top Bracket, 3 Tenons at 120°

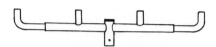

Top Bracket, 4 in Line Tenons

Bracket Arm Types

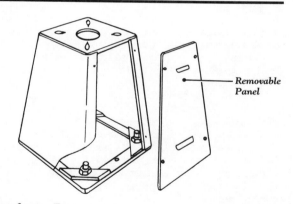

Tapered Pole

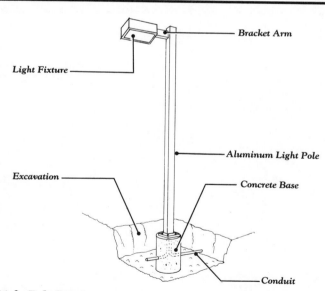

Bracket Arm

Light Fixture

Aluminum Light Pole

Excavation

Concrete Base

Conduit

Light Pole System

Removable Panel

Steel Transformer Base

Receptacles provide a convenient means of connecting portable power equipment and electrical appliances to the electrical source. These devices are available in voltage ratings of 125, 208, 250, 277, 347, 480, and 600 volts and in amperage ratings ranging from 10 to 400 amps. Receptacles are classified as either the grounded or ungrounded type. Because of the dangers of fire and shock associated with all electrical installations, care should be exercised in determining the proper size and type of receptacle and in following installation guidelines, as well as local, state, and national codes.

In accordance with the National Electrical Code, all receptacles rated at 15 and 20 amps must be classified as the grounding type. The only exception to this regulation applies to the replacement of outlets that are installed in existing ungrounded systems. Installing a grounded receptacle as a replacement in such an ungrounded system may cause a false sense of safety and security. Because of the great variation of voltage and amperage combinations and the National Electrical Code rule that calls for specific voltage and current ratings for all grounding-type receptacles, standard configurations have been adopted by NEMA (National Electrical Manufacturers Association). This standardization also prevents low-voltage caps (plugs) from being inserted into high-voltage receptacles. Configurations for general purpose, grounding-type receptacles and caps appear in the tables included in this section.

Receptacles are connected to the wiring leads in several ways, depending on the set-up of the particular receptacle or its amperage rating. Some receptacles with 15- and 20-amp ratings are equipped with terminal screws to which the lead wire is attached. After the end of the wire is stripped of its covering, it is wrapped clockwise around the terminal screw, which is then securely tightened to form a mechanical connection. Another common type of 15- and 20-amp receptacle provides a pressure-lock connection in place of the terminal screws. With this type of receptacle, the stripped end of the wire is pushed into a recessed pressure-locking contact that grips the wire to form a permanent locking connection. Receptacles rated at 30 amps and higher are equipped with set screws to form wire-to-device connections.

Several grades of receptacles exist to assure that the proper capacity and durability requirements for their use are met. Residential-grade receptacles can be installed in structures located only in noncommercial areas; specification-grade receptacles are used in office and industrial locations. Hospital-grade receptacles, which are labeled "Hospital Grade" and contain a green dot on the face of the outlet, must be able to withstand more severe damage tests than conventional receptacles.

Another safety rating required for receptacles is the ground-fault type, or GFI, which is specified by code for installations in bathrooms, attached garages, construction sites, and outdoor locations. These receptacles are equipped with a safety switch set at 5 milliamps to prevent accidental shock and with a reset button to restore power when a ground condition has been cleared.

All receptacles must be housed in some type of specified enclosure, which is determined by the installation procedure. Outdoor installations require cast weatherproof boxes and matching weathertight covers. Commercial and industrial receptacle enclosures often consist of a 4″ square box with a raised cover. In residential and office situations, where a permanent outlet must be inconspicuous, recessed single-gang boxes may be used. The covers for all receptacle enclosures are usually manufactured from metal or plastic materials.

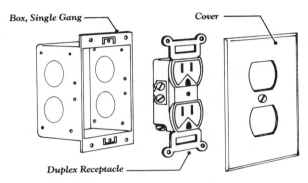

Receptacle, Including Box and Cover

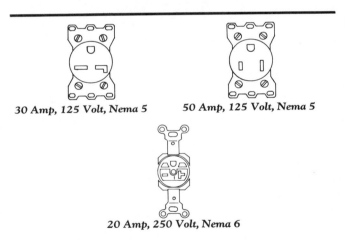

30 Amp, 125 Volt, Nema 5   **50 Amp, 125 Volt, Nema 5**

20 Amp, 250 Volt, Nema 6

Receptacles

Man-hours

Description	m/hr	Unit
Receptacle 20A 250V	.290	Ea.
Receptacle 30A 250V	.530	Ea.
Receptacle 50A 250V	.720	Ea.
Receptacle 60A 250V	1.000	Ea.
Box, 4″ Square	.400	Ea.
Box, Single Gang	.290	Ea.
Box, Cast Single Gang	.660	Ea.
Cover, Weatherproof	.120	Ea.
Cover, Raised Device	.150	Ea.
Cover, Brushed Brass	.100	Ea.

Receptacles, Connectors, and Matching Cap Configurations

NEMA No.	15A	20A	30A	50A	60A
1 125V					
2 250V					
5 125V					
6 250V					
7 277V ,AC					
10 125/250V					
11 3 Phase 250V					
14 125/250V					
15 3 Phase 250V					
18 3 Phase,Y 120/208V					

Locking Configurations

NEMA No.	15A	20A	30A	NEMA No.	15A	20A	30A
L 1 125V				**L 13** 3 Phase 600V			
L 2 250V		15A		**L 14** 125/250V			
L 5 125V				**L 15** 3 Phase 250V			
L 6 250V				**L 16** 3 Phase 480V			
L 7 277V ,AC				**L 17** 3 Phase 600V			
L 8 480V				**L 18** 3 phase, Y 120/208V			
L 9 600V				**L 19** 3 Phase,Y 277/480V			
L 10 125/250V				**L 20** 3 Phase,Y 347/600V			
L 11 3 Phase 250V				**L 21** 3 Phase,Y 120/208V			
L 12 3 Phase 480V				**L 22** 3 Phase,Y 277/480V			
				L 23 3 Phase,Y 347/600V			

9 ELECTRICAL
UNDERCARPET POWER SYSTEMS

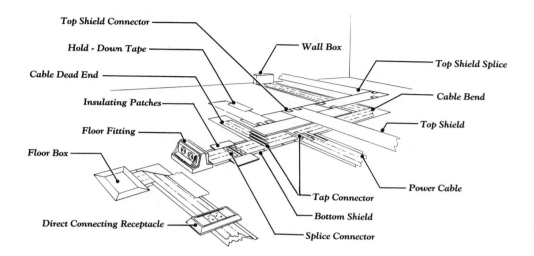

Top Shield Connector

Hold - Down Tape

Cable Dead End

Insulating Patches

Floor Fitting

Floor Box

Direct Connecting Receptacle

Wall Box

Top Shield Splice

Cable Bend

Top Shield

Power Cable

Tap Connector

Bottom Shield

Splice Connector

An undercarpet power system is an alternative to conventional round cable for wiring of commercial and industrial offices. It is a method of distributing power almost anywhere on the floors without having to channel through underfloor ducts walls or ceilings.

The flat, low profile design of this system allows for its installation directly on top of wood, concrete, composition or ceramic floors. It is then covered with carpet squares up to a maximum of 30″ square. The undercarpet system cannot be used outdoors or in wet locations, or where it would be subject to corrosive vapors. Nor is it appropriate for residential use, schools or hospital buildings.

An undercarpet power system is a branch circuit wiring method. It uses a three-layer construction consisting of a plastic bottom shield, a flat flexible cable with separate conductor legs, and a protective zinc plate steel top shield.

Insulating patches and cable tapping and splicing connectors are available to provide flexibility in wiring. The connectors are terminated with an undercarpet crimping tool.

The transition from flat conductor cable to round wire is made using an insulating, piercing transition block; this block can be installed using only a screwdriver for termination. Fittings are provided to protect the power source connections to the circuit and for mounting single and double duplex receptacles.

The flat conductor cable is available in sizes corresponding to #12 (20 amp rating) or #10 (30 amp rating) American Wire Gauge (AWG) round wire.

The number of conductors may be 3, 4, or 5, depending on the number of outlets to be serviced.

Flat conductor cable is rated at 300 volts and the three conductor cable is designed for use on 120V AC branch circuits.

The two and three circuits which utilize four and five conductor flat cable have been designed for use with 240/120V single phase and 208/120V three phase systems.

The top shield is made from a corrosive-resistant, zinc-coated steel sheet metal. The shield is a fully grounded component covering the entire system and providing physical protection for the cable. It is grounded to prevent fault and to minimize hazards. A spray adhesive and tape are used to attach the top shield to the floor surface.

The bottom shield is made of a nonconductive vinyl film. The shield is designed to protect the cable from floor moisture, chemical reaction, and abrasion.

The tap and splice connectors are made from copper alloy with insulation-piercing prongs. They are designed to connect two flat cable conductors. This connection is performed with a hand-crimping tool.

Transition fittings (wall and floor box) are made from zinc-plated steel, shaped into a box-like configuration. This box houses the transition block and provides protection where round supply cable connects to flat conductor cable.

Transition blocks include a plastic block, copper straps and mounting screws. The blocks are designed to connect round conductors to flat conductor cable - both within the transition box and at floor fittings.

Floor fittings are available in standard types and sizes to accommodate power, telephone and data connections under a single pedestal. The floor fitting is designed to provide protection for the round conductor and the flat conductor connections. It also serves as the pedestal for the connection of equipment to the power source.

The insulating patch is made from a polyester film coated on one side with a sealant gel. Two insulating patches are required when a cable is spliced, tapped or ended. The patches are designed to provide protection for the splice and for a watertight seal around exposed conductors.

The direct connect receptacle includes a steel mounting plate, a plastic housing and covers, copper terminals, stainless steel bonding clips, and the hardware required for attaching. The receptacle is terminated directly to the flat cable by tightening screws. In this way, the terminal prongs are able to pierce the cable insulation and engage the conductors.

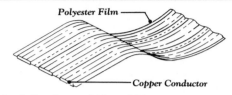

Flat Cable, 3 Conductor #12 or #10

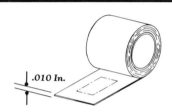

Top Shield

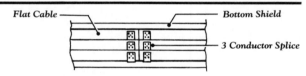

3 Conductor, Flat Cable Splice

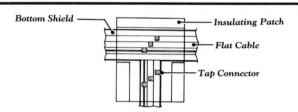

3 Conductor, Flat Cable Tap

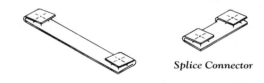

Connectors

Hand Crimping Tool

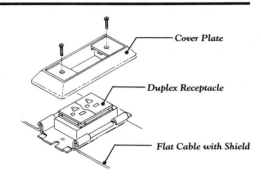

Direct Connect Receptacle

Pedestal Duplex Receptacle

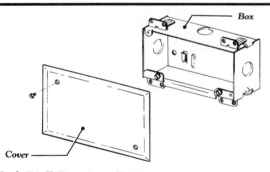

Flush Wall Transition Fitting

Man-hours

Description	m/hr	Unit
Cable 3 Conductor #12 w/Bottom Shield	.008	lf
Cable 5 Conductor #12 w/Bottom shield	.010	lf
Splice 3 Conductor w/Insulating Patch	.334	Ea.
Splice 5 Conductor w/Insulating Patch	.334	Ea.
Tap 3 Conductor w/Insulating Patch	.367	Ea.
Tap 5 Conductor w/Insulating Patch	.367	Ea.
Receptacle w/Floor Box Pedestal Type	.500	Ea.
Receptacle Direct Connect	.320	Ea.
Top Shield	.005	lf
Transition Block	.104	Ea.
Transition Box, Flush Mount w/Cover	.400	Ea.

9 ELECTRICAL
MOTOR CONTROL SYSTEM

A motor control system consists of the various electrical equipment necessary to operate and protect motors in an industrial or manufacturing facility. The system is comprised of three major components: a disconnecting device to control the circuit between incoming power and the controller; controller, or motor starter, to stop and start the motor and to provide overload protection; and a motor control device to energize or de-energize the starter.

Several types of disconnecting devices are available for motor control systems, depending on the size and use of the motors within the facility. Circuit breakers open and close the incoming circuit manually, or automatically at a predetermined value of current. Circuit breakers are usually installed in systems with low horsepower motors. A single breaker may protect on or several motors within the system. According to NEC standards, the maximum rating or setting for AC motor protection with a magnetic circuit breaker is 700% of the full load current. Fused disconnects, which are more commonly used than circuit breakers, are available in standard and heavy duty classifications in a wide range of ratings. Some commonly installed sizes are 30A, 60A, 100A, 200A, 400A, and 600A, at 250V and 600V.

The fuses themselves are manufactured in three types: dual element, single element, and current limiting. Magnetic-type starters are also used as disconnecting devices and provide overload protection. They are basically electromagnetic switches which start and stop the motor when voltage is applied to or disengaged from a magnetic coil. Magnetic starters can be used to operate motors for most types of machinery, including compressors, pumps, conveyors, and presses.

Because motors are rated by voltage, horsepower, and amperage, starters are correspondingly rated. Eleven sizes of starters are presently manufactured to control motors rated to 1600 horsepower. Some motors may draw as little as 1A and others as much as 1200A; consequently, the motor controllers are rated to carry continuous current in a specific range from 9A to 1225A. The sizes and corresponding ampere ratings of starters used with the various motor controllers, as established by NEMA, are: 00/9A, 0/18A, 1/27A, 2/45A, 3/90A, 4/150A, 5/270A, 6/540A, 7/810A, 8/1215A, and 9/2250A. More detailed descriptions of these starter sizes can be found in the official NEMA listings.

In systems using magnetic starters, the current flows from the power source through the closed power contacts and the overload thermal units to the motor. Motor control devices are contained in this control circuit, which is that portion of the wiring that conducts power to the coil of the starter. This circuit also contains the overload contact, which opens when excess current is sensed by the thermal unit within the starter power circuit.

The control power can be arranged in either "common" or "separate" control format. A common control circuit is tapped directly from L1 and L2 of the power circuit, and therefore the voltage of the control circuit is the same as that between L1 and L2 in the power circuit. The common control arrangement is normally used for low voltage

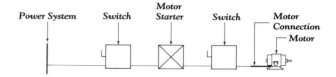

systems, but systems which require high voltage, such as 480V, in their power circuit are usually set up in separate control format. The reason for the separate control arrangement is to protect the operator from coming into contact with unnecessarily high voltage while operating the system.

Separate control arrangements contain two distinct circuits, each with a special voltage. The high voltage power circuit is connected directly to the motor, and the low voltage (usually 120V) control circuit is wired from a separate source to the control device. Another way of setting up a low voltage control circuit is to install a fused control transformer between the line power and control coil. This method eliminates the need for a separate source for the control circuit and reduces the control voltage to only 24V.

Some of the devices used to control the power to the starter coil are start-stop buttons, selector switches, limit switches, pressure switches, foot switches, pilot devices, electrical interlocks, and timers. Special control modules which can be combined with pilot indicator lights for complex motor control systems are also available. Most of these control devices can be either integrally mounted or remotely located.

Man-hours

Description	m/hr	Unit
Heavy Duty Fused Disconnect 30 Amps	2.500	Ea.
60 Amps	3.480	Ea.
100 Amps	4.210	Ea.
200 Amps	6.150	Ea.
600 Amps	13.330	Ea.
1200 Amps	20.000	Ea.
Starter 3-Pole 2 HP Size 00	2.290	Ea.
5 HP Size 0	3.480	Ea.
10 HP Size 1	5.000	Ea.
25 HP Size 2	7.270	Ea.
50 HP Size 3	8.890	Ea.
100 HP Size 4	13.330	Ea.
200 HP Size 5	17.780	Ea.
400 HP Size 6	20.000	Ea.
Control Station Stop/Start	1.000	Ea.
Stop/Start, Pilot Light	1.290	Ea.
Hand/Off/Automatic	1.290	Ea.
Stop/Start/Reverse	1.510	Ea.

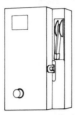

Combination Starter and Circuit Breaker Disconnect   *Circuit Breaker - NEMA 1*   *Circuit Breaker, Explosionproof - NEMA 7*

Circuit Breakers

Manual Motor Starter   *Magnetic Motor Starter*   *Motor Starter and Control, Magnetic*

Control Station - NEMA 1   *Control Station, Watertight - NEMA 4*

Motor Starters and Control Stations

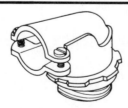

Straight Connector   *Angle Connector*

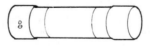

Cartridge, Nonrenewable   *Dual Element, Class R*

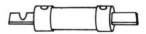

Dual Element, Class J

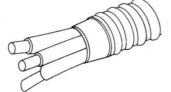

Flexible Conduit

Fuses

Motor Connections

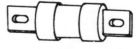

9 ELECTRICAL
MOTOR CONTROL CENTER

A motor control center is an assembly of various motor control equipment and feeders housed within a series of steel-clad enclosures. It is used to provide a compact and convenient centralization of the electrical components which service a given area within a building. Because of the modular arrangement, the components within the control center can be easily serviced or replaced without disturbing the operation of other equipment within the enclosure. Control centers were originally designed to provide for the mounting or housing of motor control combination starters and their respective circuit breakers or fused disconnects. In industry today, control centers still house motor control equipment, but they also serve as enclosures for lighting transformers and distribution panels within the area serviced by the unit.

Control centers are usually made up of a series of 20″-wide by 90″-high by 12″- to 20″- deep modular steel structures that can be bolted together in various configurations to form a continuous control line-up. Enclosures for larger-sized starters are available in 30″ to 36″ widths and in depths greater than 20″. The most common arrangement of the modules is a straight line with front-mount equipment; the second most common configuration is a straight line with equipment mounted front and back. The front-mounted arrangement measures 12″ to 14″ in depth; the front and back, or back-to-back, configuration measures 20″ in depth.

Various sizes of starters require modules of different heights. Starters of sizes 2, 3, and 4 can be enclosed in the 24″- and 36″- high modules; sizes 5 and 6 can be housed in the modules measuring 42″, 48″, 54″, 60″, and 72″ in height. Many types and sizes of starters are available, including nonreversing, reversing, multi-speed full voltage control, primary resistor, autotransformer, part-winding, and reduced voltage control models.

Several types of NEMA enclosures may be used for housing the components within the motor control center, but the NEMA 1 enclosure is the most commonly employed. This type of housing covers all control and bus work and minimizes dust and dirt penetration. If greater dust protection is required, a gasketed NEMA 1 or an even more dust resistant NEMA 12 enclosure may be utilized. NEMA 3 and 4 enclosures may also be used in outdoor locations or where resistance to dust and water are mandated.

Within the enclosure is a common three-phase horizontal bus to which is attached a three-phase vertical bus for supplying power to the various components in the vertical structures. Each vertical structure contains three vertical bus bars of approximately 6′ in length. Control components are usually assembled in 12″ modular height units, or longer, and are plugged into the vertical bus by means of spring tab fingers. Typically sized units are 12″, 18″, 24″, 36″, and 48″ in height. The incoming line cables enter the center section of the enclosure from access holes in the top, bottom, front, rear, or sides. These supply cables may be single or multiple conductor per phase in sizes up to 1,000 MCM. A main breaker or fused disconnect, which may be separately mounted or located within the enclosure in 600V switchgear, it is necessary to protect the bus of each motor control center.

Motor control centers are available in two standard classes, with several types of units making up each class. These classifications and types are defined as follows:

CLASS 1

Type A: This type consists of a control unit with a circuit breaker or fusible disconnect wired to the line side of the starter only.

Type B: This type is the same as Type A, but the control circuit leads are wired to a fixed terminal block on the control unit.

Type C: This type is the same as Type B, but the leads are brought to the control unit terminal boards which are located at the top or bottom of the motor control center.

Interwiring and interlocking do not exist between starters or cubicles in any type in this class of unit.

CLASS 2

Type B: This type is the same as Class 1, Type B, but it contains wiring between control units in the same or adjacent cubicles.

Type C: This type is the same as Class 2, Type B, but it provides interwiring from the master terminal boards at the top and bottom of the control center.

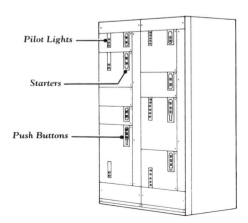

Pilot Lights

Starters

Push Buttons

Man-hours

Description	m/hr	Unit
Structures 300 Amps 72″ High	10.00	Ea.
Structures 300 Amps 72″ High Back to Back Type	13.300	Ea.
Starters Class 1 Type B Size 1	3.000	Ea.
Size 2	4.000	Ea.
Size 3	8.000	Ea.
Size 4	11.400	Ea.
Size 5	16.000	Ea.
Size 6	20.000	Ea.
Pilot Light Wiring in Starter	.500	Ea.
Push Button Wiring in Starter	.500	Ea.
Auxilliary Contacts in Starter	.500	Ea.

A safety switch provides a manual method of disconnecting power, from electrical equipment, which cannot be overridden by automatic controls. If the switch is fused, it provides branch circuit protection. The safety switch itself is mounted inside a sheet metal or cast iron enclosure and controlled by a handle connected to the mechanism and located on the outside of the enclosure. The switch mechanism is manufactured as either single- or double-break type, in knife-blade or butt-contact format. Because these switches usually conduct large amounts of power, the enclosure doors are designed so that they cannot be opened if the switch handle is set in the "up" or "on" position.

Safety switches are manufactured in general-duty and heavy-duty grades and in both fused and unfused types. General-duty switches are available in two- and three-pole design for up to 240 volts, with ratings of 30, 60, 100, 200, 400, and 600 amps. Heavy-duty switches are also manufactured in two- and three-pole formats, with voltage ratings of 250 and 600 volts and amperage ratings of 30, 60, 100, 200, 400, 600, 800, and 1,200 amps.

During installation, line side power is connected to the terminal lugs, which are located at the top of the switch before the blade mechanism. Power to a piece of electrical equipment is tapped off the load side of the switch mechanism. Accidental feeding of the load terminals, which would cause the blades of the switch to become "hot" even when the switch is in the open position, can be avoided by following precisely correct line and load installation rules.

The enclosures for safety switches are classified by the NEMA according to their particular interior or exterior safety function. Several of the NEMA enclosure types classified for interior use include NEMA 1, NEMA 2, NEMA 7, and NEMA 12. The NEMA 1 enclosure type is intended for indoor use where no unusual service conditions exist. Its primary function is to protect personnel from accidental contact with live electrical equipment. NEMA 2 is designed with a drip-proof housing to protect the enclosed equipment against falling dirt and noncorrosive liquids. NEMA 7 is used to protect equipment in hazardous locations where indoor atmospheres containing volatile gas and vapors may cause explosion. This type of switch enclosure is marked to show the class and group letter designation. NEMA 12, which is designed for indoor industrial use, features dust- and drip-tight seals. Because it contains no conduit openings or knockouts, access may be gained only through field-installed holes with oil resistant gaskets. An additional safety feature of this type of enclosure is that a special tool is required to open its cover.

Two commonly used exterior NEMA enclosure classifications include NEMA 3 and NEMA 3R. NEMA 3 is a rain-tight, dust-tight, and ice-resistant enclosure that protects equipment from penetration by water, ice, and wind-blown dust. NEMA 3R is designed to protect against rain, sleet, and snow. This type of enclosure is also equipped with a conduit hub to assure weathertight connections for conduits, which are fed into the top of the unit.

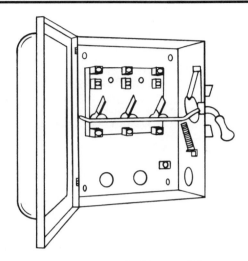

NEMA 1, Non-fusible, 600 Volt

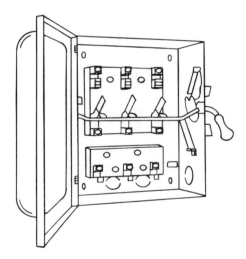

NEMA 1, Fusible, 600 Volt

Safety Switches

Man-hours

Description	m/hr	Unit
Safety Switch NEMA 1 600V 3P 200 Amps	6.150	Ea.
NEMA 3R	6.670	Ea.
NEMA 7	10.000	Ea.
NEMA 12	6.67	Ea.

Electrical enclosures serve two basic purposes: they protect people from accidental contact with enclosed electrical devices and connections, and they protect the enclosed devices and connections from specified external conditions. The National Electrical Manufacturers Association (NEMA) has established the following standards. Because these descriptions are not intended to be complete representations of NEMA listings, consultation of NEMA literature is advised for detailed information.

The following definitions and descriptions pertain to NONHAZARDOUS locations:

NEMA Type 1: General purpose, indoor enclosures which are intended for use indoors, primarily to prevent accidental contact of personnel with the enclosed equipment in areas where unusual conditions do not exist.

NEMA Type 2: Dripproof, indoor enclosures which are intended for use indoors to protect the enclosed equipment against dripping noncorrosive liquids and falling dirt.

NEMA Type 3: Dustproof, raintight and sleet-resistant (ice-resistant) outdoor enclosures which are intended for use outdoors to protect the enclosed equipment against wind-blown dust, rain, sleet, and external ice formation.

NEMA Type 3R: Rainproof and sleet-resistant (ice-resistant) outdoor enclosures which are intended for use outdoors to protect the enclosed equipment against rain and are constructed so that the accumulation and melting of sleet (ice) will not damage the enclosure and its internal mechanisms.

NEMA Type 3S: Outdoor enclosures which are intended for outdoor use to provide limited protection against wind-blown dust, rain, sleet (ice) and to allow operation of external mechanisms when ice-laden.

NEMA Type 4: Watertight and dust-tight indoor and outdoor enclosures which are intended for use indoors and outdoors to protect the enclosed equipment against splashing water, seepage of water, falling or hose-directed water, and severe external condensation.

NEMA Type 4X: Watertight, dust-tight, and corrosion-resistant indoor and outdoor enclosures which feature the same provisions as Type 4 enclosures, as well as corrosion resistance.

NEMA Type 5: Indoor enclosures which are intended for indoor use primarily to provide limited protection against dust and falling dirt.

NEMA Type 6: Indoor and outdoor enclosures which are intended for indoor and outdoor use primarily to provide limited protection against the entry of water during occasional temporary submersion at a limited depth.

NEMA Type 6R: Indoor and outdoor enclosures which are intended for indoor and outdoor use primarily to provide limited protection against the entry of water during prolonged submersion at a limited depth.

NEMA Type 11: Indoor enclosures which are intended for indoor use primarily to provide, by means of oil immersion, limited protection to enclosed equipment against the corrosive effects of liquids and gases.

NEMA Type 12: Dust-tight and driptight indoor enclosures which are intended for use indoors in industrial locations to protect the enclosed equipment against fibers, flyings, lint, dust, and dirt, as well as light splashing, seepage, dripping and external condensation of noncorrosive liquids.

NEMA Type 13: Oiltight and dust-tight indoor enclosures which are intended for use indoors primarily to house pilot devices, such as limit switches, foot switches, push buttons, selector switches, pilot lights, etc., and to protect these devices against lint and dust, seepage, external condensation, and sprayed water, oil, and noncorrosive coolant.

The following definitions and descriptions pertain to HAZARDOUS, or CLASSIFIED, locations:

NEMA Type 7: Enclosures are intended for use in indoor locations classified as Class 1, Groups A, B, C, or D, as defined in the National Electrical Code.

NEMA Type 9: Enclosures which are intended for use in indoor locations classified as Class 2, Groups E, F, or G, as defined in the National Electrical Code.

Man-hours

Description	m/hr	Unit
NEMA 1		
12"L x 12"W x 4"D	1.330	Ea.
NEMA 3R		
12"L x 12"W x 6"D	1.600	Ea.
NEMA 4		
12"L x 12"W x 6"D	4.000	Ea.
NEMA 7		
12"L x 12"W x 6"D	8.000	Ea.
NEMA 9		
12"L x 12"W x 6"D	5.000	Ea.
NEMA 12		
12"L x 14"W x 6"D	1.510	Ea.

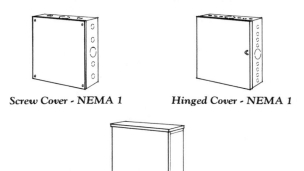

Screw Cover - NEMA 1   *Hinged Cover - NEMA 1*

Rainproof and Weatherproof, Screw Cover - NEMA 3R

Sheet Metal Pull Boxes

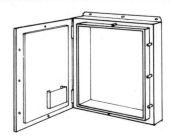

Watertight and Dust-tight, Hinged Cover - NEMA 4

Explosionproof, Screw Cover - NEMA 7

Cast Iron Pull Boxes

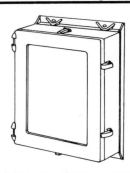

Enclosure, Quick Release Latch Door - NEMA 4X

Wiring Box, Dust-tight and Driptight - NEMA 12

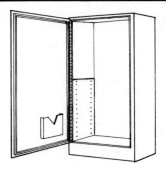

Electronic Rack Enclosure - NEMA 12

Double Door Cabinet - NEMA 12

Standard, Oiltight - NEMA 13

Sloping Front, Oiltight - NEMA 13

Pushbutton Enclosures

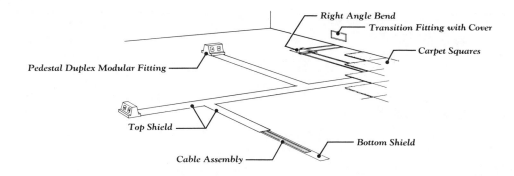

Undercarpet telephone systems employ flat conductor cabling to interconnect telephone devices to a distribution closet in open office situations. This method of interior cabling not only allows for flexibility in both installation and operation but also eliminates the access problems associated with cables in underfloor ducts. The basic elements of undercarpet telephone systems include three groupings of components: specialized flat, low profile cable; transition fittings, which house the round-to-flat conductor connections at the supply end of the cable; and floor fittings, which house the flat-to-round connections and provide various access configurations at the other end of the cable.

Undercarpet telephone cable is available in two commonly used types: low pair count cable and 25 pair cable. Low pair count cable (2,3 and 4 pair) consists of 26 AWG (American Wire Gauge) round solid copper conductors; 25 pair cable contains flat copper conductors that are equivalent to 26 AWG. In both types, a PVC core encases and insulates the conductors, and a protective polyester film covers the entire cable. The conductors within low pair count cable can be separated and terminated without exposing the bare copper wire. Both types of cable include factory-prepared termination connectors at both ends and are manufactured in harnesses which measure in 5' increments up to 100' in length. Low pair count cable (3 and 4 pair) is available in bulk so that custom lengths can be installed; however, when the bulk cable is used, termination connectors must be installed at the site.

Undercarpet cable installation involves the laying of the low pair count or 25 pair cable between two layers of material, bottom shield and top shield, which provide a secure path for the cable and protects it from wear and other harm. Bottom shield is a nonconductive abrasion resistant vinyl film which is used exclusively with 25 pair harnesses to protect the cable from moisture, chemical reaction, and abrasion. Two-inch wide telephone tape is installed as bottom shield for low pair count cable harnesses. Top shield is an adhesive-backed vinyl plastic telephone tape which is used to protect both 25 pair and low pair count cable systems from wear and abrasion. Top shield tape covers the entire width and length of the undercarpet cable. Both types of cable are secured to the floor and held in position under the carpet with a highly adhesive tape which is applied across the run of the cable.

The fittings which are employed in undercarpet telephone systems consist of transition fittings, which are used to house the round-to-flat cable connectors at the supply end of the undercarpet cable, and floor fittings, which are installed at the other end to house flat-to-round cable connectors and to provide the necessary access fixtures for telephone devices. Transition fittings are employed to protect the connections where round supply cable joins the flat undercarpet cable. Various transition fittings with individualized covers are available for wall, floor, and column mountings. Floor fittings, in addition to providing protection for the flat-to-round conductor connections, serve as pedestals for connecting telephone and other data and signal systems. Floor fitting kits are manufactured in many types and configurations, including duplex jacks, 25 pair kits, and call directors, as well as telephone, power, and data connections under a single pedestal. The thermoplastic covers used for the floor fittings are also manufactured in a wide range of options to match the fittings, including blank, duplex, slotted, and low profile models.

The hardware, which is used with the many unique installation situations possible with undercarpet telephone systems, is also available in a variety of configurations. Line assignment adaptors, line assignment plugs, modular plug connector kits, and surface jacks are some of these commonly employed hardware configurations. Specialized crimping devices and other hand tools, which are utilized during the installation process, are available from undercarpet cable manufacturers.

Man-hours

Description	m/hr	Unit
Cable Assembly 25/pr w/connectors 50 ft	.670	Ea.
Cable Assembly 3/pr w/connectors 50 ft	.340	Ea.
Cable Assembly 4/pr w/connectors 50 ft	.350	Ea.
Cable (Bulk) 3/pr	.006	lf
Cable (Bulk) 4/pr	.007	lf
Bottom Shield for 25/pr Cable	.005	lf
Bottom Shield for 3-4 pr Cable	.005	lf
Top Shield for all Cable	.005	lf
Transition Box, Flush Mount	.330	Ea.
In Floor Service Box	2.000	Ea.
Floor Fitting w/Duplex Jack and Cover	.380	Ea.
Floor Fitting Miniature w/Duplex Jack	.150	Ea.
Floor Fitting w/25 pr Kit	.380	Ea.
Floor Fitting Call Director Kit	.420	Ea.

2 Pair

3 Pair

4 Pair

Cable Assembly

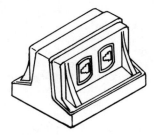

Floor Fitting with Duplex Jack and Cover

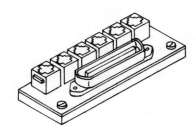

PC Board Floor Fitting

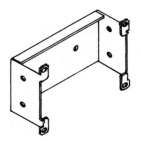

Wall Transition Fitting

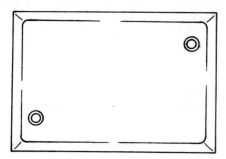

Transition Fitting Cover

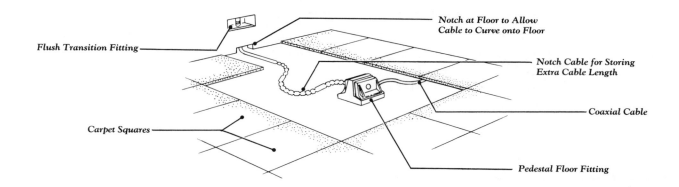

Flush Transition Fitting

Notch at Floor to Allow Cable to Curve onto Floor

Notch Cable for Storing Extra Cable Length

Coaxial Cable

Carpet Squares

Pedestal Floor Fitting

Undercarpet data systems are designed to interconnect remote data processing terminals to the main computer by means of an undercarpet cable installation. These systems provide great flexibility in open office areas because they are relatively easy to install, eliminate the access problems associated with underfloor ducts, and remove from walls and floors unsightly data poles and exposed cabling. Like undercarpet power and telephone systems, undercarpet data systems are not allowed in schools and hospitals. The wiring for undercarpet data systems includes three groupings of components: flat, low profile cable; transition fittings at the supply end of the cable; and floor fittings at the access end of the cable.

Flat undercarpet data cable is manufactured in 50, 75, and 93 OHM coaxial assemblies which may be procured with factory-prepared pre-terminated connectors at each end or purchased in bulk roll for field termination. The cable is composed of a round, solid tin-plated copper drain wire, an aluminum foil shield, two nylon cords, and a PVC protective jacket to resist moisture and abrasion damage. Pre-terminated coaxial cable assemblies include BNC series dual crimp connectors with heat-shrinking tubing to relieve stress. Field-terminated coaxial cable is available in bulk rolls with dual leads of 50 and 70 OHMS. A 75 OHM dual lead coaxial cable assembly is also available with pre-terminated BNC or TNC connectors.

The installation of undercarpet data cable does not require bottom or top shield, but cross taping must be applied at intervals along the cable run to fasten it to the floor and hold it in position under the carpet. A special crimping tool is required during the installation process of field-terminated cable. This tool assists in all aspects of the terminating procedure which involve the coaxial cable, its connectors, and the heat-shrink tubing. Because data cable cannot be folded or creased, another specialized hand tool is used for notching the flat cable where turns must be made along the run. To complete a 90° turn, eight notches must be placed in each side of the cable; to complete a 180° turn, sixteen notches are required per side.

The fittings which are employed at the ends of the under-carpet data cable include transition fittings at the supply end and floor fittings at the access end. The transition fittings for undercarpet data cable, which are the same as those used for under-carpet power systems, house and

protect the round-to-flat coaxial connectors. They are available in a variety of models for wall and floor mountings. The floor fittings are the same in design as those employed for undercarpet telephone and power cable systems. They house and protect the flat-to-round coaxial cable mating connectors and provide access connections for CRT devices in a variety of pedestal configurations with suitable blank or slotted thermpolastic covers. Highly adhesive tape should be applied over the mating connectors within the fittings to prevent interference from the building's electrical grounding system.

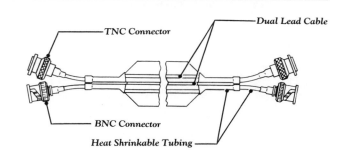

TNC Connector

Dual Lead Cable

BNC Connector

Heat Shrinkable Tubing

Dual Lead Coaxial Cable Assembly

Man-hours

Description	m/hr	Unit
Cable Assembly Single Lead w/Connectors 40 ft	.360	Ea.
Cable Assembly Dual Lead w/Connectors 40 ft	.380	Ea.
Cable (bulk) Single Lead	.010	lf
Cable (bulk) Dual Lead	.010	lf
Cable Notching 90 Degree	.080	Ea.
Cable Notching 180 Degree	.130	Ea.
Connectors BNC Coax	.200	Ea.
Connectors TNC Coax	.200	Ea.
Transition Box, Flush Mount	.330	Ea.
In Floor Service Box	2.000	Ea.
Floor Fitting w/Slotted Cover	.380	Ea.
Floor Fitting w/Blank Cover	.380	Ea.

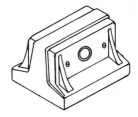

Pedestal Floor Fitting

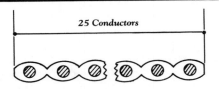

25 Conductors

Cable 25 Conductor

Single Lead

Dual Lead

Flat Cable

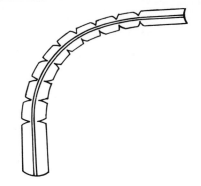

Data Cable Notching, 90°

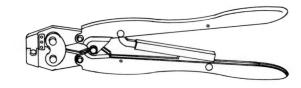

Hand Crimping Tool

9 ELECTRICAL
EMERGENCY/STANDBY POWER SYSTEM

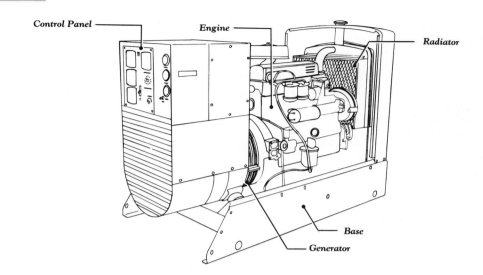

Control Panel — Engine — Radiator

Base

Generator

Although the equipment used for emergency and standby power systems is basically the same, the function of each system is markedly different, as defined by the National Electrical Code. The purpose of emergency systems is to provide the power and illumination essential for the safety to occupants as well as the power needed for the operation of equipment necessary to assist the occupants during emergencies, such as elevators, fire alarm systems, and fire pumps. These requirements must be observed with particular care in hospitals, health care facilities, and other buildings in which the occupants are dependent on electrically powered devices. Emergency power systems are required by law in these and other structures where loss of power may cause physical harm to or endanger the lives of their occupants. Standby systems, which are not required by law, are installed for the purpose of providing power to selected loads other than those classified as safety-oriented or emergency systems. Where the function of emergency power systems is to protect lives, the purpose of standby power systems is to protect property against financial loss or inconvenience.

Emergency and standby power systems are classified in two equipment categories: battery equipment and engine generator sets. Battery-powered systems consist of battery units which supply limited amounts of emergency power for illumination. The battery units may be installed at a central location with power distribution feeders serving the emergency lighting fixtures throughout the building: or the batteries may be installed separately as battery-light package units at selected locations. In both types of installations, the battery system must include automatic charging equipment.

The most commonly used battery types include lead-acid, nickle-cadmium, lead-antimony, and lead-calcium, with voltage ratings of 24 to 125 volts dc for central systems and 6 to 12 volts dc for package units. Both systems are required by law to maintain loads for a minimum of 1-1/2 hours.

Engine generator systems, which are sized from several hundred watts to several hundred kilowatts, include three groups of components within the engine generator set: the engine and control, the exhaust system, and the fuel system. The engines may be powered by gas/gasoline or diesel fuel and are rated according to the type of engine and the system's output. The most common sizes for gas/gasoline-operated 3 phase 4 wire 120/208 or 277/480 volt systems are: 7.5KW, 10KW, 15KW, 30KW, 70KW, 85KW, 115KW, and 170KW. For diesel-operated systems, the most common sizes are 50KW, 175KW, 200KW, 300KW, 400KW, 500KW, 725KW, and 1,000KW.

The generator engines are generally four-cycle units of one to six cylinders, depending on the capacity of the generator unit. The engines are cooled by fan-forced air or circulating water systems. They are controlled by manual, remote, or automatic means, with ignition for larger engines supplied by storage batteries.

The engine generator set may be a stationary or mobile unit, depending on its size and particular function. Stationary units of 10KW and lower should be mounted on a concrete pad with preset anchor bolts. Larger stationary units are equipped with steel skids and do not require separate mounting pads. Engine generator units range in size from 47″ long by 27″ wide by 27″ high, with a weight of 20,438 lbs., for a 1,000KW unit. The exhaust piping, which is constructed from wrought iron, cast iron, or steel, must terminate in the open air away from doors, windows, and other building openings.

The fuel supply for engine generator sets is stored in underground tanks outside of the building, with the exception of a small amount of fuel which is stored in the generator room in a day tank. This small, ready supply ensures immediate access to fuel during emergencies until fuel from the main storage tank is injected into the fuel lines. The run duration of engine generator systems depends on the quantity of stored fuel and the unit's rate of consumption. A 10KW unit consumes about 2.30 gallons of fuel per hour; a 170KW unit, 21.00 gallons per hour; a 500KW unit, 40.0″ gallons per hour; and a 1,000 KW unit, 83.50 gallons per hour.

The transfer of building power to emergency of standby power is controlled through a load transfer switch, which may be manually or automatically activated. The manual control type of switch is double throw in design and must be activated by hand after the generator has started. The automatic transfer type of switch not only starts and stops the generator but also transfers the load via relays without the attention of an operator.

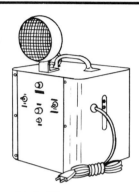

Battery Emergency Light Unit

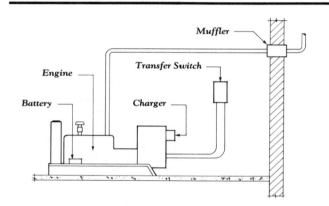

Emergency Generator System

Man-hours

Description	m/hr	Unit
Battery Light Unit 6 Volt Lead Battery and 2 Lights	2.000	Ea.
Battery Light Unit 12 Volt Nickel Cadmium 2 Lights	2.000	Ea.
Remote Mount Sealed Beam Light, 25W, 6 Volts	.300	Ea.
Self Contained Fluorescent Lamp Pack	.800	Ea.
Engine Generator 10KW Gas/Gasoline 277/480V Complete	34.000	System
Engine Generator 170KW Gas/Gasoline 277/480V Comp	96.000	System
Engine Generator 500KW Diesel 277/480V Complete	133.000	System
Engine Generator 1000KW Diesel 277/480V Compete	180.000	System

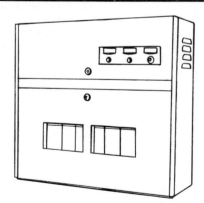

Central Battery Unit for Emergency Light System

Electric baseboard heaters are normally installed under windows and on outside walls to provide a perimeter heating system. They are usually surface-mounted convector units with fin tube or cast aluminum heating elements which are rated by their watt density. They are manufactured in varying lengths and levels of output for noncommercial and commercial installations.

Noncommercial baseboard heaters are available in 2', 3', 4', 5', 6', 7', 8', and 10' lengths with voltage ratings of 120, 208, 240, and 277 volts. Watt density ratings of these units are classified in three levels: low watt density units, which are rated at approximately 225 watts per linear foot; and high density units, approximately 275 watts per linear foot. The heating units are manufactured with knockouts in the bottom, sides, and back of the enclosure to provide access for the electrical supply which may be brought into the unit in rigid conduit, flexible metallic cable, or nonmetallic sheathed cable. Various accessories for the heating units are also available, including inside/outside corners for wall-to-wall installations, end caps, and blank sections.

The baseboard heating units are controlled by four types of thermostat systems: integral line thermostats: thermostat sections; line voltage remote-mounted thermostats; or low voltage thermostats which control line voltage contractors or relays. The number of heaters allowed on each circuit depends on the total wattage divided by the line voltage and the rating of the wire and circuit breaker servicing the circuit. Most feeder circuits are double-pole 20 amp 208/240 volt systems which are fed with #12 AWG wire.

The number of linear feet of electric baseboard needed to heat efficiently a given room or area is determined by the results of a heat loss survey. Generally, the rate of heat loss for average residential or office space may be estimated at about 10 watts per square foot of exposed area, but precise heat loss surveys can be computed to determine the linear footage of baseboard by following this five-step formula:

1. Determine the heat loss factor in watts per square foot by the R value. This data can be found in the NEMA Electric Heat Guide.
2. Multiply the total square feet of all exposed area, minus the area of glass and doors, by this factor.
3. Multiply the volume in cubic feet by the infiltration or air change factor.
4. Add the above projects to determine the total heat loss of the area.
5. Divide the total heat loss of the area by the watt density of the baseboard type being installed. The resulting figure represents the total linear feet of baseboard required.

Commercial electric baseboard heaters are similar to noncommercial units, but they are designed to provide greater heating output. Their capacities range from 300 to 1,200 watts at 120, 208, 240, and 277 volts, single phase. Commercial baseboard heater units are available in 2', 5', and 8' lengths which are designed for surface mounting at 2" above the finished floor.

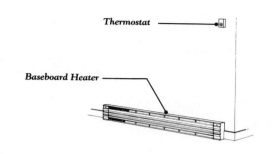

Thermostat Wire, Twisted, No Jacket

Romex Cable, #12, 2 Wire, with Ground

Armored Cable, (BX), #12 2 Wire, with Ground

Electric Supply Wiring

Man-hours

Description	m/hr	Unit
Baseboard Heater 2' Long	1.000	Ea.
4' Long	1.190	Ea.
6' Long	1.600	Ea.
8' Long	2.000	Ea.
10' Long	2.420	Ea.
Thermostat, Integral 1-Pole	.500	Ea.
2-Pole	.500	Ea.
Thermostat, Low Voltage 1-Pole	1.000	Ea.
Circuit Breaker 1-Pole 120V	.800	Ea.
2-Pole 240V	1.000	Ea.
Romex #12-2w/Ground	3.200	CLF
Armored Cable, (BX) #12-2 w/Ground	3.480	CLF

9 ELECTRICAL
LIGHTNING PROTECTION SYSTEM

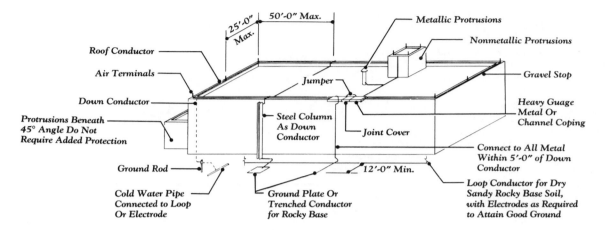

Because lightning is an uncontrollable natural force which has the potential to cause severe heat, mechanical damage, and personal injury or loss of life, lightning protection systems have been developed to protect buildings by grounding the lightning stroke. The basic operative principle of all lightning protection systems is the same: to provide a metallic path for the lightning stroke from the point of contact to the ground and, therefore, to prevent the stroke from passing through ungrounded portions of the building. A lightning protection system includes three groupings of components: rods, points, and bases, which attract the lightning stroke; cables, which conduct the stroke from the points to the ground rods; and ground rods and plates, which discharge and dissipate the stroke at the ground.

Before any lightning protection system is installed, precautions should be taken to ensure the proper grounding of a building's structural elements and any sensitive electrical equipment within it. All equipment used in the fundamental grounding system within and on the structure should be UL (Underwriters Laboratory) approved and completely installed to proper electrical standards. A partially or improperly grounded building may create a more severe electrical danger than one that has no grounding protection at all. One of the guidelines regarding basic grounding protection in concrete structures, for example, is to make sure that all reinforcing bars are bonded. If they are not bonded, the insulating gaps between the rods may break the grounding system during a lightning strike and cause severe damage to the structure. Another guideline is to place a lightning arrestor on all aerial service conductors, whether their voltage is high or low. Also, any piece of electronic equipment that is sensitive to voltage surges should be individually protected by surge arrestors. These and other guidelines must be followed in collaboration with the installation of a lightning protection system if it is to function safely and effectively.

The uppermost components of a lightning protection system are the points, or air terminals, which attract the lightning, and the various types of bases which support them and assist in the conducting process. Although aluminum can be used, the points are usually constructed

from solid copper rod, which is drawn to a sharp point with a nickle-plated tip on one end and threaded to fit its base on the other. Standard stock sizes measures 3/8″, 1/2″ and 5/8″ in diameter and 1″, 12″, 18″, 24″, 30″, 36″, 48″, and 60″ in length. Special points are also available in 3/4″ and 1″ diameters. Point lengths may be increased up to 12′ by combining two rods of any length with a coupling which is manufactured in 3/8″ to 1″ diameters.

The bases, which support the points and attach them to the building structure and the protection system's cable network, are manufactured in many different configurations. In most cases the threaded end of the point is simply fastened vertically into a fixed receptacle in the base, but some bases are equipped with swivels to alter the direction and angle of the point adjust for slanted surface mountings. Some of the commonly installed base formats and their special applications include: straight point base, for pole tops or chimneys; flat point base, for flat roofs or surfaces; and strap ridge point base, for roll ridge roofs. These configurations are just three of the hundreds of formats available to suit any installation surface of condition.

From the point and base components, the lightning stroke is conducted to the ground rods via the main conductor cable and its downleads. Cable attachments to the system's components, the building's structural steel, other cables, and grounding electrodes may be accomplished by clamping hardware, field weld, or thermoweld connections. Main conductor cable is manufactured from copper or aluminum it two classfications: cable for buildings under 75′ in height and cable for buildings over 75′ in height. Cable size is **not** determined by AWG (American Wire Gauge) ratings, but by pounds per 1,000 linear feet. The industry standard for buildings under 75′ high is cable sized at 220 pounds per 1,000 linear feet. This size cable contains 32 strands of 17 gauge bare copper wire; and it has a circular mil cross-section of 65,600 (approx #1 AWG). It is available on 500′ reels. The industry standard for structures over 75′ in height is cable sized at 375 pounds per 1,000 linear feet. This size cable contains 28 strands of 14 gauge bare copper wire; and it has a circular mil cross-section of 122,000 (approx. #2/0 AWG). Main conductor cable is placed around the roof's perimeter and secured to the soffit at intervals of 2'. All electrical equipment located within 5' of the main conductor cable must be ground connected to it with smaller branch conductor cable. Downleads are fastened to the main conductor cable at maximum 50' intervals along the main run and then attached vertically to the face of the building or imbedded in concrete columns. Steel columns may also be used as down conductors. Downleads and other down conductors, regardless or the type or method of installation must be attached to the ground rods or plates.

The lightning stroke is discharged and dissipated into the earth via the ground rods which are connected to the downleads after being driven into the ground. Ground rods which are manufactured from copperclad steel, galvanized steel or stainless steel, are available in diameters of 1/2", 5/8", 3/4", and 1" and in lengths of 8', 10' and 12'.

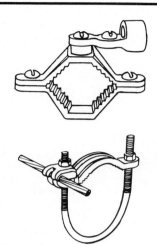

Water Pipe Clamps

Ground Rod Clamp

Sectional Ground Rod

Ground Rod Coupling

Ground Rod Driving Stud

Grounding Accessories

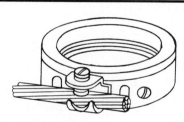

Conduit Grounding Bushing

Man-hours

Description	m/hr	Unit
Points/Air Terminals 3/8" and 1/2" to 12"	1.000	Ea.
Cable Copper 220 lb per M (Under 75')	.020	lf
Cable Copper 375 lb per M (Over 75')	.030	lf
Cable Aluminum 101 lb per M (Under 75')	.020	lf
Cable Aluminum 199 lb per M (Over 75')	.030	lf
Arrestor 175 Volt AC to Ground	1.000	Ea.
Arrestor 650 Volt AC to Ground	1.200	Ea.
Ground Rod 8' Long 5/8" Diameter	1.450	Ea.
Ground Rod 10' Long 5/8" Diameter	1.750	Ea.
Ground Clamps	1.000	Ea.
Thermowelds	1.150	Ea.
Brazed Connections	1.000	Ea.

9 ELECTRICAL
HEAT TRACE SYSTEMS

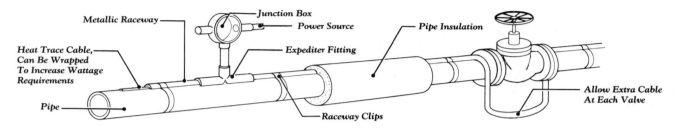

A heat trace system is designed to provide freeze protection or to maintain constant temperatures in liquids contained in pipes, valves, and storage tanks. The basic component of the system is parallel resistance heating cable, which is composed of thermally stable nichrome wire. Most heat trace systems contain a series of heating zones and are designed to produce a constant watt per hour output to ensure even distribution of heat to the protected pipes of vessels. The choice or the type of cable to be used in the system is determined by matching the fixed power output of the heat trace cable to the heat loss characteristics of the surface to be protected. The manufacturers of different cable types usually supply information on design procedures unique to their product.

Many different types of heat trace cable with various watt per foot ratings and insulation capabilities are available. Ten standard cable types, each color coded to identify its output, range from 2.0 watts per linear foot at 110 volts to 14.5 watts per linear foot at 277 volts. Heat trace cable can be used with voltages of 110, 115, 120, 208, 220, 230, 240, 277, 440, and 480 volts. Common designations for insulation jackets include EL cable (Elexar), which is recommended for use in freeze protection systems with temperature ranges of +40°F (+4°C) to +90°F (+32°C), and FP cable (Fluorinated Ethylene Propylene), which is employed to maintain higher process fluid tempratures between +40°F (+4°C) and +250°F (121°C). The method of attachment of either type of cable to the process piping and/or pipeline equipment is also a factor in determining its insulating capacity.

Various methods may be employed to attach heat trace cable to the surfaces of the process piping or storage tanks. For freeze protection, the cable is simply attached to the pipe with polyester tape. In another method, parallel heat trace cable is attached to the surface with a continuous coverage of 2″ wide aluminum tape to increase heat transfer and temperature distribution around the pipe. This method also allows higher watt densities and elimimates the need of spiral wrapping. Parallel cable may also be effectively attached with factory-extruded heat transfer cement and then covered with a metallic raceway. This method of attachment maximizes heat transfer and allows temperature maintenance of up to +250°F (121°C).

In situations where surface attachment methods cannot be employed or are not desirable, alternative means of attachment are used. For the heating of underground and pre-insulated piping or the foundations of storage vessels, heat trace cable may be run through a conduit which is adjacent to the piping or imbedded in the concrete of the foundation. With this method of installation, the cable can be placed and removed without damage to the piping insulation or foundation. High temperature steam lines may be protected against freezing by installing heat trace cable between the layers of pipe insulation.

The process of connecting and attaching sections of heat trace cable at the junctions of process piping is accomplished with a fitting called an expediter, which is clamped over both the cable and the process pipe. The heat trace cable is brought up from the bottom of the expediter and then channeled through a nipple and into a junction box where the cable splices are made. The junction box, which is attached to the expediter fitting, is constructed from cast aluminum, die cast aluminum, or stainless steel with a NEMA rating of 4 or 7.

The methods of temperature control of heat trace systems vary according to the nature of the sensing mechanism used in the system. Pipewall sensing is the most commonly applied method of control for freeze protection, as it provides for individual control of heating circuits and prevents the overheating of FRP (Fiberglass Reinforced Plastic), PVC, and reinforced fiberglass piping. The thermostat sensing bulb is attached to the pipe wall, and the thermostat itself is placed either in the junction box attached to the expediter fitting or in its own NEMA 4 or 7 enclosure. Temperature settings may be preset at the factory or determined and made at the site of installation.

For more complex applications of temperature control, such as temperature maintenance of process fluids, ambient sensing may be employed. When the thermostat senses a drop in the ambient temperature, it acts as a pilot device to energize a contactor and applies power to a distribution panel which channels power into the cable to raise the temperature. When the desired temperature of the process liquid rises to the preset level, the thermostat opens the contact until the ambient senses that the temperature be raised again. Care should be exercised when planning any process temperature control system to assure that a full engineering specification and design are completed before the system is installed.

Man-hours

Description	m/hr	Unit
Cable 400 degree 2.5 Watts per Linear Foot	.010	lf
5.0 Watts per Linear Foot	.010	lf
10.0 Watts per Linear Foot	.010	lf
Cable Metallic Raceway	.040	lf
Snap Band, Clamp	.080	Ea.
Expediter Fitting	.720	Ea.
Thermostat, NEMA 4, 30 Amp Single Pole	1.000	Ea.
Thermostat, NEMA 7, 30 Amp Single Pole	1.150	Ea.
Thermostat, NEMA 4, 30 Amp Double Pole	1.150	Ea.
Thermostat, NEMA 7, 30 Amp Double Pole	1.350	Ea.
Contactor/Thermostat Combination		
75 Amp 4 Pole	3.500	Ea.

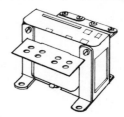

Control Transformer

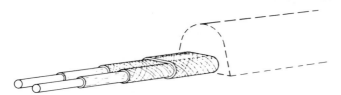

Heat Trace Cable

Single Pole Thermostat - NEMA 4

Heat Trace Raceway

Thermostat/Contactor Combination - NEMA 4

Thermostats

Expediter Fitting

DIVISION
10
EQUIPMENT

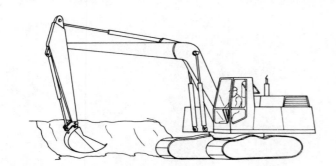

Backhoe - Crawler Type

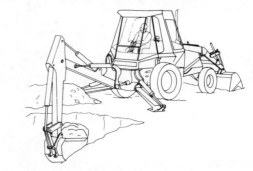

Backhoe/Loader - Wheel Type

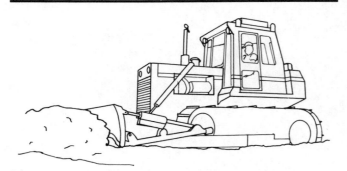

Tractor - Crawler Type

Tractor Loader - Wheel Type

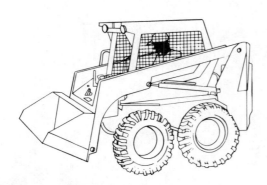

Tractor Loader - Wheel Type, Small

Dump Truck

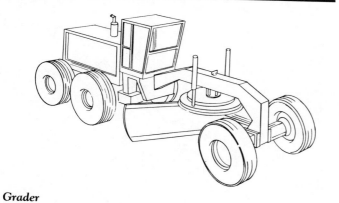

Grader

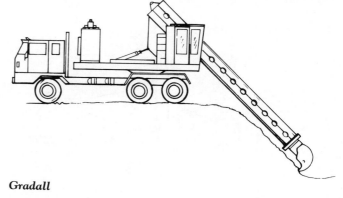

Gradall

Roller

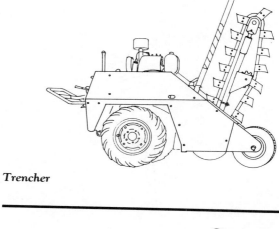

Trencher

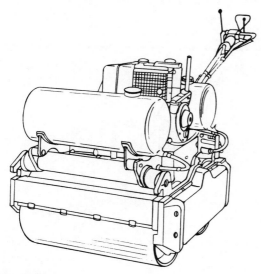

Compactor Roller

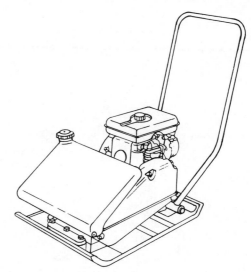

Compactor - Vibratory Plate

Ready Mix Truck

Concrete Pump

Concrete Bucket

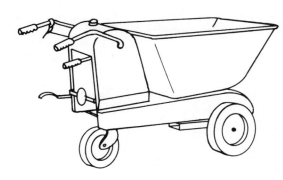

Concrete Cart

Concrete Mixer

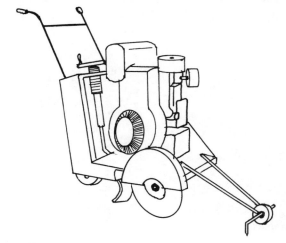

Concrete Saw

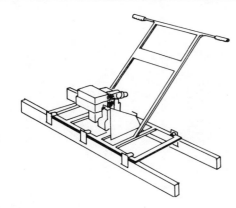

Power Screed

Concrete Finisher

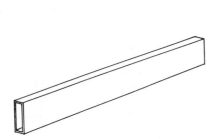

Magnesium Screed

Magnesium Darby

Bull Float

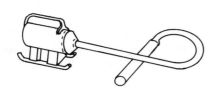

Concrete Vibrator

Finishing Broom

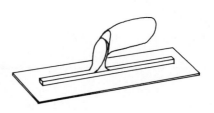

Steel Trowel

Wood Float

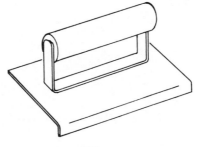

Edger

Bronze Groover

Hand Finishing Tools

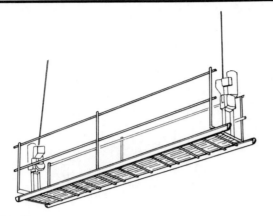

Swing Staging

Fixed Scaffold - Heavy Duty

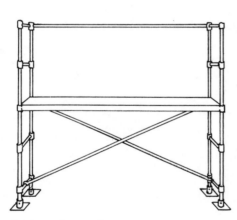

Fixed Scaffold - Lightweight

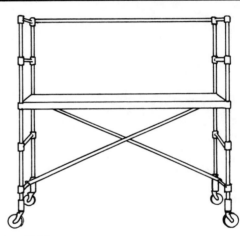

Rolling Scaffold

Rolling Ladder

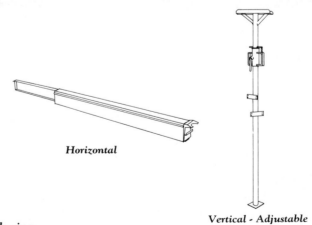

Horizontal

Shoring

Vertical - Adjustable

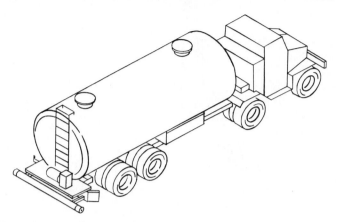

Distributor - Asphalt or Slurry

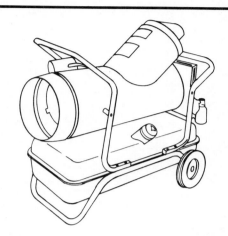

Space Heater

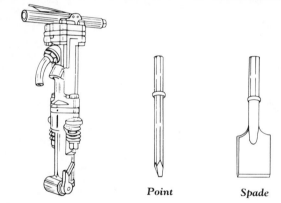

Point *Spade*

Air Hammer

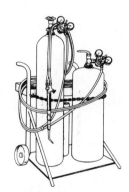

Oxygen/Acetylene Cutting Outfit

Welder on Trailer

Air Compressor on Trailer

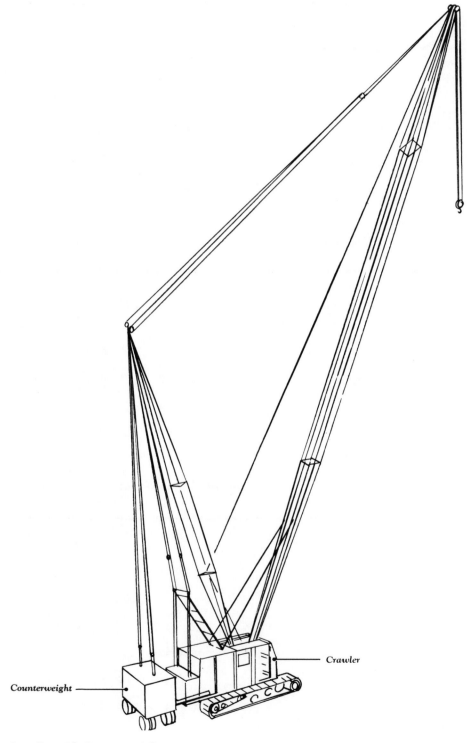

Counterweight

Crawler

Crane - Crawler with Counterweight

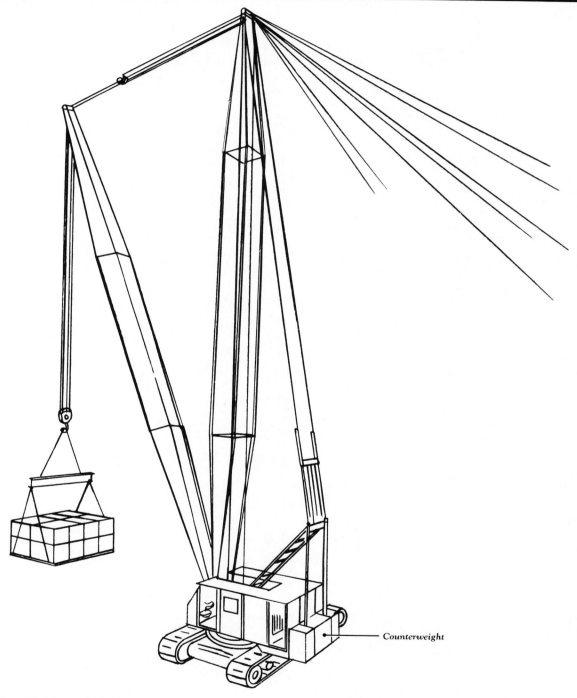

Counterweight

Crane - Crawler, Heavy Capacity

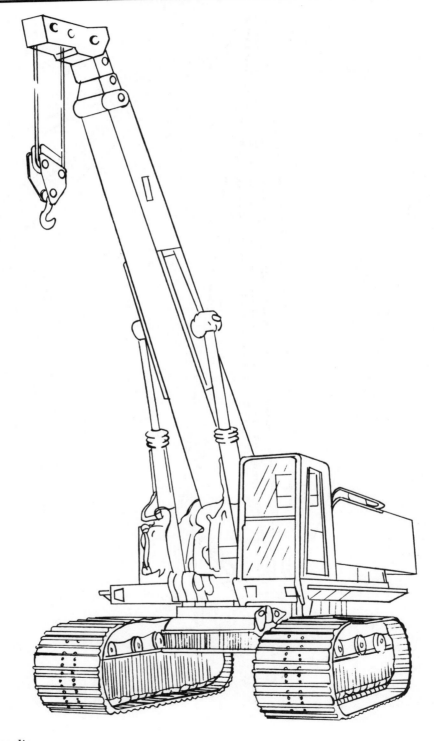

Crane - Crawler, Hydraulic

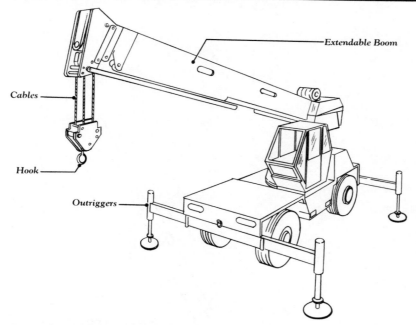

Extendable Boom

Cables

Hook

Outriggers

Crane, Truck Mounted, Fixed Cab - Hydraulic

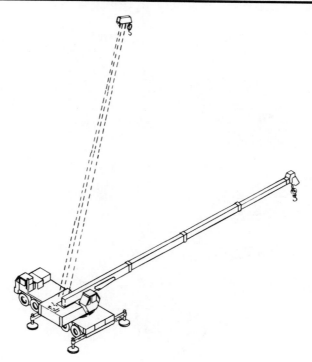

Crane, Truck Mounted - Hydraulic

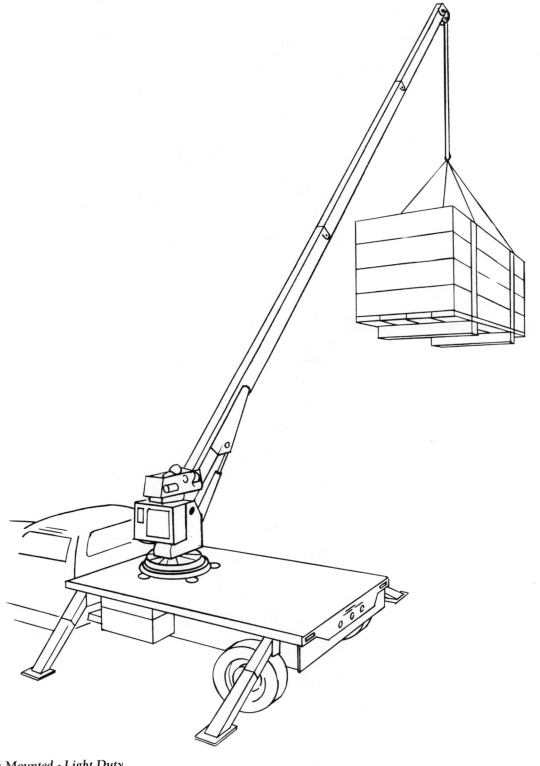

Crane, Truck Mounted - Light Duty

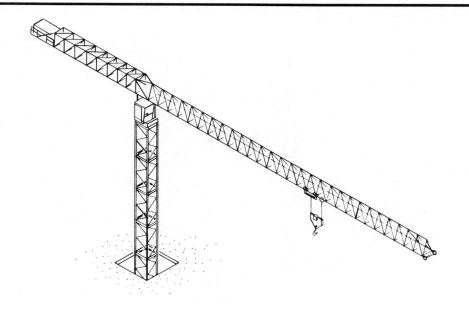

Climbing Crane

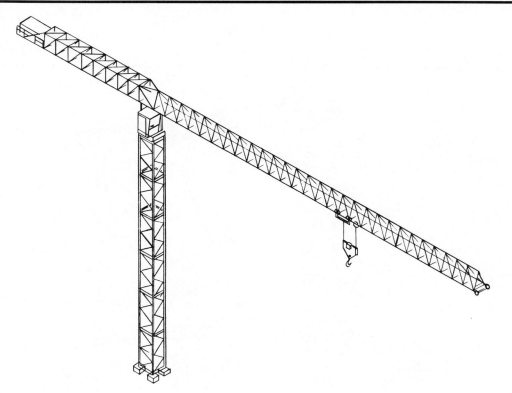

Crane Tower, Static

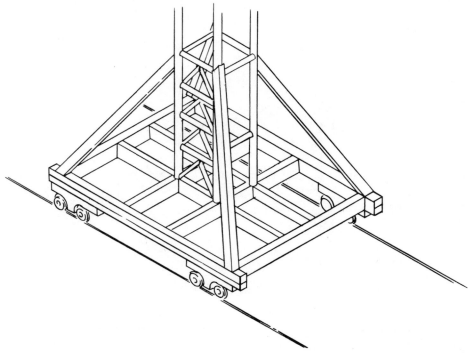

Crane Tower, Static - Rail Mounted

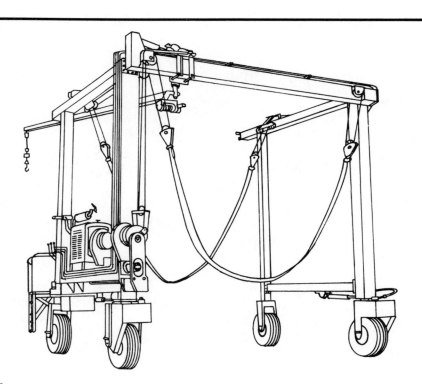

Travel Lift - Wheel Type

DIVISION
11
SPECIALTIES

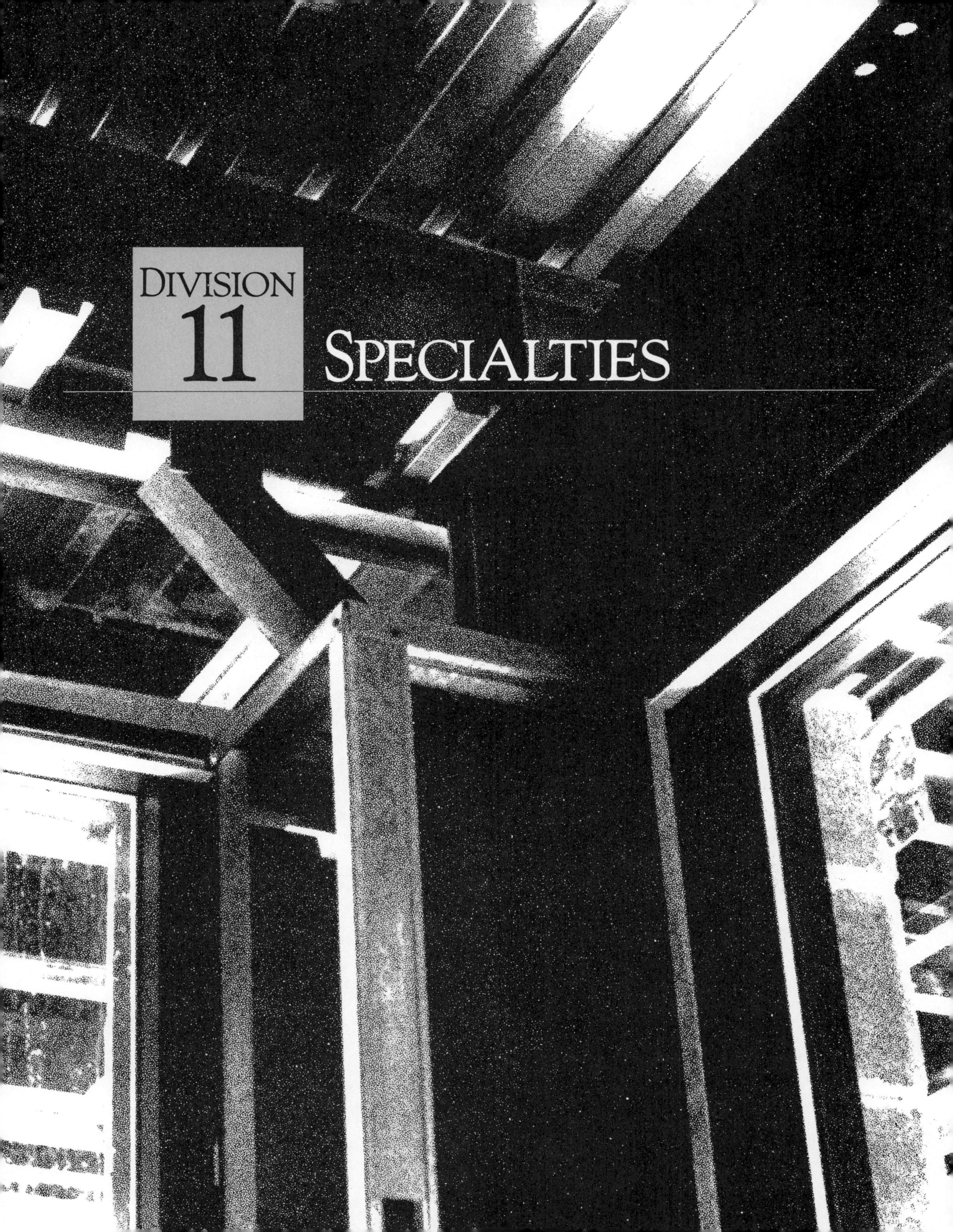

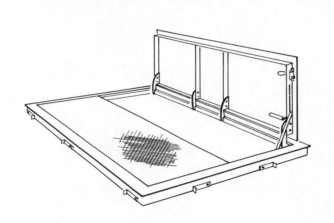

Canopy

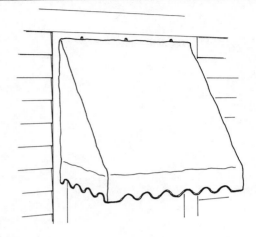

Canvas Awning

Double Leaf Floor Hatch

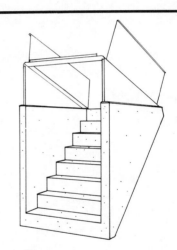

Precast Basement Entrance

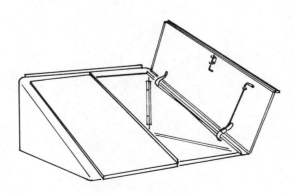

Bulkhead Basement Doors

Disappearing Stair - Folding

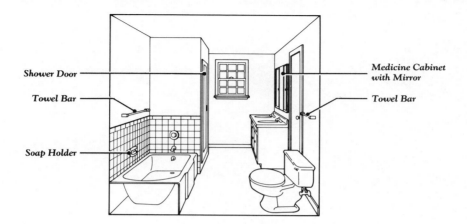

Bathroom Accessories

Woodburning Stove

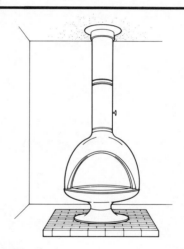

Freestanding Fireplace

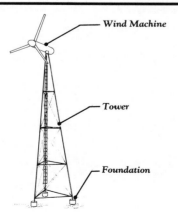

Wind Energy - Freestanding Tower

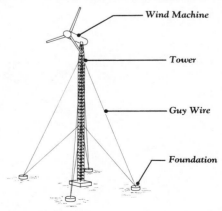

Wind Energy - Guyed Tower

385

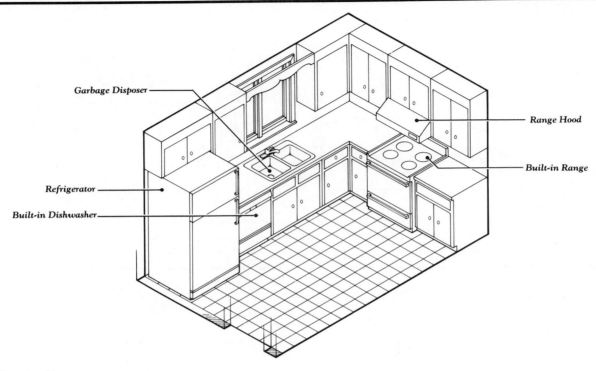

Garbage Disposer

Range Hood

Built-in Range

Refrigerator

Built-in Dishwasher

Kitchen Appliances

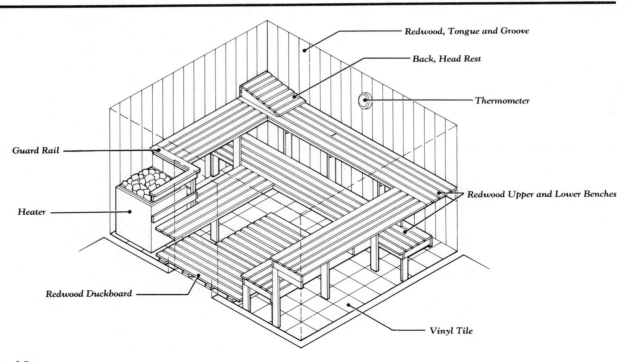

Redwood, Tongue and Groove

Back, Head Rest

Thermometer

Guard Rail

Redwood Upper and Lower Benches

Heater

Redwood Duckboard

Vinyl Tile

Prefabricated Sauna

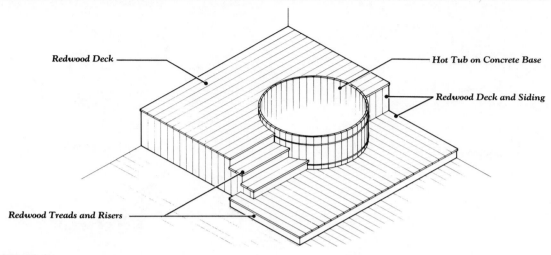

Redwood Deck ———

Hot Tub on Concrete Base

Redwood Deck and Siding

Redwood Treads and Risers ———

Redwood Hot Tub

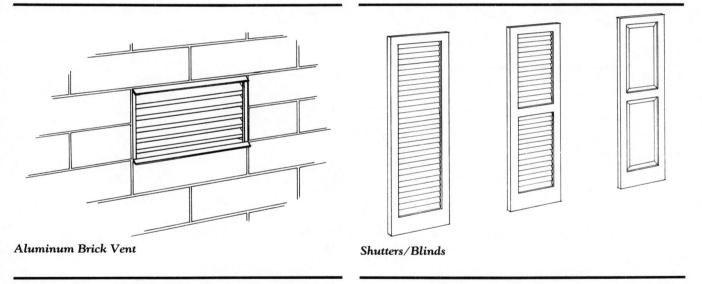

Aluminum Brick Vent

Shutters/Blinds

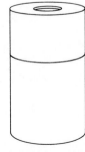

Trash Receptacle

Ash Urn

Ash/Trash Receivers

Fixed Seating, Four Seat

Free Standing Upholstered Seat and Back

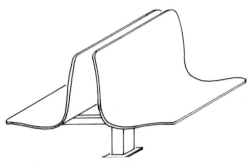

Mounted in Floor, Double Booth

Arm Chair with Cane Seat and Back

Crome Stool with Upholstered Seat and Back **Wood Stool with Swivel Upholstered Seat** **Wood Stool with Upholstered Seat and Back**

Restaurant Seating

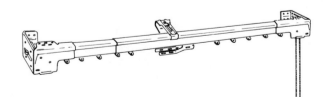

Traverse Rod, Adjustable

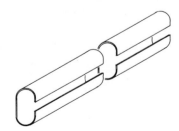

Stationary Drapery Rod

Single Pedestal Desk

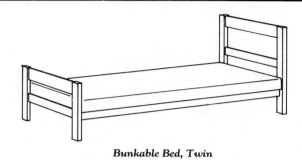

Bunkable Bed, Twin

Dormitory Furniture

Free Standing Headboard, Twin Beds

Sleep Sofa, Full

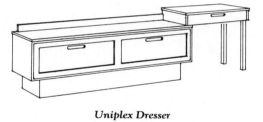

Uniplex Dresser

Hotel Furniture

Single Face Carrel

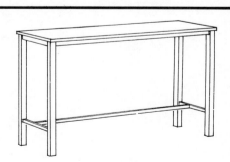

Card Catalog Reference Table

Library Furniture

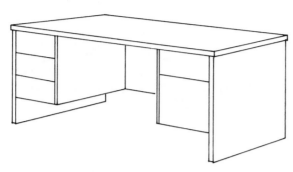

Wood Double Pedestal Desk

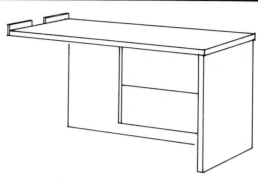

Wood Secretarial Return

Executive Office Chair

Secretarial Office Chair

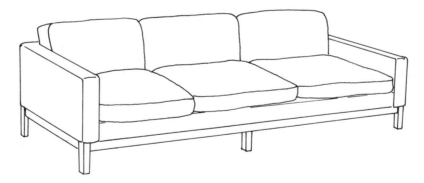

Three Seat Loose Cushion Sofa

Office Furniture

Chair, Integral Table Arm, Molded Plastic

Table, Plastic Laminate Top

School Furniture

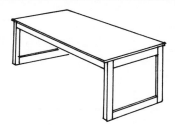

Sled Base Coffee Table

Round Designer Table

Rectangular, Folding

Round, Folding

Tables

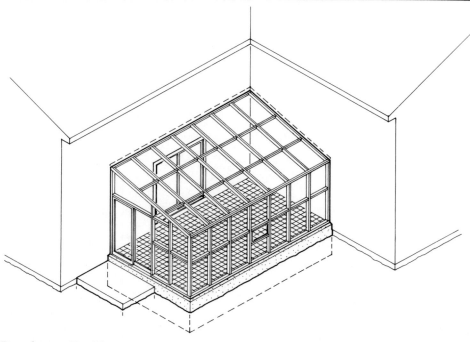

Large Sunroom/Greenhouse - Prefabricated

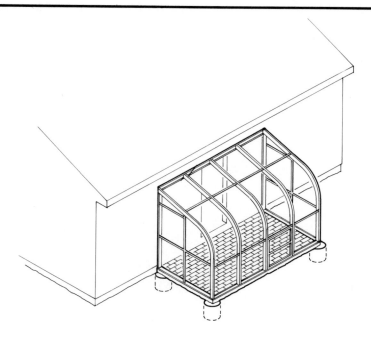

Small Lean-to Greenhouse

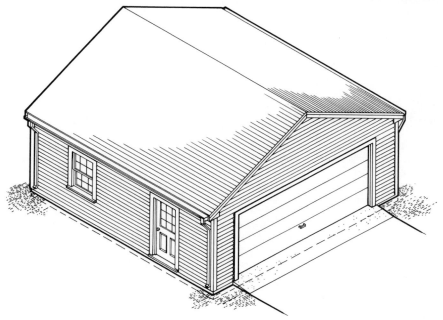

Prefabricated Two Car Garage

Large Octagonal Gazebo - Redwood

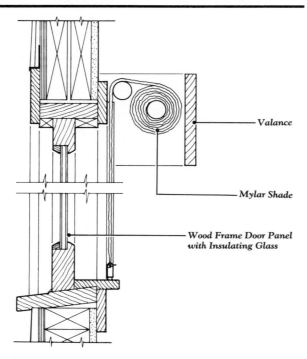

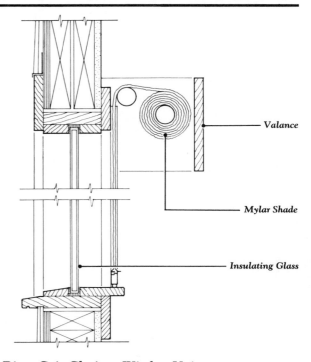

Valance

Mylar Shade

**Wood Frame Door Panel
with Insulating Glass**

Valance

Mylar Shade

Insulating Glass

Solar Direct Gain Glazing - Door Panel

Solar Direct Gain Glazing - Window Unit

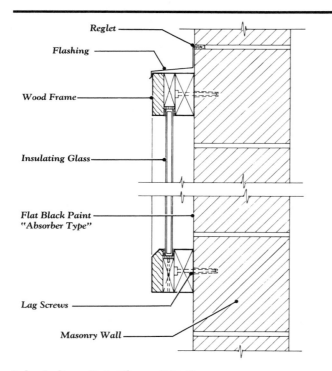

Reglet

Flashing

Wood Frame

Insulating Glass

**Flat Black Paint
"Absorber Type"**

Lag Screws

Masonry Wall

Solar Indirect Gain Thermal Wall

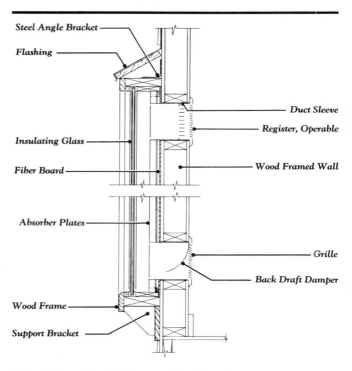

Steel Angle Bracket

Flashing

Insulating Glass

Fiber Board

Absorber Plates

Wood Frame

Support Bracket

Duct Sleeve

Register, Operable

Wood Framed Wall

Grille

Back Draft Damper

Solar Indirect Gain Thermosyphon Panel

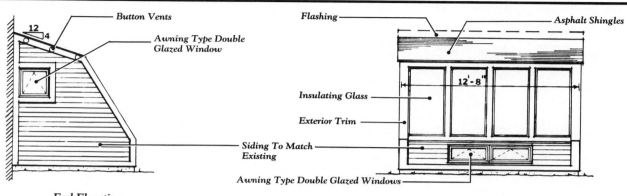

End Elevation

Front Elevation

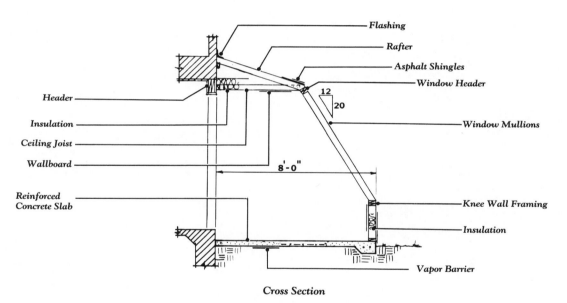

Cross Section

Solar Attached Sunspace

DIVISION 12

SITE WORK

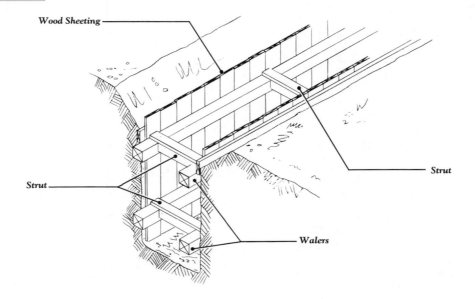

Wood Sheeting

Strut

Strut

Walers

Because of the ever present danger of cave-ins in deep, open trenches, trench shoring should be used to protect workers and equipment. OSHA regulations also mandate that shoring be provided for deep, open trenches. Three types of trench shoring are commonly used to hold trench walls intact and to ensure protection against cave-ins: wood sheeting, steel sheet piling, and the trench box.

Wood sheeting is often used as a shoring method. When the possibility of caving is slight, the sheeting planks are placed at only 5' to 15' intervals. Sheeting jacks force the vertically placed planks against the trench walls to allow for the inserting of bracing struts at each location. If the potential for caving increases and if stability requirements mandate that the sheeting be continuous, the bracing struts then support full-length horizontal walers rather than just the individual vertical sheeting planks. In very deep trenches, this system of walers and braces may be employed for every 2' to 5' of additional depth. Under normal installation and removal conditions, wood shoring materials can be reused for up to ten subsequent shoring operations.

Special lightweight sections of steel sheet piling may also be used for trench shoring. Because it is stronger than wood sheeting and shaped to handle more stress, steel sheet piling allows for longer spans, greater heights with fewer walers and braces and has more reuses than wood sheeting.

The trench box, used by many utility contractors, is a portable shoring system. The box, a prefabricated unit with steel double-wall sidewalls and spreader bars, is moved along the trench as the construction proceeds. Sizes of the assembled box range from 6' to 10' in height, from 12' to 24' in length, and from 2' to 10' in spreading capacity. Trench box units are modular and designed to allow stacking.

Man-hours

Description	m/hr	Unit
Wood Sheeting, Trench Jacks at 4' On Center		
8' Deep	.030	sf
12' Deep	.034	sf
15' Deep	.040	sf

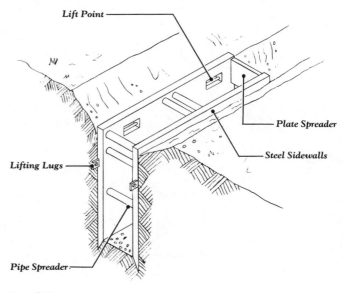

Lift Point

Plate Spreader

Steel Sidewalls

Lifting Lugs

Pipe Spreader

Trench Box

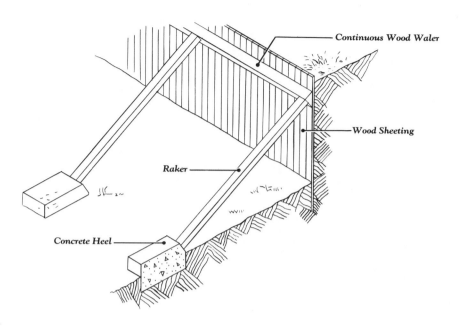

Wood Sheet Piling System

Sheet piling is installed to shore earth and reduce groundwater flow into open excavations. Two types of sheet piling are commonly used: wood sheet piling and steel sheet piling.

Wood sheet piling usually consists of planks, 3″ or more in thickness, which are placed side-by-side and driven vertically against the open cut of the excavation. Adjacent planks may simply be butted together or connected by tongue-and-groove joints. Walers, usually heavy timbers or steel channels, are placed horizontally across the sheeting planks. The spans between the walers are spaced to resist bending stresses in the sheeting material.

Rakers and tiebacks are two types of braces commonly used for supporting wood sheet piling. Rakers are diagonal braces which carry the horizontal thrust of the walers to the ground. They are abutted to heels or other suitable supports on the ground to prevent horizontal displacement. Rakers are usually installed where space requirements are not critical within the excavation. Tiebacks may be a more suitable bracing system when space is limited within the excavation or when access is required around its perimeter. Tieback tendons are grouted in place into the soil or rock behind the sheet piling and then posttensioned to support the horizontal thrust on the waler. Tiebacks are not uncommon in deep excavations of up to 60′.

Steel sheet piling consists of rolled U- or Z-shaped interlocking sections which are driven, facing in opposite directions, to form a structural wall. The horizontal pressure of the shored earth forces the interlocking piles together and forms a seal which prevents water from entering the excavation. In the case of cofferdams, the exterior water pressure forces the sheet piling to interlock. Under normal conditions, the corrugated shape and interlocking structure of steel sheet piling may produce a

sufficiently stiff, freestanding wall. However, in walls with hydraulic loading or heavy surcharge, bracing may be required with walers similar to those used for wood sheet piling.

Steel sheet piling is installed by driving the sections with an impact vibratory hammer. Alignment is maintained during the driving procedure with a temporary guide structure. The sheet piles are driven at least 2′ into firm soil at the bottom of the excavation, or deeper where soft conditions exist. Steel sheet piling is removed with vibratory extractors.

Man-hours

Description	m/hr	Unit
Wood Sheeting Including Wales Braces and Spacers		
8′ Deep Excavation Pull and Salvage	.121	sf
Left in Place	.091	sf
12′ Deep Pull and Salvage	.148	sf
Left in Place	.111	sf
16′ Deep Pull and Salvage	.167	sf
Left in Place	.125	sf
20′ Deep Pull and Salvage	.190	sf
Left in Place	.143	sf
Steel Sheet Piling		
15′ Deep Excavation Pull and Salvage	.098	sf
Left in Place	.065	sf
20′ Deep Pull and Salvage	.100	sf
Left in Place	.067	sf
25′ Deep Pull and Salvage	.096	sf
Left in Place	.064	sf
Tieback, Based on Total Length		
Minimum	.553	lf
Maximum	1.250	lf

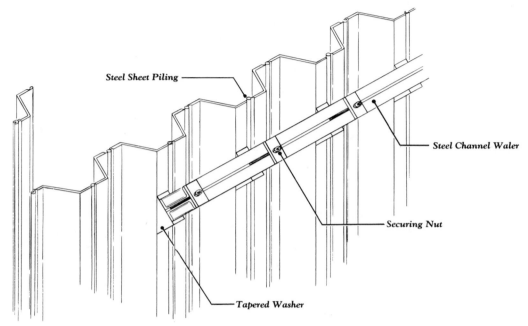

Steel Sheet Piling System

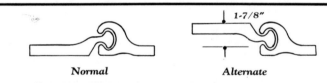

Normal   Alternate

Steel Sheet Piling Interlocking Connections

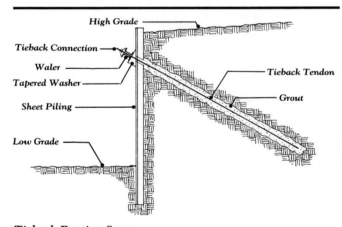

Tieback Bracing System

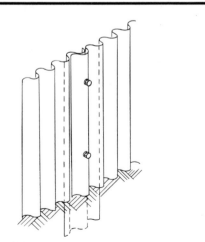

Corrugated Sheet Piling

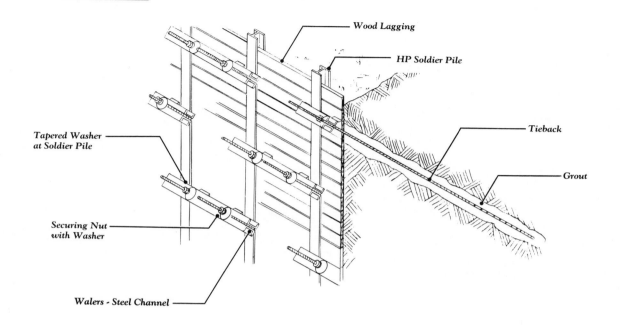

Wood Lagging

HP Soldier Pile

Tieback

Grout

Tapered Washer
at Soldier Pile

Securing Nut
with Washer

Walers - Steel Channel

Soldier piles and lagging are used to shore earth around deep excavations and along shorelines. The individual piles are manufactured from heavy steel or precast concrete sections that are driven at intervals and then spanned with horizontal lagging. The soldier piles are driven into bedrock and intermediately supported by walers and tiebacks, as in the sheet piling system. Soldier piles are driven at intervals, however, whereas sheet piling is driven continuously. The soldier pile and lagging system, therefore, reduces the overall driving time required for shoring installations.

Steel soldier piles, which are rolled sections designated by the letters HP, range between 8″ and 14″ deep in cross section. Special points attached to the driven end are recommended for use during installation to expedite the driving process and to prevent damage to and twisting of the pile while it is being driven.

The lagging used to span the steel soldier piles consists of wood planking that is placed horizontally between piles as the depth of the excavation increases. The size of the lagging, usually 3″ or more in thickness, depends upon the spacing of the soldier piles and the depth of the earth being retained. For example, an excavation 40′ in depth may require 4″ lagging between soldier piles that have been spaced at 8′ intervals with four rows of walers and tiebacks.

Precast concrete T beams may also be used as soldier piles for some marine bulkhead installations. The driving process for concrete piles is similar to that used for steel piles. Precast concrete panels are substituted for wood lagging to span between the piles.

Man-hours

Description	m/hr	Unit
Soldier Piles and Lagging, 15′ Deep		
1 Line of Braces		
Pull and Salvage	.206	sf
Left in Place	.176	sf
23′ to 35′ Deep, 3 Lines of Braces		
Pull and Salvage	.345	sf
Left in Place	.267	sf
Lagging Only, 3″ Thick Planks		
Minimum	.120	sf
Maximum	.192	sf
Tiebacks Only		
Minimum	.553	lf
Maximum	1.250	lf
HP Piles Only		
8″ x 8″	.100	vlf
14″ x 14″	.125	vlf

Excavation dewatering is the process of removing surface and ground water which may interfere with the excavation and construction of a structure. Methods used for removal of surface water are ditching, or gravity flow, and pumping.

The use of ditches to remove surface water by gravity flow may be economical where the angle of repose of the earth is steep and the amount of required excavation of the site is minimal. Where surface water is a common feature in the geographic location, the dewatering ditch may be planned as part of the final landscaping or topographical design of the site.

The removal of surface water by pumping is the alternative to the ditching method. Depending on the condition and the amount of the water to be removed, diaphragm or centrifugal pumps may be used. Where the water contains large amounts of solids, the diaphragm type is recommended. The more efficient centrifugal pumps should be used in cases which require removal of large volumes of surface water. Regardless of the type of pump selected, it should be located as close to the water as possible to minimize lift. Ideally, it should rest in a sump at a low point of the excavation. The inlet to the pump should be fitted with a trash screen to protect the pump and be set in a mesh enclosure to prevent clogging. Placing the inlet in a wooden box may keep it from settling in mud and sludge.

Water may be removed prior to excavation by lowering the water table around the site with the installation of well points or deep wells and pumping. A well point is a section of perforated pipe that is jetted vertically into place below the ground water level and then connected by riser pipes to a horizontal header at the ground surface. Well points and their accompanying risers are spaced along the header at intervals ranging from 2-1/2' to 10' on center. A pump is then connected to the header pipe, and the ground water is drawn from the water table, which, in time, is lowered to the depth of the well points (usually a maximum of 15'). Deeper well points are installed by locating headers at successively lower levels and operating the pump at the lowest of these levels. The flow of ground water into the well points depends on the porosity of the soil surrounding them. Where the soil is comprised of clay or consists of material which restricts its porosity, sand fill may be placed around the well points to facilitate the flow of water into them. The deep well system is a more complex and costly operation than the well point system because of the depth of installation involved and the drilling and lining of the shafts. Deep wells can be driven to depths of 50' and deeper and are spaced at wider intervals than well points.

Because all types of dewatering operations may require continuous pumping, a standby pump and extra operating equipment should be readily available to cope with breakdowns. To save on overtime wages for uninterrupted pumping operations that last for a week or longer, the assignment of four men rotating six-hour shifts computes to a total of only eight hours overtime wages per man per week.

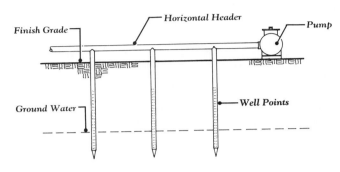

Man-hours

Description	m/hr	Unit
Excavate Drainage Trench, 2' Wide		
2' Deep	.178	cu yd
3' Deep	.160	cu yd
Sump Pits, By Hand		
Light Soil	1.130	cu yd
Heavy Soil	2.290	cu yd
Pumping 8 Hours, Diaphragm or Centrifugal Pump		
Attended 2 hours per Day	3.000	day
Attended 8 hours per Day	12.000	day
Pumping 24 Hours, Attended 24 Hours, 4 Men at 6 Hour Shifts, 1 Week Minimum	25.140	day
Relay Corrugated Metal Pipe, Including Excavation, 3' Deep		
12" Diameter	.209	lf
18" Diameter	.240	lf
Sump Hole Construction, Including Excavation, with 12" Gravel Collar		
Corrugated Pipe		
12" Diamter	.343	lf
18" Diameter	.480	lf
Wood Lining, Up to 4' x 4'	.080	sfca
Well-point System, Single Stage, Install and Remove, per Length of Header		
Minimum	.750	lf
Maximum	2.000	lf
Wells, 10' to 20' Deep with Steel Casing		
2' Diameter		
Minimum	.145	vlf
Average	.245	vlf
Maximum	.490	vlf

12 SITE WORK
WATER DISTRIBUTION SYSTEMS

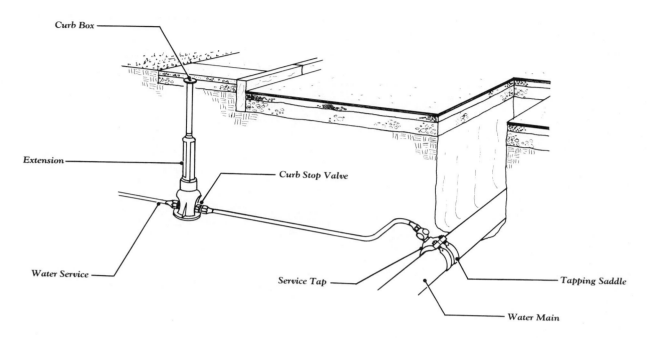

Curb Box

Extension

Curb Stop Valve

Water Service

Service Tap

Tapping Saddle

Water Main

Water distribution systems are comprised of two primary elements: the mains, which are pipes of several inches in diameter, and the piping service, which consists of pipes as small as 3/4″ diameter. When water is distributed to a structure, it is conducted from the larger existing mains, or newly installed laterals, into the smaller elements of the service piping. The installation of valves at various points assures control of the individual piping lines within the distribution system.

Water mains are usually manufactured from ductile iron, but reinforced concrete or plastic piping is also used. Laterals may be tied into the mains with Y's or T's. Wherever possible, a valve should be installed at the junction of the laterals to provide the ability to isolate a section of piping in case of a break or need for repair or maintenance. When tying a new lateral into an existing main that is still under pressure, the "wet" tap method should be used. This method allows the main to be tapped and the new valve inserted without disrupting the water service in the existing pressurized main.

Each smaller service line, which consists of copper or plastic piping, is usually provided with its own valve. In cases where the service lines supply a large commercial user, or where fire hydrants are to be installed along the line, a gate valve should be installed. For domestic service lines, a smaller curb stop valve can be used. Each service valve must be accompanied by its own gate box or curb box to provide accessibility to the valve so that the water supply to the individual line may be controlled at ground level. Service lines may be tapped into existing pressurized mains with the "wet" tap method, but without the necessity of adding a valve. The materials used in the service tap usually include a fitting, for the service line, that connects to a saddle mounted on the water main.

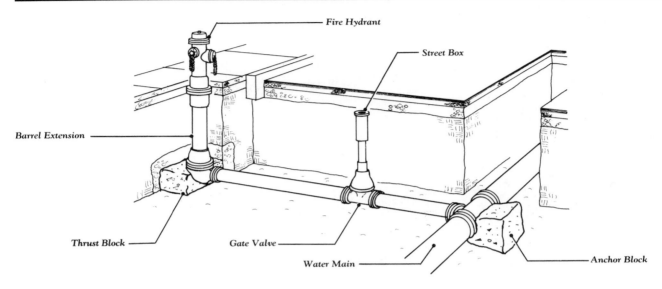

Fire Hydrant System

Man-hours

Description	m/hr	Unit
Water Distribution Piping, Not Including		
Excavation and Backfill		
Mains, Ductile Iron, 4″ Diameter	.167	lf
6″ Diameter	.190	lf
8″ Diameter	.259	lf
12″ Diameter	.389	lf
16″ Diameter	.609	lf
Polyvinyl Chloride, 4″ Diameter	.120	lf
6″ Diameter	.133	lf
8″ Diameter	.175	lf
12″ Diameter	.280	lf
Concrete, 10″ Diameter	.104	lf
12″ Diameter	.112	lf
16″ Diameter	.156	lf
Fittings For Mains, Ductile Iron, Bend,		
4″ Diameter	.649	Ea.
8″ Diameter	1.143	Ea.
16″ Diameter	2.000	Ea.
Wye, 4″ Diameter	.960	Ea.
8″ Diameter	1.714	Ea.
16″ Diameter	3.500	Ea.
Increaser, 4″ x 6″	2.000	Ea.
6″ x 16″	4.670	Ea.
Flange, 4″ Diameter	1.600	Ea.
8″ Diameter	3.080	Ea.
12″ Diameter	4.000	Ea.
Polyvinyl Chloride, Bend, 4″ Diameter	.240	Ea.
8″ Diameter	.300	Ea.

Man-hours (cont.)

Description	m/hr	Unit
12″ Diameter	.800	Ea.
Wye, 4″ Diameter	.267	Ea.
8″ Diameter	.343	Ea.
12″ Diameter	1.200	Ea.
Concrete, Bend, 12″ Diameter	1.170	Ea.
16″ Diameter	1.560	Ea.
Wye, 12″ Diameter	1.560	Ea.
16″ Diameter	2.800	Ea.
Service, Copper, Type K, 3/4″ Diameter	.050	lf
1″ Diameter	.060	lf
2″ Diameter	.080	lf
4″ Diameter	.150	lf
Polyvinyl Chloride, 1-1/2″ Diameter	.080	lf
2-1/2″ Diameter	.096	lf
Fittings for Service, Copper, General,		
3/4″ Diameter	.421	Ea.
2″ Diameter	.727	Ea.
Wye, 3/4″ Diameter	.667	Ea.
2″ Diameter	1.140	Ea.
Curb Box, 3/4″ Diameter Service	.667	Ea.
2″ Diameter Service	1.000	Ea.
Valves for Mains, 4″ Diameter	4.000	Ea.
8″ Diameter	7.000	Ea.
12″ Diameter	9.330	Ea.
Valves for Service, Curb Stop, 3/4″ Diameter	.421	Ea.
1″ Diameter	.500	Ea.
2″ Diameter	.727	Ea.

Trench drains are used to remove surface water and to serve as a linear boundary to retain surface drainage. Situations where trench drains may be called for include: a sloped driveway entering a building, a stairway intersecting a plaza, or a sloping landscape where surface water may collect. Trench drains are often constructed from formed concrete with a cast iron grating cover, but other natural and man-made materials may also be used.

Trench drains are designed for open-channel flow. The cover of the drain is recessed to the level of the surrounding grade to allow for unobstructed crossing by foot or wheeled traffic. The cover can be manufactured from light- or heavy-duty material, depending on the amount of support required by the expected traffic flow. A framing angle is embedded into the perimeter of the trench to receive the cover.

Prefabricated concrete trench drains can be installed much faster than poured-in-place concrete drains. One type, for example, is manufactured from polymer concrete with sections that snap together for relatively easy and quick installation. The polymer concrete material used in the drain also resists chemicals and is not affected by the freeze-thaw cycle. Modular catch basins are also available with this prefabricated drainage system.

Materials other than concrete are used occasionally to create trench drains in lawns and/or other landscaped or site areas where concrete drains are not desired for aesthetic or other reasons. One alternative method of trench drain installation is to place drainage stone in an envelope of geotextile fabric, usually 4- to 6-ounce nonwoven polyester. The fabric is unrolled into an open trench, filled with the stone, and then overlapped. Additional stone or suitable landscaping materials may then be used to cover the drain.

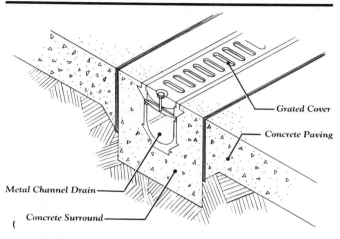

Embedded Trench Drain

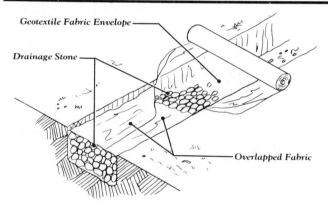

Stone Trench Drain

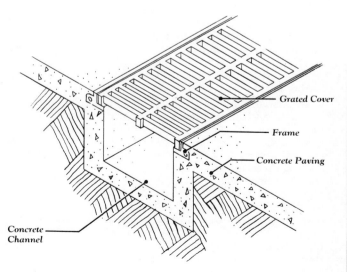

Man-hours

Description	m/hr	Unit
Trench Forms		
1 Use	.200	sfca
4 Uses	.173	sfca
Reinforcing	15.240	ton
Concrete		
Direct Chute	.320	cu yd
Pumped	.492	cu yd
With Crane and Bucket	.533	cu yd
Trench Cover, Including Angle Frame		
To 18″ Wide	.400	lf
Cover Frame Only		
For 1″ Grating	.178	lf
For 2″ Grating	.229	lf
Geotextile Fabric in Trench		
Ideal Conditions	.007	sq yd
Adverse Conditions	.010	sq yd
Drainage Stone		
3/4″	.092	cu yd
Pea Stone	.092	cu yd

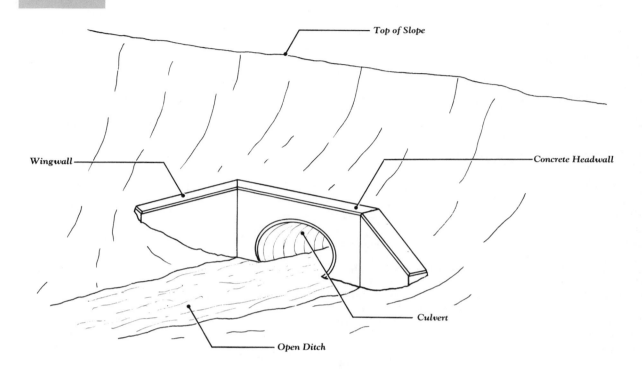

The flow of storm water runoff on the surfaces of slopes and the embankments of swales and ditches must be controlled to prevent erosion and the undermining of site structures and pavements. Where the vegetative cover is insufficient to control surface erosion, transverse water diversion ditches should be placed at the top of the slope. Where the end of the intercepting ditch becomes too steep or must follow the slope, a paved spillway section may be installed to conduct the flow. Paving may be required where the flow in a ditch or swale exceeds the runoff limits for bare earth or turf grass. The paving materials commonly used for these installations include stone rubble, asphalt, and concrete. Concrete drop structures, which are similar in construction to catch basins, may also be installed at critical locations along the ditch to reduce the volume of the flow and to allow some of the runoff to be carried by pipe to safe remote discharge locations.

Culverts may also be employed to conduct storm water runoff, especially under roads, highways, and railroad beds. Small culverts of only a few feet in diameter may be constructed from sections of reinforced concrete pipe, corrugated metal, or plastic. Large culverts require more complex installations which may be constructed from reinforced concrete pipe of up to 8' in diameter, precast box culverts of up to 12' in height, corrugated steel or aluminum arches of up to 6' in radius, or multi-plate steel arches. Poured-in-place concrete bridges may even be required for extremely large-scale culvert installations. Regardless of the size of the culvert, the structure, wherever possible, should be placed at right angles to the roadway or rail bed to minimize the length of the structure and to reduce its cost.

Care should be exercised in planning and constructing the entrances and exits of culverts so that appropriately designed and installed headwalls will protect and retain the surrounding fill. For small culverts which are correctly aligned on slight slopes, a simple straight headwall constructed perpendicular to the flow serves as adequate protection. For larger culverts and moderate to steep slopes, wing walls must be placed alongside the straight headwall to prevent erosion of the sloped fill and undermining of the culvert itself. In situations where the flow changes abruptly at a culvert, the wing walls must be skewed in a direction to coincide with the natural stream. In addition to the headwall, a paved apron may also be required where high intake or exit flow velocities can cause further erosion or undermining.

The materials used for the construction of culverts, headwalls and culvert piping vary according to the size of the structure and the conditions of its placement. Large field constructed headwalls are normally constructed from poured-in-place concrete or from building stones set in mortar. Precast culvert boxes may be ordered with custom end walls at an additional cost. Some types of small culvert piping can be ordered with special sections with flared ends which can function as headwalls. Concrete culvert pipe may be ordered with a vitreous lining which greatly improves the pipe's hydraulic characteristics. Corrugated metal pipe also achieves greater hydraulic efficiency with the addition of a specially ordered paved invert.

Man-hours

Description	m/hr	Unit
Paving, Asphalt, Ditches	.185	sq yd
Concrete, Ditches	.360	sq yd
Filter Stone Rubble	.258	cu yd
Paving, Ashpalt, Aprons	.320	sq yd
Concrete, Aprons	.620	sq yd
Drop Structure	8.000	Ea.
Culverts, Reinforced Concrete, 12" Diameter	.162	lf
24" Diameter	.183	lf
48" Diameter	.280	lf
72" Diameter	.431	lf
96" Diameter	.560	lf
Flared Ends, 12" Diameter	1.080	Ea.
24" Diameter	1.750	Ea.
Corrugated Metal, 12" Diameter	.114	lf
24" Diameter	.175	lf
48" Diameter	.560	lf
72" Diameter	1.240	lf
Reinforced Plastic, 12" Diameter	.280	lf
Precast Box Culvert, 6' x 3'	.343	lf
8' x 8'	.480	lf
12' x 8'	.716	lf
Aluminum Arch Culvert, 17" x 11"	.150	lf
35" x 24"	.300	lf
57" x 38"	.800	lf
Multi-plate Arch, Steel	.014	lb

Corrugated Metal Culvert

Man-hours (cont.)

Description	m/hr	Unit
Headwall, Concrete, 30" Diameter Pipe,		
3' Wing Walls	28.750	Ea.
4'-3" Wing Walls	33.500	Ea.
60" Diameter Pipe, 5'-6" Wing Walls	63.250	Ea.
8'-0" Wing Walls	76.650	Ea.
Stone, 30" Diameter Pipe, 3' Wing Walls	12.650	Ea.
4'-3" Wing Walls	14.800	Ea.
60" Diameter Pipe, 5'-6" Wing Walls	30.200	Ea.
8'-0" Wing Walls	37.200	Ea.

12 SITE WORK
SEWAGE AND DRAINAGE COLLECTION SYSTEMS

Sewage and drainage usually flow by gravity from the source of collection, through service lines, into mains, and eventually to a point of treatment. However, force mains are not uncommonly substituted for gravity mains in sewage systems in which the fluid head must be augmented with pumping stations to assist the drainage flow. Manholes are spaced at regular intervals along the main lines to provide access for repair and maintenance. Sewage and drainage systems operate in the same manner, but their components vary in the types of materials used, because their functions are different.

Sewage-collection systems conduct biological waste to a treatment facility. Gravity sewer mains are usually manufactured from reinforced concrete, but plastic piping can also be used. Asbestos cement piping, at one time the most commonly used form of main piping material, has declined in use due to the hazards of handling asbestos. Force sewer mains (under pressure) are usually manufactured from ductile iron or reinforced plastic pipe. Service lines to individual users consist of vitrified clay or plastic piping. The fittings used in tying the service lines into the main should be made of the same material as that of the service line.

Drainage systems are used to collect stormwater and surface drainage from roadways and parking areas and to conduct the flow away to a suitable outfall. Drainage piping is usually manufactured from reinforced concrete or corrugated metal, although plastic piping has become increasingly popular. Catch basins, which are located at the sump, or low point, of the area to be drained, are sized to handle the maximum volume of rainfall for that area. A catch basin is protected from clogging by a cast iron grating that is removed for maintaining and periodic cleaning. Because the basin outlet pipe is located several feet above the lowest level of the basin, silt and debris settle and accumulate at the bottom of the structure. This material must be cleaned out at regular intervals to prevent buildup and the eventual clogging of the outlet pipe.

Although manholes may require the installation of steps, and may differ from catch basins in function, size, and type of cover, both structures are commonly constructed in the same way and of the same materials. Both structures are fabricated from brick, concrete blocks, c.i.p. concrete or precast concrete sections that are assembled on a poured-in-place concrete slab or on a precast base. A typical manhole or catch basin is a cylinder 4′ ± in diameter and 6′ high (internal dimensions) at the lower section. The cylinder tapers into an upper section, 2′ in diameter by 2′ high, that receives the cover. The cover, and the frame, range from 4″ to 10″ in depth and may be adjusted to grade with brick.

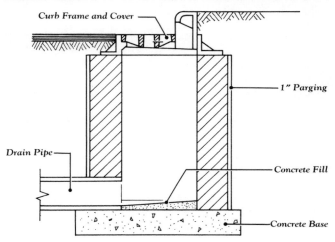

Curb Inlet

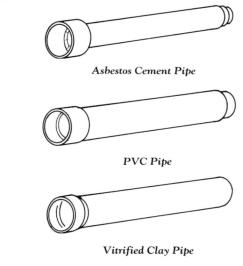

Asbestos Cement Pipe

PVC Pipe

Vitrified Clay Pipe

Drainage and Sewage Pipe Types

Catch Basin

Man-hours

Description	m/hr	Unit
Catch Basins or Manholes, Not Including Excavation and Backfill, Frame and Cover		
Brick, 4' I.D., 6' Deep	23.881	Ea.
8' Deep	32.000	Ea.
10' Deep	44.444	Ea.
Concrete Block, 4' I.D., 6' Deep	16.000	Ea.
8' Deep	21.333	Ea.
10' Deep	29.630	Ea.
Precast Concrete, 4' I.D., 6' Deep	3.000	Ea.
8' Deep	4.000	Ea.
10' Deep	6.000	Ea.
Cast in Place Concrete, 4' I.D., 6' Deep	10.667	Ea.
8' Deep	16.000	Ea.
10' Deep	32.000	Ea.
Frames and Covers, 18" Square, 160 lbs	2.400	Ea.
270 lbs	2.791	Ea.
24" Square, 220 lbs	2.667	Ea.
400 lbs	3.077	Ea.
26" D Shape, 600 lbs	3.429	Ea.
Roll Type Curb, 24" Square, 400 lbs	3.077	Ea.
Light Traffic, 18" Diameter, 100 lbs	2.526	Ea.
24" Diameter, 300 lbs	2.759	Ea.
36" Diameter, 900 lbs	4.138	Ea.
Heavy Traffic, 24" Diameter, 400 lbs	3.077	Ea.
36" Diameter, 1150 lbs.	8.000	Ea.
Raise Frame and Cover 2", for Resurfacing,		
20" to 26" Frame	3.640	Ea.
30" to 36" Frame	4.440	Ea.

Man-hours

Description	m/hr	Unit
Drainage and Sewage Piping, Not Including Excavation and Backfill		
Asbestos Cement, 6" to 8" Diameter	.073	lf
12" Diameter	.100	lf
16" Diameter	.140	lf
24" Diameter	.261	lf
Concrete, up to 8" Diameter	.140	lf
10" to 18" Diameter	.168	lf
21" to 24" Diameter	.183	lf
30" Diameter	.212	lf
36" Diameter	.250	lf
Corrugated Metal, 8" Diameter	.073	lf
12" Diameter	.114	lf
18" Diameter	.147	lf
24" Diameter	.175	lf
30" Diameter	.233	lf
36" Diameter	.280	lf
Ductile Iron, 6" Diameter	.190	lf
8" Diameter	.259	lf
12" Diameter	.389	lf
16" Diameter	.609	lf
Polyvinyl Chloride, 4" Diameter	.064	lf
8" Diameter	.072	lf
12" Diameter	.088	lf
15" Diameter	.147	lf
Vitrified Clay, 4" Diameter	.063	lf
6" Diameter	.071	lf
8" Diameter	.093	lf
12" Diameter	.147	lf

12 SITE WORK
SEPTIC SYSTEMS

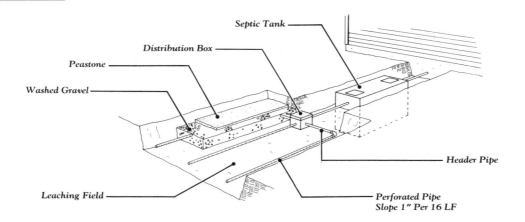

- Septic Tank
- Distribution Box
- Peastone
- Washed Gravel
- Leaching Field
- Header Pipe
- Perforated Pipe Slope 1″ Per 16 LF

Self-contained individual septic systems dispose of sewage by percolating the waste fluids through a gravel base and the surrounding soil. The basic components of a typical septic system include a septic tank, distribution box, header pipe, perforated field piping, and septic, or leaching, field. Overflow from the tank is carried by the header pipe to a junction box and then distributed to the leaching field by a network of perforated field pipes.

The leaching field normally consists of a large area of washed gravel base on which the perforated pipes are laid, usually at a slope of 1″ per 16 lf. After the perforated pipes are laid, more gravel is added between and around the piping to a level flush with the top of the pipes. Filtering paper can be placed over the gravel base to prevent the penetration of dirt and silt into the leaching field.

Because the liquids which drain out through the holes in the perforated pipes percolate through the washed gravel and, eventually, the subsoil, the porosity of the existing soil is a critical factor in the system's operation. Generally, the size of the system is determined by the porosity factor and the anticipated volume of fluid to be disposed of. In cases where the natural soil is not porous enough to absorb the flow, the entire leaching area must be excavated and suitable subsoil hauled in to replace it. Whenever possible, the leaching field should be placed in an untravelled, open, well-ventilated area, which is located at a safe distance from ponds, streams, wells, and reservoirs. Local codes should be consulted to determine the minimum allowable distance between the leaching field and these natural water sources.

Most localities require that a percolation test be performed on the existing soil to determine its porosity and rate of absorption of fluids. The test is performed by first digging several test holes to the anticipated depth of the leaching field, usually 2′ to 3′ below grade. The holes are then filled at least twice with water and allowed to drain. After they have been filled and drained at least twice, they are filled again, and the rate of absorption is measured. The results of the test are based on the amount of time required for the water level to drop 1″ in the final filling.

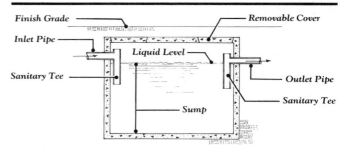

- Finish Grade
- Inlet Pipe
- Sanitary Tee
- Liquid Level
- Sump
- Removable Cover
- Outlet Pipe
- Sanitary Tee

Septic Tank System

Man-hours

Description	m/hr	Unit
Septic Tanks Not Including Excavation or Piping, Precast, 1000 Gallon	3.500	Ea.
2000 Gallon	5.600	Ea.
5000 Gallon	16.471	Ea.
Fiberglass 1000 Gallon	4.667	Ea.
1500 Gallon	7.000	Ea.
Excavation, 3/4 cu yd Backhoe	.110	cu yd
Distribution Boxes, Concrete, 5 Outlets	1.000	Ea.
12 Outlets	2.000	Ea.
Leaching Field Chambers, 13′ x 3′-7″ x 1′-4″, Standard	3.500	Ea.
Heavy Duty, 8′ x 4′ x 1′-6″	4.000	Ea.
13′ x 3′-9″ x 1′-6″	4.667	Ea.
20′ x 4′ x 1′-6″	11.200	Ea.
Leaching Pit, Precast Concrete, 3′ Pit	3.500	Ea.
6′ Pit	5.957	Ea.
Disposal Field		
Excavation, 4′ Trench, 3/4 cu yd Backhoe	.048	lf
Crushed Stone	.160	cu yd
Piping, Asbestos Cement, Perforated, 4″ Diameter	.062	lf
6″ Diameter	.063	lf
Bituminous Fiber Perforated, 4″ Diameter	.042	lf
6″ Diameter	.047	lf
Vitrified Clay, Perforated, 4″ Diameter	.060	lf
6″ Diameter	.076	lf
PVC, Perforated, 4″ Diameter	.056	lf
6″ Diameter	.060	lf

12 SITE WORK
EROSION CONTROL

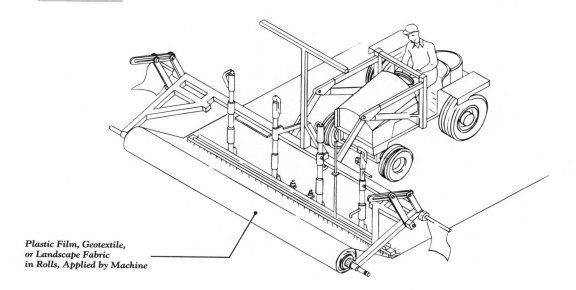

Plastic Film, Geotextile,
or Landscape Fabric
in Rolls, Applied by Machine

Erosion control may be used when top soil, subsoil, or fill tends to run off with the rapid flow of water down a sloped surface. The type of erosion control selected varies for different situations and is determined by the size and severity of the erosion problem, the amount of time available to establish the control system, cost factors, and the final landscaping design of the site. Natural controls, mulches of all types, fiber grids, and rip rap stone are among the many erosion control materials used to keep soil in place.

For moderate slopes and swales, the planting of natural materials may serve as an adequate method for slowing runoff. The root systems of shrubbery, ground cover, and, especially, grass provide a constantly growing and expanding source of natural erosion control. This method of control, however, often takes a long time to establish and may require regular maintenance of mowing and pruning after the vegetation takes root. More immediate controls, even if they are temporary, can be implemented to hold the soil in place while grass or other vegetation gets established.

Included among these faster methods of erosion control are many different types of organic and stone mulches which can be spread over large areas to control runoff. Most organic mulches are relatively inexpensive and can be easily distributed by wheelbarrow or light-duty loader. Wood chips, often a by-product of site clearing, can be placed to a depth of 1″ to 2″. Straw can be placed by itself or as a cover to protect seeded banks. Blankets of excelsior, held together with biodegradable plastic netting, can be unrolled and stapled into place on the slope. For rapid installation, mulch may also be applied by the hydro-spraying method from a distribution truck. This slurry consists of water and wood cellulose alone or mixed with grass seed and fertilizer. Expensive organic mulches, which include peat and bark in shredded or nugget form, should be installed where the aesthetic impact of the site is a prime consideration.

Stone mulches provide permanent protection against slope erosion, but they are harder to install and more expensive than most organic erosion control materials. Quarry stone is the most economical type of stone mulch, but other types of decorative stone are available at costs of ten to thirty times that of quarry stone. The size of stone mulch ranges from pea stone to boulders.

To control weeds and other unwanted growth which may establish themselves beneath and between the mulch, several methods may be employed. For example, two mil black plastic film can be installed on the slope prior to the spreading of the mulch. Geotextile and landscaping fabric can be substituted for the plastic film if a porous material is needed to allow water penetration into the slope. Lightweight polypropylene fabric can be installed prior to the placing of small stone mulch, but heavier stones may require a heavier fabric or a light fabric with a granular bedding layer.

Several types of organic and synthetic fiber grids, nettings, and fabric materials may also be used alone or in conjunction with mulches or with ground covers and plantings to control slope erosion. Recycled tobacco cloth, which is used for shading tobacco fields, is available from landscaping wholesalers in many parts of the country. Jute mesh and woven polyethylene mesh cost a little more than tobacco cloth, but they can serve as an adequate substitute in areas where tobacco cloth is not available. Polypropylene, which is often used in the manufacture of geotextiles, has been adapted for use in erosion control as a flat, loosely bonded film that allows air and moisture to pass freely while inhibiting weed growth. Flexible three-dimensional matting, which is fabricated from bonded vinyl, nylon, or polypropylene filaments, becomes permanently interlocked in vegetation growth after it has been placed.

For more severe erosion control situations, such as shorelines, the banks of channels, and very steep slopes, heavier materials are required. Rip rap stone, which ranges in size from 8″ to 1/3 of a cubic yard and larger, can be installed by a variety of methods. Small rip rap may be distributed by truck or loader and then spread or placed by hand or machine. Large rip rap must be placed a stone at a time, often with a crane. The stone may be set in a gravel bedding or a geotextile base, grouted in place with concrete, or dry grouted by filling voids with smaller stone.

Several other materials and methods besides rip rap can be employed for severe erosion control. Precast interlocking concrete blocks, which are installed like rip rap, can be employed as a heavy-duty paving material on eroded slopes. A unique concrete forming system for interconnected modular blocks is also available. This system consists of a heavy fabric which can be unrolled in place and then pumped full of concrete. Gabion revetment mats can be set on banks, filled with stone, and then coated with an asphaltic slurry.

A final method of severe erosion control includes the installation of a barrier or a series of barriers set at right angles to the direction of flow to interrupt the runoff. These barriers consist of materials which filter out and retain the sediment in the runoff but allow the water to pass. Hay bales supported by wood stakes comprise the most commonly installed barrier of this type. Another type of barrier, called a silt fence, is constructed from sections of woven polypropylene strip fabric that are reinforced and drawn between posts located at intervals of 8′ to 10′.

Man-hours

Description	m/hr	Unit
Mulch		
Hand Spread		
Wood Chips, 2″ Deep	.004	sf
Oat Straw, 1″ Deep	.002	sf
Excelsior w/Netting	.001	sf
Polyethylene Film	.001	sf
Shredded Bark, 3″ Deep	.009	sf
Pea Stone	.643	cu yd
Marble Chips	2.400	cu yd
Polypropylene Fabric	.001	sf
Jute Mesh	.001	sf
Machine Spread		
Wood Chips, 2″ Deep	1.970	MSF
Oat Straw, 1″ Deep	.089	MSF
Shredded Bark, 3″ Deep	2.960	MSF
Pea Stone	.047	cu yd
Hydraulic Spraying		
Wood Cellulose	.200	MSF
Rip Rap		
Filter Stone, Machine Placed	.258	cu yd
1/3 cu yd Pieces, Crane Set Grouted	.700	sq yd
18″ Thick, Crane Set, Not Grouted	1.060	sq yd
Gabion Revetment Mats, Stone Filled, 12″ Deep	.366	sq yd
Precast Interlocking Concrete Block Pavers	.078	sf

Pavement and slab base stabilization may be employed to improve the bearing and flexural capacities of the supporting base soil. Several methods can be used to accomplish the stabilization, including the application of an additive by direct mechanical mixing or the installation of a layer of geotextile fabric. The type of material—cement, asphalt cutback, or lime—mixed into the base soil in the additive method depends on the amount of silt and clay found in the soil.

Cement is added to base soil which contains no less than 10% nor more than 35% combined silt and clay. Cement should be added at a rate of between 4% and 14% by volume, and distributed over the graded base by individual bags or in bulk with a mechanical spreader. After the cement has been placed, it is combined with the soil by a rotary or traveling plant mixer which simultaneously adds water and deposits the stabilized mixture. The base is then compacted, first by sheepsfoot and then by rubber tired rollers. The application of a concrete curing compound completes the process, and the stabilized base is left to set up and cure.

Base soil which contains up to 45% silt and clay can be stabilized with asphalt cutback. The type of soil dictates both the amount of cutback to be added, from 1/3 to 2/3 of a gallon per square yard per inch of depth, and the rate of cure, from slow to rapid. The asphalt cutback is introduced into the soil as it is being mixed in a process similar to the one in which water is added to the soil-cement operation described in the previous paragraph. The compacting methods employed for the cement-soil stabilization process are also used for the asphalt cutback-soil procedure. The stabilized deposit should then be allowed to cure for about a week before being sealed with a layer of appropriate coating material.

The use of lime as an additive helps to stabilize base soils that contain large amounts of clay in the range of 2% to 6% by weight. The process of installing the lime is identical to that used for soil-cement stabilization, with the addition of a second mixing of the deposit after a two-day preliminary curing period. After the second mixing, the stabilization requires an additional three to seven days of curing time. During the final curing period, the base must be continuously sprinkled with water or be covered by a single bituminous prime coat membrane.

Stabilization fabric may be installed to stabilize base soil as an alternative to the various additive methods. The material normally used in this procedure is woven polypropylene geotextile in weights which usually range between 3 and 7 ounces per square yard. For extreme cases where heavier material is required, fabric of 20 ounces per square yard may be called for. The fabric is usually placed over the sub-base with a roller bar mounted on a front-end loader, and then the pavement base is installed on the layer of fabric. In addition to increasing the strength of the sub-base, the fabric prevents the mixing of the layers.

Man-hours

Description	m/hr	Unit
Cement Stabilization		
6″ Deep	.058	sq yd
12″ Deep	.067	sq yd
Asphalt Stabilization		
3″ Deep	.007	sq yd
6″ Deep	.008	sq yd
Lime Stabilization		
For Base		
6″ Deep	.036	sq yd
12″ Deep	.041	sq yd
For Sub-base, Initial Cure and Remixing		
6″ Deep	.058	sq yd
12″ Deep	.067	sq yd
Stabilization Fabric	.002	sq yd

Asphalt pavements transfer and distribute traffic loads to the subgrade. The pavement is made up of two layers of material, the wearing course and the base course. The wearing course consists of two layers: the thin surface course and the thicker binder course that bonds the surface course to the heavy base layer underneath. The base consists of one layer that varies in thickness, type of material, and design, according to the bearing value of the subgrade material. If the subgrade material has a low-bearing value, either the thickness or the flexural strength of the base must be increased to spread the load over a larger area.

Increasing the flexural strength of the base course can be accomplished in two ways, one of which is to mix asphalt with the base material. The addition of the asphalt doubles the load distributing ability of a conventional granular base. A second method of increasing flexural strength is to add a layer of geotextile stabilization fabric between the base and the subgrade. This fabric not only adds tensile strength to the base, but also prevents the subgrade material from pumping up and contaminating the base during load cycles.

The thicknesses of the various courses within the pavement vary with each layer, according to the intended use of the pavement and, as noted earlier for the base course, the bearing value of the subgrade. For example, pavement may contain a 1″ surface course, a 2″-3″ binder course, and a 5″ or more base course. Standard commercial parking lot pavement may consist of 2-1/2″ of wearing surface and 8″ of granular base, while a highway may require a full-depth pavement with 5-1/2″ of asphalt base and 4″ of wearing surface. Granular base courses usually range from 6″ to 18″ in thickness. The cost of the asphalt mix for a pavement depends on the quality of the aggregate used in the various layers. Generally, the surface mixes contain smaller, higher-grade aggregates and, therefore, cost more than the heavier base mixes.

The installation of pavement material involves several steps and requires highly specialized equipment. Asphalt mix is delivered by truck to the hopper on the front end of the paver and is carried back to the spreading screws by bar feeders. The screws deposit a continuous flow of mix in front of the screed unit, which controls the thickness of the course being spread by the machine. The width of the mat may be increased with extensions added to the screed unit. Automatic sensors, which follow a previously set string line or a ski that rides on an adjacent grade, guide the screed unit to maintain the correct grade of the paved surface. After the pavement has been spread, steel-wheeled or pneumatic-tired rollers compact the asphalt in three separate rollings: the first, to achieve the desired density; the second, to seal the surface; and the third, to remove compactor roller marks and to smooth the wearing surface.

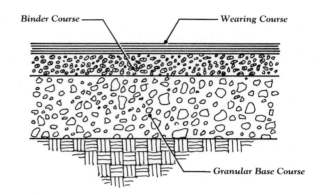

Asphalt Pavement

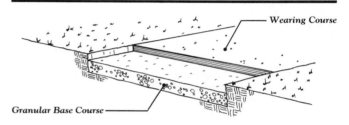

Bituminous Sidewalk

Man-hours

Description	m/hr	Unit
Subgrade, Grade and Roll		
Small Area	.024	sq yd
Large Area	.011	sq yd
Base Course		
Bank Run Gravel, Spread Compact		
6″ Deep	.004	sq yd
18″ Deep	.013	sq yd
Crushed Stone, Spread Compact		
6″ Deep	.016	sq yd
18″ Deep	.029	sq yd
Asphalt Concrete Base		
4″ Thick	.053	sq yd
8″ Thick	.089	sq yd
Stabilization Fabric, Polypropylene, 6 oz./sq yd	.002	sq yd
Asphalt Pavement, Wearing Course		
1-1/2″ Thick	.026	sq yd
3″ Thick	.052	sq yd

The process of recycling used asphalt pavement provides a means of reclaiming and rejuvenating old pavement materials so that they can be respread. Generally, all of the methods of recycling involve the tearing up or breaking of the old material, pulverizing it into uniform consistency at a treatment plant or with on-site machines, combining it with new materials, and reinstalling the paving mixture.

The reclaiming part of the recycling process includes the loosening and removing of all or part of the worn-out asphalt pavement. The method used for removal of the existing paved surface depends on the thickness of the pavement to be recycled and the nature of the underlying base. If the base is concrete and the entire layer of pavement above it is to be removed, a ripper or scarifier, which has been mounted on a dozer, loader, or motor grader, can be used. If the base is composed of gravel, the total pavement can be removed with a pavement breaker. Another method, which can be implemented with any type of base material, is to sawcut the pavement into manageable sized strips that can be removed with a backhoe or gradall. Regardless of the method used in tearing up or breaking the old pavement, the loosened, variously sized pieces must be loaded for stockpiling and eventual pulverization.

The pulverization of the old asphalt pavement material can be accomplished on the site with three types of machines designed for this procedure: the stabilizer, the cold planer, or the miller. The actual pulverizing of the old material is performed by a rotating drum with inlaid carbide cutting teeth, called the "cutter," within the planer unit. If the planer is track-mounted, material with depths of just above 0" to a maximum of 5" to 12" can be processed; for wheel-mounted machines, the maximum falls between 3" and 7" in depth. The width of the cutter varies.

After the material has been pulverized, several methods of handling it can be employed, depending on how it is to be used when respread. The recycled pavement may be blended with the base as it is being pulverized to upgrade the strength of the base; it may be left behind in a windrow; or it may be deposited into a truck with an elevating loader. The pulverized material may also be deposited into a traveling hammermill or portable crushing plant and processed into a uniformly graded mixture that is suitable for respreading.

Two operations may be implemented for rejuvenating the pulverized old pavement: cold-mix recycling and hot-mix recycling. Cold-mix recycling involves the combining-in-place of asphalt emulsions or cutbacks with the reclaimed pavement materials at the installation site. If, as often happens, the reclaimed pavement has been respread over the roadway after pulverization, asphaltic emulsion is sprayed on the material from a distribution tanker and then blade-mixed with motor graders. In another cold-mix method, the emulsion is added during the mixing with a stabilizer connected to the tanker. Some planers are also equipped with blending and asphalt emulsion spray bars, and therefore perform three tasks—pulverization, addition of emulsion, and mixing—and function as self-contained, mobile recycling plants.

The hot-mix recycling process is carried out at an asphalt recycling facility, rather than at the removal and/or installation site. In the hot-mix recycling operation, stockpiled recycled pavement must be crushed and screened before being mixed with new aggregate and asphalt. The proportion of old to new material depends on the new mix design and type of plant in which the recycling operation is taking place. In a batch plant, the new mix usually consists of 30% old material and 70% new; in a continuous mix plant, the new mix is usually composed of 70% old material and 30% new. The reason for the superior efficiency of continuous mix plants is that both the recycled material and the new aggregate are directly heated. In the batch plant operation, the recycled pavement is heated only mixing it with superheated new aggregate. Additional savings in both operations can be realized by utilizing recycled aggregate materials in addition to the recycled pavement.

Man-hours

Description	m/hr	Unit
Asphalt Pavement Demolition		
Hydraulic Hammer	.035	sq yd
Ripping Pavement, Load and Sweep	.007	sq yd
Crush and Screen, Traveling Hammermill		
3" Deep	.004	sq yd
6" Deep	.007	sq yd
12" Deep	.012	sq yd
Pulverizing, Crushing, and Blending into Base		
4" Pavement		
Over 15,000 sq yd	.027	sq yd
5,000 to 15,000 sq yd	.029	sq yd
8" Pavement		
Over 15,000 sq yd	.029	sq yd
5,000 to 15,000 sq yd	.032	sq yd
Remove, Rejuvenate and Spread, Mixer-Paver	.026	sq yd
Profiling, Load and Sweep		
1" Deep	.002	sq yd
3" Deep	.006	sq yd
6" Deep	.011	sq yd
12" Deep	.019	sq yd

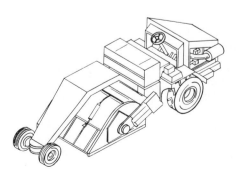

Pulverizer

Curbs are set along the sides of roadways to protect the unpaved roadside and to provide erosion-free means of drainage for the road surface. Commonly used curb materials include cast-in-place concrete, precast concrete, bituminous concrete, asphalt, and stone (usually granite). The methods of curb installation vary with the type of curbing material.

Cast-in-place concrete curbs can be formed and cast or placed monolithically with a form-and-place machine. In the latter method, the concrete is placed into a large retaining bin on the machine, formed into the shape of the curb, and then placed as the machine moves along at a set pace. Bituminous concrete curbing can be formed and placed in the same way. In both the cast-in-place and monolithic methods, gutters can be placed as an integral part of the curb.

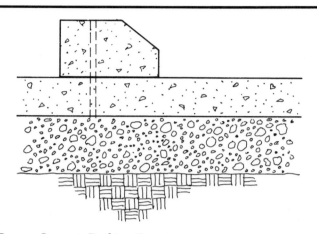

Precast Concrete Parking Bumper

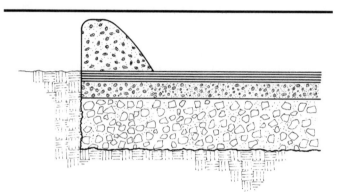

Bituminous Curb

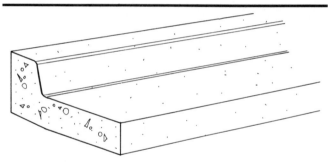

Cast-in-Place Concrete Curb and Gutter

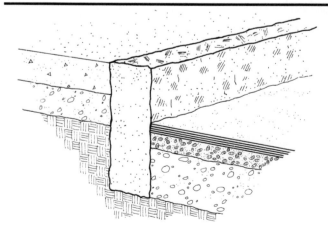

Granite Curb

Man-hours

Description	m/hr	Unit
Curbs, Bituminous, Plain, 8″ Wide, 6″ High, 50 lf/ton	.032	lf
8″ Wide, 8″ High, 44 lf/ton	.036	lf
Bituminous Berm, 12″ Wide, 3″ to 6″ High, 35 lf/ton, before Pavement	.046	lf
12″ Wide, 1-1/2″ to 4″ High, 60 lf/ton, Laid with Pavement	.030	lf
Concrete, 6″ x 18″, Cast-in-place, Straight	.096	lf
6″ x 18″ Radius	.107	lf
Precast, 6″ x 18″, Straight	.160	lf
6″ x 18″ Radius	.172	lf
Granite, Split Face, Straight, 5″ x 16″	.112	lf
6″ x 18″	.124	lf
Radius Curbing, 6″ x 18″, Over 10′ Radius	.215	lf
Corners, 2′ Radius	.700	Ea.
Edging, 4-1/2″ x 12″, straight	.187	lf
Curb inlets, (guttermouth) straight	1.366	Ea.
Monolithic concrete curb and gutter, cast in place with 6′ high curb and 6″ thick gutter		
24″ wide, .055 cy per lf	.128	lf
30″ wide, .066 cy per lf	.141	lf

12 SITE WORK
BRICK, STONE AND CONCRETE PAVING

Brick and stone provide durable, weather-resistant surfaces for exterior pavements. Many types of hard brick and stone are available to meet the practical and cosmetic needs of a given surface. The patterns in which the brick or stone is placed should be determined according to the surface's use and its desired appearance.

Both brick and stone may be set in a sand or concrete bed and grouted with mortar or watered and tamped sand. Regardless of the type of bed material, the subbase must first be leveled and thoroughly compacted to prevent cracking and settling of the finished surface. Wire mesh reinforcing may be required in concrete beds that are large in area or subjected to heavy traffic.

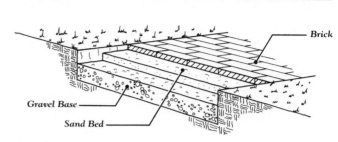

Brick Sidewalk

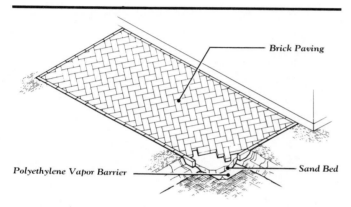

Brick Paving on Sand Bed

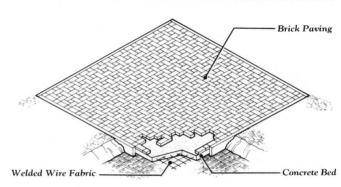

Brick Paving on Concrete Bed

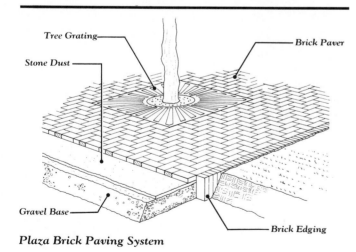

Plaza Brick Paving System

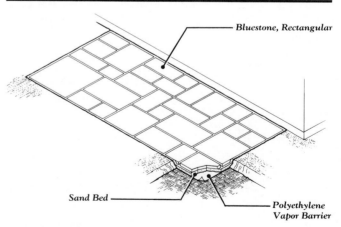

Stone Paving on Sand Bed

417

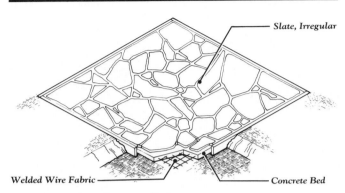

Stone Paving on Concrete Bed

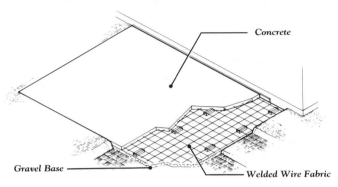

Concrete Paving

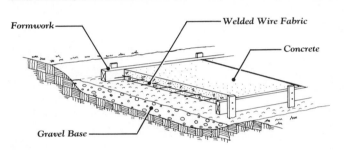

Concrete Sidewalk

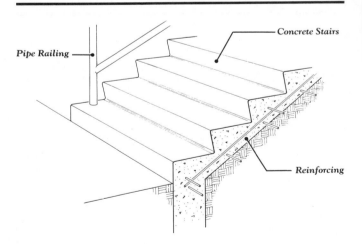

Concrete Stairs

Man-hours

Description	m/hr	Unit
Brick Paving without Joints (4.5 Brick/sf)	.145	sf
Grouted, 3/8″ Joints (3.9 Brick/sf)	.178	sf
Sidewalks		
Brick on 4″ Sand Bed		
Laid on Edge (7.2/sf)	.229	sf
Flagging		
Bluestone, Irregular, 1″ Thick	.198	sf
Snapped Randon Rectangular 1″ Thick	.174	sf
1-1/2″ Thick	.188	sf
2″ Thick	.193	sf
Slate		
Natural Cleft, Irregular, 3/4″ Thick	.174	sf
Random Rectangular, Gauged, 1/2″ Thick	.152	sf
Random Rectangular, Butt Joint, Gauged,		
1/4″ Thick	.107	sf
Granite Blocks, 3-1/2″ x 3-1/2″ x 3-1/2″	.174	sf
4″ to 12″ Long, 3″ to 5″ Wide, 3″ to 5″ Thick	.163	sf
6″ to 15″ Long, 3″ to 6″ Wide, 3″ to 5″ Thick	.152	sf

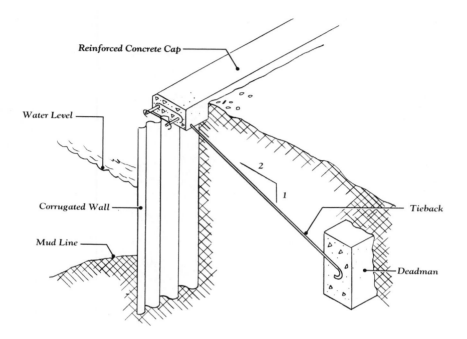

Bulkhead retaining walls are usually used for canal bank and shoreline protection. Sheet piling is installed with a continuous cap that is anchored at regular intervals with tiebacks connected to a deadman buried in the retained bank. Three types of sheet piling are used for these walls: asbestos-cement, aluminum, and steel. The choice of anchor material to be used depends on a number of variables, including the amount of force to be withstood by the wall, its height, and climate considerations. Aluminum and steel sheet piling can be installed with either a pile driving hammer or jetting; asbestos-cement can be erected only with jetting.

Asbestos-cement sheeting, because it does not have the strength of aluminum or steel sheeting, is restricted in its use to sheltered waterways with tidal variations of less than 2'. It is available in panels measuring 3'-6" in width and 3' to 10' in length. The maximum allowable exposed face is 5' in height or less, with a maximum embedment of 6'-6". It resists corrosion, but it cannot be used in climates that experience a regular freeze-thaw cycle. After the piling is installed and anchored with tie rods at spacings of 10', a reinforced concrete cap is formed and placed to maintain wall alignment. Asbestos-cement sheeting can also be used as a knee wall to protect the toe of a larger bulkhead where severe tidal and erosion forces must be resisted.

Aluminum sheet piling, because it is stronger than asbestos-cement piling, can be used in waterways and bays where wave action is considerable. Also, this sheeting is not restricted by climate or corrosive environments. It is available in panels measuring 5' in width and 13' in length which can be installed with a maximum exposed face of 8'. Longer aluminum sheet piling is available in interlocking Z and U shaped sections which can be used in conjunction with walers when the exposed face must be 12' or more in

height. After the piling is installed and secured, a concrete cap, similar to the cap used for asbestos-cement walls, or an aluminum cap, is installed to maintain alignment and to anchor one end of the tieback.

Steel sheet piling possesses the greatest strength of the three types. With appropriate use of walers, steel sheeting can be used to heights of 30' and can withstand severe tidal and wave forces. Galvanized or aluminum coatings can be applied to provide corrosion protection when needed. The sheeting profile is only 3" deep, as compared with 9" to 16" for Z shaped steel sheeting. The thickness ranges from 5 to 12 gauge to suit the specific application. Successive panels interlock to simplify alignment during driving. Steel sheeting is superior to other types when difficult driving conditions are encountered.

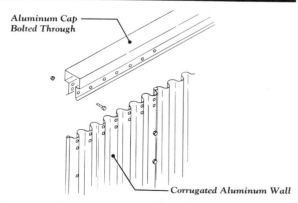

Aluminum Bulkhead Retaining Wall - Aluminum Cap

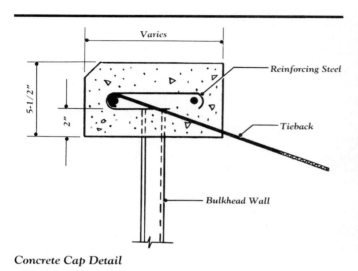

Concrete Cap Detail

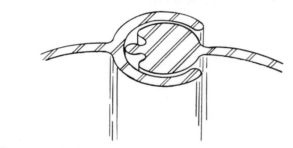

Connection Detail

Description	m/hr	Unit
Asbestos-Cement Sheeting, by Jetting, Including Cap and Anchors, Coarse Compact Sand,		
2'-6" Embedment	.267	lf
5'-6" Embedment	.533	lf
Loose Silty Sand		
2'-6" Embedment	.200	lf
5'-6" Embedment	.400	lf
Aluminum Panel Sheeting, Vibratory Hammer Driven, Including Cap and Anchors, Coarse Compact Sand,		
2'-0" Embedment	.320	lf
5'-6" Embedment	.674	lf
Loose Silty Sand,		
3'-0" Embedment	.312	lf
5'-6" Embedment	.492	lf
Steel Sheet Piling, Exposed Face		
Shore Driven	.067	sf
Barge Driven	.116	sf

Concrete retaining walls are freestanding structures used to retain earth. They take up much less space than bin- or crib-type wall structures and can be used in situations where very high retaining walls are needed. When soil conditions are normal and the backfill is level, the forces of earth pressure on the wall vary with the height. However, normal earth pressure forces can be increased several times by special conditions such as water, sloped backfill, and building, highway, or railroad bed surcharges. Concrete retaining walls must be built to resist the tendency for sliding and overturning which these forces generate.

There are two basic types of concrete retaining walls: gravity and cantilever. A gravity wall resists sliding and overturning by its mass alone. The cross-sectional shape is usually trapezoidal—narrowest at the top and widest at the base where the forces of the retained earth are greatest. Since the width at the base is greater than 50% of the wall's height, the volume of concrete and related cost become prohibitive if the wall rises above 10′.

A cantilever wall consists of two segments: a vertical wall stem and a horizontal base slab. The wall stem acts as a cantilever fixed at the base. Vertical reinforcing bars resist bending in the stem, and horizontal bars resist bending in the base. Cantilever wall thickness may be constant for structures of less than 10′ in height. Higher walls may require a section that increases in width toward the base by means of a 1:12 slope on the inside face. This slope may also be included on the outside face for aesthetic effect. The wall stem is usually keyed into the base, and the base may be keyed into the underlying, undisturbed earth to resist sliding.

Because retaining walls are not usually designed to withstand hydraulic pressure, provisions must be made to remove any groundwater that might collect behind the structure. This condition can be prevented by placing a blanket of drainage stone backfill behind the wall and placing weep holes at about 10′ intervals at the bottom of the wall. To remove the water entirely, perforated pipe, surrounded by drainage stone, can be installed along the inside face at the base of the wall prior to backfilling.

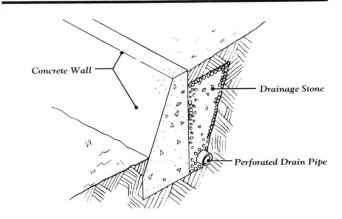

Gravity Retaining Wall

Man-hours

Description	m/hr	Unit
Concrete Retaining Wall		
Footing Formwork	.066	sfca
Wall Formwork, Under 8′ High	.049	sfca
Under 16′ High	.079	sfca
Wall Formwork, Battered to 16′ High	.150	sfca
Footing Reinforcing Bars	15.240	ton
Wall Reinforcing Bars	10.670	ton
Footing Concrete, Direct Chute	.400	cu yd
Pumped	.640	cu yd
Crane and Bucket	.711	cu yd
Wall Concrete, Direct Chute	.480	cu yd
Pumped	.674	cu yd
Crane and Bucket	.711	cu yd
Perforated, Clay, 4″ Diameter	.060	lf
6″ Diameter	.076	lf
8″ Diameter	.083	lf
Bituminous, 4″ Diameter	.032	lf
6″ Diameter	.035	lf
Porous Concrete, 4″ Diameter	.072	lf
6″ Diameter	.076	lf
8″ Diameter	.090	lf
Drainage Stone, 3/4″ Diameter	.092	cu yd
Backfill, Dozer	.010	cu yd
Compaction, Vibrating Plate, 12″ Lifts	.044	cu yd

Cantilever Retaining Wall

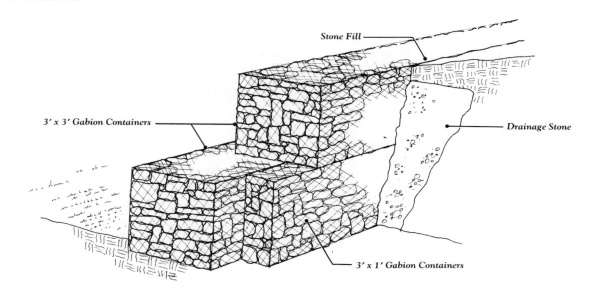

Stone Fill

3′ x 3′ Gabion Containers

Drainage Stone

3′ x 1′ Gabion Containers

Gabion retaining walls are designed and built to hold back earth with their weight and mass. Although they have the greatest mass of retaining wall types, they are among the least expensive to erect. The basic components of the walls include individual gabion containers, wire or plastic, and the 4″ to 8″ stone with which the containers are filled. The containers are 3′ in width and range from 1′ to 3′ in height and 6′ to 12′ in length.

The process of installing the walls involves several steps. Empty containers are arranged, one course at a time and one or more deep, and then filled. During the installation and filling process, care must be taken not to damage the galvanized or plastic coating on the wire mesh. The stone on the exposed face of the wall may be set by hand with the balance of each of the containers filled by machine. Voids within the container average about 35% of its total volume.

Special construction techniques may add to the life and effectiveness of gabion walls. Either face of the wall may be stepped, for example, and added stability may be realized by tilting the wall into the retained earth at a 1:6 slope. In clay soils, additional gabions, spaced from 13′ to 30′ on center, should run traverse to the wall line, from the outside face to beyond the slip plane of the retained bank. If drainage is a problem, especially in instances where the retained earth supports a highway or railroad bed, a blanket of drainage stone should be placed behind the wall. This drainage system helps in preventing fines from washing through and undermining the wall's backfill.

Man-hours

Description	m/hr	Unit
Gabion with stone fill		
3′ x 6′ x 1′	.280	Ea.
3′ x 6′ x 3′	1.020	Ea.
3′ x 9′ x 1′	.431	Ea.
3′ x 9′ x 3′	1.510	Ea.
3′ x 12′ x 1′	.560	Ea.
3′ x 12′ x 3′	2.240	Ea.
Drainage stone, 3/4″ diameter	.092	cu yd
Backfill Dozer	.010	cu yd
Compaction, roller, 12″ lifts	.014	cu yd

Masonry retaining walls may be constructed of block, brick or stone.

Brick or block walls are usually placed on a concrete footing that acts as a leveling pad and distributes imposed loads to the subsoil. Both the wall and the footing may be reinforced. Voids in the masonry are usually filled with mortar or grout, and the wall capped with a suitable material. Solid masonry walls should include porous backfill against the back of the wall and weep holes or drainage piping to eliminate hydrostatic head.

Stone retaining walls may be constructed dry or mortar set with or without a suitable concrete footing. All masonry retaining walls should be placed a sufficient depth below grade to eliminate the danger of frost heave. Mortar set walls should include an adequate drainage system.

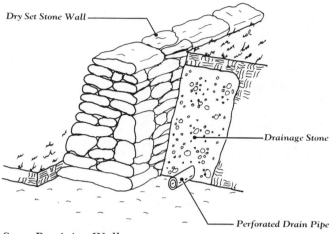

Stone Retaining Wall

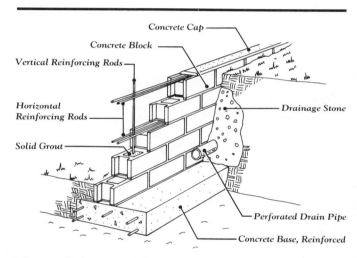

Masonry Retaining Wall

Man-hours

Description	m/hr	Unit
Masonry Retaining Wall, 8″ Thick, 4′ High		
Continous Wall Footing	.140	lf
Concrete Block Wall Including Reinforcing		
and Grouting	.610	lf
Fill in Trench Crushed Bank Run	.010	lf
Perforated Asbestos Cement Drain,		
4″ Diameter	.062	lf
Masonry Retaining Wall, 10″ Thick, 6′ High		
Continuous Wall Footing	.218	lf
Concrete Block Wall Including Reinforcing		
and Grouting	.957	lf
Fill in Trench Crushed Bank Run	.015	lf
Perforated Asbestos Cement Drain,		
4″ Diameter	.062	lf
Masonry Retaining Wall, 12″ Thick, 8′ High		
Continuous Wall Footing	.278	lf
Concrete Block Wall Including Reinforcing		
and Grouting	1.548	lf
Fill in Trench Crushed Bank Run	.020	lf
Perforated Asbestos Cement Drain,		
4″ Diameter	.062	lf
Stone Retaining Wall, 3′ Above Grade, Dry Set	2.742	lf
Mortar Set	2.400	lf
6′ Above Grade, Dry Set	4.114	lf
Mortar Set	3.600	lf

Note: Units are per lf of wall.

Wood retaining walls are usually constructed of redwood, cedar, pressure-treated lumber, or creosoted lumber.

The three systems normally used to construct these walls are wood post, post-and-board, and wood tie.

Wood post walls consist of posts placed side by side and anchored a suitable distance below the subgrade in order to resist overturning.

Post-and-board walls are erected of spaced, anchored posts, horizontal wood sheathing, and a wood cap.

Wood tie walls are constructed of railroad or landscape ties. These are anchored by steel rods driven thru holes in the ties.

Wood retaining walls are normally used for their pleasing aesthetic effect and are not suitable for high cuts or heavy surcharges.

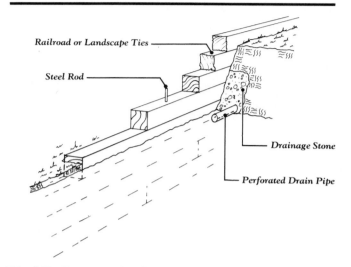

Wood Tie Retaining Wall

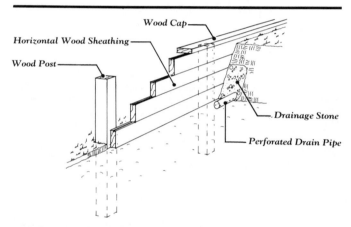

Wood Post and Board Retaining Wall

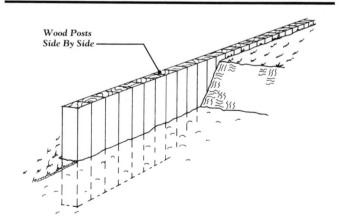

Wood Post Retaining Wall

Man-hours

Description	m/hr	Unit
Wood Post and Board Retaining Wall		
4' High		
Redwood Posts, Plank and Cap	.664	lf
Porous Backfill	.010	lf
Perforated Pipe	.062	lf
6' High		
Redwood Posts, Plank and Cap	.742	lf
Porous Backfill	.015	lf
Perforated Pipe	.062	lf
8' High		
Redwood Posts, Plank and Cap	1.307	lf
Porous Backfill	.020	lf
Perforated Pipe	.062	lf
Wood Post Retaining Wall		
2' High Redwood Posts	.358	lf
Porous Fill	.005	lf
Perforated Pipe	.062	lf
4' High Redwood Posts	1.074	lf
Porous Fill	.010	lf
Perforated Pipe	.062	lf
6' High Redwood Posts	2.505	lf
Porous Fill	.015	lf
Perforated Pipe	.062	lf
Wood Tie Retaining Wall		
4' High Redwood Ties	.689	lf
Porous Fill	.010	lf
Perforated Pipe	.062	lf
6' High Redwood Ties	1.034	lf
Porous Fill	.015	lf
Perforated Pipe	.062	lf

12 SITE WORK
FENCING

Fences are usually composed of a series of vertical posts set into the earth and spanned with rails, panels, or stretched metal fencing fabric. They are installed for many reasons, including privacy, safety, security, weather protection, cosmetics, and mandate of local codes. Fence posts may be installed and secured by direct driving or by setting them in concrete or compacted fill in predug post holes. The most commonly used fencing materials are wood (usually cedar or redwood), galvanized steel, aluminum, and plastic-coated steel.

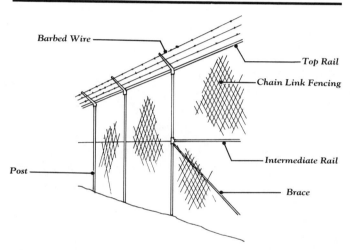

Chain Link Fence, Industrial

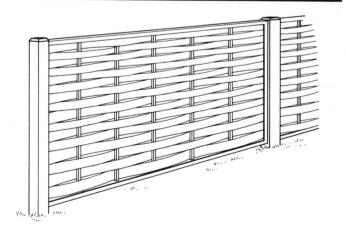

Basketweave Fence

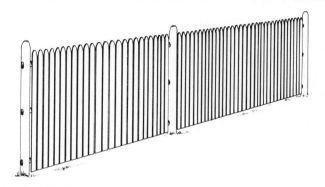

Stockade Fence

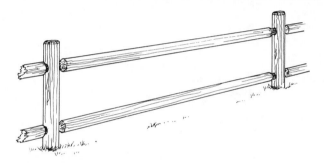

Open Rail Rustic Fence

Man-hours

Description	m/hr	Unit
Fence, Chain Link, Industrial Plus 3 Strands Barbed Wire, 2″ Line Post @ 10′ on Center		
1-5/8″ Top Rail, 6′ High	.096	lf
Corners, Add	.600	Ea.
Braces, Add	.300	Ea.
Gate, Add	.686	Ea.
Residential, 11 Gauge Wire, 1-5/8″ Line Post @ 10′ on Center 1-3/8″		
Top Rail, 3′ High	.048	lf
4′ High	.060	lf
Gate, Add	.400	Ea.
Tennis Courts, 11 Gauge Wire, 1-3/4″ Mesh, 2-1/2″ Line Posts, 1-5/8″ Top Rail,		
10′ High	.155	lf
12′ High	.185	lf
Corner Posts, 3″ Diameter, Add	.800	Ea.
Fence, Security, 12′ High	.960	lf
16′ High	1.200	lf
Fence, Wood, Cedar Picket, 2 Rail, 3′ High	.150	lf
Gate, 3′-6″, Add	.533	Ea.
Cedar Picket, 3 Rail, 4′ High	.160	lf
Gate, 3′-6″, Add	.585	Ea.
Open Rail Rustic, 2 Rail, 3′ High	.150	lf
Stockade, 6′ High	.150	lf
Board, Shadow Box, 1″ x 6″ Treated Pine, 6′ High	.150	lf

12 SITE WORK
SITE IRRIGATION

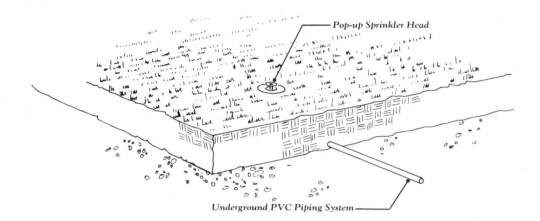

Pop-up Sprinkler Head

Underground PVC Piping System

Site irrigation is accomplished with sprinkler heads that are attached to a permanent underground piping system, usually of PVC pipe or flexible polyethylene tubing. The system may be controlled at a main valve, either manually or automatically, or at each individual sprinkler head.

Several types of sprinkler head operations are available: spray, impact, or rotary. Any of these heads may accommodate one of several types of mountings: pop-up, riser mount, or quick connect. The mountings are made of plastic or brass in both economy and heavy-duty models.

The area covered by a single sprinkler head can range in a circular or arc pattern from 10' to 100' in radius, or in a quadrilateral pattern, such as a 24' square or a 40' by 6' rectangle. Radius pattern heads can be adjusted for irrigating areas along property boundaries. Since large radius sprays require great amounts of water, the local water pressure must be sufficient. If this is not the case, then booster pumps may be required to provide the necessary pressure. Generally, the larger-coverage sprinklers provide a more economical system.

The site served is usually divided into zones determined by the nature of the landscaping (turf or shrub), the amount of irrigation needed (sunny or shady), and the coverage of the sprinkler head (small or large). Each zone is controlled by a separate valve. In an automated system, each of these valves and the main valve from the water supply are electrically operated. The electrical service to each of these valves is connected to and controlled from a remote panel.

Man-hours

Description	m/hr	Unit
Sprinkler System, Golf Course, Fully Automatic	.600	9 holes
12' Radius Heads, 15' Spacing		
Minimum	.343	head
Maximum	.600	head
30' Radius Heads, Automatic		
Minimum	.857	head
Maximum	1.040	head
Sprinkler Heads		
Minimum	.267	head
Maximum	.320	head
Trenching, Chain Trencher, 12 hp		
4" Wide, 12" Deep	.010	lf
6" Wide, 24" Deep	.015	lf
Backfill and Compact		
4" Wide, 12" Deep	.010	lf
6" Wide, 24" Deep	.030	lf
Trenching and Backfilling, Chain Trencher, 40 hp		
6" Wide, 12" Deep	.007	lf
8" Wide, 36" Deep	.010	lf
Compaction		
6" Wide, 12" Deep	.003	lf
8" Wide, 36" Deep	.005	lf
Vibrating Plow		
8" Deep	.004	lf
12" Deep	.006	lf
Automatic Valves, Solenoid		
3/4" Diameter	.363	Ea.
2" Diameter	.500	Ea.
Automatic Controllers		
4 Station	1.500	Ea.
12 Station	2.000	Ea.

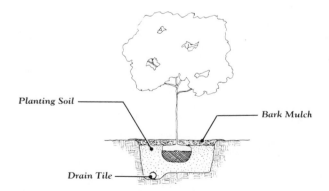

Planting Soil

Bark Mulch

Drain Tile

Tree/Shrub Irrigation and Drainage

APPENDIX

APPENDIX 1

REFERENCE AIDS

The following charts, graphs, and tables are a quick, convenient reference for construction planning and design. These references are particularly useful for loading checks, heat loss, and the appropriate building code applications.

Minimum Design Live Loads in Pounds Per SF for Various Building Codes

Occupancy	Description	Minimum Live Loads, Pounds Per SF				
		BOCA	ASBC	UBC	Chicago	New York
Armories150		150			150	
Assembly	Fixed seats	60	60	50	60	60
	Movable seats	100	100	100	100	100
	Platforms or stage floors	100	100	125	150	
Commercial & Industrial	Light manufacturing	125	125	75	100	100
	Heavy manufacturing	250	250	125	100	100
	Light storage	125	125	75	100	100
	Heavy storage	250	250	100	100	150
	Stores, retail, first floor	100	100	75	100	100
	Stores, retail, upper floors	75	75			75
	Stores, wholesale	125	125	100	100	100
Court rooms		100				
Dance halls	Ballrooms	100	100			100
Dining rooms	Restaurants	100	100			100
Fire escapes	Other than below	100	100			100
	Multi or single family residential	40	40			40
Garages	Passenger cars only	50	50	50	50	50
Gymnasiums	Main floors and balconies	100	100			100
Hospitals	Operating rooms, laboratories	60	60			60
	Private room	40	40			40
	Wards	40	40			40
	Corridors, above first floor	80	80			
Libraries	Reading rooms	60	60			60
	Stack rooms	150	150	125		
	Corridors, above first floor	80				
Marquees		75	75			
Office Buildings	Offices	50	50	50	50	50
	Lobbies	100	100			100
	Corridors, above first floor	80	80	100	100	75
Residential	Multi family private apartments	40	40	40	40	
	Multi family, public rooms	100	100			
	Multi family, corridors	80	80	100	100	100
	Dwellings, first floor	40	40	40	40	40
	Dwellings, second floor & habitable attics	30	30			30
	Dwellings, uninhabitable attics	20	20			20
	Hotels, guest rooms	40	40	40	40	40
	Hotels, public rooms	100	100			
	Hotels, corridors serving public rooms	100	100			
	Hotels, corridors	80	80			
Roofs	Flat	12-20	12-20	20	25	40
	Pitched	12-16	12-16			
Schools	Classrooms	40	40	40	40	40
	Corridors	80	80	100	100	100
Sidewalks	Driveways, etc. subject to trucking	250	250	250		600
Stairs	Exits	100	100	100	100	100
Theaters	Aisles, corridors and lobbies	100	100			
	Orchestra floors	60	60			
	Balconies	60	60			
	Stage floors	150	150		150	
Yards	Terrace, Pedestrian	100	100			

BOCA = Building Officials & Code Administration International, Inc. National Building Code 1985
ANSI = Standard A58.1-1982
UBC = Uniform Building Code, International Conference of Building Officials, 1967
Chicago = Chicago, Ill., amended to July 1, 1967
New York = New York City, amended to 1970
* Corridor loading equal to occupancy loading

Design Weight Per SF For Walls and Partitions

Type	Wall Thickness	Description	Weight Per SF	Type	Wall Thickness	Description	Weight Per SF
Brick	4"	Clay brick, high absorption	34 lb.	Clay tile	2"	Split terra cotta furring	10 lb.
		Clay brick, medium absorption	39			Non load bearing clay tile	11
		Clay brick, low absorption	46		3"	Split terra cotta furring	12
		Sand-lime brick	38			Non load bearing clay tile	18
		Concrete brick, heavy aggregate	46		4"	Non load bearing clay tile	20
		Concrete brick, light aggregate	33			Load bearing clay tile	24
	8"	Clay brick, high absorption	69		6"	Non load bearing clay tile	30
		Clay brick, medium absorption	79			Load bearing clay tile	36
		Clay brick, low absorption	89		8"	Non load bearing clay tile	36
		Sand-lime brick	74			Load bearing clay tile	42
		Concrete brick, heavy aggregate	89		12"	Non load bearing clay tile	46
		Concrete brick, light aggregate	68			Load bearing clay tile	58
	12"	Common brick	120	Gypsum block	2"	Hollow gypsum block	9.5
		Pressed brick	130			Solid gypsum block	12
		Sand-lime brick	105		3"	Hollow gypsum block	10
		Concrete brick, heavy aggregate	130			Solid gypsum block	18
		Concrete brick, light aggregate	98		4"	Hollow gypsum block	15
	16"	Clay brick, high absorption	134			Solid gypsum block	24
		Clay brick, medium absorption	155		5"	Hollow gypsum block	18
		Clay brick, low absorption	173		6"	Hollow gypsum block	24
		Sand-lime brick	138	Structural facing tile	2"	Facing tile	15
		Concrete brick, heavy aggregate	174		4"	Facing tile	25
		Concrete brick, light aggregate	130		6"	Facing tile	38
Concrete block	4"	Solid conc. block, stone aggregate	45	Glass	4"	Glass block	18
		Solid conc. block, lightweight	34		1"	Structural glass	15
		Hollow conc. block, stone aggregate	30	Plaster	1"	Gypsum plaster (1 side)	5
		Hollow conc. block, lightweight	20			Cement plaster (1 side)	10
	6"	Solid conc. block, stone aggregate	50			Gypsum plaster on lath	8
		Solid conc. block, lightweight	37			Cement plaster on lath	13
		Hollow conc. block, stone aggregate	42	Plaster partition (2 finished faces)	2"	Solid gypsum on metal lath	18
		Hollow conc. block, lightweight	30			Solid cement on metal lath	25
	8"	Solid conc. block, stone aggregate	67			Solid gypsum on gypsum lath	18
		Solid conc. block, lightweight	48			Gypsum on lath & metal studs	18
		Hollow conc. block, stone aggregate	55		3"	Gypsum on lath & metal studs	19
		Hollow conc. block, lightweight	38		4"	Gypsum on lath & metal studs	20
	10"	Solid conc. block, stone aggregate	84		6"	Gypsum on lath & wood studs	18
		Solid conc. block, lightweight	62	Concrete	6"	Reinf. concrete, stone aggregate	75
		Hollow conc. block, stone aggregate	55			Reinf. concrete, lightweight	36-60
		Hollow conc. block, lightweight	38		8"	Reinf. concrete, stone aggregate	100
	12"	Solid conc. block, stone aggregate	108			Reinf. concrete, lightweight	48-80
		Solid conc. block, lightweight	72		10"	Reinf. concrete, stone aggregate	125
		Hollow conc. block, stone aggregate	85			Reinf. concrete, lightweight	60-100
		Hollow conc. block, lightweight	55		12"	Reinf. concrete stone aggregate	150
Drywall	6"	Drywall on wood studs	10			Reinf. concrete, lightweight	72-120

Design Weight Per SF for Roof Coverings

Type		Description	Weight Per SF	Type		Description	Weight Per SF
Sheathing	Gypsum	1" thick	4	Metal	Aluminum	Corr. & ribbed, .024" to .040"	.4-.8
	Wood	¾" thick	3		Copper	or tin	1.0
Insulation	Per 1"	Loose	.5		Steel	Corrugated, 29 ga to 12 ga	.6-5.0
		Poured in place	2	Shingles	Asphalt	Strip shingles	3
		Rigid	1.5		Clay	Tile	9-20
Built-up	Tar & gravel	3 ply felt	5.5		Slate	¼" thick	10
		5 ply felt			Wood		2

Design Weight Per SF for Floor Fills and Finishes

Type		Description	Weight Per SF	Type		Description	Weight Per SF
Floor fill	Per 1"	Cinder fill	5	Wood	Single 7/8"	On sleepers, light concrete fill	16
		Cinder concrete	9			On sleepers, stone concrete fill	25
		Lightweight concrete	7		Double 7/8"	On sleepers, light concrete fill	19
		Stone concrete	12			On sleepers, stone concrete fill	28
		Sand	8		3"	Wood block on mastic, no fill	10
		Gypsum	4			Wood block on ½" mortar	16
Terrazzo	1"	Terrazzo, 2" stone concrete	32		Per 1"	Hardwood flooring	4
Marble	and mortar	on stone concrete fill	33			Underlayment	3
Resilient	1/16-1/4	Linoleum, asphalt, vinyl tile	2	Asphalt	1-1/2"	Mastic flooring	18
Tile	3/4"	Ceramic or quarry	10		2"	Block on ½" mortar	30

Design Weight for Various Materials

Type		Description	Weight Per CF	Type		Description	Weight Per CF
Bituminous	Coal	Anthracite	97	Masonry	Ashlar	Granite	168
		Bituminous	84			Limestone, crystalline	168
		Peat, turf, dry	47			Limestone, oolitic	135
		Coke	75			Marble	173
	Petroleum	Unrefined	54			Sandstone	144
		Refined	50		Rubble, in mortar	Granite	153
		Gasoline	42			Limestone, crystalline	147
	Pitch		69			Limestone, oolitic	138
	Tar	Bituminous	75			Marble	156
Concrete	Plain	Stone aggregate	144			Sandstone	137
		Slag aggregate	132		Brick	Pressed	140
		Expanded slag aggregate	100			Common	120
		Haydite (burned clay agg.)	90			Soft	100
		Vermiculite & perlite, load bearing	70-105		Cement	Portland, loose	90
		Vermiculite & perlite, nonload bear	25-50			Portland set	183
	Rein-forced	Stone aggregate	150		Lime	Gypsum, loose	53-64
		Slag aggregate	138		Mortar	Set	103
		Lightweight aggregates	30-106	Metals	Aluminum	Cast, hammered	165
Earth	Clay	Dry	63		Brass	Cast, rolled	534
		Damp, plastic	110		Bronze	7.9 to 14% Sn	509
		and gravel, dry	100		Copper	Cast, rolled	556
	Dry	Loose	76		Iron	Cast, pig	450
		Packed	95			Wrought	485
	Moist	Loose	78		Lead		710
		Packed	96		Monel		556
	Mud	Flowing	108		Steel	Rolled	490
		Packed	115		Tin	Cast, hammered	459
	Riprap	Limestone	80-85		Zinc	Cast rolled	440
		Sandstone	90	Timber	Cedar	White or red	22
		Shale	105		Fir	Douglas	32
	Sand & gravel	Dry, loose	90-105			Eastern	25
		Dry, packed	100-120		Maple	Hard	43
		Wet	118-120			White	33
Gases	Air	0°C., 760 mm.	.0807		Oak	Red or Black	41
	Gas	Natural	.0385			White	46
Liquids	Alcohol	100%	49		Pine	White	26
	Water	4°C., maximum density	62.5			Yellow, long leaf	44
		Ice	56			Yellow, short leaf	38
		Snow, fresh fallen	8		Redwood	California	26
		Sea water	64		Spruce	White or Black	27

Design Weight Per SF for Structural Floor and Roof Systems

Type Slab	Description		Weight In Pounds Per SF — Slab Depth In Inches											
Concrete Slab	Reinforced	Stone aggregate	1"	12.5	2"	25	3"	37.5	4"	50	5"	62.5	6"	75
		Lightwgt sand aggregate		9.5		19		28.5		38		47.5		57
		All lightwgt aggregate		9.0		18		27.0		36		45.0		54
	Plain, non-reinforced	Stone aggregate		12.0		24		36.0		48		60.0		72
		Lightwgt sand aggregate		9.0		18		27.0		36		45.0		54
		All lightwgt aggregate		8.5		17		25.5		34		42.5		51
Concrete Waffle	19" x 19"	5" wide ribs @ 24" O.C.	6+3	77	8+3	92	10+3	100	12+3	118				
			6+4-1/2	96	8+4-1/2	110	10+4-1/2	119	12+4-1/2	136				
	30" x 30"	6" wide ribs @ 36" O.C.	8+3	83	10+3	95	12+3	109	14+3	118	16+3	130	20+3	155
			8+4-1/2	101	10+4-1/2	113	12+4-1/2	126	14+4-1/2	137	16+4-1/2	149	20+4-1/2	173
Concrete Joist	20" wide form	5" wide rib	8+3	60	10+3	67	12+3	74	14+3	81				
		6" wide rib		63		70		78		86	16+3	94	20+3	111
		7" wide rib										99		118
		5" wide rib	8+4-1/2	79	10+4-1/2	85	12+4-1/2	92	14+4-1/2	99				
		6" wide rib		82		89		97		104				
		7" wide rib									16+4-1/2	113	20+4-1/2	130
												118		136
	30" wide form	5" wide rib	8+3	54	10+3	58	12+3	63	14+3	68				
		6" wide rib		56		61		67		72	16+3	78	20+3	91
		7" wide rib										83		96
		5" wide rib	8+4-1/2	72	10+4-1/2	77	12+4-1/2	82	14+4-1/2	87				
		6" wide rib		75		80		85		91				
		7" wide rib									16+4-1/2	97	20+4-1/2	109
												101		115
Wood Joists	Incl. Sub-floor	12" O.C.	2x6	6	2x8	6	2x10	7	2x12	8	3x8	8	3x12	11
		16" O.C.		5		6		6		7		7		9

Design Loads for Structures for Wind Load

Wind Loads: Structures are designed to resist the wind force from any direction. Usually 2/3 is assumed to act on the windward side, 1/3 on the leeward side.

For more than 1/3 openings, add 10 psf for internal wind pressure or 5 psf for suction, whichever is critical.

For buildings and structures, use psf values from Table.

For glass over 4 SF use values in the Table below after determining 30′ wind velocity.

Type Structure	Height Above Grade	Horizontal Load in Lb per SF
Buildings	Up to 50 ft	15
	50 to 100 ft	20
	Over 100 ft	20+.025 per ft
Ground signs & towers	Up to 50 ft	15
	Over 50 ft	20
Roof structures		30
Glass	See Table below	

Design Wind Loads in PSF for Glass at Various Elevations

Height From Grade	Velocity in Miles Per Hour and Design Load in Pounds Per SF																				
	Vel.	PSF	Vel.	PSF	Vel.	PSF	Vel.	PSF	Vel.	PSF	Vel.	PSF	Vel.	PSF	Vel.	PSF	Vel.	PSF	Vel.	PSF	Vel.	PSF
To 10 feet	42	6	46	7	49	8	52	9	55	10	59	11	62	12	66	14	69	15	76	19	83	22
10-20	52	9	58	11	61	11	65	14	70	16	74	18	79	20	83	22	87	24	96	30	105	35
20-30	60	12	67	14	70	16	75	18	80	20	85	23	90	26	95	29	100	32	110	39	120	46
30-60	66	14	74	18	77	19	83	22	88	25	94	28	99	31	104	35	110	39	121	47	132	56
60-120	73	17	82	21	85	12	92	27	98	31	104	35	110	39	116	43	122	48	134	57	146	68
120-140	81	21	91	26	95	29	101	33	108	37	115	42	122	48	128	52	135	48	149	71	162	84
240-480	90	26	100	32	104	35	112	40	119	45	127	51	134	57	142	65	149	71	164	86	179	102
480-960	98	31	110	39	115	42	123	49	131	55	139	62	148	70	156	78	164	86	180	104	197	124
Over 960	98	31	110	39	115	42	123	49	131	55	139	62	148	70	156	78	164	86	180	104	197	124

Design Wind Velocity at 30 Feet Above Ground

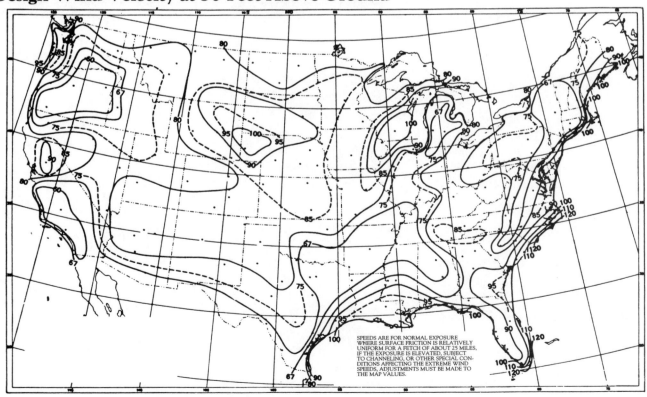

SPEEDS ARE FOR NORMAL EXPOSURE WHERE SURFACE FRICTION IS RELATIVELY UNIFORM FOR A FETCH OF ABOUT 25 MILES. IF THE EXPOSURE IS ELEVATED, SUBJECT TO CHANNELING, OR OTHER SPECIAL CONDITIONS AFFECTING THE EXTREME WIND SPEEDS, ADJUSTMENTS MUST BE MADE TO THE MAP VALUES.

Snow Load in Pounds Per Square Foot on the Ground

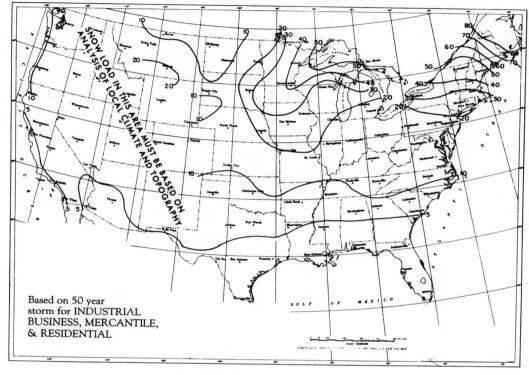

Based on 50 year
storm for INDUSTRIAL
BUSINESS, MERCANTILE,
& RESIDENTIAL

Snow Load in Pounds Per Square Foot on the Ground

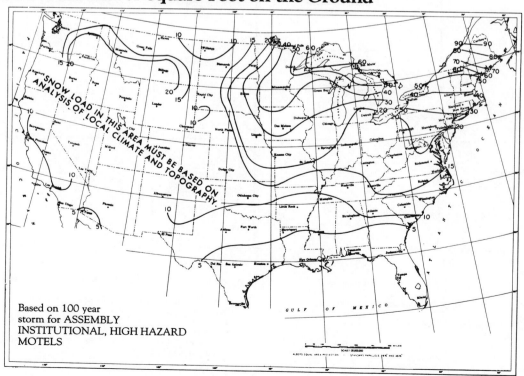

Based on 100 year
storm for ASSEMBLY
INSTITUTIONAL, HIGH HAZARD
MOTELS

To convert the ground snow loads on the previous page to roof snow loads, the ground snow loads should be multiplied by the following factors depending upon the roof characteristics.

Snow Loads

Description	Sketch	Formula	X	Conversion Factor = C_f	
				Sheltered	Exposed
Simple flat and shed roofs	Load Diagram	For $\alpha > 30°$ $C_f = 0.8 - \dfrac{\alpha - 30}{50}$	0 to 30° 40° 50° 60° 70° to 90°	0.8 0.6 0.4 0.2 0	0.6 0.45 0.3 0.15 0
				Case I	Case II
Simple gable and hip roofs	Case I Load Diagram Case II Load Diagram	$C_f = 0.8 - \dfrac{\alpha - 30}{50}$ $C_f = 1.25 \left(0.8 - \dfrac{\alpha - 30}{50}\right)$	10° 20° 30° 40° 50° 60°	0.8 0.8 0.8 0.6 0.4 0.2	— — 1.0 0.75 0.5 0.25
Valley areas of Two span roofs	Case I Case II	$\beta = \dfrac{\alpha_1 + \alpha_2}{2}$ $C_f = 0.8 - \dfrac{\alpha - 30}{50}$	$\beta \leqslant 10°$ use Case I only $\beta > 10°$ $\beta < 20°$ use Case I & II $\beta \geqslant 20°$ use Case I, II & III		
Lower level of multi level roofs (or on an adjacent building not more then 15 ft. away		$C_f = 15\dfrac{h}{g}$ h=difference in roof height in feet g=ground snow load in psf w=width of drift For h < 5, w = 10 h > 15, w = 30	When $15\dfrac{h}{g} < .8$, use 0.8 When $15\dfrac{h}{g} > 3.0$, use 3.0		

Summary of above:

1. For flat roofs or roofs up to 30°, use 0.8 x ground snow load for the roof snow load.
2. For roof pitches in excess of 30°, conversion factor becomes lower than 0.8.
3. For exposed roofs there is a further 25% reduction of conversion factor.
4. For steep roofs a more highly loaded half span must be considered.
5. For shallow roof valleys conversion factor is 0.8.
6. For moderate roof valleys conversion factor is 1.0 for half the span.
7. For steep roof valleys conversion factor is 1.5 for one quarter of the span.
8. For roofs adjoining vertical surfaces the conversion factor is up to 3.0 for part of the span.
9. If snow load is less than 30 psf, use water load on roof for clogged drain condition.

Floor Area Ratios: commonly used gross to net area and net to gross area ratios expressed in % for various building types.

Building Type	Gross to Net Ratio	Net to Gross Ratio	Building Type	Gross to Net Ratio	Net to Gross Ratio
Apartment	156	64	School's (campus Type)		
Bank	140	72	Admistrative	150	67
Church	142	70	Auditorium	142	70
Courthouse	162	61	Biology	161	62
Department Store	123	81	Chemistry	170	59
Garage	118	85	Classroom	152	66
Hospital	183	55	Dining Hall	138	72
Hotel	158	63	Dormitory	154	65
Laboratory	171	58	Engineering	164	61
Library	132	76	Fraternity	160	63
Office	135	75	Gymnasium	142	70
Restaurant	141	70	Science	167	60
Warehouse	108	93	Service	120	83
			Student Union	172	59

The gross area of a building is the total floor area based on outside dimensions.

The net area of a building is the usable floor area for the function intended and excludes such items as stairways, corridors and mechanical rooms. In the case of a commercial building, it might be considered as the "leasable area."

Partition/Door Density

Building Type		Stories	Partition/ Density	Doors	Description of Partition
Apartments		1 story	9 SF/LF	90 SF/door	Plaster, wood doors & trim
		2 story	8 SF/LF	80 SF/door	Drywall, wood studs, wood doors & trim
		3 story	9 SF/LF	90 SF/door	Plaster, wood studs, wood doors & trim
		5 story	9 SF/LF	90 SF/door	Plaster, wood studs, wood doors & trim
		6-15 story	8 SF/LF	80 SF/door	Drywall, wood studs, wood doors & trim
Bakery		1 story	50 SF/LF	500 SF/door	Conc. block, paint, door & drywall, wood studs
		2 story	50 SF/LF	500 SF/door	Conc. block, paint, door & drywall, wood studs
Bank		1 story	20 SF/LF	200 SF/door	Plaster, wood studs, wood doors & trim
		2-4 story	15 SF/LF	150 SF/door	Plaster, wood studs, wood doors & trim
Bottling plant		1 story	50 SF/LF	500 SF/door	Conc. block, drywall, wood studs, wood trim
Bowling Alley		1 story	50 SF/LF	500 SF/door	Conc. block, wood & metal doors, wood trim
Bus Terminal		1 story	15 SF/LF	150 SF/door	Conc. block, ceramic tile, wood trim
Cannery		1 story	100/SF/LF	1000 SF/door	Drywall on metal studs
Car Wash		1 story	18 SF/LF	180 SF/door	Concrete block, painted & hollow metal door
Dairy Plant		1 story	30 SF/LF	300 SF/door	Concrete block, glazed tile, insulated cooler doors
Department Store		1 story	60 SF/LF	600 SF/door	Drywall, wood studs, wood doors & trim
		2-5 story	60 SF/LF	600 SF/door	30% concrete block, 70% drywall, wood studs
Dormitory		2 story	9 SF/LF	90 SF/door	Plaster, concrete block, wood doors & trim
		3-5 story	9 SF/LF	90 SF/door	Plaster, concrete block, wood doors & trim
		6-15 story	9 SF/LF	90 SF/door	Plaster, concrete block, wood doors & trim
Funeral Home		1 story	15 SF/LF	150 SF/door	Plaster on concrete block & wood studs, paneling
		2 story	14 SF/LF	140 SF/door	Plaster, wood studs, paneling & wood doors
Garage Sales & Service		1 story	30 SF/LF	300 SF/door	50% conc. block, 50% drywall, wood studs
Hotel		3-8 story	9 SF/LF	90 SF/door	Plaster, conc. block, wood doors & trim
		9-15 story	9 SF/LF	90 SF/door	Plaster, conc. block, wood doors & trim
Laundromat		1 story	25 SF/LF	250 SF/door	Drywall, wood studs, wood doors & trim
Medical Clinic		1 story	6 SF/LF	60 SF/door	Drywall, wood studs, wood doors & trim
		2-4 story	6 SF/LF	60 SF/door	Drywall, wood studs, wood doors & trim
Motel		1 story	7 SF/LF	70 SF/door	Drywall, wood studs, wood doors & trim
		2-3 story	7 SF/LF	70 SF/door	Concrete block, drywall on wood studs, wood paneling
Movie theater	200-600 seats	1 story	18 SF/LF	180 SF/door	Concrete block, wood, metal, vinyl trim
	601-1400 seats		20 SF/LF	200 SF/door	Concrete block, wood, metal, vinyl trim
	1401-2200 seats		25 SF/LF	250 SF/door	Concrete block, wood, metal, vinyl trim
Nursing Home		1 story	8 SF/LF	80 SF/door	Drywall, wood studs, wood doors & trim
		2-4 story	8 SF/LF	80 SF/door	Drywall, wood studs, wood doors & trim
Office		1 story	20 SF/LF	200-500 SF/door	30% concrete block, 70% drywall on wood studs
		2 story	20 SF/LF	200-500 SF/door	30% concrete block, 70% drywall on wood studs
		3-5 story	20 SF/LF	200-500 SF/door	30% concrete block, 70% movable partitions
		6-10 story	20 SF/LF	200-500 SF/door	30% concrete block, 70% movable partitions
		11-20 story	20 SF/LF	200-500 SF/door	30% concrete block, 70% movable partitions
Parking Ramp (Open)		2-8 story	60 SF/LF	600 SF/door	Stair and elevator enclosures only
Parking Garage		2-8 story	60 SF/LF	600 SF/door	Stair and elevator enclosures only
Pre-Engineered	Steel	1 story	0		
	Store	1 story	60 SF/LF	600 SF/door	Drywall on wood studs, wood doors & trim
	Office	1 story	15 SF/LF	150 SF/door	Concrete block, movable wood partitions
	Shop	1 story	15 SF/LF	150 SF/door	Movable wood partitions
	Warehouse	1 story	0		
Radio & TV Broadcasting		1 story	25 SF/LF	250 SF/door	Concrete block, metal and wood doors
& TV Transmitter		1 story	40 SF/LF	400 SF/door	Concrete block, metal and wood doors
Self Service Restaurant		1 story	15 SF/LF	150 SF/door	Concrete block, wood and aluminum trim
Cafe & Drive-In Restaurant		1 story	18 SF/LF	180 SF/door	Drywall, wood studs, ceramic & plastic trim
Restaurant with seating		1 story	25 SF/LF	250 SF/door	Concrete block, paneling, wood studs & trim
Supper Club		1 story	25 SF/LF	250 SF/door	Concrete block, paneling, wood studs & trim
Bar or Lounge		1 story	24 SF/LF	240 SF/door	Plaster or gypsum lath, wooded studs
Retail Store or Shop		1 story	60 SF/LF	600 SF/door	Drywall wood studs, wood doors & trim
Service Station	Masonry	1 story	15 SF/LF	150 SF/door	Concrete block, paint, door & drywall, wood studs
	Metal panel	1 story	15 SF/LF	150 SF/door	Concrete block paint door & drywall, wood studs
	Frame	1 story	15 SF/LF	150 SF/door	Drywall, wood studs, wood doors & trim
Shopping Center	(strip)	1 story	30 SF/LF	300 SF/door	Drywall, wood studs, wood doors & trim
	(group)	1 story	40 SF/LF	400 SF/door	50% concrete block, 50% drywall, wood studs
		2 story	40 SF/LF	400 SF/door	50% concrete block, 50% drywall, wood studs
Small Food Store		1 story	30 SF/LF	300 SF/door	Concrete block drywall, wood studs, wood trim
Store/Apt. above	Masonry	2 story	10 SF/LF	100 SF/door	Plaster, wood studs, wood doors, & trim
	Frame	2 story	10 SF/LF	100 SF/door	Plaster, wood studs, wood doors, & trim
	Frame	3 story	10 SF/LF	100 SF/door	Plaster, wood studs wood doors & trim
Supermarkets		1 story	40 SF/LF	400 SF/door	Concrete block, paint, drywall & porcelain panel
Truck Terminal		1 story	0		
Warehouse		1 story	0		

438

Occupancy Determinations

Description		SF Required Per Person*			
		BBC	BOCA	SBC	UBC
Assembly Areas	Fixed Seats	6	**	6	7
	Movable Seats	15		15	15
	Concentrated		7		
	Unconcentrated		15		
	Standing Space		3		
Educational	Unclassified	40			
	Classrooms		20	40	20
	Shop Areas		50	100	50
Institutional	Unclassified	150		125	
	In-Patient Areas		240		
	Sleeping Areas		120		
Mercantile	Basement	30	30	30	20
	Ground Floor	30	30	30	30
	Upper Floors	60	60	60	50
Office		100	100	100	100

* BBC=Basic Building Code
BOCA=Building Officials & Code Administrators
SBC=Southern Building Code
UBC=Uniform Building Code

** The occupancy load for assembly area with fixed seats shall be determined by the number of fixed seats installed.

Length of Exitway Access Travel (Feet)

Use Group	Without Fire Suppression System	With Fire Suppression System
Assembly	150	200
Business	200	300
Factory and industrial	200	300
High hazard	—	75
Institutional	100	200
Mercantile	100	150
Residential	100	150
Storage, low hazard	300	400
Storage, moderate hazard	200	300

Note: The maximum length of exitway access travel in unlimited area buildings shall be 400'.

Capacity Per Unit Egress Width*

Use Group	Without Fire Suppression System Number of Occupants		With Fire Suppression System Number of Occupants	
	Stairways	Doors, Ramps and Corridors	Stairways	Doors, Ramps and Corridors
Assembly	75	100	113	150
Business	60	100	90	150
Factory & industrial	60	100	90	150
High hazard	—	—	60	100
Institutional	22	30	33	45
Mercantile	60	100	90	150
Residential	75	100	113	150
Storage	60	100	90	150

*A Unit of Egress Width is 22". Add 1/2 Unit For Each Additional 12".

"U" Values for Type "A" Building

Type A buildings shall include:

A1 Detached one and two family dwellings

A2 All other residential buildings, three stories or less, including but not limited to: multi-family dwellings, hotels and motels.

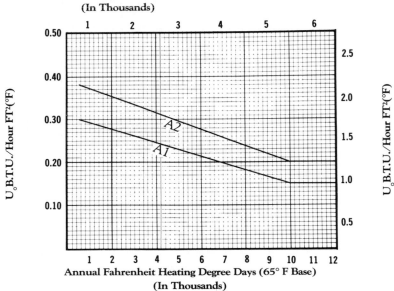

Annual Celsius Heating Degree Days (18° C Base)
(In Thousands)

"U" Values for Type "B" Buildings

For all Buildings Not Classified Type "A"

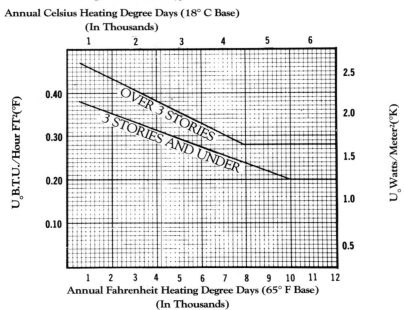

Annual Celsius Heating Degree Days (18° C Base)
(In Thousands)

Resistances ("R") of Building and Insulating Materials

Material	Wt. Lbs Per CF	R Per Inch	R Listed Size
Air Spaces and Surfaces			
Enclosed non-reflective spaces,			
E=0.82,			
50°F mean temp., 30°/10°F diff.			
.5"			.90/.91
.75"			.94/1.01
1.50"			.90/1.02
3.50"			.91/1.01
Inside vert. surface (still air)			0.68
Outside vert. surface (15 mph wind)			0.17
Building Boards			
Asbestos cement, 0.25" thick	120		0.06
Gypsum or plaster, 0.5" thick	50		0.45
Hardboard regular	50	1.37	
Tempered	63	1.00	
Laminated paper	30	2.00	
Particle board	37	1.85	
	50	1.06	
	63	0.85	
Plywood (Douglas Fir), 0.5" thick	34		0.62
Shingle backer, .375" thick	18		0.94
Sound deadening board, 0.5" thick	15		1.35
Tile and lay-in panels, plain or			
acoustical, 0.5" thick	18		1.25
Vegetable fiber, 0.5" thick	18		1.32
	25		1.14
Wood, hardwoods	48	0.91	
Softwoods	32	1.25	
Flooring Carpet with fibrous pad			2.08
With rubber pad			1.23
Cork tile, 1/8" thick			0.28
Terrazzo			0.08
Tile, resilient			0.05
Wood, hardwood, 0.75" thick			0.68
Subfloor, 0.75" thick			0.94
Glass			
Insulation, 0.50" air space			2.04
Single glass			0.91
Insulation Blanket or batt, mineral, glass			
or rock fiber, approximate thickness			
3.0" to 3.5" thick			11
3.5" to 4.0" thick			13
6.0" to 6.5" thick			19
6.5" to 7.0" thick			22
8.5" to 9.0" thick			30
Boards			
Cellular glass	8.5	2.63	
Fiberboard, wet felted			
Acoustical tile	21	2.70	
Roof insulation	17	2.94	
Fiberboard, wet molded			
Acoustical tile	23	2.38	
Mineral fiber with resin binder	15	3.45	
Polystyrene, extruded,			
cut cell surface	1.8	4.00	
smooth skin surface	2.2	5.00	
	3.5	5.26	
Bead boards	1.0	3.57	
Polyurethane	1.5	6.25	
Wood or cane fiberboard, 0.5" thick			25

Material	Wt. Lbs Per CF	R Per Inch	R Listed Size
Insulation Loose Fill			
Cellulose	2.3	3.13	
	3.2	3.70	
Mineral fiber, 3.75" to 5" thick	2-5		11
6.5" to 8.75" thick			19
7.5" to 10" thick			22
10.25" to 13.75" thick			30
Perlite	5-8	2.70	
Vermiculite	4-6	2.27	
Wood fiber	2-3.5	3.33	
Masonry Brick, Common	120	0.20	
Face	130	0.11	
Cement mortar	116	0.20	
Clay tile, hollow			
1 cell wide, 3" width			0.80
4" width			1.11
2 cells wide, 6" width			1.52
8" width			1.85
10" width			2.22
3 cells wide, 12" width			2.50
Concrete, gypsum fiber	51	0.60	
Lightweight	120	0.19	
	80	0.40	
	40	0.86	
Perlite	40	1.08	
Sand and gravel or stone	140	0.08	
Concrete block, lightweight,			
3 cell units, 4"-15 lbs. ea.			1.68
6"-23 lbs. ea.			1.83
8"-28 lbs. ea.			2.12
12"-40 lbs. ea.			2.62
Sand and gravel aggregates,			
4"-20 lbs. ea.			1.17
6"-33 lbs. ea.			1.29
8"-38 lbs. ea.			1.46
12"-56 lbs. ea.			1.81
Plastering Cement Plaster, sand aggregate	116	0.20	
Gypsum plaster, Perlite aggregate	45	0.67	
Sand aggregate	105	0.18	
Vermiculite aggregate	45	0.59	
Roofing			
Asphalt, felt, 15 lb.			0.06
Rolled roofing	70		0.15
Shingles	70		0.44
Built-up roofing .375" thick	70		0.33
Cement shingles	120		0.21
Vapor-permeable felt			0.06
Vapor seal, 2 layers of mopped 15 lb. felt			0.12
Wood, shingles 16"-7.5" exposure			0.87
Siding			
Aluminum or steel (hollow backed)			
oversheathing			0.61
With .375" insulating backer board			1.82
Foil backed			2.96
Wood siding, beveled, ½" x 8"			0.81

Weather Data and Design Conditions (winter design @ 97.5% - summer design @ 2.5%)

City	Latitude (1) 0	1'	Winter Temperatures (1) Med. of Annual Extremes	99%	97½%	Winter Degree Days (2)	Summer (Design Dry Bulb) Temperatures and Relative Humidity 1%	2½%	5%
UNITED STATES									
Albuquerque, NM	35	0	6	12	16	4,400	96/61	94/61	92/61
Atlanta, GA	33	4	14	17	22	3,000	95/74	92/74	90/73
Baltimore, MD	39	2	12	14	17	4,600	94/75	92/75	89/74
Birmingham, AL	33	3	17	17	21	2,600	97/75	94/75	93/74
Bismarck, ND	46	5	-31	-23	-19	8,800	95/68	91/68	88/67
Boise, ID	43	3	0	3	10	5,800	96/65	93/64	91/64
Boston, MA	42	2	-1	6	9	5,600	91/73	88/71	85/70
Burlington, VT	44	3	-18	-12	-7	8,200	88/72	85/70	83/69
Charleston, WV	38	2	1	7	11	4,400	92/74	90/73	88/72
Charlotte, NC	35	1	13	18	22	3,200	96/74	94/74	92/74
Casper, WY	42	5	-20	-11	-5	7,400	92/58	90/57	87/57
Chicago, IL	41	5	-5	-3	2	6,600	94/75	91/74	88/73
Cincinnati, OH	39	1	2	1	6	4,400	94/73	92/72	90/72
Cleveland, OH	41	2	-2	1	5	6,400	91/73	89/72	86/71
Columbia, SC	34	0	16	20	24	2,400	98/76	96/75	94/75
Dallas, TX	32	5	14	18	22	2,400	101/75	99/75	97/75
Denver, CO	39	5	-9	-5	1	6,200	92/59	90/59	89/59
Des Moines, IA	41	3	-13	-10	-5	6,600	95/75	92/74	89/73
Detroit, MI	42	2	0	3	6	6,200	92/73	88/72	85/71
Great Falls, MT	47	3	-29	-21	-15	7,800	91/60	88/60	85/59
Hartford, CT	41	5	-4	3	7	6,200	90/74	88/73	85/72
Houston, TX	29	5	24	28	33	1,400	96/77	94/77	92/77
Indianapolis, IN	39	4	-2	-2	2	5,600	93/74	91/74	88/73
Jackson, MS	32	2	17	21	25	2,200	98/76	96/76	94/76
Kansas City, MO	39	1	-2	2	6	4,800	100/75	97/74	94/74
Las Vegas, NV	36	1	18	25	28	2,800	108/66	106/65	104/65
Lexington, KY	38	0	0	3	8	4,600	94/73	92/72	90/72
Little Rock, AR	34	4	13	15	20	3,200	99/76	96/77	94/77
Los Angeles, CA	34	0	38	41	43	2,000	94/70	90/70	87/69
Memphis, TN	35	0	11	13	18	3,200	98/77	96/76	94/76
Miami, FL	25	5	39	44	47	200	92/77	90/77	89/77
Milwaukee, WI	43	0	-11	-8	-4	7,600	90/74	87/73	84/71
Minneapolis, MN	44	5	-19	-16	-12	8,400	92/75	89/73	86/71
New Orleans, LA	30	0	29	29	33	1,400	93/78	91/78	90/77
New York, NY	40	5	6	11	15	5,000	94/74	91/73	88/72
Norfolk, VA	36	5	18	20	22	3,400	94/77	91/76	89/76
Oklahoma City, OK	35	2	4	9	13	3,200	100/74	97/74	95/73
Omaha, NE	41	2	-12	-8	-3	6,600	97/76	94/75	91/74
Philadelphia, PA	39	5	7	10	14	4,400	93/75	90/74	87/72
Phoenix, AZ	33	3	25	31	34	1,800	108/71	106/71	104/71
Pittsburgh, PA	40	3	1	3	7	6,000	90/72	88/71	85/70
Portland, ME	43	4	-14	-6	-1	7,600	88/72	85/71	81/69
Portland, OR	45	4	17	17	23	4,600	89/68	85/67	81/65
Portsmouth, NH	43	1	-8	-2	2	7,200	88/73	86/71	83/70
Providence, RI	41	4	0	5	9	6,000	89/73	86/72	83/70
Rochester, NY	43	1	-5	1	5	6,800	91/73	88/71	85/70
Salt Lake City, UT	40	5	-2	3	8	6,000	97/62	94/62	92/61
San Francisco, CA	37	5	38	38	40	3,000	80/63	77/62	83/61
Seattle, WA	47	4	22	22	27	5,200	81/68	79/66	76/65
Sioux Falls, SD	43	4	-21	-15	-11	7,800	95/73	92/72	89/71
St. Louis, MO	38	4	1	3	8	5,000	96/75	94/75	92/74
Tampa, FL	28	0	32	36	40	680	92/77	91/77	90/76
Trenton, NJ	40	1	7	11	14	5,000	92/75	90/74	87/73
Washington, DC	38	5	12	14	17	4,200	94/75	92/74	90/74
Wichita, KS	37	4	-1	3	7	4,600	102/72	99/73	96/73
Wilmington, DE	39	4	6	10	14	5,000	93/74	93/74	20/73
ALASKA									
Anchorage	61	1	-29	-23	-18	10,800	73/59	70/58	67/56
Fairbanks	64	5	-59	-51	-47	14,280	82/62	78/60	75/59
CANADA									
Edmonton, Alta.	53	3	-30	-29	-25	11,000	86/66	83/65	80/63
Halifax, N.S.	44	4	-4	1	5	8,000	83/66	80/65	77/64
Montreal, Que.	45	3	-20	-16	-10	9,000	88/73	86/72	84/71
Saskatoon, Sask.	52	1	-35	-35	-31	11,000	90/68	86/66	83/65
St. Johns, Nwf.	47	4	1	3	7	8,600	79/66	77/65	75/64
Saint John, N.B.	45	2	-15	-12	-8	8,200	81/67	79/65	77/64
Toronto, Ont.	43	4	-10	-5	-1	7,000	90/73	87/72	85/71
Vancouver, B.C.	49	1	13	15	19	6,000	80/67	78/66	76/65
Winnipeg, Man.	49	5	-31	-30	-27	10,800	90/73	87/71	84/70

(1) Handbook of Fundamentals, ASHRAE, Inc., NY 1972/1985
(2) Local Climatological Annual Survey, USDC Env. Science Services Administration, Ashville, NC

Maximum Depth of Frost Penetration in Inches

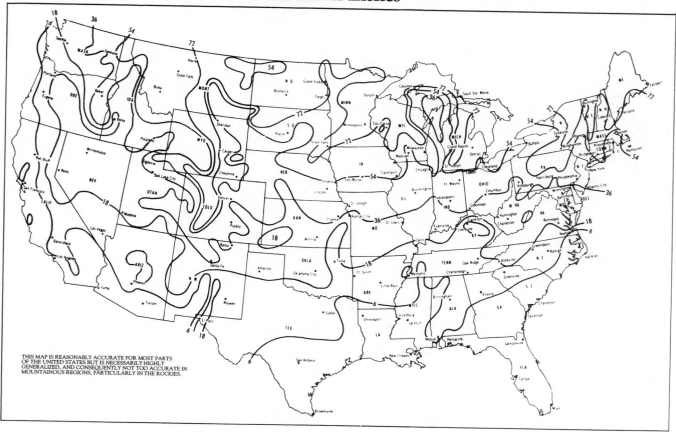

THIS MAP IS REASONABLY ACCURATE FOR MOST PARTS
OF THE UNITED STATES BUT IS NECESSARILY HIGHLY
GENERALIZED, AND CONSEQUENTLY NOT TOO ACCURATE IN
MOUNTAINOUS REGIONS, PARTICULARLY IN THE ROCKIES.

Fire Resisting Ratings of Structural Elements (In hours)

Description of the Structural Element	No. 1 Fireproof		No. 2 Non Combustible			No. 3 Exterior Masonry Wall			No. 4 Frame	
			Protected		Unpro-tected	Heavy Timber	Ordinary		Pro-tected	Unpro-tected
							Pro-tected	Unpro-tected		
	1A	1B	2A	2B	2C	3A	3B	3C	4A	4B
Exterior, Bearing Walls	4	3	2	1½	1	2	2	2	1	1
Nonbearing Walls	2	2	1½	1	1	2	2	2	1	1
Interior Bearing Walls and Partitions	4	3	2	1	0	2	1	0	1	0
Fire Walls and Party Walls	4	3	2	2	2	2	2	2	2	2
Fire Enclosure of Exitways, Exit Hallways and Stairways	2	2	2	2	2	2	2	2	1	1
Shafts other than Exitways, Hallways and Stairways	2	2	2	2	2	2	2	2	1	1
Exitway access corridors and Vertical separation of tenant space	1	1	1	1	0	1	1	0	1	0
Columns, girders, trusses (other than roof trusses) and framing: Supporting more than one floor	4	3	2	1	0	—	1	0	1	0
Supporting one floor only	3	2	1½	1	0	—	1	0	1	0
Structural members supporting wall	3	2	1½	1	0	1	1	0	1	0
Floor construction including beams	3	2	1½	1	0	—	1	0	1	0
Roof construction including beams, trusses & framing arches & roof deck 15' or less in height to lowest member	2	1½	1	1	0	—	1	0	1	0

Note:
a. Codes include special requirements and exceptions that are not included in the table above.
b. Each type of construction has been divided into sub-types which vary according to the degree of fire-resistance required. Sub-types (A) requirements are more severe than those for sub-types (B).
c. Protected construction means all structural members are chemically treated, covered or protected so that the unit has the required fire-resistance specified.

Type No. 1, Fireproof Construction — Buildings and structures of fireproof construction are those in which the walls, partitions, structural elements, floors, ceilings, and roofs, and the exitways are protected with approved noncombustible materials to afford the fire-resistance rating specified. Fire-resistant treated wood may be used as specified.

Type No. 2, Noncombustible Construction —Buildings and structures of noncombustible construction are those in which the walls, partitions, structural elements, floors, ceilings, roofs and the exitways are approved noncombustible materials meeting the fire-resistance rating requirements. Fire-retardant treated wood may be used as specified.

Type No. 3, Exterior Masonry Wall Construction — Buildings and structures of exterior masonry wall construction are those in which the exterior, fire and party walls are masonry or other approved noncombustible materials of the required fire-resistance rating and structural properties. The floors, roofs, and interior framing are wholly or partly wood or metal or other approved construction. The fire and party walls are ground-supported; except that girders and their supports, carrying walls of masonry shall be protected to afford the same degree of fire-resistance rating of the supported walls.

Type No. 4, Frame Construction — Buildings and structures of frame construction are those in which the exterior walls, bearing walls, partitions, floor and roof construction are wholly or partly of wood stud and joist assemblies with a minimum nominal dimension of two inches or of other approved combustible materials. Fire stops are required at all vertical and horizontal draft openings in which the structural elements have required fire-resistance ratings.

Fire Hazard for Fire Walls

The degree of fire hazard of buildings relating to their intended use is defined by "Fire Grading" the buildings. Such a grading system is listed in the table below. This type of grading determines the requirements for fire walls and fire separation walls (exterior fire exposure). For mixed use occupancy, use the higher Fire Grading requirement of the components.

Fire Grading of Building in Hours for Fire Walls and Fire Separation Walls

Building Type	Hours	Building Type	Hours
Businesses	2	Recreation Centers	2
Churches	1½	Residential Hotels	2
Factories	3	Residential, Multi-family Dwellings	1½
High Hazard*	4	Residential, 1 & 2 Family Dwellings	¾
Industrial	3	Restaurants	2
Institutional, Incapacitated Occupants	2	Schools	1½
Institutional, Restrained Occupants	3	Storage, Low Hazard*	2
Lecture Halls	2	Storage, Moderate Hazard*	3
Mercantile	3	Terminals	2
Night Club	3	Theatres	3

Note: *The difference in "Fire Hazards" is determined by their occupancy and use.

High Hazard: Industrial and storage buildings in which the combustible contents might cause fires to be unusually intense or where explosives, combustible gases or flammable liquids are manufactured or stored.

Moderate Hazard: Mercantile buildings, industrial and storage buildings in which combustible contents might cause fires of moderate intensity.

Low Hazard: Business buildings that ordinarily do not burn rapidly.

Flame Spread for Interior Finishes

The flame spreadability of a material is the burning characteristic of the material relative to the fuel contributed by its combustion and the density of smoke developed. The flame spread classification of a material is based on a ten minute test on a scale of 0 to 100. Cement asbestos board is assigned a rating of 0 and select red oak flooring a rating of 100. The four classes are listed in the table. The flame spread ratings for interior finish walls and ceilings shall not be greater than the Class listed in the chart below.

Interior Finish Classification

Class of Material	Surface Burning Characteristics
I	0 to 25
II	26 to 75
III	76 to 200
IV	201 to 5000

Interior Finish Requirements by Class

Building Type	Vertical Exitways and Passageways	Corridors Providing Exitways	Rooms or Enclosed Spaces	Building Type	Vertical Exitways and Passageways	Corridors Providing Exitways	Rooms or Enclosed Space
Assembly Halls	I	I	II	Mercantile Walls	I	II	III
Businesses	I	II	III	Night Clubs	I	I	II
Churches	I	I	II	Residential Hotels	I	II	III
Factory	I	II	III	Residential, Multi-family	I	II	III
High Hazard	I	II	II	Residential, 1 & 2 Family	IV	IV	IV
Industrial	I	II	III	Restaurants	I	I	II
Institutional Incapacitated	I	II	I	Storage, Low Hazard	I	II	III
Institutional, Restrained	I	I	I	Storage, Moderate Hazard	I	II	III
Mercantile Ceilings	I	II	II	Terminals	I	I	II

Description: This table is primarily for converting customary U.S. units in the left hand column to SI metric units in the right hand column. In addition, conversion factors for some commonly encountered Canadian and non-SI metric units are included.

Metric Conversion Factors

	If You Know		Multiply By		To Find
Length	Inches	x	25.4[a]	=	Millimeters
	Feet	x	0.3048[a]	=	Meters
	Yards	x	0.9144[a]	=	Meters
	Miles (statute)	x	1.609	=	Kilometers
Area	Square inches	x	645.2	=	Square millimeters
	Square feet	x	0.0929	=	Square meters
	Square yards	x	0.8361	=	Square meters
Volume (Capacity)	Cubic inches	x	16,387	=	Cubic millimeters
	Cubic feet	x	0.02832	=	Cubic meters
	Cubic yards	x	0.7646	=	Cubic meters
	Gallons (U.S. liquids)[b]	x	0.003785	=	Cubic meters[c]
	Gallons (Canadian liquid)[b]	x	0.004546	=	Cubic meters[c]
	Ounces (U.S. liquid)[b]	x	29.57	=	Milliliters[c,d]
	Quarts (U.S. liquid)[b]	x	0.9464	=	Liters[c,d]
	Gallons (U.S. liquid)[b]	x	3.785	=	Liters[c,d]
Force	Kilograms force[d]	x	9.807	=	Newtons
	Pounds force	x	4.448	=	Newtons
	Pounds force	x	0.4536	=	Kilograms force[d]
	Kips	x	4448	=	Newtons
	Kips	x	453.6	=	Kilograms force[d]
Pressure, Stress, Strength (Force per unit area)	Kilograms force per square centimeter[d]	x	0.09807	=	Megapascals
	Pounds force per square inch (psi)	x	0.006895	=	Megapascals
	Kips per square inch	x	6.895	=	Megapascals
	Pounds force per square inch (psi)	x	0.07031	=	Kilograms force per square centimeter[d]
	Pounds force per square foot	x	47.88	=	Pascals
	Pounds force per square foot	x	4.882	=	Kilograms force per square meter[d]
Bending Moment Or Torque	Inch-pounds force	x	0.01152	=	Meter-kilograms force[d]
	Inch-pounds force	x	0.1130	=	Newton-meters
	Foot-pounds force	x	0.1383	=	Meter-kilograms force[d]
	Foot-pounds force	x	1.356	=	Newton-meters
	Meter-kilograms force[d]	x	9.807	=	Newton-meters
Mass	Ounces (avoirdupois)	x	28.35	=	Grams
	Pounds (avoirdupois)	x	0.4536	=	Kilograms
	Tons (metric)	x	1000[a]	=	Kilograms
	Tons, short (2000 pounds)	x	907.2	=	Kiloprams
	Tons, short (2000 pounds)	x	0.9072	=	Megagrams[e]
Mass per Unit Volume	Pounds mass per cubic foot	x	16.02	=	Kilograms per cubic meter
	Pounds mass per cubic yard	x	0.5933	=	Kilograms per cubic meter
	Pounds mass per gallon (U.S. liguid)[b]	x	119.8	=	Kilograms per cubic meter
	Pounds mass per gallon (Canadian liquid)[b]	x	99.78	=	Kilograms per cubic meter
Temperature	Degrees Fahrenheit	(F-32)/1.8		=	Degrees Celsius
	Degrees Fahrenheit	(F+459.67)/1.8		=	Degrees Kelvin
	Degrees Celsius	C+273.15		=	Degrees Kelvin

[a]The factor given is exact

[b]One U.S. gallon = 0.8327 Canadian gallon

[c]1 liter = 1000 milliliters = 1000 cubic centimeters

 1 cubic decimeter = 0.001 cubic meter

[d]Metric but not SI unit

[e]Called "tonne" in England and "metric ton" in other metric countries

2 ABBREVIATIONS

In the front portion of this book, we have identified abbreviations that were used in writing *Means Graphic Construction Standards.* On the following pages, you will find an expanded list of abbreviations commonly used throughout industry.

447

a acre
A area
A&E architect-engineer
AAMA Architectural Aluminum Manufacturers Association
ABC aggregate base course, Associated Builders and Contractors.
ABS acrylonitrile butadiene styrene
ABT air blast transformer, (about)
ac, a-c, a.c. alternating current
a.c. asphaltic (a.c. paving)
AC air conditioning, alternating current (on drawings), armored cable (on drawings), asbestos cement
ACB asbestos-cement board, air circuit breaker
ACC accumulator
Access. accessory
ACD automatic closing device
ACI American Concrete Institute
ACM asbestos-covered metal
ACS American Ceramic Society
ACSR aluminum cable steel reinforced, aluminum conductor steel reinforced
Acst acoustic
Actl actual
a.d. air dried
AD access door, air dried, area drain, as drawn
ADD addendum (on drawings), addition (on drawings)
Addit. additional
ADF after deducting freight (used in lumber industry)
ADH adhesive
adj adjust, adjustable, adjoining, adjacent
ADS automatic door seal
AG above grade
AGA American Gas Association
AGC Associated General Contractors.
Aggr aggregate
AGL above ground level
AH ampere hour (also, amp hr)
AIA American Institute of Architects
AIC ampere interrupting capacity
AIEE American Institute of Electrical Engineers

AIMA Acoustical and Insulating Materials Association
AISC American Institute of Steel Construction
AISI American Iron and Steel Institute
AITC American Institute of Timber Construction
AL aluminum (also, alum)
ALLOW allowance (also, Allow)
ALM alarm
ALS American Lumber Standards
ALT alternate
ALTN alteration
ALY alloy
AMB asbestos millboard
AMD air-moving device
amp ampere
ANL anneal
ANSI American National Standards Institute
AP access panel
APC acoustical plaster ceiling
APF acid-proof floor
Appd approved
Approx approximate
Apt apartment
APW Architectural Projected Window
AR as required, as rolled
ARC W, ARC/W arc weld
ARS asbestos roof shingles
ART. artificial
AS automatic sprinkler
ASA American Standards Association
asb asbestos
ASBC American Standard Building Code
ASC asphalt surface course
ASCE American Society of Civil Engineers
ASEC American Standard Elevator Codes
ASHRAE American Society of Heating, Refrigeration and Air Conditioning Engineers
ASI American National Standards Institute
ASME American Society of Mechanical Engineers
asph asphalt
ASR automatic sprinkler riser

ASSE American Society of
Sanitary Engineering
ASTM American Society for
Testing and Materials
AT asphalt tile, airtight
ATB asphalt-tile base
ATC acoustical tile ceiling,
architectural terra cotta
ATF asphalt-tile floor
atm atmosphere, atmospheric
aux auxiliary
av, ave, avg average
A/W all-weather
AW actual weight
AWG American wire gauge
A.W.W.I. American Wood
Window Institute

B1S banded one side, bead one side
B2E banded two ends
B2S banded two sides, bead two
sides, bright two sides
B2S1E banded two sides and one
end
B3E beveled on three edges
B4E beveled on four edges
B&B in the lumber industry, grade
B and better
b&cb beaded on the edge and
center
B&O back-our punch
B&S beams and stringers, bell and
spigot, Brown and Sharpe gauge
B beam, boron, brightness
BA bright annealed
bat. batten
bbl, brl barrel
BC building code
BCM broken cubic meter
BCY broken cubic yard
bd. in the lumber industry, board
BF, bd. ft. in the lumber industry,
board foot
bdl bundle
BET. between
Beth. B Bethlehem beam
bev in the lumber industry, beveled
bev sid beveled siding
BFP backflow preventer
bg bag
BG below ground
Bh Brinell hardness
Bhn Brinell hardness number

B

BHP brake horsepower
BL building line
B/L bill of lading
bldg building
blk block, black
BLKG blocking
BLO blower
BLR boiler
blt built, borrowed light
B/M, BOM bill of materials
b.m. in the lumber industry, board
measure
BM bench mark, beam
BMEP brake mean effective
measure
b of b back of board
BP blueprint, baseplate, bearing
pile, building paper
bpd barrels per day
BPG beveled plate glass
BR bedroom
brc brace
brcg bracing
BRG bearing
BRK brick
BRKT, bkt bracket
BRS brass
Br Std, BS British Standard
BRZ bronze
BRZG brazing
BSMT basement
BSR building space requirements
BTB bituminous treated base
Btr., btr better
Btu British Thermal Unit
BTUH Btu per hour
but. buttress
BW butt weld
BX interlocked armored cable

C

1/C single conductor
2/C two conductors
c candle, cathode, cycle, channel
C carbon, centigrade, Celsius
C&Btr. grade C and better (used in
lumber industry)
CAB cement-asbestos board,
cabinet
cal calorie
cap. capacity
CAT. catalog, catalogue
CATW catwalk

CB catch basin
CB1S center beam one side
CB2S center beam two sides
CBR California bearing ratio
cc cubic centimeter
CCW counterclockwise
cd candela
ceil ceiling
cem. fin. cement finish
cem. m cement mortar
cent. central
cer ceramic
CF centrifugal force, cost and freight, cooling fan
cfm, CFM cubic feet per minute
CFS cubic feet per second
CG center of gravity, coarse grain, ceiling grille, corner guard
CG2E center groove two edges
CHIM chimney
CHU centigrade heat unit
CI cast iron, certificate of insurance
CIP cast iron pipe
CIR circle, circuit
CIRC circumference
CL center line
cm centimeter
CM construction management, center matched
CMP corrugated metal pipe
CMPA corrugated metal pipe arch
CND conduit
CO change order, certificate of occupancy, cleanout, cutout
coef coefficient
col column
com common
COMB. combination
comp compensate, component, composition
COMPF composition floor
COMPR composition roof, compress, compressor
conc concrete
cond conductivity
const constant, construction
constr construction
CONTR contractor
conv convector
cop. coping
corb corbeled
corn. cornice

corr corrugated
CP cesspool
CPFF cost plus fixed fee
CPM Critical Path Method, cycles per minute
crib. cribbing
CRN cost of reproduction/replacement new
CRP controlled rate of penetration
CRT cathode ray tube
CS cast stone
CSI Construction Specifications Institute
CSK countersink
c/s cycles per second
ct coat, coats
CTB cement treated base
c to c center to center
ctr center
cu cubic
cu. ft. cubic feet
cu. in. cubic inch
cu. yd. cubic yard
cw clockwise
CV1S center vee one side
CV2S center vee two sides
C.W. pt. cold water point
cwt hundred weight
cyl cylinder
CYL L cylinder lock
cyp cypress

D

d degree, density, penny (nail size)
D diameter, dimensional, deep, depth, discharge
D&CM dressed and center matched
D&M dressed and matched
D&MB dressed and matched beaded
D&SM dressed and standard matched
D1s dressed one side
D2s dressed two sides
D2S&CM dressed two sides and center matched
D2S&SM dressed two sides and standard matched
D4S dressed on four sides
DAD. double acting door
dB decibel
dBA a unit of sound level (as from

the A-scale of a sound-level meter)

DB. Clg double-headed ceiling
DBL double
DBT dry-bulb temperature
DEC decimal
DEG degree, degrees
DEL delineation
DEPT department
DET detail, detached, double end trimmed
DF drinking fountain, drainage free, direction finder
dflct deflection
d.f.u. drainage fixture unit
DHW double-hung window
DIA diameter
DIAG diagonal
DIM. dimension
DIN Dutch Industry Normal (German industry standard)
DIV division
DL dead load, deadlight
DN down
DO ditto
DOZ dozen
DP dew point, double pitched, degree of polymerization
DPC dampproof course
DR drain, dressing room, dining room, driver
DRG drawing
drn drain, drainage
drwl drywall
DS downspout
DSGN design
DT drum trap
DT&G double tongue and groove
DU disposal unit
DUP duplicate
DVTL dovetail
DWG, dwg drawing
DWV drain, waste, and vent

E

e eccentricity, erg
E Modulus of Elasticity, Engineer
ea. each
EA Exhaust Air
E and OE Errors and Omissions Excepted
EB1S edge bead one side
Econ economy

EDP Electronic Data Processing
EDR Equivalent Direct Radiation
EE eased edges, electrical engineer, errors expected
EEO Equal Opportunity Employer
eff efficiency
EG edge (vertical) grain
ehf extremely high frequency
EHP Effective Horse Power, Electric Horsepower
elec electric or electrical
elev, EL elevation, elevator
EM end matched
EMF Electromotive Force
enam enamelled
encl enclosure
eng engine
engr engineer
EPDM Ethylene Propylene Diene Monomer
eq equal
equip equipment
equiv equivalent
erec erection
ERW Electric Resistance Welding
est estimate
esu electrostatic unit
EV electron volt
evap evaporate
EV1S edge vee one side
EW each way
EWT Entering Water Temperature
ex extra, example
exc excavation, except
excav excavation
exh exhaust
exp expansion
exp bt expansion bolt
ext exterior
extg extracting
extru extrusion
exx examples

F

f fine, focal length, force, frequency
F Fahrenheit, fluorine
FA fresh air, fire alarm
fab fabricate
fac facsimile
FAI fresh air intake
F.A.I.A. Fellow of the American Institute of Architects
FAO finish all over

FAR floor-area ratio
FAS free alongside ship, firsts and seconds
FBM foot board measure
f.c./F.C. footcandle
f.c.c. face-centered cubic
FDB forced-draft blower
FDC fire-department connection
fdn/fdtn/fds/FDN foundation, foundations
fdry foundry
Fe ferrum (iron)
FE fire escape
FEA Federal Energy Administration
FFA full freight allowance
FG fine grain, flat grain
F.G. finished grade
FHA Federal Housing Administration
FHC fire-hose cabinet
Fig. figure
fill. filling
Fin. finish
Fixt fixture
fl floor, fluid
FL floorline, floor, flashing
flash. flashing
FLG flooring
fl oz fluid ounce
Flr floor
FLUOR fluorescent
fm fathom
FM frequency modulation, Factory Mutual
FMT flush metal threshold
FMV fair market value
fndtn. foundation
FOB free on board
FOC free of charge
FOHC free of heart centers
FOK free of knots
Fount. fountain
fp fireplace, freezing point
f.pfg. fireproofing
fpm feet per minute
FPRF fireproof
fps feet per second
fr frame
Fr. fire rating
F.R. fire rating
frmg framing, forming

FRP fiber reinforced plastic
frt freight
frwy freeway
FS Federal specifications
FST flat seam tin
ft foot, feet
ftc footcandle
ftg footing
fth fathom
ft lb foot pound
Furn furnish(ed)
fus fusible
fv face velocity
FW flash welding
fwd forward

g gram, gravity, guage, girth, gain
G gas
ga gauge
gal. gallon
galv galvanized
gar garage
GB glass block
GC general contractor
GCF greatest common factor
gen general
GI galvanized iron
gl glass, glazing
GM grade marked
GMV gram molecular volume
Goth Gothic
gov/govt government
gpd gallons per day
gph gallons per hour
gpm gallons per minute
gr grade, gravity, gross, grains
G/R grooved roofing
gran granular
gr.fl. ground floor
gr.fl.ar ground floor area
grnd ground
gr.wt gross weight
GT gross ton
gtd guaranteed
g.u.p. grading under pavement
GYP gypsum

h harbor, hard, height, hours, house, hundred
H "head" on drawings, high, high strength bar joist, Henry, hydrogen

HA hour angle
HC, H.C. high capacity
HD, H.D. heavy duty, high density
H.D.O. High Density Overlaid
Hdr header
Hdwe. Hardware
hdwr hardware
He. helium
HE. high explosive
Help. Helper average
hem. hemlock
HEPA high efficiency particulate air
hex hexagon
hf half, high-frequency
H.F. hot finished
hg hectogram
Hg mercury
hgt height
HI height of instrument
hip. hipped (roof)
hl hectoliter
hm hectometer
HM hollow metal
HO high output
hor, horiz horizontal
hp horsepower
HP high pressure, steel pile section
H.P. Horsepower
H.P.F. High Power Factor
hr hour
Hrs./Day Hours Per Day
HSC High Short Circuit
ht, Ht. height
HT high-tension
htg, Htg. heating
Htrs. Heaters
hv, HV high voltage
HVAC heating, ventilating, and air conditioning
hvy, Hvy. heavy
HW high water, Hot Water
HWM high-water mark
hwy highway
hyd hydraulics, hydrostatics
Hyd, Hydr. Hydraulic
hydraul hydraulics, hydrostatics
hyp, hypoth hypothesis, hypothetical
hz hertz (cycles)

I

I moment of inertia
IC interrupting capacity, ironclad, incense cedar
ID inside dimension, inside diameter, identification
IF inside frosted
ihp indicated horsepower
IMC intermediate metal conduit
imp imperfect
in. inch
inc included, including, incorporated, increase, incoming
incan incandescent
incl included, including
Ins insulate, insurance
inst installation
insul insulation, insulate
int intake, interior, internal
IP iron pipe
IPS iron pipe size
IPT iron pipe threaded
IR inside radius

J

J jack, joule
jct junction
J.I.C Joint Industrial Council
jour journeyman
JP jet propulsion
jt/jnt joint
junc junction
jsts joists

K

k kilo, knot
K Kalium
Ka cathode
kc kilocycle
kcal kilocalorie
kc/s kilocycles per second
KD kiln dried
KDN knocked down
kg keg, kilogram
KIT. kitchen
kl kiloliter
KLF kips per lineal foot
km kilometer
kmps kilometers per second
kn knot
Kr krypton
kv kilovolt
kvar kilovar
kw kilowatt
kwhr/kwh kilowatt hour

L

l labor only, left, length, liter, long, lumen
L Lambert, large
L&CN lime and cement mortar
L&H light and heat
L&L latch and lock
L&O lead and oil (paint)
L&P lath and plaster
Lab. labor
LAG lagging
LAM laminated
LAT latitude, lattice
Lath lather
LAV lavatory
lb, lbs pound, pounds
lb/hr pounds per hour
lbf/sq in pound-force per square inch
lb/LF pounds per linear foot
Lbr lumber
LCL less-than-carload lot
LCM least common multiple, loose cubic meter
LCY loose cubic yard
LDG landing
ld load
LE leading edge
LECA light expanded clay aggregate
LEMA Lighting Equipment Manufacturers' Association
lf lightface, low frequency, lineal foot, linear foot
LG liquid gas
lg large, length, long
lgr longer
lgt lighting
lgth length
LH left hand, long-span, high strength bar joist
LIC license
lin linear, lineal
lin ft linear foot, linear feet
lino linoleum
LJ obsolete designation for long span standard strength bar joist
LL live load
LL&B latch, lock, and bolt
LLD lamp lumen depreciation
lm lumen
LM lime mortar
lm/sf lumen per square foot

lm/W lumen per watt
lng, Lng lining
LOA length over all
log logarithm
LP liquid petroleum, low pressure
LPF low power factor
LPG liquid petroleum gas
LR living room
LS (1) left side, (2) loudspeaker
LT long ton, light
Lt Ga light gauge
LTL less than truckload lot
Lt Wt lightweight
LV low voltage
LW low water
LWC lightweight concrete
LWM low water mark

M

m meter
M thousand, bending moment (on drawings)
ma milliampere
MA mechanical advantage
mach machine, machinist
mag magazine, magneto
MAN manual
manuf manufacture
mas masonry
mat, matl material
max maximum
mb millibar
MBH 1,000 BTU's per hour
MBM, M.b.m. thousand feet board measure
MC moisture content, metal-clad, mail chute
me marbled edges
ME mechanical engineer
meas measure
mech mechanic, mechanical
med medium
memb member
mep mean effective pressure
MER mechanical equipment room
met metallurgy
mezz mezzanine
mf mill finish
mfg manufactured
Mg magnesium
MG motor generator
mgt management
MH manhole

MHW mean high water
mi mile
mid middle
min minimum, minor, minute
misc miscellaneous
mix, mixt mixture
mks meter-kilogram-second
ML, ml material list
Mldg, mldg molding
MLW mean low water
MMF magnetomotive force
Mn manganese
MN magnetic North, main
MO month
Mo molybdenum
MOD model
mod, modif modification
MOL maximum overall length
MOT motor
mp melting point
mpg miles per gallon
mph miles per hour
MRT mean radiant temperature
MSF per 1,000 square feet
msl mean sea level
mtg, mtge mortgage
mult multiple, multiplier
mun, munic municipal
mxd mixed

N

n noon, number
N North, nail, nitrogen, normal
Na sodium
NAAMM National Association of Architectural Metal Manufacturers
NAT natural
NBC National Building Code
NBS National Bureau of Standards
NC noise criterion
NCM noncorrosive metal
NEC National Electric Code
NEMA National Electrical Manufacturers Association
NESC National Electrical Safety Code
NFC National Fire Code
Ni nickel
NIC not in contract
NOM nominal
NOP not otherwise provided for
norm normal

NPS nominal pipe size
nr near
NRC noise reduction coefficient
NS not specified
ntp normal temperature and pressure
NTS not to scale
nt. wt.; n.wt. net weight
num numeral
N1E nosed one edge (used in lumber industry)
N2E nosed two edges (used in lumber industry)

O

O oxygen
OA overall
O/A on approval
OAI outside air intake
O.B.M. ordinance bench mark
OBS (1) obsolete (2) open back strike
OC, o.c. on center
OCT octagon
OD outside diameter
OFF. office
OG, o.g. ogee
O/H overhead
OHS oval-headed screw
O.J.T. on-the-job-training
opp opposite
opt optional
OR. (1) outside radius (2) owner's risk
ord (1) order (2) ordnance
ORIG original
OSHA (1) Occupational Safety and Health Administration, Department of Labor (2) Occupational Safety and Health Act
OVHD overhead
OZ, oz ounce

P

P&G post and girder
P&T post and timbers
P1E planed one edge
P1S planed one side
1S2E planed one side and two edges
P4S planed four sides
p part, per, pint, pipe, pitch, pole, post, port, power
P phosphorus, pressure, pole, page

PA particular average, power amplifier, purchasing agent, public address system
pan. panel
par. parapet
p.a.r planed all round
PAR. paragraph
part. partition
partn partition
PASS. passenger
pat. patent
Pat. pattern
PAX private automatic (telephone) exchange
Pb lead
pc piece
pc/pct percent
PC portland cement
PCE pyrometric cone equivalent
pcf pounds per cubic foot
pcs pieces
pd paid
Pd palladium
PD per diem, potential difference
p.e. plain edged
PE professional engineer, probable error, plain end, polyethylene
P.E. professional engineer
pecky cyp pecky cypress
PEP Public Employment Program
per. perimeter, by the, period
PERF perforate
PERM permanent
PERP perpendicular
PERT project evaluation and review technique
PF power factor
PFA pulverized fuel ash
PFD preferred
ph phase, phot
Ph phenyl
PH phase, Phillips head
1PH single phase
3PH three phase
pil pilaster
piv pivoted
pk park, peak, plank
pk.fr. plank frame
pkwy parkway
pl place, plate
PL pile, plate, plug, power line, pipe line, private line

P/L plastic laminate
platf platform
PLG piling
plmb, plb, PLMB plumbing
PLYWD plywood
pmh production man-hour
PNEU pneumatic
PNL panel
PO purchase order
POL polish
PORC porcelain
PORT. CEM portland cement
pos, POS positive
pot. potential
PP-AC air-conditioning power panel
ppd prepaid
PPGL polished plate glass
ppm parts per million
ppt, pptn precipitate, precipitation
PR payroll, pair
PRCST precast
preb prebend
prec preceding
PREFAB. prefabricated
prelim preliminary
prin principal
prod. production
proj project, projection
PROJ project
prop. property
prov provisional
prs pairs
PRV Pressure Regulating Valve
ps pieces
p.s.e. planed and square-edged
psf pounds per square foot
psi pounds per square inch
p.s.j. planed and square-jointed
pt paint, pint, payment, port, point
PT part, point
P.T. pipe thread
ptfe polytetrafluorethylene
p.t.g. planed, tongued, and grooved
PTN partition
PU pickup
PUD pickup and delivery
pur purlins
PVA polyvinyl acetate
PVC polyvinyl chloride
PWA Public Works Administration

pwr power
pwt pennyweight

q quart
qda quantity discount agreement
QF quick firing
qr quarter
QR quarter-round
qs quarter-sawn
qt quart
QTR (1) quarry-tile roof. (2) On drawings, abbr. for quarter.
quad. quadrant
QUAD. quadrangle
QUAL quality
quar quarterly

r rain, range, rare, red, river, roentgen, run
R radius, right
Ra radium
R.A. registered architect
rab rabbeted
RAB rabbet
rad radiator
raft. rafter
RBM reinforced brick masonry
RC, R/C reinforced concrete
RC asphalt rapid-curing asphalt
RCP reinforced concrete pipe
1/4 RD quarter-round
1/2 RD half-round
rd road, rod, round
RD roof drain, round
rebar reinforcing bar
recap. recapitulation
recd received
recip reciprocal
RECP receptacle
rec. room recreation room
rect rectangle, rectified
red. reduce, reduction
ref reference, refining
REF refer, reference
REFR refractory, refrigerate
refrig refrigeration
reg registered
Reg regular
REG register, regulator
rein., reinf reinforced
REINF reinforce, reinforcing
REM removable

remod remodel
rent. rental
rep, REP repair
repl, REPL replace, replacement
REPRO reproduce
reqd, REQD required
res resawn
ret retain, retainage
RET. return
rev revenue, reverse, revised
REV revise
rf roof
RF roof, radio frequency
Rfg roofing
RFP request for proposal
rgh, Rgh rough
Rh rhodium, Rockwell hardness
RH relative humidity
RHN Rockwell hardness number
RI refractive index
R.I.B.A. Royal Institute of British Architects
rib. gl. ribbed glass
riv river
RJ road junction
R/L random lengths
rm ream, room
RM room
r. mld. raised mold
rms root mean square
rnd round
ROP record of production
ROPS roll-over protection system
rot. rotating, rotation
rpm revolutions per minute
RRGCP reinforced rubber gasket concrete pipe
RRS railroad siding
RSJ rolled steel joist
rt right
RT raintight
Rub., rub. Ruberoid, rubble
r.w. redwood, roadway, right-of-way
R/W right-of-way
R/W&L random widths and lengths
rwy, ry railway

S

S side, south, southern, seamless, subject, sulphur
S&E surfaced one side and edge
S&G studs and girts
S&H staple and hasp
S&M surfaced and matched
S/A shipped assembled
SAE Society of Automotive Engineers
SAF safety
SAN sanitary
sanit sanitation
sat. saturate, saturation
sch school
SCH schedule
scp spherical candlepower
SD sea-damaged, standard deviation
S/D shop drawings
SDA specific dynamic action
Sdg siding
Se selenium
S/E square-edged
SE&S square edge and sound
sec second
SECT section
sed sediment, sedimentation
sel select, selected
Sel select
sep, SEP separate
SERV service
SE Sdg, S.E. Sdg. square-edge siding
SEW. sewer
Sftwd. softwood
sf surface foot
sfu supply fixture unit
SGD sliding glass door
sh shingles
SH sheet, shower, single-hung
shf superhigh frequency
shp shaft horsepower
sht sheet, sheath
Si silicon
SIC Standard Industrial Classification
sid siding
SIM similar
sk sack
SK sketch
sky. skylights
S/L, S/LAP shiplap

SL&C shipper's load and count
slid. sliding
SM standard matched, surface measure
s. mld. stuck mold
SMS sheet-metal screw
so. south
SO. seller's option
soln solution
SOV shutoff valve
sp specific, specimen, spirit, single pitch (roof)
SP soil pipe, standpipe, self-propelled, single pole
SPEC specification
sp. gr., SP GR specific gravity
sp. ht. specific heat
SPKR loudspeaker
spl spline
SPL special
spr spruce
SPT Standard Penetration Test
sp. vol. specific volume
sq. square
sq. e. square edge
sq. E&S square edge and sound
sq. ft. square foot
sq. in. square inch
sq. yd. square yard
SR sedimentation rate
ss single strength (glass)
SS, S/S stainless steel
sst standing seam tin (roof)
SST stainless steel
st stairs, stone, street
ST steam, street
STC sound transmission class
std, STD standard
Std. M standard matched
STG storage
STK stock
STL steel
STP standard temperature and pressure
Stpg stepping
str stringers
Str. structural
STR. strike
Struc structural
st. sash. steel sash
ST W storm water
sty. story

sty. hgt. story height
SUB. substitute
sub. fl. subfloor
subpar subparagraph
subsec subsection
sup supplementary, supplement
SUP supply
supp supplement
SUPSD supersede
supt, SUPT superintendent
SUPV supervise
supvr supervisor
sur, SUR surface
surv survey, surveying, surveyor
svc service
sw switch
SW switch, seawater, southwest
SWBD switchboard
SWG, S.W.G. standard wire gauge
sy jet syphon jet (water closet)
SYM symmetrical
SYN synthetic
SYS system
syst system
S1E surfaced one edge
S1S surfaced one side
S1S1E surfaced one side and
   one edge
S1S2E surfaced one side and
   two edges
S2E surfaced two edges
S2S surfaced two sides
S2S&CM surfaced two sides and
   center matched
S2S&SL surfaced two sides and
   shiplapped
S2S1E surfaced two sides and
   one edge
S4S surfaced four sides
S4S&CS surfaced four sides and
   caulking seam

T

t temperature, time, ton
T tee, township, true, thermostat
Ta tantalum
t.b. turnbuckle
TB through-bolt
TC terra-cotta
Te tellurium
TE table of equipment, trailing edge
tech technical
TEL telephone

T.E.M. Total Energy Management
temp temperature, temporary
TEMP temperature
TER terrazzo
t.f. tar felt
T&G, T and G tongue-and-groove
TG&B tongued, grooved, and
   beaded
t.g.&d. tongued, grooved, and
   dressed
TH true heading
therm thermometer
THERMO thermostat
thou thousand
THK thick
thp thrust horsepower
THRU through
Ti titanium
TL transmission loss
tlr trailer
TM technical manual
tn ton, town, train
TN true north
tnpk/tpk turnpike
t.o. take off (estimate)
TOL tolerance
tonn tonnage
topog/topo topography
TOT. total
tp title page, township, tar paper
tps townships
tr tread
trans transom
TRANS transformer
transp transportation
trf tuned radio frequency
trib tributary
trib. ar tributary area
ts tensile strength
TU trade union, transmission unit
TUB. tubing
TV terminal velocity
twp township
TYP typical

U

u unit
U uranium
UBC Uniform Building Code
UDC universal decimal
   classification
U/E unedged
uhf ultrahigh frequency

UL Underwriters' Laboratories, Inc.
ult ultimate
unins. uninsurable
uns unsymmetrical
up upper
ur, UR urinal
USASI American National Standards Institute
USG United States gauge
UV ultraviolet

V

V volt, valve, vacuum, v-groove
val value, valuation
van vanity
VAP vapor
var variation, varnished
VAR visual-aural range, volt-ampere reactive
VAT vinyl-asbestos tile
VD vapor density
vel velocity
ven veneer
vent. ventilator
VENT. ventilate
vert, VERT vertical
VF video frequency
VG vertical grain
vhf very high frequency
vic vicinity
VIF verify in field
vil village
vis visibility, visual
VIT vitreous
vit. ch. vitreous china
v.j. V-joint
vlf very low frequency
VLR very long range
vol, VOL volume
vou voussoirs
VP vent pipe
VS versus, vent stack, vapor seal
VT vacuum tube, variable time
VU volume unit
V1S vee one side

W

w water, watt, weight, wicket, wide, width, work, with
W watt, west, western, width
W/ with
WA with average
WAF wiring around frame
WB welded base, water ballast, waybill

WBT wet-bulb temperature
WC, W.C. water closet
wd wood, window
Wdr wider
WF wide flange
wfl waffle
wg wing, wire gauge
WG wire gauge
wh watt-hour
WH water heater
WHP water horsepower
whr watt-hour
whse, WHSE warehouse
WI wrought iron
WK week, work
wm wattmeter
WM wire mesh
W/M weight or measurement
w/o water-in-oil, without
W/O without
WP waterproof, weatherproof, white phosphorus
wpc watts per candle
w proof waterproofing
wrt wrought
WS weather strip
wsct/wains wainscoting
wt., Wt. weight
WT watertable, watertight
ww white wash
WWM welded wire mesh

X

X experimental
XBAR crossbar
XH, X HVY extra heavy
XL extra large
xr without rights
X STR extra strong
xw without warrants
XXH double extra heavy

Y

y yard
Y yttrium, wye, Y-branch
yd yard
y.p. yellow pine
YP yield paint
YR year
YS yield strength

Z

z zero, zone
Z modulus of section
ZI zone of interior
Zn azimuth, zinc

3 OSHA EXCERPTS

The following **excerpts** from the OSHA code show the sections mainly concerned with construction. These should help clarify the requirements necessary to comply with the Occupational Safety and Health Act. The entire document may be obtained from the Superintendent of Documents, U.S. Government Printing Office, Washington, D.C. 20402.

Sanitation.

Toilets at construction jobsites. (1) Toilets shall be provided for employees according to the following table:

Table D-1

Number of employees	Minimum number of facilities
20 or less	1.
20 or more	1 toilet seat and 1 urinal per 40 workers.
200 or more	1 toilet seat and 1 urinal per 50 workers

(2) Under temporary field conditions, provisions shall be made to assure not less than one toilet facility is available.

(3) Job sites, not provided with a sanitary sewer, shall be provided with one of the following toilet facilities unless prohibited by local codes:

(i) Privies (where their use will not contaminate ground or surface water);

(ii) Chemical toilets;

(iii) Recirculating toilets;

(iv) Combustion toilets.

(4) The requirements of this paragraph (c) for sanitation facilities shall not apply to mobile crews having transportation readily available to nearby toilet facilities.

(d) *Food handling.* All employees' food service facilities and operations shall meet the applicable laws, ordinances, and regulations of the jurisdictions in which they are located.

Table D-3 – Minimum Illumination Intensities in Foot-Candles

Foot-candles	Area or operation
5	General construction area lighting.
3	General construction areas, concrete placement, excavation and waste areas, accessways, active storage areas, loading platforms, refueling, and field maintenance areas.
5	Indoors: warehouses, corridors, hallways, and exitways.
5	Tunnels, shafts, and general underground work areas: (Exception: minimum of 10 footcandles is required at tunnel and shaft heading during drilling, mucking and scaling. Bureau of Mines approved cap lights shall be acceptable for use in the tunnel heading.)
10	General construction plant and shops (e.g., batch plants, screening plants, mechanical and electrical equipment rooms, carpenter shops, rigging lofts and active storerooms, barracks or living quarters, locker or dressing rooms, mess halls, and indoor toilets and workrooms).
30	First aid stations, infirmaries, and offices.

(e) *Temporary sleeping quarters.* When temporary sleeping quarters are provided, they shall be heated, ventilated, and lighted.

(f) *Washing facilities.* The employer shall provide adequate washing facilities for employees engaged in the application of paints, coating, herbicides, or insecticides, or in other operations where contaminants may be harmful to the employees. Such facilities shall be in near proximity to the worksite and shall be so equipped as to enable employees to remove such substances.

Illumination.

(a) *General.* Construction areas, ramps, runways, corridors, offices, shops, and storage areas shall be lighted to not less than the minimum illumination intensities listed in Table D-3 while any work is in progress:

(b) *Other areas.* For areas or operations not covered above, refer to the American National Standard A11.1-1965, R1970, Practice for Industrial Lighting, for recommended values of illumination.

Fire Protection and Prevention

Fire protection.

(a) *General requirements.* (1) The employer shall be responsible for the development of a fire protection program to be followed throughout all phases of the construction and demolition work, and he shall provide for the firefighting equipment as specified in this subpart. As fire hazards occur, there shall be no delay in providing the necessary equipment.

(2) Access to all available firefighting equipment shall be maintained at all times.

(3) All firefighting equipment, provided by the employer, shall be conspicuously located.

(4) All firefighting equipment shall be periodically inspected and maintained in operating condition. Defective equipment shall be immediately replaced.

(5) As warranted by the project, the employer shall provide a trained and equipped firefighting organization (Fire Brigade) to assure adequate protection to life.

(b) *Water supply.* (1) A temporary or permanent water supply, of sufficient volume, duration, and pressure, required to properly operate the firefighting equipment shall be made available as soon as combustible materials accumulate.

(2) Where underground water mains are to be provided, they shall be installed, completed, and made available for use as soon as practicable.

(c) *Portable firefighting equipment* — (1) *Fire extinguishers and small hose lines.* (i) A fire extinguisher, rated not less than 2A, shall be provided for each 3,000 square feet of the protected building area, or major fraction thereof. Travel distance from any point of the protected area to the nearest fire extinguisher shall not exceed 100 feet.

(ii) One 55-gallon open drum of water with two fire pails may be substituted for a fire extinguisher having a 2A rating.

(iii) A ½-inch diameter garden-type hose line, not to exceed 100 feet in length and equipped with a nozzle, may be substituted for a 2A-rated fire extinguisher, providing it is capable of discharging a minimum of 5 gallons per minute with a minimum hose stream range of 30 feet horizontally. The garden-type hose lines shall be mounted on conventional racks or reels. The number and location of hose racks or reels shall be such that at least one hose stream can be applied to all points in the area.

(iv) One or more fire extinguishers, rated not less than 2A, shall be provided on each floor. In multistory buildings, at least one fire extinguisher shall be located adjacent to stairway.

(v) Extinguishers and water drums, subject to freezing, shall be protected from freezing.

(vi) A fire extinguisher, rated not less than 10B, shall be provided within 50 feet of wherever more than 5 gallons of flammable or combustible liquids or 5 pounds of flammable gas are being used on the jobsite. This requirement does not apply to the integral fuel tanks of motor vehicles.

(vii) Carbon tetrachloride and other toxic vaporizing liquid fire extinguishers are prohibited.

(viii) Portable fire extinguishers shall be inspected periodically and maintained in accordance with Maintenance and Use of Portable Fire Extinguishers, NFPA No. 10A-1970.

(ix) Fire extinguishers which have been listed or approved by a nationally recognized testing laboratory, shall be used to meet the requirements of this subpart.

(2) *Fire hose and connections.* (i) One hundred feet, or less, of 1½-inch hose, with a nozzle capable of discharging water at 25 gallons or more per minute, may be substituted for a fire extinguisher rated not more than 2A in the designated area provided that the hose line can reach all points in the area.

(ii) If fire hose connections are not compatible with local firefighting equipment, the contractor shall provide adapters, or equivalent, to permit connections.

(iii) During demolition involving combustible materials, charged hose lines, supplied by hydrants, water tank trucks with pumps, or equivalent, shall be made available.

(d) *Fixed firefighting equipment* — (1) *Sprinkler protection.* (i) If the facility being constructed includes the installation of automatic sprinkler protection, the installation shall closely follow the construction and be placed in service as soon as applicable laws permit following completion of each story.

(ii) During demolition or alterations, existing automatic sprinkler installations shall be retained in service as long as reasonable. The operation of sprinkler control valves shall be permitted only by properly authorized persons. Modification of sprinkler systems to permit alterations or additional demolition should be expedited so that the automatic protection may be returned to service as quickly as possible. Sprinkler control valves shall be checked daily at close of work to ascertain that the protection is in service.

(2) *Standpipes.* In all structures in which standpipes are required, or where standpipes exist in structures being altered, they shall be brought up as soon as applicable laws permit, and shall be maintained as construction progresses in such a manner that they are always ready for fire protection use. The standpipes shall be provided with Siamese fire department connections on the outside of the structure, at the street level, which shall be conspicuously marked. There shall be at least one standard hose outlet at each floor.

(e) *Fire alarm devices.* (1) An alarm system, e.g., telephone system, siren, etc., shall be established by the employer whereby employees on the site and the local fire department can be alerted for an emergency.

(2) The alarm code and reporting instructions shall be conspicuously posted at phones and at employee entrances.

(f) *Fire cutoffs.* (1) Fire walls and exit stairways, required for the completed buildings, shall be given construction priority. Fire doors, with automatic closing devices, shall be hung on openings as soon as practicable.

(2) Fire cutoffs shall be retained in buildings undergoing alterations or demolition until operations necessitate their removal.

General requirements for storage.

(a) *General.* (1) All materials stored in tiers shall be stacked, racked, blocked, interlocked, or otherwise secured to prevent sliding, falling or collapse.

(2) Maximum safe load limits of floors within buildings and structures, in pounds per square foot, shall be conspicuously posted in all storage areas, except for floor or slab on grade. Maximum safe loads shall not be exceeded.

(3) Aisles and passageways shall be kept clear to provide for the free and safe movement of material handling equipment or employees. Such areas shall be kept in good repair.

(4) When a difference in road or working levels exist, means such as ramps, blocking, or grading shall be used to ensure the safe movement of vehicles between the two levels.

(b) *Material storage.* (1) Material stored inside buildings under construction shall not be placed within 6 feet of any hoistway or inside floor openings, nor within 10 feet of an exterior wall which does not extend above the top of the material stored.

(2) Employees required to work on stored material in silos, hoppers, tanks, and similar storage areas shall be equipped with lifelines and safety belts.

(3) Noncompatible materials shall be segregated in storage.

(4) Bagged materials shall be stacked by stepping back the layers and crosskeying the bags at least every 10 bags high.

(5) Materials shall not be stored on scaffolds or runways in excess of supplies needed for immediate operations.

(6) Brick stacks shall not be more than 7 feet in height. When a loose brick stack reaches a height of 4 feet, it shall be tapered back 2 inches in every foot of height above the 4-foot level.

(7) When masonry blocks are stacked higher than 6 feet, the stack shall be tapered back one-half block per tier above the 6-foot level.

(8) Lumber:

(i) Used lumber shall have all nails withdrawn before stacking.

(ii) Lumber shall be stacked on level and solidly supported sills.

(iii) Lumber shall be so stacked as to be stable and self-supporting.

(iv) Lumber piles shall not exceed 20 feet in height provided that lumber to be handled manually shall not be stacked more than 16 feet high.

(9) Structural steel, poles, pipe, bar stock, and other cylindrical materials, unless racked, shall be stacked and blocked so as to prevent spreading or tilting.

Rigging equipment for material handling.

(a) *General.* (1) Rigging equipment for material handling shall be inspected prior to use on each shift and as necessary during its use to ensure that it is safe. Defective rigging equipment shall be removed from service.

(2) Rigging equipment shall not be loaded in excess of its recommended safe working load.

(3) Rigging equipment, when not in use, shall be removed from the immediate work area so as not to present a hazard to employees.

(4) Special custom design grabs, hooks, clamps, or other lifting accessories, for such units as modular panels, prefabricated structures and similar materials, shall be marked to indicate the safe working loads and shall be proof-tested prior to use to 125 percent of their rated load.

(b) *Alloy steel chains.* (1) Welded alloy steel chain slings shall have permanently affixed durable identification stating size, grade, rated capacity, and sling manufacturer.

(2) Hooks, rings, oblong links, pear-shaped links, welded or mechanical coupling links, or other attachments, when used with alloy steel chains, shall have a rated capacity at least equal to that of the chain.

(3) Job or shop hooks and links, or makeshift fasteners, formed from bolts, rods, etc., or other such attachments, shall not be used.

(4) Rated capacity (working load limit) for alloy steel chain slings shall conform to the values shown in Table H-1.

(c) *Wire Rope* (2) Protruding ends of strands in splices on slings and bridles shall be covered or blunted.

(3) Wire rope shall not be secured by knots, except on haul back lines on scrapers.

(4) The following limitations shall apply to the use of wire rope:

(i) An eye splice made in any wire rope shall have not less than three full tucks. However, this requirement shall not operate to preclude the use of another form of splice or connection which can be shown to be as efficient and which is not otherwise prohibited.

(ii) Except for eye splices in the ends of wires and for endless rope slings, each wire rope used in hoisting or lowering, or in pulling loads, shall consist of one continuous piece without knot or splice.

(iii) Eyes in wire rope bridles, slings, or bull wires shall not be formed by wire rope clips or knots.

(iv) Wire rope shall not be used if, in any length of eight diameters, the total number of visible broken wires exceeds 10 percent of the total number of wires, or if the rope shows other signs of excessive wear, corrosion, or defect.

(5) When U-bolt wire rope clips are used to form eyes, Table H-20 shall be used to determine the number and spacing of clips.

(i) When used for eye splices, the U-bolt shall be applied so that the "U" section is in contact with the dead end of the rope.

(2) All splices in rope slings provided by the employer shall be made in accordance with fiber rope manufacturers recommendations.

(i) In manila rope, eye splices shall contain at least three full tucks, and short splices shall contain at least six full tucks (three on each side of the centerline of the splice).

(ii) In layed synthetic fiber rope, eye splices shall contain at least four full tucks, and short splices shall contain at least eight full tucks (four on each side of the centerline of the splice).

(iii) Strand end tails shall not be trimmed short (flush with the surface of the rope) immediately adjacent to the full tucks. This precaution applies to both eye and short splices and all types of fiber rope. For fiber ropes under 1-inch diameter, the

Table H-1 – Rated Capacity (Working Load Limit), For Alloy Steel Chain Slings[1]

Rated Capacity (Working Load Limit), Pounds
[Horizontal angles shown in parentheses] (2)

Chain size (inches)	Single branch sling – 90° loading	Double sling vertical angle (1)			Triple and quadruple sling vertical angle (1)		
		30° (60°)	45° (45°)	60° (30°)	30° (60°)	45° (45°)	60° (30°)
1/4	3,250	5,560	4,550	3,250	8,400	6,800	4,900
3/8	6,600	11,400	9,300	6,600	17,000	14,000	9,900
1/2	11,250	19,500	15,900	11,250	29,000	24,000	17,000
5/8	16,500	28,500	23,300	16,500	43,000	35,000	24,500
3/4	23,000	39,800	32,500	23,000	59,500	48,500	34,500
7/8	28,750	49,800	40,600	28,750	74,500	61,000	43,000
1	38,750	67,100	54,800	38,750	101,000	82,000	58,000
1-1/8	44,500	77,000	63,000	44,500	115,500	94,500	66,500
1-1/4	57,500	99,500	81,000	57,500	149,000	121,500	86,000
1-3/8	67,000	116,000	94,000	67,000	174,000	141,000	100,500
1-1/2	80,000	138,000	112,500	80,000	207,000	169,000	119,500
1-3/4	100,000	172,000	140,000	100,000	258,000	210,000	150,000

[1]Other grades of proof tested steel chain include Proof Coil, BBB Coil and Hi-Test Chain. These grades are not recommended for overhead lifting and therefore are not covered by this code.

(1) Rating of multileg slings adjusted for angle of loading measured as the included angle between the inclined leg and the vertical.

(2) Rating of multileg slings adjusted for angle of loading between the inclined leg and the horizontal plane of the load.

Table H-2 – Maximum Allowable Wear At Any Point Of Link

Chain size (inches)	Maximum allowable wear (inch)
1/4	3/64
3/8	5/64
1/2	7/64
5/8	9/64
3/4	5/32
7/8	11/64
1	3/16
1-1/8	7/32
1-1/4	1/4
1-3/8	9/32
1-1/2	5/16
1-3/4	11/32

Table H-3 – Rated Capacities For Single Leg Slings
6 X 19 and 6 X 37 Classification Improved Plow Steel Grade Rope with Fiber Core (FC)

Rope		Rated capacities, tons (2,000 lb.)								
Dia. (inches)	Constr.	Vertical			Choker			Vertical basket[1]		
		HT	MS	S	HT	MS	S	HT	MS	S
1/4	6 X 19	0.49	0.51	0.55	0.37	0.38	0.41	0.99	1.0	1.1
5/16	6 X 19	0.76	0.79	0.85	0.57	0.59	0.64	1.5	1.6	1.7
3/8	6 X 19	1.1	1.1	1.2	0.80	0.85	0.91	2.1	2.2	2.4
7/16	6 X 19	1.4	1.5	1.6	1.1	1.1	1.2	2.9	3.0	3.3
1/2	6 X 19	1.8	2.0	2.1	1.4	1.5	1.6	3.7	3.9	4.3
9/16	6 X 19	2.3	2.5	2.7	1.7	1.9	2.0	4.6	5.0	5.4
5/8	6 X 19	2.8	3.1	3.3	2.1	2.3	2.5	5.6	6.2	6.7
3/4	6 X 19	3.9	4.4	4.8	2.9	3.3	3.6	7.8	8.8	9.5
7/8	6 X 19	5.1	5.9	6.4	3.9	4.5	4.8	10.0	12.0	13.0
1	6 X 19	6.7	7.7	8.4	5.0	5.8	6.3	13.0	15.0	17.0
1-1/8	6 X 19	8.4	9.5	10.0	6.3	7.1	7.9	17.0	19.0	21.0
1-1/4	6 X 37	9.8	11.0	12.0	7.4	8.3	9.2	20.0	22.0	25.0
1-3/8	6 X 37	12.0	13.0	15.0	8.9	10.0	11.0	24.0	27.0	30.0
1-1/2	6 X 37	14.0	16.0	17.0	10.0	12.0	13.0	28.0	32.0	35.0
1-5/8	6 X 37	16.0	18.0	21.0	12.0	14.0	15.0	33.0	37.0	41.0
1-3/4	6 X 37	19.0	21.0	24.0	14.0	16.0	18.0	38.0	43.0	48.0
2	6 X 37	25.0	28.0	31.0	18.0	21.0	23.0	49.0	55.0	62.0

[1]These values only apply when the D/d ratio for HT slings is 10 or greater, and for MS and S Slings is 20 or greater where:
D = Diameter of curvature around which the body of the sling is bent. d = Diameter of rope.
HT = Hand Tucked Splice and Hidden Tuck Splice. For hidden tuck splice (IWRC) use values in HT columns.
MS = Mechanical Splice.
S = Swaged or Zinc Poured Socket.

Table H-4 – Rated Capacities For Single Leg Slings
6 X 19 and 6 X 37 Classification Improved Plow Steel Grade Rope With Independent Wire Rope Core (IWRC)

Rope		Rated capacities, tons (2,000 lb.)								
Dia. (inches)	Constr.	Vertical			Choker			Vertical Basket[1]		
		HT	MS	S	HT	MS	S	HT	MS	S
1/4	6 X 19	0.53	0.56	0.59	0.40	0.42	0.44	1.0	1.1	1.2
5/16	6 X 19	0.81	0.87	0.92	0.61	0.65	0.69	1.6	1.7	1.8
3/8	6 X 19	1.1	1.2	1.3	0.86	0.93	0.98	2.3	2.5	2.6
7/16	6 X 19	1.5	1.7	1.8	1.2	1.3	1.3	3.1	3.4	3.5
1/2	6 X 19	2.0	2.2	2.3	1.5	1.6	1.7	3.9	4.4	4.6
9/16	6 X 19	2.5	2.7	2.9	1.8	2.1	2.2	4.9	5.5	5.8
5/8	6 X 19	3.0	3.4	3.6	2.2	2.5	2.7	6.0	6.8	7.2
3/4	6 X 19	4.2	4.9	5.1	3.1	3.6	3.8	8.4	9.7	10.0
7/8	6 X 19	5.5	6.6	6.9	4.1	4.9	5.2	11.0	13.0	14.0
1	6 X 19	7.2	8.5	9.0	5.4	6.4	6.7	14.0	17.0	18.0
1-1/8	6 X 19	9.0	10.0	11.0	6.8	7.8	8.5	18.0	21.0	23.0
1-1/4	6 X 37	10.0	12.0	13.0	7.9	9.2	9.9	21.0	24.0	26.0
1-3/8	6 X 37	13.0	15.0	16.0	9.6	11.0	12.0	25.0	29.0	32.0
1-1/2	6 X 37	15.0	17.0	19.0	11.0	13.0	14.0	30.0	35.0	38.0
1-5/8	6 X 37	18.0	20.0	22.0	13.0	15.0	17.0	35.0	41.0	44.0
1-3/4	6 X 37	20.0	24.0	26.0	15.0	18.0	19.0	41.0	47.0	51.0
2	6 X 37	26.0	30.0	33.0	20.0	23.0	25.0	53.0	61.0	66.0

[1]These values only apply when the D/d ratio for HT slings is 10 or greater, and for MS and S Slings is 20 or greater where:

D = Diameter of curvature around which the body of the sling is bent. d = Diameter of rope.

HT = Hand Tucked Splice: For hidden tuck splice (IWRC) use Table H-3 values in HT column.

MS = Mechanical Splice.

S = Swaged or Zinc Poured Socket.

Table H-5 – Rated Capacities For Single Leg Slings

Cable Laid Rope – Mechanical Splice Only
7 X 7 X 7 and 7 X 7 X 19 Construction Galvanized Aircraft Grade Rope
7 X 6 X 19 IWRC Construction Improved Plow Steel Grade Rope

Rope		Rated capacities, tons (2,000 lb.)			Rope		Rated capacities, tons (2,000 lb.)		
Dia. (inches)	Constr.	Vertical	Choker	Vertical basket[1]	Dia. (inches)	Constr.	Vertical	Choker	Vertical basket[1]
1/4	7 X 7 X 7	0.50	0.38	1.0	1-1/8	7 X 7 X 19	8.2	6.2	16.0
3/8	7 X 7 X 7	1.1	0.81	2.2	1-1/4	7 X 7 X 19	9.9	7.4	20.0
1/2	7 X 7 X 7	1.8	1.4	3.7	3/4	[2]7 X 6 X 19	3.8	2.8	7.6
5/8	7 X 7 X 7	2.8	2.1	5.5	7/8	[2]7 X 6 X 19	5.0	3.8	10.0
3/4	7 X 7 X 7	3.8	2.9	7.6	1	[2]7 X 6 X 19	6.4	4.8	13.0
5/8	7 X 7 X 19	2.9	2.2	5.8	1-1/8	[2]7 X 6 X 19	7.7	5.8	15.0
3/4	7 X 7 X 19	4.1	3.0	8.1	1-1/4	[2]7 X 6 X 19	9.2	6.9	18.0
7/8	7 X 7 X 19	5.4	4.0	11.0	1-5/16	[2]7 X 6 X 19	10.0	7.5	20.0
1	7 X 7 X 19	6.9	5.1	14.0	1-3/8	[2]7 X 6 X 19	11.0	8.2	22.0
					1-1/2	[2]7 X 6 X 19	13.0	9.6	26.0

[1]These values only apply when the D/d ratio is 10 or greater where:
D = Diameter of curvature around which the
body of the sling is bent.
d = Diameter of rope.
[2]IWRC.

[1]These values only apply when the D/d ratio is 10 or greater where:
D = Diameter of curvature around which the
body of the sling is bent.
d = Diameter of rope.
[2]IWRC.

Table H-6 – Rated Capacities For Single Leg Slings
8-Part and 6-Part Braided Rope
6 X 7 and 6 X 19 Construction Improved Plow Steel Grade Rope
7 X 7 Construction Galvanized Aircraft Grade Rope

Component ropes		Rated capacities, tons (2,000 lb.)					
		Vertical		Choker		Basket vertical to 30°[1]	
Diameter (inches)	Constr.	8-Part	6-Part	8-Part	6-Part	8-Part	6-Part
3/32	6 X 7	0.42	0.32	0.32	0.24	0.74	0.55
1/8	6 X 7	0.76	0.57	0.57	0.42	1.3	0.98
3/16	6 X 7	1.7	1.3	1.3	0.94	2.9	2.2
3/32	7 X 7	0.51	0.39	0.38	0.29	0.89	0.67
1/8	7 X 7	0.95	0.71	0.71	0.53	1.6	1.2
3/16	7 X 7	2.1	1.5	1.5	1.2	3.6	2.7
3/16	6 X 19	1.7	1.3	1.3	0.98	3.0	2.2
1/4	6 X 19	3.1	2.3	2.3	1.7	5.3	4.0
5/16	6 X 19	4.8	3.6	3.6	2.7	8.3	6.2
3/8	6 X 19	6.8	5.1	5.1	3.8	12.0	8.9
7/16	6 X 19	9.3	6.9	6.9	5.2	16.0	12.0
1/2	6 X 19	12.0	9.0	9.0	6.7	21.0	15.0
9/16	6 X 19	15.0	11.0	11.0	8.5	26.0	20.0
5/8	6 X 19	19.0	14.0	14.0	10.0	32.0	24.0
3/4	6 X 19	27.0	20.0	20.0	15.0	46.0	35.0
7/8	6 X 19	36.0	27.0	27.0	20.0	62.0	47.0
1	6 X 19	47.0	35.0	35.0	26.0	81.0	61.0

[1]These values only apply when the D/d ratio is 20 or greater where: D = Diameter of curvature around which the body of the sling is bent. d = Diameter of component rope.

Table H-7 – Rated Capacities for 2-Leg and 3-Leg Bridle Slings
6 X 19 and 6 X 37 Classification Improved Plow Steel Grade Rope With Fiber Core (FC)

Rope		Rated capacities, tons (2,000 lb.)											
		2-leg bridle slings						3-leg bridle slings					
Dia. (inches)	Constr.	30°¹(60°)²		45° angle		60°¹(30°)²		30°¹(60°)²		45° angle		60°¹(30°)²	
		HT	MS	HT	MS	HT	MS	HT	MS	HT	MS	HT	MS
1/4	6 X 19	0.85	0.88	0.70	0.72	0.49	0.51	1.3	1.3	1.0	1.1	0.74	0.7
5/16	6 X 19	1.3	1.4	1.1	1.1	0.76	0.79	2.0	2.0	1.6	1.7	1.1	1.2
3/8	6 X 19	1.8	1.9	1.5	1.6	1.1	1.1	2.8	2.9	2.3	2.4	1.6	1.7
7/16	6 X 19	2.5	2.6	2.0	2.2	1.4	1.5	3.7	4.0	3.0	3.2	2.1	2.3
1/2	6 X 19	3.2	3.4	2.6	2.8	1.8	2.0	4.8	5.1	3.9	4.2	2.8	3.0
9/16	6 X 19	4.0	4.3	3.2	3.5	2.3	2.5	6.0	6.5	4.9	5.3	3.4	3.7
5/8	6 X 19	4.8	5.3	4.0	4.4	2.8	3.1	7.3	8.0	5.9	6.5	4.2	4.6
3/4	6 X 19	6.8	7.6	5.5	6.2	3.9	4.4	10.0	11.0	8.3	9.3	5.8	6.6
7/8	6 X 19	8.9	10.0	7.3	8.4	5.1	5.9	13.0	15.0	11.0	13.0	7.7	8.9
1	6 X 19	11.0	13.0	9.4	11.0	6.7	7.7	17.0	20.0	14.0	16.0	10.0	11.0
1-1/8	6 X 19	14.0	16.0	12.0	13.0	8.4	9.5	22.0	24.0	18.0	20.0	13.0	14.0
1-1/4	6 X 37	17.0	19.0	14.0	16.0	9.8	11.0	25.0	29.0	21.0	23.0	15.0	17.0
1-3/8	6 X 37	20.0	23.0	17.0	19.0	12.0	13.0	31.0	35.0	25.0	28.0	18.0	20.0
1-1/2	6 X 37	24.0	27.0	20.0	22.0	14.0	16.0	36.0	41.0	30.0	33.0	21.0	24.0
1-5/8	6 X 37	28.0	32.0	23.0	26.0	16.0	18.0	43.0	48.0	35.0	39.0	25.0	28.0
1-3/4	6 X 37	33.0	37.0	27.0	30.0	19.0	21.0	49.0	56.0	40.0	45.0	28.0	32.0
2	6 X 37	43.0	48.0	35.0	39.0	25.0	28.0	64.0	72.0	52.0	59.0	37.0	41.0

HT = Hand Tucked Splice.
MS = Mechanical Splice.
¹Vertical angles.
²Horizontal angles.

Table H-8 — Rated Capacities for 2-Leg and 3-Leg Bridle Slings
6 X 19 and 6 X 37 Classification Improved Plow Steel Grade Rope
With Independent Wire Rope Core (IWRC)

Rope		Rated capacities, tons (2,000 lb.)											
		2-leg bridle slings						3-leg bridle slings					
Dia. (inches)	Constr.	30°¹(60°)²		45° angle		60°¹(30°)²		30°¹(60°)²		45° angle		60°¹(30°)²	
		HT	MS	HT	MS	HT	MS	HT	MS	HT	MS	HT	MS
1/4	6 X 19	0.92	0.97	0.75	0.79	0.53	0.56	1.4	1.4	1.1	1.2	0.79	0.84
5/16	6 X 19	1.4	1.5	1.1	1.2	1.81	1.87	2.1	2.3	1.7	1.8	1.2	1.3
3/8	6 X 19	2.0	2.1	1.6	1.8	1.1	1.2	3.0	3.2	2.4	2.6	1.7	1.9
7/16	6 X 19	2.7	2.9	2.2	2.4	1.5	1.7	4.0	4.4	3.3	3.6	2.3	2.5
1/2	6 X 19	3.4	3.8	2.8	3.1	2.0	2.2	5.1	5.7	4.2	4.6	3.0	3.3
9/16	6 X 19	4.3	4.8	3.5	3.9	2.5	2.7	6.4	7.1	5.2	5.8	3.7	4.1
5/8	6 X 19	5.2	5.9	4.2	4.8	3.0	3.4	7.8	8.8	6.4	7.2	4.5	5.1
3/4	6 X 19	7.3	8.4	5.9	6.9	4.2	4.9	11.0	13.0	8.9	10.0	6.3	7.3
7/8	6 X 19	9.6	11.0	7.8	9.3	5.5	6.6	14.0	17.0	12.0	14.0	8.3	9.9
1	6 X 19	12.0	15.0	10.0	12.0	7.2	8.5	19.0	22.0	15.0	18.0	11.0	13.0
1-1/8	6 X 19	16.0	18.0	13.0	15.0	9.0	10.0	23.0	27.0	19.0	22.0	13.0	16.0
1-1/4	6 X 37	18.0	21.0	15.0	17.0	10.0	12.0	27.0	32.0	22.0	26.0	16.0	18.0
1-3/8	6 X 37	22.0	25.0	18.0	21.0	13.0	15.0	33.0	38.0	27.0	31.0	19.0	22.0
1-1/2	6 X 37	26.0	30.0	21.0	25.0	15.0	17.0	39.0	45.0	32.0	37.0	23.0	26.0
1-5/8	6 X 37	31.0	35.0	25.0	29.0	18.0	20.0	46.0	53.0	38.0	43.0	27.0	31.0
1-3/4	6 X 37	35.0	41.0	29.0	33.0	20.0	24.0	53.0	61.0	43.0	50.0	31.0	35.0
2	6 X 37	46.0	53.0	37.0	43.0	26.0	30.0	68.0	79.0	56.0	65.0	40.0	46.0

HT = Hand Tucked Splice.
MS = Mechanical Splice.
¹Vertical angles.
²Horizontal angles.

Table H-9 – Rated Capacities for 2-Leg and 3-Leg Bridle Slings
Cable Laid Rope – Mechanical Splice Only
7 X 7 X 7 and 7 X 7 X 19 Constructions Galvanized Aircraft Grade Rope
7 X 6 X 19 IWRC Construction Improved Plow Steel Grade Rope

Rope		Rated capacities, tons (2,000 lb.)					
Dia. (inches)	Constr.	2-leg bridle sling			3-leg bridle sling		
		30°[1] (60°)[2]	45° angle	60°[1] (30°)[2]	30°[1] (60°)[2]	45° angle	60°[1] (30°)[2]
1/4	7 X 7 X 7	0.87	0.71	0.50	1.3	1.1	0.75
3/8	7 X 7 X 7	1.9	1.5	1.1	2.8	2.3	1.6
1/2	7 X 7 X 7	3.2	2.6	1.8	4.8	3.9	2.8
5/8	7 X 7 X 7	4.8	3.9	2.8	7.2	5.9	4.2
3/4	7 X 7 X 7	6.6	5.4	3.8	9.9	8.1	5.7
7/8	7 X 7 X 19	5.0	4.1	2.9	7.5	6.1	4.3
3/4	7 X 7 X 19	7.0	5.7	4.1	10.0	8.6	6.1
7/8	7 X 7 X 19	9.3	7.6	5.4	14.0	11.0	8.1
1	7 X 7 X 19	12.0	9.7	6.9	18.0	14.0	10.0
1-1/8	7 X 7 X 19	14.0	12.0	8.2	21.0	17.0	12.0
1-1/4	7 X 7 X 19	17.0	14.0	9.9	26.0	21.0	15.0
3/4	7 X 6 X 19 IWRC	6.6	5.4	3.8	9.9	8.0	5.7
7/8	7 X 6 X 19 IWRC	8.7	7.1	5.0	13.0	11.0	7.5
1	7 X 6 X 19 IWRC	11.0	9.0	6.4	17.0	13.0	9.6
1-1/8	7 X 6 X 19 IWRC	13.0	11.0	7.7	20.0	16.0	11.0
1-1/4	7 X 6 X 19 IWRC	16.0	13.0	9.2	24.0	20.0	14.0
1-5/16	7 X 6 X 19 IWRC	17.0	14.0	10.0	26.0	21.0	15.0
1-3/8	7 X 6 X 19 IWRC	19.0	15.0	11.0	28.0	23.0	16.0
1-1/2	7 X 6 X 19 IWRC	22.0	18.0	13.0	33.0	27.0	19.0

[1]Vertical angles.

[2]Horizontal angles.

Table H-10 – Rated Capacities for 2-Leg and 3-Leg Bridle Slings
8-Part and 6-Part Braided Rope
6 X 7 and 6 X 19 Construction Improved Plow Steel Grade Rope
7 X 7 Construction Galvanized Aircraft Grade Rope

Rope		Rated capacities, tons (2,000 lb.)											
		2-leg bridle slings						3-leg bridle slings					
		30°1(60°)2		45° angle		60°1(30°)2		30°1(60°)2		45° angle		60°1(30°)2	
Dia. (inches)	Constr.	8-Part	6-Part	8-Part	6-Part	8-Part	6-Part	8-Part	6-Part	8-Part	6-Part	8-Part	6-Part
3/32	6 X 7	0.74	0.55	0.60	0.45	0.42	0.32	1.1	0.83	0.90	0.68	0.64	0.48
1/8	6 X 7	1.3	0.98	1.1	0.80	0.76	0.57	2.0	1.5	1.6	1.2	1.1	0.85
3/16	6 X 7	2.9	2.2	2.4	1.8	1.7	1.3	4.4	3.3	3.6	2.7	2.5	1.9
3/32	7 X 7	0.89	0.67	0.72	0.55	0.51	0.39	1.3	1.0	1.1	0.82	0.77	0.58
1/8	7 X 7	1.6	1.2	1.3	1.0	0.95	0.71	2.5	1.8	2.0	1.5	1.4	1.1
3/16	7 X 7	3.6	2.7	2.9	2.2	2.1	1.5	5.4	4.0	4.4	3.3	3.1	2.3
3/16	6 X 19	3.0	2.2	2.4	1.8	1.7	1.3	4.5	3.4	3.7	2.8	2.6	1.9
1/4	6 X 19	5.3	4.0	4.3	3.2	3.1	2.3	8.0	6.0	6.5	4.9	4.6	3.4
5/16	6 X 19	8.3	6.2	6.7	5.0	4.8	3.6	12.0	9.3	10.0	7.6	7.1	5.4
3/8	6 X 19	12.0	8.9	9.7	7.2	6.8	5.1	18.0	13.0	14.0	11.0	10.0	7.7
7/16	6 X 19	16.0	12.0	13.0	9.8	9.3	6.9	24.0	18.0	20.0	15.0	14.0	10.0
1/2	6 X 19	21.0	15.0	17.0	13.0	12.0	9.0	31.0	23.0	25.0	19.0	18.0	13.0
9/16	6 X 19	26.0	20.0	21.0	16.0	15.0	11.0	39.0	29.0	32.0	24.0	23.0	17.0
5/8	6 X 19	32.0	24.0	26.0	20.0	19.0	14.0	48.0	36.0	40.0	30.0	28.0	21.0
3/4	6 X 19	46.0	35.0	38.0	28.0	27.0	20.0	69.0	52.0	56.0	42.0	40.0	30.0
7/8	6 X 19	62.0	47.0	51.0	38.0	36.0	27.0	94.0	70.0	76.0	57.0	54.0	40.0
1	6 X 19	81.0	61.0	66.0	50.0	47.0	35.0	22.0	91.0	99.0	74.0	70.0	53.0

[1] Vertical angles.
[2] Horizontal angles.

Table H·11 Rated Capacities for Strand Laid Grommet – Hand Tucked
Improved Plow Steel Grade Rope

Rope body		Rated capacities, tons (2,000 lb.)		
Dia. (inches)	Constr.	Vertical	Choker	Vertical basket[1]
1/4	7 X 19	0.85	0.64	1.7
5/16	7 X 19	1.3	1.0	2.6
3/8	7 X 19	1.9	1.4	3.8
7/16	7 X 19	2.6	1.9	5.2
1/2	7 X 19	3.3	2.5	6.7
9/16	7 X 19	4.2	3.1	8.4
5/8	7 X 19	5.2	3.9	10.0
3/4	7 X 19	7.4	5.6	15.0
7/8	7 X 19	10.0	7.5	20.0
1	7 X 19	13.0	9.7	26.0
1-1/8	7 X 19	16.0	12.0	32.0
1-1/4	7 X 37	18.0	14.0	37.0
1-3/8	7 X 37	22.0	16.0	44.0
1-1/2	7 X 37	26.0	19.0	52.0

[1]These values only apply when the D/d ratio is 5 or greater where:
D = Diameter of curvature around which rope is bent.
d = Diameter of rope body.

Table H 12 – Rated Capacities for Cable Laid Grommet – Hand Tucked
7 X 6 X 7 and 7 X 6 X 19 Constructions Improved Plow Steel Grade Rope
7 X 7 X 7 Construction Galvanized Aircraft Grade Rope

Cable body		Rated capacities, tons (2,000 lb.)		
Dia. (inches)	Constr.	Vertical	Choker	Vertical basket[1]
3/8	7 X 6 X 7	1.3	0.95	2.5
9/16	7 X 6 X 7	2.8	2.1	5.6
5/8	7 X 6 X 7	3.8	2.8	7.6
3/8	7 X 7 X 7	1.6	1.2	3.2
9/16	7 X 7 X 7	3.5	2.6	6.9
5/8	7 X 7 X 7	4.5	3.4	9.0
5/8	7 X 6 X 19	3.9	3.0	7.9
3/4	7 X 6 X 19	5.1	3.8	10.0
15/16	7 X 6 X 19	7.9	5.9	16.0
1-1/8	7 X 6 X 19	11.0	8.4	22.0
1-5/16	7 X 6 X 19	15.0	11.0	30.0
1-1/2	7 X 6 X 19	19.0	14.0	39.0
1-11/16	7 X 6 X 19	24.0	18.0	49.0
1-7/8	7 X 6 X 19	30.0	22.0	60.0
2-1/4	7 X 6 X 19	42.0	31.0	84.0
2-5/8	7 X 6 X 19	56.0	42.0	112.0

[1]These values only apply when the D/d ratio is 5 or greater where:
D = Diameter of curvature around which cable body is bent.
d = Diameter of cable body.

Table H·13 – Rated Capacities for Strand Laid Endless Slings – Mechanical Joint
Improved Plow Steel Grade Rope

Rope body		Rated capacities, tons (2,000 lb.)		
Dia. (inches)	Constr.	Vertical	Choker	Vertical basket[1]
1/4	[2]6 X 19	0.92	0.69	1.8
3/8	[2]6 X 19	2.0	1.5	4.1
1/2	[2]6 X 19	3.6	2.7	7.2
5/8	[2]6 X 19	5.6	4.2	11.0
3/4	[2]6 X 19	8.0	6.0	16.0
7/8	[2]6 X 19	11.0	8.1	21.0
1	[2]6 X 19	14.0	10.0	28.0
1-1/8	[2]6 X 19	18.0	13.0	35.0
1-1/4	[2]6 X 37	21.0	15.0	41.0
1-3/8	[2]6 X 37	25.0	19.0	50.0
1-1/2	[2]6 X 37	29.0	22.0	59.0

[1]These values only apply when the D/d ratio is 5 or greater where:
D = Diameter of curvature around which rope is bent.
d = Diameter of rope body.
[2]IWRC.

Table H·14 Rated Capacities for Cable Laid Endless Slings – Mechanical Joint
7 X 7 X 7 and 7 X 7 X 19 Constructions Galvanized Aircraft Grade Rope 7 X 6 X 19 IWRC Construction Improved Plow Steel Grade Rope

Cable body		Rated capacities, tons (2,000 lb.)		
Dia. (inches)	Constr.	Vertical	Choker	Vertical basket[1]
1/4	7 X 7 X 7	0.83	0.62	1.6
3/8	7 X 7 X 7	1.8	1.3	3.5
1/2	7 X 7 X 7	3.0	2.3	6.1
5/8	7 X 7 X 7	4.5	3.4	9.1
3/4	7 X 7 X 7	6.3	4.7	12.0
3/8	7 X 7 X 19	4.7	3.5	9.5
3/4	7 X 7 X 19	6.7	5.0	13.0
7/8	7 X 7 X 19	8.9	6.6	18.0
1	7 X 7 X 19	11.0	8.5	22.0
1-1/8	7 X 7 X 19	14.0	10.0	28.0
1-1/4	7 X 7 X19	17.0	12.0	33.0
3/4	[2]7 X 6 X 19	6.2	4.7	12.0
7/8	[2]7 X 6 X 19	8.3	6.2	16.0
1	[2]7 X 6 X 19	10.0	7.9	21.0
1-1/8	[2]7 X 6 X 19	13.0	9.7	26.0
1-1/4	[2]7 X 6 X 19	16.0	12.0	31.0
1-3/8	[2]7 X 6 X 19	18.0	14.0	37.0
1-1/2	[2]7 X 6 X 19	22.0	16.0	43.0

[1]These values only apply when the D/d value is 5 or greater where:
D = Diameter of curvature around which cable body is bent.
d = Diameter of cable body.
[2]/WRC.

Table H·15 — Manila Rope Slings
[Angle of rope to vertical shown in parentheses]

Rope diameter nominal in inches	Nominal weight per 100 ft. in pounds	Minimum breaking strength in pounds	Rated capacity in pounds (safety factor = 5)											
			Eye and eye sling						Endless sling					
			Vertical hitch	Choker hitch	Basket hitch; angle of rope to horizontal				Vertical hitch	Choker hitch	Basket hitch; angle of rope to horizontal			
					90° (0°)	60° (30°)	45° (45°)	30° (60°)			90° (0°)	60° (30°)	45° (45°)	30° (60°)
1/2	7.5	2,650	550	250	1,100	900	750	550	950	500	1,900	1,700	1,400	950
9/16	10.4	3,450	700	350	1,400	1,200	1,000	700	1,200	600	2,500	2,200	1,800	1,200
5/8	13.3	4,400	900	450	1,800	1,500	1,200	900	1,600	800	3,200	2,700	2,200	1,600
3/4	16.7	5,400	1,100	550	2,200	1,900	1,500	1,100	2,000	950	3,900	3,400	2,800	2,000
13/16	19.5	6,500	1,300	650	2,600	2,300	1,800	1,300	2,300	1,200	4,700	4,100	3,300	2,300
7/8	22.5	7,700	1,500	750	3,100	2,700	2,200	1,500	2,800	1,400	5,600	4,800	3,900	2,800
1	27.0	9,000	1,800	900	3,600	3,100	2,600	1,800	3,200	1,600	6,500	5,600	4,600	3,200
1-1/16	31.3	10,500	2,100	1,100	4,200	3,600	3,000	2,100	3,800	1,900	7,600	6,600	5,400	3,800
1-1/8	36.0	12,000	2,400	1,200	4,800	4,200	3,400	2,400	4,300	2,200	8,600	7,500	6,100	4,300
1-1/4	41.7	13,500	2,700	1,400	5,400	4,700	3,800	2,700	4,900	2,400	9,700	8,400	6,900	4,900
1-5/16	47.9	15,000	3,000	1,500	6,000	5,200	4,300	3,000	5,400	2,700	11,000	9,400	7,700	5,400
1-1/2	59.9	18,500	3,700	1,850	7,400	6,400	5,200	3,700	6,700	3,300	13,500	11,500	9,400	6,700
1-5/8	74.6	22,500	4,500	2,300	9,000	7,800	6,400	4,500	8,100	4,100	16,000	14,000	11,500	8,000
1-3/4	89.3	26,500	5,300	2,700	10,500	9,200	7,500	5,300	9,500	4,800	19,000	16,500	13,500	9,500
2	107.5	31,000	6,200	3,100	12,500	10,500	8,800	6,200	11,000	5,600	22,500	19,500	16,000	11,000
2-1/3	125.0	36,000	7,200	3,600	14,500	12,500	10,000	7,200	13,000	6,500	26,000	22,500	18,500	13,000
2-1/4	146.0	41,000	8,200	4,100	16,500	14,000	11,500	8,200	15,000	7,400	29,500	25,500	21,000	15,000
2-1/2	166.7	46.500	9,300	4,700	18,500	16,000	13,000	9,300	16,500	8,400	33,500	29,000	23,500	16,500
2-5/8	190.8	52,000	10,500	5,200	21,000	18,000	14,500	10,500	18,500	9,500	37,500	32,500	26,500	18,500

Table H·16 — Nylon Rope Slings
[Angle of rope to vertical shown in parentheses]

			Rated capacity in pounds (safety factor = 9)												
			Eye and eye sling						Endless sling						
Rope diameter nominal in inches	Nominal weight per 100 ft. in pounds	Minimum breaking strength in pounds	Vertical hitch	Choker hitch	Basket hitch; angle of rope to horizontal				Vertical hitch	Choker hitch	Basket hitch; angle of rope to horizontal			
					90° (0°)	60° (30°)	45° (45°)	30° (60°)			90° (0°)	60° (30°)	45° (45°)	30° (60°)
1/2	6.5	6,080	700	350	1,400	1,200	950	700	1,200	600	2,400	2,100	1,700	1,200
9/16	8.3	7,600	850	400	1,700	1,500	1,200	850	1,500	750	3,000	2,600	2,200	1,500
5/8	10.5	9,880	1,100	550	2,200	1,900	1,600	1,100	2,000	1,000	4,000	3,400	2,800	2,000
3/4	14.5	13,490	1,500	750	3,000	2,600	2,100	1,500	2,700	1,400	5,400	4,700	3,800	2,700
12/16	17.0	16,150	1,800	900	3,600	3,100	2,600	1,800	3,200	1,600	6,400	5,600	4,600	3,200
7/8	20.0	19,000	2,100	1,100	4,200	3,700	3,000	2,100	3,800	1,900	7,600	6,600	5,400	3,800
1	26.0	23,750	2,600	1,300	5,300	4,600	3,700	2,600	4,800	2,400	9,500	8,200	6,700	4,800
1-1/16	29.0	27,360	3,000	1,500	6,100	5,300	4,300	3,000	5,500	2,700	11,000	9,500	7,700	5,500
1-1/8	34.0	31,350	3,500	1,700	7,000	6,000	5,000	3,500	6,300	3,100	12,500	11,000	8,900	6,300
1-1/4	40.0	35,625	4,000	2,000	7,900	6,900	5,600	4,000	7,100	3,600	14,500	12,500	10,000	7,100
1-5/16	45.0	40,850	4,500	2,300	9,100	7,900	6,400	4,500	8,200	4,100	16,500	14,000	12,000	8,200
1-1/2	55.0	50,350	5,600	2,800	11,000	9,700	7,900	5,600	10,000	5,000	20,000	17,500	14,000	10,000
1-5/8	68.0	61,750	6,900	3,400	13,500	12,000	9,700	6,900	12,500	6,200	24,500	21,500	17,500	12,500
1-3/4	83.0	74,100	8,200	4,100	16,500	14,500	11,500	8,200	15,000	7,400	29,500	27,500	21,000	15,000
2	95.0	87,400	9,700	4,900	19,500	17,000	13,500	9,700	17,500	8,700	35,000	30,500	24,500	17,500
2-1/8	109.0	100,700	11,000	5,600	22,500	19,500	16,000	11,000	20,000	10,000	40,500	35,000	28,500	20,000
2-1/4	129.0	118.750	13,000	6,600	26,500	23,000	18,500	13,000	24,000	12,000	47,500	41,000	33,500	24,000
2-1/2	149.0	133,000	15,000	7,400	29,500	25,500	21,000	15,000	26,500	13,500	53,000	46,000	37,500	26,500
2-5/8	168.0	153.900	17,100	8,600	34,000	29,500	24,000	17,000	31,000	15,500	61,500	53,500	43,500	31,000

Table H·17 — Polyester Rope Slings
[Angle of rope to vertical shown in parentheses]

Rope diameter nominal in inches	Nominal weight per 100 ft. in pounds	Minimum breaking strength in pounds	Eye and eye sling						Endless sling					
			Vertical hitch	Choker hitch	Basket hitch; angle of rope to horizontal				Vertical hitch	Choker hitch	Basket hitch; angle of rope to horizontal			
					90° (0°)	60° (30°)	45° (45°)	30° (60°)			90° (0°)	60° (30°)	45° (45°)	30° (60°)
1/2	8.0	6,080	700	350	1,400	1,200	950	700	1,200	600	2,400	2,100	1,700	1,200
9/16	10.2	7,600	850	400	1,700	1,500	1,200	850	1,500	750	3,000	2,600	2,200	1,500
5/8	13.0	9,500	1,100	550	2,100	1,800	1,500	1,100	1,900	950	3,800	3,300	2,700	1,900
3/4	17.5	11,875	1,300	650	2,600	2,300	1,900	1,300	2,400	1,200	4,800	4,100	3,400	2,400
13/16	21.0	14,725	1,600	800	3,300	2,800	2,300	1,600	2,900	1,500	5,900	5,100	4,200	2,900
7/8	25.0	17,100	1,900	950	3,800	3,300	2,700	1,900	3,400	1,700	6,800	5,900	4,800	3,400
1	30.5	20.900	2,300	1,200	4,600	4,000	3,300	2,300	4,200	2,100	8,400	7,200	5,900	4,200
1-1/16	34.5	24,225	2,700	1,300	5,400	4,700	3,800	2,700	4,800	2,400	9,700	8,400	6,900	4,800
1-1/8	40.0	28,025	3,100	1,600	6,200	5,400	4,400	3,100	5,600	2,800	11,000	9,700	7,900	5,600
1-1/4	46.3	31,540	3,500	1,800	7,000	6,100	5,000	3,500	6,300	3,200	12,500	11,000	8,900	6,300
1-5/16	52.5	35,625	4,000	2,000	7,900	6,900	5,600	4,000	7,100	3,600	14,500	12,500	10,000	7,100
1-1/2	66.8	44,460	4,900	2,500	9,900	8,600	7,000	4,900	8,900	4,400	18,000	15,500	12,500	8,900
1-5/8	82.0	54,150	6,000	3,000	12,000	10,400	8,500	6,000	11,000	5,400	21,500	19,000	15,500	11,000
1-3/4	98.0	64,410	7,200	3,600	14,500	12,500	10,000	7,200	13,000	6,400	26,000	22,500	18,000	13,000
2	118.0	76,000	8,400	4,200	17,000	14,500	12,000	8,400	15,000	7,600	30,500	26,500	21,500	15,000
2-1/8	135.0	87,400	9,700	4,900	19,500	17,000	13,500	9,700	17,500	8,700	35,000	30,500	24,500	17,500
2-1/4	157.0	101,650	11,500	5,700	22,500	19,500	16,000	11,500	20,500	10,000	40,500	35,000	29,000	20,500
2-1/2	181.0	115,900	13,000	6,400	26,000	22,500	18,000	13,000	23,000	11,500	46,500	40,000	33,000	23,000
2-5/8	205.0	130,150	14,500	7,200	29,000	25,000	20,500	14,500	26,000	13,000	52,000	45,000	37,000	26,000

Rated capacity in pounds (safety factor = 9)

Table H-18 — Polypropylene Rope Slings
[Angle of rope to vertical shown in parentheses]

			Rated capacity in pounds (safety factor = 6)											
			Eye and eye sling						Endless sling					
Rope diameter nominal in inches	Nominal weight per 100 ft. in pounds	Minimum breaking strength in pounds	Vertical hitch	Choker hitch	Basket hitch; angle of rope to horizontal				Vertical hitch	Choker hitch	Basket hitch; angle of rope to horizontal			
					90° (0°)	60° (30°)	45° (45°)	30° (60°)			90° (0°)	60° (30°)	45° (45°)	30° (60°)
1/2	4.7	3,990	650	350	1,300	12,00	950	650	1,200	600	2,400	2,100	1,700	1,200
9/16	6.1	4,845	800	400	1,600	1,400	1,100	800	1,500	750	2,900	2,500	2,100	1,500
5/8	7.5	5,890	1,000	500	2,000	1,700	1,400	1,000	1,800	900	3,500	3,100	2,500	1,800
3/4	10.7	8,075	1,300	700	2,700	2,300	1,900	1,300	2,400	1,200	4,900	4,200	3,400	2,400
13/16	12.7	9,405	1,600	800	3,100	2,700	2,200	1,600	2,800	1,400	5,600	4,900	4,000	2,800
7/8	15.0	10,925	1,800	900	3,600	3,200	2,600	1,800	3,300	1,600	6,600	5,700	4,600	3,300
1	18.0	13,300	2,200	1,100	4,400	3,800	3,100	2,200	4,000	2,000	8,000	6,900	5,600	4,000
1-1/6	20.4	15,200	2,500	1,300	5,100	4,400	3,600	2,500	4,600	2,300	9,100	7,900	6,500	4,600
1-1/8	23.7	17,385	2,900	1,500	5,800	5,000	4,100	2,900	5,200	2,600	10,500	9,000	7,400	5,200
1-1/4	27.0	19,950	3,300	1,700	6,700	5,800	4,700	3,300	6,000	3,000	12,000	10,500	8,500	6,000
1-5/16	30.5	22,325	3,700	1,900	7,400	6,400	5,300	3,700	6,700	3,400	13,500	11,500	9,500	6,700
1-1/2	38.5	28,215	4,700	2,400	9,400	8,100	6,700	4,700	8,500	4,200	17,000	14,500	12,000	8,500
1-5/8	47.5	34,200	5,700	2,900	11,500	9,900	8,100	5,700	10,500	5,100	20,500	18,000	14,500	10,500
1-3/4	57.0	40,850	6,800	3,400	13,500	12,000	9,600	6,800	12,500	6,100	24,500	21,000	17,500	12,500
2	69.0	49,400	8,200	4,100	16,500	14,500	11,500	8,200	15,000	7,400	29,500	25,500	21,000	15,000
2-1/8	80.0	57,950	9,700	4,800	19,500	16,500	13,500	9,700	17,500	8,700	35,000	30,100	24,500	17,500
2-1/4	92.0	65,550	11,000	5,500	22,000	19,000	15,500	11,000	19,500	9,900	39,500	34,000	28,000	19,500
2-1/2	107.0	76,000	12,500	6,300	25,500	22,000	18,000	12,500	23,000	11,500	45,500	39,500	32,500	23,000
2-5/8	120.0	85,500	14,500	7,100	28,500	24,500	20,000	14,500	25,500	13,000	51,500	44,500	36,500	25,500

tails shall project at least six rope diameters beyond the last full tuck. For fiber ropes 1-inch diameter and larger, the tails shall project at least 6 inches beyond the last full tuck. In applications where the projecting tails may be objectionable, the tails shall be tapered and spliced into the body of the rope using at least two additional tucks (which will require a tail length of approximately six rope diameters beyond the last full tuck).

(iv) For all eye splices, the eye shall be sufficiently large to provide an included angle of not greater than 60° at the splice when the eye is placed over the load or support.

(v) Knots shall not be used in lieu of splices.

(e) *Synthetic webbing (nylon, polyester, and polypropylene).* (1) The employer shall have each synthetic web sling marked or coded to show:

(i) Name or trademark of manufacturer.

(ii) Rated capacities for the type of hitch.

(iii) Type of material.

(2) Rated capacity shall not be exceeded.

(f) *Shackles and hooks.* (1) Table H-19 shall be used to determine the safe working loads of various sizes of shackles, except that higher safe working loads are permissable when recommended by the manufacturer for specific, identifiable products, provided that a safety factor of not less than 5 is maintained.

(2) The manufacturer's recommendations shall be followed in determining the safe working loads of the various sizes and types of specific and identifiable hooks. All hooks for which no applicable manufacturer's recommendations are available shall be tested to twice the intended safe working load before they are initially put into use. The employer shall maintain a record of the dates and results of such tests.

Table H-19 – Safe Working Loads For Shackles
[In tons of 2,000 pounds]

Material size (inches)	Pin diameter (inches)	Sae working load
1/2	5/8	1.4
5/8	3/4	2.2
3/4	1	3.2
7/8	1	4.3
1	1-1/8	5.6
1-1/8	1-1/4	6.7
1-1/4	1-3/8	8.2
1-3/8	1-1/2	10.0
1-1/2	1-5/8	11.9
1-3/4	2	16.2
2	2-1/4	21.2

Table H-20 – Number and Spacing of U-Bolt Wire Rope Clips

Improved plow steel, rope diameter (inches)	Number of clips		Minimum spacing (inches)
	Drop forged	Other material	
1/2	3	4	3
5/8	3	4	3-3/4
3/4	4	5	4-1/2
7/8	4	5	5-1/4
1	5	6	6
1-1/8	6	6	6-3/4
1-1/4	6	7	7-1/2
1-3/8	7	7	8-1/4
1-1/2	7	8	9

Disposal of waste materials.

(a) Whenever materials are dropped more than 20 feet to any point lying outside the exterior walls of the building, an enclosed chute of wood, or equivalent material, shall be used. For the purpose of this paragraph, an enclosed chute is a slide, closed in on all sides, through which material is moved from a high place to a lower one.

(b) When debris is dropped through holes in the floor without the use of chutes, the area onto which the material is dropped shall be completely enclosed with barricades not less than 42 inches high and not less than 6 feet back from the projected edge of the opening above. Signs warning of the hazard of falling materials shall be posted at each level. Removal shall not be permitted in this lower area until debris handling ceases above.

(c) All scrap lumber, waste material, and rubbish shall be removed from the immediate work area as the work progresses.

(d) Disposal of waste material or debris by burning shall comply with local fire regulations.

(e) All solvent waste, oily rags, and flammable liquids shall be kept in fire resistant covered containers until removed from worksite.

Tools — Hand and Power

General requirements.

(a) *Condition of tools.* All hand and power tools and similar equipment, whether furnished by the employer or the employee, shall be maintained in a safe condition.

(b) *Guarding.* (1) When power operated tools are designed to accommodate guards, they shall be equipped with such guards when in use.

(2) Belts, gears, shafts, pulleys, sprockets, spindles, drums, fly wheels, chains, or other reciprocating, rotating or moving parts of equipment shall be guarded if such parts are exposed to contact by employees or otherwise create a hazard. Guarding shall meet the requirements as set forth in American National Standards Institute, B15.1-1953 (R1958), Safety Code for Mechanical Power-Transmission Apparatus.

(c) *Personal protective equipment.* Employees using hand and power tools and exposed to the hazard of falling, flying,

478

abrasive, and splashing objects, or exposed to harmful dusts, fumes, mists, vapors, or gases shall be provided with the particular personal protective equipment necessary to protect them from the hazard.

(d) *Switches.* (1) All hand-held powered platen sanders, grinders with wheels 2-inch diameter or less, routers, planers, laminate trimmers, nibblers, shears, scroll saws, and jigsaws with blade shanks one-fourth of an inch wide or less may be equipped with only a positive "on-off" control.

(2) All hand-held powered drills, tappers, fastener drivers, horizontal, vertical, and angle grinders with wheels greater than 2 inches in diameter, disc sanders, belt sanders, reciprocating saws, saber saws, and other similar operating powered tools shall be equipped with a momentary contact "on-off" control and may have a lock-on control provided that turnoff can be accomplished by a single motion of the same finger or fingers that turn it on.

(3) All other hand-held powered tools, such as circular saws, chain saws, and percussion tools without positive accessory holding means, shall be equipped with a constant pressure switch that will shut off the power when the pressure is released.

(4) The requirements of this paragraph shall become effective on July 15, 1972.

(5) Exception: This paragraph does not apply to concrete vibrators, concrete breakers, powered tampers, jack hammers, rock drills, and similar hand operated power tools.

Hand tools.

(a) Employers shall not issue or permit the use of unsafe hand tools.

(b) Wrenches, including adjustable, pipe, end, and socket wrenches shall not be used when jaws are sprung to the point that slippage occurs.

(c) Impact tools, such as drift pins, wedges, and chisels, shall be kept free of mushroomed heads.

(d) The wooden handles of tools shall be kept free of splinters or cracks and shall be kept tight in the tool.

Power-operated hand tools.

(a) *Electric power-operated tools.* (1) Electric power operated tools shall be of the approved double-insulated type.

(2) The use of electric cords for hoisting or lowering tools shall not be permitted.

(b) *Pneumatic power tools.* (1) Pneumatic power tools shall be secured to the hose or whip by some positive means to prevent the tool from becoming accidentally disconnected.

(2) Safety clips or retainers shall be securely installed and maintained on pneumatic impact (percussion) tools to prevent attachments from being accidentally expelled.

(3) All pneumatically driven nailers, staplers, and other similar equipment provided with automatic fastener feed, which operate at more than 100 p.s.i. pressure at the tool shall have a safety device on the muzzle to prevent the tool from ejecting fasteners, unless the muzzle is in contact with the work surface.

(4) Compressed air shall not be used for cleaning purposes except where reduced to less than 30 p.s.i. and then only with effective chip guarding and personal protective equipment. The 30 p.s.i. requirement does not apply for concrete form, mill scale and similar cleaning purposes.

(5) The manufacturer's safe operating pressure for hoses, pipes, valves, filters, and other fittings shall not be exceeded.

(6) The use of hoses for hoisting or lowering tools shall not be permitted.

(7) All hoses exceeding ½-inch inside diameter shall have a safety device at the source of supply or branch line to reduce pressure in case of hose failure.

(8) Airless spray guns of the type which atomize paints and fluids at high pressures (1,000 pounds or more per square inch) shall be equipped with automatic or visible manual safety devices which will prevent pulling of the trigger to prevent release of the paint or fluid until the safety device is manually released.

(9) In lieu of the above, a diffuser nut which will prevent high pressure, high velocity release, while the nozzle tip is removed, plus a nozzle tip guard which will prevent the top from coming into contact with the operator, or other equivalent protection, shall be provided.

(c) *Fuel powered tools.* (1) All fuel powered tools shall be stopped while being refueled, serviced, or maintained.

(d) *Hydraulic power tools.* (1) The fluid is used in hydraulic powered tools shall be fire-resistant fluids approved under Schedule 30 of the U.S. Bureau of Mines, Department of the Interior, and shall retain its operating characteristics at the most extreme temperatures to which it will be exposed.

(2) The manufacturer's safe operating pressures for hoses, valves, pipes, filters, and other fittings shall not be exceeded.

(e) *Powder-actuated tools.* (1) Only employees who have been trained in the operation of the particular tool in use shall be allowed to operate a powder-actuated tool.

(2) The tool shall be tested each day before loading to see that safety devices are in proper working condition. The method of testing shall be in accordance with the manufacturer's recommended procedure.

(3) Any tool found not in proper working order, or that develops a defect during use, shall be immediately removed from service and not used until properly repaired.

(5) Tools shall not be loaded until just prior to the intended firing time. Neither loaded nor empty tools are to be pointed at any employees. Hands shall be kept clear of the open barrel end.

(6) Loaded tools shall not be left unattended.

(7) Fasteners shall not be driven into very hard or brittle materials including, but not limited to, cast iron, glazed tile, surface-hardened steel, glass block, live rock, face brick, or hollow tile.

(8) Driving into materials easily penetrated shall be avoided unless such materials are backed by a substance that will prevent the pin or fastener from passing completely through and creating a flying missile hazard on the other side.

(9) No fastener shall be driven into a spalled area caused by an unsatisfactory fastening.

(10) Tools shall not be used in an explosive or flammable atmosphere.

(11) All tools shall be used with the correct shield, guard, or attachment recommended by the manufacturer.

Abrasive wheels and tools.

(a) *Power.* All grinding machines shall be supplied with sufficient power to maintain the spindle speed at safe levels under all conditions of normal operation.

(b) *Guarding.* Grinding machines shall be equipped with safety guards in conformance with the requirements of American National Standards Institute, B7.1-1970, Safety Code for the Use, Care and Protection of Abrasive Wheels and paragraph (d) of this section.

(c) *Use of abrasive wheels.* (1) Floor stand and bench mounted abrasive wheels, used for external grinding, shall be provided with safety guards (protection hoods). The maximum angular exposure of the grinding wheel periphery and sides shall be not more than 90°, except that when work requires contact with the wheel below the horizontal plane of the spindle, the angular exposure shall not exceed 125°. In either case, the exposure shall begin not more than 65° above the horizontal plane of the spindle. Safety guards shall be strong enough to withstand the effect of a bursting wheel.

(2) Floor and bench-mounted grinders shall be provided with work rests which are rigidly supported and readily adjustable. Such work rests shall be kept at a distance not to exceed one-eighth inch from the surface of the wheel.

(3) Cup types wheels used for external grinding shall be protected by either a revolving cup guard or a band type guard in accordance with the provisions of the American National Standards Institute, B7.1-1970 Safety Code for the Use, Care, and Protection of Abrasive Wheels. All other portable abrasive wheels used for external grinding, shall be provided with safety guards (protection hoods) meeting the requirements of paragraph (c)(5) of this section, except as follows:

(i) When the work location makes it impossible, a wheel equipped with safety flanges, as described in paragraph (c)(6) of this section, shall be used;

(ii) When wheels 2 inches or less in diameter which are securely mounted on the end of a steel mandrel are used.

(4) Portable abrasive wheels used for internal grinding shall be provided with safety flanges (protection flanges) meeting the requirements of paragraph (c)(6) of this section, except as follows:

(i) When wheels 2 inches or less in diameter which are securely mounted on the end of a steel mandrel are used;

(ii) If the wheel is entirely within the work being ground while in use.

(5) When safety guards are required, they shall be so mounted as to maintain proper alignment with the wheel, and the guard and its fastenings shall be of sufficient strength to retain fragments of the wheel in case of accidental breakage. The maximum angular exposure of the grinding wheel periphery and sides shall not exceed 180°.

(6) When safety flanges are required, they shall be used only with wheels designed to fit the flanges. Only safety flanges, of a type and design and properly assembled so as to ensure that the pieces of the wheel will be retained in case of accidental breakage, shall be used.

(7) All abrasive wheels shall be closely inspected and ring-tested before mounting to ensure that they are free from cracks or defects.

(8) Grinding wheels shall fit freely on the spindle and shall not be forced on. The spindle nut shall be tightened only enough to hold the wheel in place.

(9) All employees using abrasive wheels shall be protected by eye protection equipment, except when adequate eye protection is afforded by eye shields which are permanently attached to the bench or floor stand.

(d) *Other requirements.* All abrasive wheels and tools used by employees shall meet other applicable requirements of American National Standards Institute, B7.1-1970, Safety Code for the Use, Care and Protection of Abrasive Wheels.

Woodworking tools.

(a) *Disconnect switches.* All fixed power driven woodworking tools shall be provided with a disconnect switch that can either be locked or tagged in the off position.

(b) *Speeds.* The operating speed shall be etched or otherwise permanently marked on all circular saws over 20 inches in diameter or operating at over 10,000 peripheral feet per minute. Any saw so marked shall not be operated at a speed other than that marked on the blade. When a marked saw is retensioned for a different speed, the marking shall be corrected to show the new speed.

(c) *Self-feed.* Automatic feeding devices shall be installed on machines whenever the nature of the work will permit. Feeder attachments shall have the feed rolls or other moving parts covered or guarded so as to protect the operator from hazardous points.

(d) *Guarding.* All portable, power-driven circular saws shall be equipped with guards above and below the base plate or shoe. The upper guard shall cover the saw to the depth of the teeth, except for the minimum arc required to permit the base to be tilted for bevel cuts. The lower guard shall cover the saw to the depth of the teeth, except for the minimum arc required to allow proper retraction and contact with the work. When the tool is withdrawn from the work, the lower guard shall automatically and instantly return to the covering position.

(f) *Other requirements.* All woodworking tools and machinery shall meet other applicable requirements of American National Standards Institute, 01.1-1961, Safety Code for Woodworking Machinery.

Jacks — lever and ratchet, screw, and hydraulic.

(a) *General requirements.* (1) the manufacturer's rated capacity shall be legibly marked on all jacks and shall not be exceeded.

(2) All jacks shall have a positive stop to prevent overtravel.

(b) *Lift slab construction.* (1) Hydraulic jacks used in lift slab construction shall have a safety device which will cause the jacks to support the load in any position in the event the jack malfunctions.

(2) If lifts slabs are automatically controlled, a device shall be installed which will stop the operation when the ½-inch leveling tolerance is exceeded.

(c) *Blocking.* When it is necessary to provide a firm foundation, the base of the jack shall be blocked or cribbed. Where there is a possibility of slippage of the metal cap of the jack, a wood block shall be placed between the cap and the load.

Welding and Cutting

Gas welding and cutting.

(a) *Transporting, moving, and storing compressed gas cylinders.*

(1) Valve protection caps shall be in place and secured.

(2) When cylinders are hoisted, they shall be secured on a cradle, slingboard, or pallet. They shall not be hoisted or transported by means of magnets or choker slings.

(3) Cylinders shall be moved by tilting and rolling them on their bottom edges. They shall not be intentionally dropped, struck, or permitted to strike each other violently.

(4) When cylinders are transported by powered vehicles, they shall be secured in a vertical position.

(5) Valve protection caps shall not be used for lifting cylinders from one vertical position to another. Bars shall not be used under valves or valve protection caps to pry cylinders loose when frozen. Warm, not boiling, water shall be used to thaw cylinders loose.

(6) Unless cylinders are firmly secured on a special carrier intended for this purpose, regulators shall be removed and valve protection caps put in place before cylinders are moved.

(7) A suitable cylinder truck, chain, or other steadying device shall be used to keep cylinders from being knocked over while in use.

(8) When work is finished, when cylinders are empty, or when cylinders are moved at any time, the cylinder valve shall be closed.

(9) Compressed gas cylinders shall be secured in an upright position at all times except, if necessary, for short periods of time while cylinders are actually being hoisted or carried.

(b) *Placing cylinders.* (1) Cylinders shall be kept far enough away from the actual welding or cutting operation so that sparks, hot slag, or flame will not reach them. When this is impractical, fire resistant shields shall be provided.

(2) Cylinders shall be placed where they cannot become part of an electrical circuit. Electrodes shall not be struck against a cylinder to strike an arc.

(3) Fuel gas cylinders shall be placed with valve end up whenever they are in use. They shall not be placed in a location where they would be subject to open flame, hot metal, or other sources of artificial heat.

(4) Cylinders containing oxygen or acetylene or other fuel gas shall not be taken into confined spaces.

(c) *Treatment of cylinders.* (1) Cylinders, whether full or empty, shall not be used as rollers or supports.

(2) No person other than the gas supplier shall attempt to mix gases in a cylinder. No one except the owner of the cylinder or person authorized by him, shall refill a cylinder. No one shall use a cylinder's contents for purposes other than those intended by the supplier. All cylinders used shall meet the Department of Transportation requirements published in 49 CFR Part 178, Subpart C, Specification for Cylinders.

(3) No damaged or defective cylinder shall be used.

(d) *Use of fuel gas.* The employer shall thoroughly instruct employees in the safe use of fuel gas, as follows:

(1) Before a regulator to a cylinder valve is connected, the valve shall be opened slightly and closed immediately. (This action is generally termed "cracking" and is intended to clear the valve of dust or dirt that might otherwise enter the regulator.) The person cracking the valve shall stand to one side of the outlet, not in front of it. The valve of a fuel gas cylinder shall not be cracked where the gas would reach welding work, sparks, flame, or other possible sources of ignition.

(2) The cylinder valve shall always be opened slowly to prevent damage to the regulator. For quick closing, valves on fuel gas cylinders shall not be opened more than 1½ turns. When a special wrench is required, it shall be left in position on the stem of the valve while the cylinder is in use so that the fuel gas flow can be shut off quickly in case of an emergency. In the case of manifolded or coupled cylinders, at least one such wrench shall always be available for immediate use. Nothing shall be placed on top of a fuel gas cylinder, when in use, which may damage the safety device or interfere with the quick closing of the valve.

(3) Fuel gas shall not be used from cylinders through torches or other devices which are equipped with shutoff valves without reducing the pressure through a suitable regulator attached to the cylinder valve or manifold.

(4) Before a regulator is removed from a cylinder valve, the cylinder valve shall always be closed and the gas released from the regulator.

(5) If, when the valve on a fuel gas cylinder is opened, there is found to be a leak around the valve stem, the valve shall be closed and the gland nut tightened. If this action does not stop the leak, the use of the cylinder shall be discontinued, and it shall be properly tagged and removed from the work area. In the event that fuel gas should leak from the cylinder valve, rather than from the valve stem, and the gas cannot be shut off, the cylinder shall be properly tagged and removed from the work area. If a regulator attached to a cylinder valve will effectively stop a leak through the valve seat, the cylinder need not be removed from the work area.

(6) If a leak should develop at a fuse plug or other safety device, the cylinder shall be removed from the work area.

(e) *Fuel gas and oxygen manifolds.* (1) Fuel gas and oxygen manifolds shall bear the name of the substance they contain in letters at least 1-inch high which shall be either painted on the manifold or on a sign permanently attached to it.

(2) Fuel gas and oxygen manifolds shall be placed in safe, well ventilated, and accessible locations. They shall not be located within enclosed spaces.

(3) Manifold hose connections, including both ends of the supply hose that lead to the manifold, shall be such that the hose cannot be interchanged between fuel gas and oxygen manifolds and supply header connections. Adapters shall not be used to permit the interchange of hose. Hose connections shall be kept free of grease and oil.

(4) When not in use, manifold and header hose connections shall be capped.

(5) Nothing shall be placed on top of a manifold, when in use, which will damage the manifold or interfere with the quick closing of the valves.

(f) *Hose.* (1) Fuel gas hose and oxygen hose shall be easily distinguishable from each other. The contrast may be made by different colors or by surface characteristics readily distinguishable by the sense of touch. Oxygen and fuel gas hoses shall not be interchangeable. A single hose having more than one gas passage shall not be used.

481

(2) When parallel sections of oxygen and fuel gas hose are taped together, not more than 4 inches out of 12 inches shall be covered by tape.

(3) All hose in use, carrying acetylene, oxygen, natural or manufactured fuel gas, or any gas or substance which may ignite or enter into combustion, or be in any way harmful to employees, shall be inspected at the beginning of each working shift. Defective hose shall be removed from service.

(4) Hose which has been subject to flashback, or which shows evidence of severe wear or damage, shall be tested to twice the normal pressure to which it is subject, but in no case less than 300 p.s.i. Defective hose, or hose in doubtful condition, shall not be used.

(5) Hose couplings shall be of the type that cannot be unlocked or disconnected by means of a straight pull without rotary motion.

(6) Boxes used for the storage of gas hose shall be ventilated.

(7) Hoses, cables, and other equipment shall be kept clear of passageways, ladders and stairs.

(g) *Torches.* (1) Clogged torch tip openings shall be cleaned with suitable cleaning wires, drills, or other devices designed for such purpose.

(2) Torches in use shall be inspected at the beginning of each working shift for leaking shutoff valves, hose couplings, and tip connections. Defective torches shall not be used.

(3) Torches shall be lighted by friction lighters or other approved devices, and not by matches or from hot work.

(h) *Regulators and gauges.* Oxygen and fuel gas pressure regulators, including their related gauges, shall be in proper working order while in use.

(i) *Oil and grease hazards.* Oxygen cylinders and fittings shall be kept away from oil or grease. Cylinders, cylinder caps and valves, couplings, regulators, hose, and apparatus shall be kept free from oil or greasy substances and shall not be handled with oily hands or gloves. Oxygen shall not be directed at oily surfaces, greasy clothes, or within a fuel oil or other storage tank or vessel.

(j) *Additional rules.* For additional details not covered in this subpart, applicable technical portions of American National Standards Institute, Z49.1-1967, Safety in Welding and Cutting, shall apply.

Arc welding and cutting.

(a) *Manual electrode holders.* (1) Only manual electrode holders which are specifically designed for arc welding and cutting, and are of a capacity capable of safely handling the maximum rated current required by the electrodes, shall be used.

(2) Any current-carrying parts passing through the portion of the holder which the arc welder or cutter grips in his hand, and the outer surfaces of the jaws of the holder, shall be fully insulated against the maximum voltage encountered to ground.

(b) *Welding cables and connectors.* (1) All arc welding and cutting cables shall be of the completely insulated, flexible type capable of handling the maximum current requirements of the work in progress, taking into account the duty cycle under which the arc welder or cutter is working.

(2) Only cable free from repair or splices for a minimum distance of 10 feet from the cable end to which the electrode holder is connected shall be used, except that cables with standard insulated connectors or with splices whose insulating quality is equal to that of the cable are permitted.

(3) When it becomes necessary to connect or splice lengths of cable one to another, substantial insulated connectors of a capacity at least equivalent to that of the cable shall be used. If connections are effected by means of cable lugs, they shall be securely fastened together to give good electrical contact, and the exposed metal parts of the lugs shall be completely insulated.

(4) Cables in need of repair shall not be used. When a cable, other than the cable lead referred to in paragraph (b)(2) of this section, becomes worn to the extent of exposing bare conductors, the portion thus exposed shall be protected by means of rubber and friction tape or other equivalent insulation.

(c) *Ground returns and machine grounding.* (1) A ground return cable shall have a safe current carrying capacity equal to or exceeding the specified maximum output capacity of the arc welding or cutting unit which it services. When a single ground return cable services more than one unit, its safe current-carrying capacity shall equal or exceed the total specified maximum output capacities of all the units which it services.

(2) Pipelines containing gases or flammable liquids, or conduits containing electrical circuits, shall not be used as a ground return. For welding on natural gas pipelines, the technical portions of regulations issued by the Department of Transportation, Office of Pipeline Safety, 49 CFR Part 192, Minimum Federal Safety Standards for Gas Pipelines, shall apply.

(3) When a structure or pipeline is employed as a ground return circuit, it shall be determined that the required electrical contact exists at all joints. The generation of an arc, sparks, or heat at any point shall cause rejection of the structures as a ground circuit.

(4) When a structure or pipeline is continuously employed as a ground return circuit, all joints shall be bonded, and periodic inspections shall be conducted to ensure that no condition of electrolysis or fire hazard exists by virtue of such use.

(5) The frames of all arc welding and cutting machines shall be grounded either through a third wire in the cable containing the circuit conductor or through a separate wire which is grounded at the source of the current. Grounding circuits, other than by means of the structure, shall be checked to ensure that the circuit between the ground and the grounded power conductor has resistance low enough to permit sufficient current to flow to cause the fuse or circuit breaker to interrupt the current.

(6) All ground connections shall be inspected to ensure that they are mechanically strong and electrically adequate for the required current.

(d) *Operating instructions.* Employers shall instruct employees in the safe means of arc welding and cutting as follows:

(1) When electrode holders are to be left unattended, the electrodes shall be removed and the holders shall be so placed or protected that they cannot make electrical contact with employees or conducting objects.

(2) Hot electrode holders shall not be dipped in water; to do so may expose the arc welder or cutter to electric shock.

(3) When the arc welder or cutter has occasion to leave his work or to stop work for any appreciable length of time, or when the arc welding or cutting machine is to be moved, the

power supply switch to the equipment shall be opened.

(4) Any faulty or defective equipment shall be reported to the supervisor.

(5) Other requirements, as outlined in Article 630, National Electrical Code, NFPA 70-1971; ANSI C1-1971 (Rev. of 1968), Electric Welders, shall be used when applicable.

(e) *Shielding.* Whenever practicable, all arc welding and cutting operations shall be shielded by noncombustible or flameproof screens which will protect employees and other persons working in the vicinity from the direct rays of the arc.

Fire prevention.

(a) When practical, objects to be welded, cut, or heated shall be moved to a designated safe location or, if the objects to be welded, cut, or heated, cannot be readily moved, all movable fire hazards in the vicinity shall be taken to a safe place, or otherwise protected.

(b) If the object to be welded, cut, or heated cannot be moved and if all the fire hazards cannot be removed, positive means shall be taken to confine the heat, sparks, and slag, and to protect the immovable fire hazards from them.

(c) No welding, cutting, or heating shall be done where the application of flammable paints, or the presence of other flammable compounds, or heavy dust concentrations creates a hazard.

(d) Suitable fire extinguishing equipment shall be immediately available in the work area and shall be maintained in a state of readiness for instant use.

(e) When the welding, cutting, or heating operation is such that normal fire prevention precautions are not sufficient, additional personnel shall be assigned to guard against fire while the actual welding, cutting, or heating operation is being performed, and for a sufficient period of time after completion of the work to ensure that no possibility of fire exists. Such personnel shall be instructed as to the specific anticipated fire hazards and how the firefighting equipment provided is to be used.

(f) When welding, cutting, or heating is performed on walls, floors, and ceilings, since direct penetration of sparks or heat transfer may introduce a fire hazard to an adjacent area, the same precautions shall be taken on the opposite side as are taken on the side on which the welding is being performed.

(g) For the elimination of possible fire in enclosed spaces as a result of gas escaping through leaking or improperly closed torch valves, the gas supply to the torch shall be positively shut off at some point outside the enclosed space whenever the torch is not to be used or whenever the torch is left unattended for a substantial period of time, such as during the lunch period. Overnight and at the change of shifts, the torch and hose shall be removed from the confined space. Open end fuel gas and oxygen hoses shall be immediately removed from enclosed spaces when they are disconnected from the torch or other gas-consuming device.

(h) Except when the contents are being removed or transferred, drums, pails, and other containers which contain or have contained flammable liquids shall be kept closed. Empty containers shall be removed to a safe area apart from hot work operations or open flames.

(i) Drums, containers, or hollow structures which have contained toxic or flammable substances shall, before welding, cut-

ting, or heating is undertaken on them, either be filled with water or thoroughly cleaned of such substances and ventilated and tested. For welding, cutting and heating on steel pipelines containing natural gas, the pertinent portions of regulations issued by the Department of Transportation, Office of Pipeline Safety, 49 CFR Part 192, Minimum Federal Safety Standards for Gas Pipelines, shall apply.

(j) Before heat is applied to a drum, container, or hollow structure, a vent or opening shall be provided for the release of any built-up pressure during the application of heat.

Ladders and Scaffolding

Ladders.

(a) *General requirements.* (1) Except where either permanent or temporary stairways or suitable ramps or runways are provided, ladders described in this subpart shall be used to give safe access to all elevations.

(2) The use of ladders with broken or missing rungs or steps, broken or split side rails, or other faulty or defective construction is prohibited. When ladders with such defects are discovered, they shall be immediately withdrawn from service. Inspection of metal ladders shall include checking for corrosion of interiors of open end hollow rungs.

(3) Manufactured portable wood ladders provided by the employer shall be in accordance with the provisions of the American National Standards Institute, A 14.1-1968, Safety Code for Portable Wood Ladders.

(4) Portable metal ladders shall be of strength equivalent to that of wood ladders. Manufactured portable metal ladders provided by the employer shall be in accordance with the provisions of the American National Standards Institute, A 14.2-1956, Safety Code for Portable Metal Ladders.

(5) Fixed ladders shall be in accordance with the provisions of the American National Standards Institute, A 14.3-1956, Safety Code for Fixed Ladders.

(6) Portable ladder feet shall be placed on a substantial base, and the area around the top and bottom of the ladder shall be kept clear.

(7) Portable ladders shall be used at such a pitch that the horizontal distance from the top support to the foot of the ladder is about one-quarter of the working length of the ladder (the length along the ladder between the foot and the top support). Ladders shall not be used in a horizontal position as platforms, runways, or scaffolds.

(8) Ladders shall not be placed in passageways, doorways, driveways, or any location where they may be displaced by activities being conducted on any other work, unless protected by barricades or guards.

(9) The side rails shall extend not less than 36 inches above the landing. When this is not practical, grab rails, which provide a secure grip for an employee moving to or from the point of access, shall be installed.

(10) Portable ladders in use shall be tied, blocked, or otherwise secured to prevent their being displaced.

(11) Portable metal ladders shall not be used for electrical work or where they may contact electrical conductors.

(b) *Job-made ladders.* (1) Job-made ladders shall be constructed for intended use. If a ladder is to provide the only means of access or exit from a working area for 25 or more employees, or simultaneous two-way traffic is expected, a double cleat ladder shall be installed.

(2) Double cleat ladders shall not exceed 24 feet in length.

(3) Single cleat ladders shall not exceed 30 feet in length between supports (base and top landing). If ladders are to connect different landings, or if the length required exceeds this maximum length, two or more separate ladders shall be used, offset with a platform between each ladder. Guardrails and toe-boards shall be erected on the exposed sides of the platforms.

Table L-1 – Average Densities of Various Species of Wood for Use in Ladders

Group 1

Species	Density (lbs./ft.³)
White ash	41
Beech	43
Birch	44
Rock elm	43
Hickory	50
Locust	47
Hard maple	42
Red maple	36
Red oak	43
White oak	46
Pecan	46
Persimmon	50

Group 2

Douglas fir (coast region)	34
Western larch	38
Southern yellow pine	37

Group 3

Red alder	28
Oregon ash	38
Pumpkin ash	37
Alaska cedar	31
Port Orford cedar	30
Cucumber	34
Cypress	32
Soft elm	36
Douglas fir (Rocky Mountain type)	30
Noble fir	27
Gum	34
West Coast hemlock	30
Magnolia	35
Oregon maple	34
Norway pine	31
Poplar	28
Redwood	25
Eastern spruce	28
Sitka spruce	28
Sycamore	35
Tamarack	37
Tupelo	35

Group 4

Aspen	27
Basswood	25
Buckeye	25
Butternut	27
Incense cedar	25
Western red cedar	23
Black cottonwood	24
White fir	26
Hackberry	37
Eastern hemlock	28
Holly	39
Soft maple	33
Lodgepole pine	29
Idaho white pine	28
Northern white pine	25
Ponderosa pine	28
Sugar pine	26

(4) The width of single cleat ladders shall be at least 15 inches, but not more than 20 inches, between rails at the top.

(5) Side rails shall be parallel or flared top to bottom by not more than one-quarter of an inch for each 2 feet of length.

(6) Wood side rails of ladders having cleats shall be not less than 1½ inches thick and 3½ inches deep (2 inches by 4 inches nominal) when made of Group 2 or Group 3 woods (see Table L-1). Wood side rails of Group 4 woods (see Table L-1) may be used in the same cross-section of dimensions for cleat ladders up to 20 feet in length.

Table L-2

Length of cleat (inches)	Thickness (inches)	Width (inches)
Up to and including 20	3/4	3
Over 20 and up to and including 30	3/4	3-3/4

(7) It is preferable that side rails be continuous. If splicing is necessary to attain the required length, however, the splice must develop the full strength of a continuous side rail of the same length.

(8) 2-inch by 4-inch lumber shall be used for side rails of single cleat ladders up to 16 feet long; 3-inch by 6-inch lumber shall be used for single cleat ladders from 16 to 30 feet in length.

(9) 2-inch by 4-inch lumber shall be used for side and middle rails of double cleat ladders up to 12 feet in length; 2-inch by 6-inch lumber for double cleat ladders from 12 to 24 feet in length.

(10) Wood cleats shall have the following minimum dimensions when made of Group 1 woods (see Table L-1):

(11) Cleats may be made of species of any other group of wood (see Table L-1) provided equal or greater strength is maintained.

(12) Cleats shall be inset into the edges of the side rails one-half inch, or filler blocks shall be used on the rails between the cleats. The cleats shall be secured to each rail with three 10d common wire nails or other fasteners of equivalent strength. Cleats shall be uniformly spaced, 12 inches top-to-top.

Scaffolding.

(a) *General requirements.* (1) Scaffolds shall be erected in accordance with requirements of this section.

(2) The footing or anchorage for scaffolds shall be sound, rigid, and capable of carrying the maximum intended load without settling or displacement. Unstable objects such as barrels, boxes, loose brick, or concrete blocks, shall not be used to support scaffolds or planks.

(3) No scaffold shall be erected, moved, dismantled, or altered except under the supervision of competent persons.

(4) Guardrails and toeboards shall be installed on all open sides and ends of platforms more than 10 feet above the ground or floor, except needle beam scaffolds and floats (see paragraphs (p) and (w) of this section). Scaffolds 4 feet to 10 feet in height, having a minimum horizontal dimension in either direction of less than 45 inches, shall have standard guardrails installed on all open sides and ends of the platform.

(5) Guardrails shall be 2 x 4 inches, or the equivalent, approximately 42 inches high, with a midrail, when required. Supports shall be at intervals not to exceed 8 feet. Toeboards shall be a minimum of 4 inches in height.

(6) Where persons are required to work or pass under the scaffold, scaffolds shall be provided with a screen between the toeboard and the guardrail, exending along the entire opening, consisting of No. 18 gauge U.S. Standard wire ½-inch mesh, or the equivalent.

(7) Scaffolds and their components shall be capable of supporting without failure at least 4 times the maximum intended load.

(8) Any scaffold including accessories such as braces, brackets, trusses, screw legs, ladders, etc. damaged or weakened from any cause shall be immediately repaired or replaced.

(9) All load-carrying timber members of scaffold framing shall be a minimum of 1,500 fiber (Stress Grade) construction grade lumber. All dimensions are nominal sizes as provided in the American Lumber Standards, except that where rough sizes are noted, only rough or undressed lumber of the size specified will satisfy minimum requirements.

Table L-3 – Material

	Full thickness undressed lumber			Nominal thickness lumber[1]	
Working load (p.s.f.)	25	50	75	25	50
Permissible span (ft.)	10	8	6	8	6

[1]Nominal thickness lumber not recommended for heavy duty use.

(10) All planking shall be Scaffold Grades, or equivalent, as recognized by approved grading rules for the species of wood used. The maximum permissible spans for 2- x 10-inch or wider planks shall be as shown in the following:

11) The maximum permissible span for 1¼ x 9-inch or wider plank of full thickness shall be 4 feet with medium duty loading of 50 p.s.f.

(12) All planking of platforms shall be overlapped (minimum 12 inches), or secured from movement.

(13) An access ladder or equivalent safe access shall be provided.

(14) Scaffold planks shall extend over their end supports not less than 6 inches nor more than 12 inches.

(15) The poles, legs, or uprights of scaffolds shall be plumb, and securely and rigidly braced to prevent swaying and displacement.

(16) Overhead protection shall be provided for men on a scaffold exposed to overhead hazards.

(17) Slippery conditions on scaffolds shall be eliminated as soon as possible after they occur.

(18) No welding, burning, riveting, or open flame work shall be performed on any staging suspended by means of fiber or synthetic rope. Only treated or protected fiber or synthetic ropes shall be used for or near any work involving the use of corrosive substances or chemicals. Specific requirements for boatswain's chairs and float or ship scaffolds are contained in paragraphs (1) and (w) of this section.

(19) Wire, synthetic, or fiber rope used for scaffold suspension shall be capable of supporting at least 6 times the rated load.

(20) The use of shore or lean-to-scaffolds is prohibited.

(21) Lumber sizes, when used in this subpart, refer to nominal sizes except where otherwise stated.

(b) *Wood pole scaffolds.* (1) Scaffold poles shall bear on a foundation of sufficient size and strength to spread the load from the pole over a sufficient area to prevent settlement. All poles shall be set plumb.

(2) Where wood poles are spliced, the ends shall be squared and the upper section shall rest squarely on the lower section. Wood splice plates shall be provided on at least two adjacent sides and shall be not less than 4 feet in length, overlapping the abutted ends equally, and have the same width and not less than the cross-sectional area of the pole. Splice plates or other materials of equivalent strength may be used.

(3) Independent pole scaffolds shall be set as near to the wall of the building as practicable.

(4) All pole scaffolds shall be securely guyed or tied to the building or structure. Where the height or length exceeds 25 feet, the scaffold shall be secured at intervals not greater than 25 feet vertically and horizontally.

(5) Putlogs or bearers shall be set with their greater dimension vertical, long enough to project over the ledgers of the inner and outer rows of poles at least 3 inches for proper support.

(6) Every wooden putlog on single pole scaffolds shall be reinforced with a 3/16 x 2-inch steel strip, or equivalent, secured to its lower edge throughout its entire length.

(7) Ledgers shall be long enough to extend over two pole spaces. Ledgers shall not be spliced between the poles. Ledgers shall be reinforced by bearing blocks securely nailed to the side

of the pole to form a support for the ledger.

(8) Diagonal bracing shall be provided to prevent the poles from moving in a direction parallel with the wall of the building, or from buckling.

(9) Cross bracing shall be provided between the inner and outer sets of poles in independent pole scaffolds. The free ends of pole scaffolds shall be cross braced.

(10) Full diagonal face bracing shall be erected across the entire face of pole scaffolds in both directions. The braces shall be spliced at the poles. The inner row of poles on medium and heavy duty scaffolds shall be braced in a similar manner.

(11) Platform planks shall be laid with their edges close together so the platform will be tight with no spaces through which tools or fragments of material can fall.

(12) Where planking is lapped, each plank shall lap its end supports at least 12 inches. Where the ends of planks abut each other to form a flush floor, the butt joint shall be at the centerline of a pole. The abutted ends shall rest on separate bearers. Intermediate beams shall be provided where necessary to prevent dislodgment of planks due to deflection, and the ends shall be secured to prevent their dislodgment.

(13) When a scaffold materially changes its direction, the platform planks shall be laid to prevent tipping. The planks that meet the corner putlog at an angle shall be laid first, extending over the diagonally placed putlog far enough to have a good safe bearing, but not far enough to involve any danger from

Table L-4 — Minimum Nominal Size and Maximum Spacing of Members of Single Pole Scaffolds, Light Duty

| | Maximum height of scaffold | |
	20 ft.	60 ft.
Uniformly distributed load	Not to exceed 25 p.s.f.	
Poles or uprights	2 x 4 in.	4 x 4 in.
Pole spacing (longitudinal)	6 ft. 0 in.	10 ft. 0 in.
Maximum width of scaffold	5 ft. 0 in.	5 ft. 0 in.
Bearers or putlogs to 3 ft. 0 in. width	2 x 4 in.	2 x 4 in.
Bearers or putlogs to 5 ft. 0 in. width	2 x 6 in. or 3 x 4 in.	2 x 6 in. or 3 x 4 in. (rough)
Ledgers	1 x 4 in.	1¼ x 9 in.
Planking	1¼ x 9 in. (rough)	2 x 10 in.
Vertical spacing of horizontal members	7 ft. 0 in.	9 ft. 0 in.
Bracing, horizontal and diagonal	1 x 4 in.	1 x 4 in.
Tie-ins	1 x 4 in.	1 x 4 in.
Toeboards	4 in. high (minimum)	4 in. high (minimum)
Guardrail	2 x 4 in.	2 x 4 in.

All members except planking are used on edge.

Table L-5 – Minimum Nominal Size and Maximum Spacing of Members of Single Pole Scaffolds – Medium Duty

Uniformly distributed load	Not to exceed 50 p.s.f.
Maximum height of scaffold	60 ft.
Poles or upright	4 X 4 in.
Pole spacing (longitudinal)	8 ft. 0 in.
Maximum width of scaffold	5 ft. 0 in.
Bearers or putlogs	2 X 10 in. or 3 X 4 in.
Spacing of bearers or putlogs	8 ft. 0 in.
Ledgers	2 X 10 in.
Vertical spacing of horizontal members	7 ft. 0 in.
Bracing, horizontal	1 X 6 in. or 1-1/4 X 4 in.
Bracing, diagonal	1 X 4 in.
Tie-ins	1 X 4 in.
Planking	2 X 10 in.
Toeboards	4-in. high (minimum)
Guardrail	2 X 4 in.

All members except planking are used on edge.

Table L-6 – Minimum Nominal Size and Maximum Spacing of Members of Single Pole Scaffolds – Heavy Duty

Uniformly distributed load	Not to exceed 75 p.s.f.
Maximum height of scaffold	60 ft.
Poles or uprights	4 X 6 in.
Pole spacing (longitudinal)	6 ft. 0 in.
Maximum width of scaffold	5 ft. 0 in.
Bearers or putlogs	2 X 10 in. or 3 X 5 in.
Spacing of bearers or putlog	6 ft. 0 in.
Ledgers	2 X 10 in.
Vertical spacing of horizontal members	6 ft. 6 in.
Bracing, horizontal and diagonal	2 X 4 in.
Tie-ins	1 X 4 in.
Planking	2 X 10 in.
Toeboards	4-in. high (minimum)
Guardrail	2 X 4 in.

All members except planking are used on edge.

Table L-7 – Minimum Nominal Size and Maximum Spacing of Members of Independent Pole Scaffold – Light Duty

	Maximum height of scaffold	
	20 ft.	**60 ft.**
Uniformly distributed load	Not to exceed 25 p.s.f.	
Poles or uprights	2 X 4 in.	4 X 4 in.
Pole spacing (longitudinal)	6 ft. 0 in.	10 ft. 0 in.
Pole spacing (transverse)	6 ft. 0 in.	10 ft. 0 in.
Ledgers	1-1/4 X 4 in.	1-1/4 X 9 in.
Bearers to 3 ft. 0 in. span	2 X 4 in.	2 X 4 in.
Bearers to 10 ft. 0 in.	2 X 6 in. or 3 X 4 in.	2 X 10 (rough) or 3 X 8 in.
Planking	1-1/4 X 9 in.	2 X 10 in.
Vertical spacing of horizontal members	7 ft. 0 in.	7 ft. 0 in.
Bracing, horizontal and diagonal	1 X 4 in.	1 X 4 in.
Tie-ins	1 X 4 in.	1 X 4 in.
Toeboards	4 in. high	4 in. high (minimum)
Guardrail	2 X 4 in.	2 X 4 in.

All members except planking are used on edge.

tipping. The planking running in the opposite direction at an angle shall be laid so as to extend over and rest on the first layer of planking.

(14) When moving platforms to the next level, the old platform shall be left undisturbed until the new putlogs or bearers have been set in place, ready to receive the platform planks.

(15) Guardrails, made of lumber not less than 2 x 4 inches (or other material providing equivalent protection), approximately 42 inches high, with a midrail of 1 x 6 inch lumber (or other material providing equivalent protection), and toeboards, shall be installed at all open sides and ends on all scaffolds more than 10 feet above the ground or floor. Toeboards shall be a minimum of 4 inches in height. Wire mesh shall be installed in accordance with paragraph (a) (6) of this section, when required.

(16) All wood pole scaffolds 60 feet or less in height shall be constructed and erected in accordance with Tables L-4 to 10. If they are over 60 feet in height, they shall be designed by a qualified engineer competent in this field, and it shall be constructed and erected in accordance with such design.

(c) *Tube and coupler scaffolds.* (1) A light duty tube and coupler scaffold shall have all posts, bearers, runners, and bracing of nominal 2-inch O.D. steel tubing. The posts shall be spaced no more than 6 feet apart by 10 feet along the length of the scaffold. Other structural metals when used must be designed to carry an equivalent load. No dissimilar metals shall be used together.

(2) A medium duty tube and coupler scaffold shall have all posts, runners, and bracing of nominal 2-inch O.D. steel tub-

Table L-8 – Minimum Nominal Size and Maximum Spacing of Members of Independent Pole Scaffolds – Medium Duty

Uniformly distributed load	Not to exceed 50 p.s.f.
Maximum height of scaffold	60 ft.
Poles or uprights	4 X 4 in.
Pole spacing (longitudinal)	8 ft. 0 in.
Pole spacing (transverse)	8 ft. 0 in.
Ledgers	2 X 10 in.
Vertical spacing of horizontal members	6 ft. 0 in.
Spacing of bearers	8 ft. 0 in.
Bearers	2 X 10 in.
Bracing, horizontal	1 X 6 in. or 1-1/4 X 4 in.
Bracing, diagonal	1 X 4 in.
Tie-ins	1 X 4 in.
Planking	2 X 10 in.
Toeboards	4-in. high (minimum)
Guardrail	2 X 4 in.

All members except planking are used on edge.

Table L-9 – Minimum Nominal Size and Maximum Spacing of Members of Independent Pole Scaffolds – Heavy Duty

Uniformly distributed	Not to exceed 75 p.s.f.
Maximum height of scaffold	60 ft.
Poles or uprights	4 X 4 in.
Pole spacing (longitudinal)	6 ft. 0 in.
Pole spacing (transverse)	8 ft. 0 in.
Ledgers	2 X 10 in.
Vertical spacing of horizontal members	6 ft. 0 in.
Bearers	2 X 10 in. (rough)
Bracing, horizontal and diagonal	2 X 4 in.
Tie-ins	1 X 4 in.
Planking	2 X 10 in.
Toeboards	4-in. high (minimum)
Guardrail	2 X 4 in.

All members except planking are used on edge.

ing. Posts spaced not more than 6 feet apart by 8 feet along the length of the scaffold shall have bearers of nominal 2½-inch O.D. steel tubing. Posts spaced not more than 5 feet apart by 8 feet along the length of the scaffold shall have bearers of nominal 2-inch O.D. steel tubing. Other structural metals, when used, must be designed to carry an equivalent load. No dissimilar metals shall be used together.

(3) A heavy duty tube and coupler scaffold shall have all posts, runners, and bracing of nominal 2-inch O.D. steel tubing, with the posts spaced not more than 6 feet by 6 feet-6 inches. Other structural metals, when used, must be designed to carry an equivalent load. No dissimilar metals shall be used together.

(4) Tube and coupler scaffolds shall be limited in heights and working levels to those permitted in Tables L-10, 11, and 12. Drawings and specifications of all tube and coupler scaffolds above the limitations in Tables L-10, 11, and 12 shall be designed by a qualified engineer competent in this field.

(5) All tube and coupler scaffolds shall be constructed and erected to support four times the maximum intended loads, as set forth in Tables L-10, 11, and 12, or as set forth in the specifications by a licensed professional engineer competent in this field.

Table L-10 – Tube and Coupler Scaffolds – Light Duty

Uniformly distributed load	Not to exceed 25 p.s.f.
Post spacing (longitudinal)	10 ft. 0 in.
Post spacing (transverse)	6 ft. 0 in.

Working levels	Additional planked levels	Maximum height
1	8	125 ft.
2	4	125 ft.
3	0	91 ft. 0 in.

Table L-11 – Tube and Coupler Scaffolds – Medium Duty

Uniformly distributed load	Not to exceed 50 p.s.f.
Post spacing (longitudinal)	8 ft. 0 in.
Post spacing (transverse)	6 ft. 0 in.

Working levels	Additional planked levels	Maximum height
1	6	125 ft.
2	0	78 ft. 0 in.

(6)) Posts shall be accurately spaced, erected on suitable bases, and maintained plumb.

(7) Runners shall be erected along the length of the scaffold, located on both the inside and the outside posts at even height. Runners shall be interlocked to the inside and the outside posts at even heights. Runners shall be interlocked to form

Table L-12 – Tube and Coupler Scaffolds – Heavy Duty

Uniformly distributed load	Not to exceed 75 p.s.f.
Post spacing (longitudinal)	6 ft. 6 in.
Post spacing (transverse)	6 ft. 0 in.

Working levels	Additional planked levels	Maximum height
1	6	125 ft.

continuous lengths and coupled to each post. The bottom runners shall be located as close to the base as possible. Runners shall be placed not more than 6 feet-6 inches on centers.

(8) Bearers shall be installed transversely between posts and shall be securely coupled to the posts bearing on the runner coupler. When coupled directly to the runners, the coupler must be kept as close to the posts as possible.

(9) Bearers shall be at least 4 inches but not more than 12 inches longer than the post spacing or runner spacing.

(10) Cross bracing shall be installed across the width of the scaffold at least every third set of posts horizontally and every fourth runner vertically. Such bracing shall extend diagonally from the inner and outer runners upward to the next outer and inner runners.

(11) Longitudinal diagonal bracing on the inner and outer rows of poles shall be installed at approximately a 45° angle from near the base of the first outer post upward to the extreme top of the scaffold. Where the longitudinal length of the scaffold permits, such bracing shall be duplicated beginning at every fifth post. In a similar manner, longitudinal diagonal bracing shall also be installed from the last post extending back and upward toward the first post. Where conditions preclude the attachment of this bracing to the posts, it may be attached to the runners.

(12) The entire scaffold shall be tied to and securely braced against the building at intervals not to exceed 30 feet horizontally and 26 feet vertically.

(13) Guardrails, made of lumber not less than 2 x 4 inches (or other material providing equivalent protection), approximately 42 inches high, with a midrail of 1 x 6 inch lumber (or other material providing equivalent protection), and toeboard shall be installed at all open sides and ends on all scaffolds more than 10 feet above the ground or floor. Toeboards shall be a minimum of 4 inches in height. Wire mesh shall be installed in accordance with paragraph (a)(6) of this section.

(d) *Tubular welded frame scaffolds.* (1) Metal tubular frame scaffolds, including accessories such as braces, brackets, trusses, screw legs, ladders, etc., shall be designed, constructed, and erected to safely support four times the maximum rated load.

(2) Spacing of panels or frames shall be consistent with the loads imposed.

(3) Scaffolds shall be properly braced by cross bracing or diagonal braces, or both, for securing vertical members together laterally, and the cross braces shall be of such length as will automatically square and aline vertical members so that the erected scaffold is always plumb, square, and rigid. All brace connections shall be made secure.

(4) Scaffold legs shall be set on adjustable bases or plain bases placed on mud sills or other foundations adequate to support the maximum rated load.

(5) The frames shall be placed one on top of the other with coupling or stacking pins to provide proper vertical alinement of the legs.

(6) Where uplift may occur, panels shall be locked together vertically by pins or other equivalent suitable means.

(7) To prevent movement, the scaffold shall be secured to the building or structure at intervals not to exceed 30 feet horizontally and 26 feet vertically.

(8) Maximum permissible spans or planking shall be in conformity with paragraph (a)(10) of this section.

(9) Drawings and specifications for all frame scaffolds over 125 feet in height above the base plates shall be designed by a registered professional engineer.

(10) Guardrails made of lumber, not less than 2 x 4 inches (or other material providing equivalent protection), and approximately 42 inches high, with a midrail of 1 x 6 inch lumber (or other material providing equivalent protection), and toeboards, shall be installed at all open sides and ends on all scaffolds more than 10 feet above the ground or floor. Toeboards shall be a minimum of 4 inches in height. Wire mesh shall be installed in accordance with paragraph (a)(6) of this section.

(e) *Manually propelled mobile scaffolds.* (1) When free-standing mobile scaffold towers are used, the height shall not exceed four times the minimum base dimension.

(2) Casters shall be properly designed for strength and dimensions to support four times the maximum intended load. All casters shall be provided with a positive locking device to hold the scaffold in position.

(3) Scaffolds shall be properly braced by cross bracing and horizontal bracing conforming with paragraph (d)(3) of this section.

(4) Platforms shall be tightly planked for the full width of the scaffold except for necessary entrance opening. Platforms shall be secured in place.

(5) A ladder or stairway shall be provided for proper access and exit and shall be affixed or built into the scaffold and so located that when in use it will not have a tendency to tip the scaffold. A landing platform must be provided at intervals not to exceed 35 feet.

(6) The force necessary to move the mobile scaffold shall be applied near or as close to the base as practicable and provision shall be made to stabilize the tower during movement from one location to another. Scaffolds shall only be moved on level floors, free of obstructions and openings.

(7) The employer shall not allow employees to ride on manually propelled scaffolds unless the following conditions exist:

(i) The floor or surface is within 3° of level, and free from pits, holes, or obstructions;

(ii) The minimum dimension of the scaffold base when ready for rolling, is at least one-half of the height. Outriggers, if used, shall be installed on both sides of staging;

(iii) The wheels are equipped with rubber or similar resilient tires;

(iv) All tools and materials are secured or removed from the platform before the mobile scaffold is moved.

(8) Scaffolds in use by any persons shall rest upon a suitable footing and shall stand plumb. The casters or wheels shall be locked to prevent any movement.

(9) Mobile scaffolds constructed of metal members shall also conform to applicable provisions of paragraphs (b), (c), or (d) of this section, depending on the material of which they are constructed.

(10) Guardrails made of lumber, not less than 2 x 4 inches (or other material providing equivalent protection), approximately 42 inches high, with a midrail of 1 x 6 inch lumber (or other material providing equivalent protection), and toeboards, shall be installed at all open sides and ends on all scaffolds more than 10 feet above the ground or floor. Toeboards shall be a minimum of 4 inches in height. Wire mesh shall be installed in accordance with paragraph (a)(6) of this section.

(f) *Elevating and rotating work platforms.* Applicable requirements of American National Standards Institute A92.2-1969, Vehicle Mounted Elevating and Rotating Work Platforms, shall be complied with for such equipment.

(g) *Outrigger scaffolds.* (1) Outrigger beams shall extend not more than 6 feet beyond the face of the building. The inboard end of outrigger beams, measured from the fulcrum point to anchorage point, shall be not less than 1½ times the outboard end in length. The beams shall rest on edge, the sides shall be plumb, and the edges shall be horizontal. The fulcrum point of the beam shall rest on a secure bearing at least 6 inches in each horizontal dimension. The beam shall be secured in place against movement and shall be securely braced at the fulcrum point against tipping.

(2) The inboard ends of outrigger beams shall be securely anchored either by means of struts bearing against sills in contact with the overhead beams or ceiling, or by means of tension members secured to the floor joists underfoot, or by both if necessary. The inboard ends of outrigger beams shall be secured against tipping and the entire supporting structure shall be securely braced in both directions to prevent any horizontal movement.

(3) Unless outrigger scaffolds are designed by a registered professional engineer competent in this field, they shall be constructed and erected in accordance with Table L-13. Outrigger scaffolds, designed by a registered professional engineer, shall be constructed and erected in accordance with such design.

Table L-13 – Minimum Nominal Size And Maximum Spacing of Members of Outrigger Scaffolds

Maximum scaffold load	Light duty 25 p.s.f.	Medium duty 50 p.s.f.
Outrigger size	2 X 10 in.	3 X 10 in.
Maximum outrigger spacing	10 ft. 0 in.	6 ft. 0 in.
Planking	2 X 10 in.	2 X 10 in.
Guardrail	2 X 4 in.	2 X 4 in.
Guardrail uprights	2 X 4 in.	2 X 4 in.
Toeboards	4 in. (minimum)	4 in. (minimum)

(4) Planking shall be laid tight and shall extend to within 3 inches of the building wall. Planking shall be secured to the beams.

(5) Guardrails made of lumber, not less than 2 x 4 inches (or other material providing equivalent protection), approximately 42 inches high, with a midrail of 1 x 6 inch lumber (or other material providing equivalent protection), and toeboards, shall be installed at all open sides and ends on all scaffolds more than 10 feet above the ground or floor. Toeboards shall be a minimum of 4 inches in height. Wire mesh shall be installed in accordance with paragraph (a)(6) of this section.

(h) *Masons' adjustable multiple-point suspension scaffolds.* (1) The scaffold shall be capable of sustaining a working load of 50 pounds per square foot and shall not be loaded in excess of that figure.

(2) The scaffold shall be provided with hoisting machines that meet the requirements of Underwriters' Laboratories or Factory Mutual Engineering Corporation.

(3) The platform shall be supported by wire ropes, capable of supporting at least 6 times the intended load, suspended from overhead outrigger beams.

(4) The scaffold outrigger beams shall consist of structural metal securely fastened or anchored to the frame or floor system of the building or structure.

(5) Each outrigger beam shall be equivalent in strength to at least a standard 7-inch, 15.3-pound steel I-beam, at least 15 feet long, and shall not project more than 6 feet 6 inches beyond the bearing point.

(6) Where the overhang exceeds 6 feet 6 inches, outrigger beams shall be composed of stronger beams or multiple beams and be installed under the supervision of a competent person.

(7) All outrigger beams shall be set and maintained with their webs in a vertical position.

(8) A stop bolt shall be placed at each end of every outrigger beam.

(9) The outrigger beam shall rest on suitable wood bearing blocks.

(10) The free end of the suspension wire ropes shall be equipped with proper size thimbles and secured by splicing or other equivalent means. The running ends shall be securely attached to the hoisting drum and at least four turns of wire rope shall at all times remain on the drum. The use of fiber rope is prohibited.

(11) Where a single outrigger beam is used, the steel shackles or clevises with which the wire ropes are attached to the outrigger beams shall be placed directly over the hoisting drums.

(12) The scaffold platform shall be equivalent in strength to at least 2-inch planking. (For maximum planking spans, see paragraph (a)(11) of this section).

(13) When employees are at work on the scaffold and an overhead hazard exists, overhead protection shall be provided on the scaffold, not more than 9 feet above the platform, consisting of 2-inch planking, or material of equivalent strength, laid tight, and extending not less than the width of the scaffold.

(14) Each scaffold shall be installed or relocated under the supervision of a competent person.

(15) Guardrails made of lumber, not less than 2 x 4 inches (or other material providing equivalent protection), approximately 42 inches high, with a midrail, and toeboards, shall be installed at all open sides and ends on all scaffolds more than 10 feet above the ground or floor. Toeboards shall be a minimum of 4 inches in height. Wire mesh shall be installed in accordance with paragraph (a)(6) of this section.

(i) *(Swinging scaffolds) two-point suspension.* (1) Two-point suspension scaffold platforms shall be not less than 20 inches nor more than 36 inches wide overall. The platform shall be securely fastened to the hangers by U-bolts or by other equivalent means.

(2) The hangers of two-point suspension scaffolds shall be made of mild steel, or other equivalent materials, having a cross-sectional area capable of sustaining 4 times the maximum rated load, and shall be designed with a support for guardrail, intermediate rail, and toeboard.

(3) When hoisting machines are used on two-point suspension scaffolds, such machines shall be of a design tested and approved by Underwriters' Laboratories or Factory Mutual Engineering Corporation.

(4) The roof irons or hooks shall be of mild steel, or other equivalent material, of proper size and design, securely installed and anchored. Tiebacks of ¾-inch manila rope, or the equivalent, shall serve as a secondary means of anchorage, installed at right angles to the face of the building, whenever possible, and secured to a structurally sound portion of the building.

(5) Two-point suspension scaffolds shall be suspended by wire, synthetic, or fiber ropes capable of supporting at least 6 times the rated load. All other components shall be capable of supporting at least four times the rated load.

(6) The sheaves of all blocks, consisting of at least one double and one single block, shall fit the size and type of rope used.

(7) All wire ropes, fiber and synthetic ropes, slings, hangers, platforms, and other supporting parts shall be inspected before every installation. Periodic inspections shall be made while the scaffold is in use.

(8) On suspension scaffolds designed for a working load of 500 pounds, no more than two men shall be permitted to work at one time. On suspension scaffolds with a working load of 750 pounds, no more than three men shall be permitted to work at one time. Each employee shall be protected by an approved safety life belt attached to a lifeline. The lifeline shall be securely attached to substantial members of the structure (not scaffold), or to securely rigged lines, which will safely suspend the employee in case of a fall. In order to keep the lifeline continuously attached, with a minimum of slack, to a fixed structure, the attachment point of the lifeline shall be appropriately changed as the work progresses.

(9) Two-point suspension scaffolds shall be securely lashed to the building or structure to prevent them from swaying. Window cleaners' anchors shall not be used for this purpose.

(10) The platform of every two-point suspension scaffold shall be one of the following types:

(i) *Ladder-type platforms.* The side stringer shall be of clear straight-grained spruce or materials of equivalent strength and durability. The rungs shall be of straight-grained oak, ash, or hickory, at least 1⅛ inch in diameter, with ⅞-inch tenons mortised into the side stringers at least seven-eighths inch. The stringers shall be tied together with tie rods not less than one-quarter inch in diameter, passing through the stringers and riveted up tight against washers on both ends. The flooring strips shall be spaced not more than five-eighths inch apart except at the side rails where the space may be 1 inch. Ladder-type platforms shall be constructed in accordance with Table L-14.

of a type tested and listed by Underwriters' Laboratories or Factory Mutual Engineering Corporation.

(3) The platform shall be securely fastened to the hangers by U-bolts or other equivalent means. (For materials and spans, see subdivision (ii) of paragraph (i) (10), Plank-Type Platforms, and Table L-14 of this section.)

Table L-14 – Schedule For Ladder Type Platforms

	Length of platform (feet)				
	12	14 and 16	18 and 20	22 and 24	28 and 30
Side Stringers, minimum cross section (finished sizes):					
At ends (inches)	1-3/4 X 2-3/4	1-3/4 X 2-3/4	1-3/4 X 3	1-3/4 X 3	1-3/4 X 3-1/2
At middle (inches)	1-3/4 X 3-3/4	1-3/4 X 3-3/4	1-3/4 X 4	1-3/4 X 4-1/4	1-3/4 X 5
Reinforcing strip (minimum)	1	1	1	1	1
Rungs	2	2	2	2	2
Tie rods:					
Number (minimum)	3	4	4	5	6
Diameter (minimum)	1/4 in.	1/4 in.	1/4 in.	1/4 in.	1/4 in.
Flooring, minimum finished size (inches)	1/2 X 2-3/4	1/2 X 2-3/4	1/2 X 2-3/4	1/2 X 2-3/4	1/2 X 2-3/4

[1]A 1/8 X 7/8-inch steel reinforcing strip or its equivalent shall be attached to the side or underside, full length.

[2]Rungs shall be 1-1/8-inch minimum diameter with at least 7/8-inch diameter tenons, and the maximum spacing shall be 12 inches center to center.

(ii) *Plank-type platforms.* Plank-type platforms shall be composed of not less than nominal 2- x 10-inch unspliced planks, properly cleated together on the underside, starting 6 inches from each end; intervals in between shall not exceed 4 feet. The plank-type platform shall not extend beyond the hangers more than 12 inches. A bar or other effective means shall be securely fastened to the platform at each end to prevent its slipping off the hanger. The span between hangers for plank-type platforms shall not exceed 8 feet.

(iii) *Beam-type platforms.* Beam platforms shall have side stringers of lumber not less than 2 x 6 inches set on edge. The span between hangers shall not exceed 12 feet when beam platforms are used. The flooring shall be supported on 2- x 6-inch cross beams, laid flat and set into the upper edge of the stringers with a snug fit, at intervals of not more than 4 feet, securely nailed in place. The flooring shall be of 1- x 6-inch material properly nailed. Floor boards shall not be spaced more than one-half inch apart.

(iv) *Light metal-type platforms,* when used, shall be tested and listed according to Underwriters' Laboratories or Factory Mutual Engineering Corporation.

(11) Guardrails made of lumber, not less than 2 x 4 inches (or other material providing equivalent protection), approximately 42 inches high, with a midrail, and toeboards, shall be installed at all open sides and ends on all scaffolds more than 10 feet above the ground or floor. Toeboards shall be a minimum of 4 inches in height. Wire mesh shall be installed in accordance with paragraph (a) (6) of this section.

(j) *Stone setters' adjustable multiple-point suspension scaffolds.* (1) The scaffold shall be capable of sustaining a working load of 25 pounds per square foot and shall not be overloaded. Scaffolds shall not be used for storage of stone or other heavy materials.

(2) When used, the hoisting machine and its supports shall be

(4) The scaffold unit shall be suspended from metal outriggers, iron brackets, wire rope slings, or iron hooks.

(5) Outriggers, when used, shall be set with their webs in a vertical position, securely anchored to the building or structure and provided with stop bolts at each end.

(6) The scaffold shall be supported by wire rope capable of supporting at least 6 times the rated load. All other components shall be capable of supporting at least 4 times the rated load.

(7) The free ends of the suspension wire ropes shall be equipped with proper size thimbles, secured by splicing or other equivalent means. The running ends shall be securely attached to the hoisting drum and at least four turns of wire rope shall remain at the drum at all times.

(8) When two or more scaffolds are used on a building or structure, they shall not be bridged one to the other, but shall be maintained at even height with platforms abutting closely.

(9) Guardrails made of lumber, not less than 2 x 4 inches (or other material providing equivalent protection), approximately 42 inches high, with a midrail, and toeboards, shall be installed at all open sides and ends on all scaffolds more than 10 feet above the ground or floor. Toeboards shall be a minimum of 4 inches in height. Wire mesh shall be installed in accordance with paragraph (a) (6) of this section.

(k) *Single-point adjustable suspension scaffolds.* (1) The scaffolding, including power units or manually operated winches, shall be of a type tested and listed by Underwriters' Laboratories or Factory Mutual Engineering Corporation.

(2) The power units may be either electrically or air motor driven.

(3) All power-operated gears and brakes shall be enclosed.

(4) In addition to the normal operating brake, all power-driven units shall have an emergency brake which engages automa-

tically when the normal speed of descent is exceeded.

(5) The hoisting machines, cables, and equipment shall be regularly serviced and inspected.

(6) The units may be combined to form a two-point suspension scaffold. Such scaffold shall then comply with paragraph (i) of this section.

(7) The supporting cable shall be vertical for its entire length, and the basket shall not be swayed nor the cable fixed to any intermediate points to change the original path of travel.

(8) Suspension methods shall conform to applicable provisions of paragraphs (h) and (i) of this section.

(9) Guards, midrails, and toeboards shall completely enclose the cage or basket. Guardrails shall be no less than 2 x 4 inches or the equivalent, approximately 42 inches above the platform. Midrails shall be 1 x 6 inches or the equivalent, installed equidistant between the guardrail and the platform. Toeboards shall be a minimum of 4 inches in height.

(10) For additional details not covered in this paragraph, applicable technical portions of American National Standards Institute, A120.1-1970, Power-Operated Devices for Exterior Building Maintenance Powered Platforms, shall be used.

(1) *Boatswain's chairs.* (1) The chair seat shall not be less than 12 x 24 inches, and 1-inch thickness. The seat shall be reinforced on the underside by cleats securely fastened to prevent the board from splitting.

(2) The two fiber rope seat slings shall be of ⅝-inch diameter, reeved through the four seat holes so as to cross each other on the underside of the seat.

(3) Seat slings shall be of at least ⅜-inch wire rope when an employee is conducting a heat-producing process, such as gas or arc welding.

(4) The employee shall be protected by a safety belt and lifeline. The attachment point of the lifeline to the structure shall be appropriately changed as the work progresses.

(5) The tackle shall consist of correct size ball bearing or bushed blocks and properly spliced ⅝-inch diameter first-grade manila rope, or equivalent.

(6) The roof irons, hooks, or the object to which the tackle is anchored, shall be securely installed. Tiebacks, when used, shall be installed at right angles to the face of the building and securely fastened.

(m) *Carpenters' bracket scaffolds.* (1) The brackets shall consist of a triangular wood frame not less than 2 x 3 inches in cross section, or of metal of equivalent strength. Each member shall be properly fitted and securely joined.

(2) Each bracket shall be attached to the structure by means of one of the following:

(i) A bolt, no less than five-eighths inch in diameter, which shall extend through to the inside of the building wall;

(ii) A metal stud attachment device;

(iii) Welding to steel tanks;

(iv) Hooking over a well-secured and adequately strong supporting member.

(3) The brackets shall be spaced no more than 8 feet apart.

(4) No more than two employees shall occupy any given 8 feet

of a bracket scaffold at any one time. Tools and materials shall not exceed 75 pounds in addition to the occupancy.

(5) The platform shall consist of not less than two 2- x 10-inch nominal size planks extending not more than 12 inches or less than 6 inches beyond each end support.

(6) Guardrails made of lumber, not less than 2 x 4 inches (or other material providing equivalent protection), approximately 42 inches high, with a midrail, of 1 x 6 inch lumber (or other material providing equivalent protection), and toeboards, shall be installed at all open sides and ends on all scaffolds more than 10 feet above the ground or floor. Toeboards shall be a minimum of 4 inches in height. Wire mesh shall be installed in accordance with paragraph (a)(6) of this section.

(n) *Bricklayers' square scaffolds.* (1) The squares shall not exceed 5 feet in width and 5 feet in height.

(2) Members shall not be less than those specified in Table L-15.

Table L-15 – Minimum Dimensions For Bricklayers' Square Scaffold Members

Members	Dimensions
Bearers or horizontal members	2 X 6 in.
Legs	2 X 6 in.
Braces at corners	1 X 6 in.
Braces diagonally from center frame	1 X 8 in.

(3) The squares shall be reinforced on both sides of each corner with 1- x 6-inch gusset pieces. They shall also have diagonal braces 1 x 8 inches on both sides running from center to center of each member, or other means to secure equivalent strength and rigidity.

(4) The squares shall be set not more than 5 feet apart for medium duty scaffolds, and not more than 8 feet apart for light duty scaffolds. Bracing, 1 x 8 inches, extending from the bottom of each square to the top of the next square, shall be provided on both front and rear sides of the scaffold.

(5) Platform planks shall be at least 2-x 10-inch nominal size. The ends of the planks shall overlap the bearers of the squares and each plank shall be supported by not less than three squares.

(6) Bricklayers' square scaffolds shall not exceed three tiers in height and shall be so constructed and arranged that one square shall rest directly above the other. The upper tiers shall stand on a continuous row of planks laid across the next lower tier and be nailed down or otherwise secured to prevent displacement.

(7) Scaffolds shall be level and set upon a firm foundation.

(o) *Horse scaffolds.* (1) Horse scaffolds shall not be constructed or arranged more than two tiers or 10 feet in height.

(2) The members of the horses shall be not less than those specified in Table L-16.

Table L-16 – Minimum Dimensions for Horse Scaffold Members

Members	Dimensions
Horizontal members or bearers	3 X 4 in.
Legs	1-1/4 X 4-1/2 in.
Longitudinal brace between legs	1 X 6 in.
Gusset brace at top of legs	1 X 8 in.
Half diagonal braces	1-1/4 X 4-1/2 in.

(3) Horses shall be spaced not more than 5 feet for medium duty and not more than 8 feet for light duty.

(4) When arranged in tiers, each horse shall be placed directly over the horse in the tier below.

(5) On all scaffolds arranged in tiers, the legs shall be nailed down or otherwise secured to the planks to prevent displacement or thrust and each tier shall be substantially cross braced.

(6) Horses or parts which have become weak or defective shall not be used.

(7) Guardrails made of lumber, not less than 2 x 4 inches (or other material providing equivalent protection), approximately 42 inches high, with a midrail, of 1 x 6 inch lumber (or other material providing equivalent protection), and toeboards, shall be installed at all open sides and ends on all scaffolds more than 10 feet above the ground or floor. Toeboards shall be a minimum of 4 inches in height. Wire mesh shall be installed in accordance with paragraph (a)(6) of this section.

(p) *Needle beam scaffold.* (1) Wood needle beams shall be not less than 4 x 6 inches in size, with the greater dimension placed in a vertical direction. Metal beams or the equivalent, conforming to paragraphs (a) (8) and (10) of this section, may be used and shall not be altered or moved horizontally while they are in use.

(2) Ropes or hangers shall be provided for supports. The span between supports on the needle beam shall not exceed 10 feet for 4- x 6-inch timbers. Rope supports shall be equivalent in strength to 1-inch diameter first-grade manila rope.

(3) The ropes shall be attached to the needle beams by a scaffold hitch or a properly made eye splice. The loose end of the rope shall be tied by a bowline knot or by a round turn and a half hitch.

(4) The scaffold hitch shall be arranged so as to prevent the needle beam from rolling or becoming otherwise displaced.

(5) The platform span between the needle beams shall not exceed 8 feet when using 2-inch scaffold plank. For spans greater than 8 feet, platforms shall be designed based on design requirements for the special span. The overhang of each end of the platform planks shall not be less than 6 inches and not more than 12 inches.

(6) When needle beam scaffolds are used, the planks shall be secured against slipping.

(7) All unattached tools, bolts, and nuts used on needle beam scaffolds shall be kept in suitable containers, properly secured.

(8) One end of a needle beam scaffold may be supported by a permanent structural member conforming to paragraphs (a)(8) and (10) of this section.

(9) Each employee working on a needle beam scaffold shall be protected by a safety belt and lifeline.

(q) *Plasterers', decorators', and large area scaffolds.* (1) Plasterers', lathers', and ceiling workers' inside scaffolds shall be constructed in accordance with the general requirements set forth for independent wood pole scaffolds. (See paragraph (b) and Tables L-7, 8, and 9 of this section.)

(2) All platform planks shall be laid with the edges close together.

(3) When independent pole scaffold platforms are erected in sections, such sections shall be provided with connecting runways equipped with substantial guardrails.

(4) Guardrails made of lumber, not less than 2 x 4 inches (or other material providing equivalent protection), approximately 42 inches high, with a midrail of 1 x 6 inch lumber (or other material providing equivalent protection), and toeboards, shall be installed on all open sides and ends of all scaffolds more than 10 feet above the ground or floor. Toeboards shall be a minimum of 4 inches in height. Wire mesh shall be installed in accordance with paragraph (a)(6) of this section.

(r) *Interior hung scaffolds.* (1) An interior hung scaffold shall be hung or suspended from the roof structure or ceiling beams.

(2) The suspending wire or fiber rope shall be capable of supporting at least 6 times the rated load. The rope shall be wrapped at least twice around the supporting members and twice around the bearers of the scaffold, with each end of the wire rope secured by at least three standard wire-rope clips properly installed.

(3) For hanging wood scaffolds, the following minimum nominal size material shall be used:

(i) Supporting bearers 2 x 10 inches on edge;

(ii) Planking 2 x 10 inches, with maximum span 7 feet for heavy duty and 10 feet for light duty or medium duty.

(4) Steel tube and coupler members may be used for hanging scaffolds with both types of scaffold designed to sustain a uniform distributed working load up to heavy duty scaffold loads with a safety factor of four.

(5) Guardrails made of lumber, not less than 2 x 4 inches (or other material providing equivalent protection), approximately 42 inches high, with a midrail of 1 x 6 inch lumber (or other material providing equivalent protection), and toeboards, shall be installed at all open sides and ends on all scaffolds more than 10 feet above the ground or floor. Toeboards shall be a minimum of 4 inches in height. Wire mesh shall be installed in accordance with paragraph (a)(6) of this section.

(s) *Ladder jack scaffolds.* (1) All ladder jack scaffolds shall be limited to light duty and shall not exceed a height of 20 feet above the floor or ground.

(2) All ladders used in connection with ladder jack scaffolds shall be heavy-duty ladders and shall be designed and constructed in accordance with American National Standards Institute A 14.1-1968, Safety Code for Portable Wood Ladders, and A 14.2-1968, Safety Code for Portable Metal Ladders. Cleated ladders shall not be used for this purpose.

(3) The ladder jack shall be so designed and constructed that it will bear on the side rails in addition to the ladder rungs, or if bearing on rungs only, the bearing area shall be at least 10 inches on each rung.

(4) Ladders used in conjunction with ladder jacks shall be so placed, fastened, held, or equipped with devices so as to prevent slipping.

(5) The wood platform planks shall be not less than 2 inches nominal in thickness. Both metal and wood platform planks shall overlap the bearing surface not less than 12 inches. The span between supports for wood shall not exceed 8 feet. Platform width shall be not less than 18 inches.

(6) Not more than two employees shall occupy any given 8 feet of any ladder jack scaffold at any one time.

(t) *Window jack scaffolds.* (1) Window jack scaffolds shall be used only for the purpose of working at the window opening through which the jack is placed.

(2) Window jacks shall not be used to support planks placed between one window jack and another or for other elements of scaffolding.

(3) Window jack scaffolds shall be provided with guardrails unless safety belts with lifelines are attached and provided for employee.

(4) Not more than one employee shall occupy a window jack scaffold at any one time.

(u) *Roofing brackets.* (1) Roofing brackets shall be constructed to fit the pitch of the roof.

(2) Brackets shall be secured in place by nailing in addition to the pointed metal projections. When it is impractical to nail brackets, rope supports shall be used. When rope supports shall be used, they shall consist of first-grade manila of at least ¾-inch diameter, or equivalent.

(3) A catch platform shall be installed below the working area of roofs more than 16 feet from the ground to eaves with a slope greater than 4 inches in 12 inches without a parapet. In width, the platform shall extend 2 feet beyond the protection of the eaves and shall be provided with a guardrail, midrail, and toeboard. This provision shall not apply where employees engaged in work upon such roofs are protected by a safety belt attached to a lifeline.

(v) *Crawling boards or chicken ladders.* (1) Crawling boards shall be not less than 10 inches wide and 1 inch thick, having cleats 1 x 1½ inches. The cleats shall be equal in length to the width of the board and spaced at equal intervals not to exceed 24 inches. Nails shall be driven through and clinched on the underside. The crawling board shall extend from the ridge pole to the eaves when used in connection with roof construction, repair, or maintenance.

(2) A firmly fastened lifeline of at least ¾-inch diameter rope, or equivalent, shall be strung beside each crawling board for a handhold.

(3) Crawling boards shall be secured to the roof by means of adequate ridge hooks or other effective means.

(w) *Float or ship scaffolds.* (1) Float or ship scaffolds shall not be used to support more than three men and a few light tools, such as those needed for riveting, bolting, and welding. They shall be constructed as designed in paragraphs (w) (2) through (6) of this section unless substitute designs and materials provide equivalent strength, stability, and safety.

(2) The platform shall be not less than 3 feet wide and 6 feet long, made of ¾-inch plywood, equivalent to American Plywood Association Grade B-B, Group I, Exterior, or other similar material.

(3) Under the platform, there shall be two supporting bearers made from 2- x 4-inch, or 1- x 10-inch, rough, "selected lumber," or better, They shall be free of knots or other flaws and project 6 inches beyone the platform on both sides. The ends of the platform shall extend 6 inches beyond the outer edges of the bearers. Each bearer shall be securely fastened to the platform.

(4) An edging of wood not less than ¾ x 1½ inches or equivalent shall be placed around all sides of the platform to prevent tools from rolling off.

(5) Supporting ropes shall be 1-inch diameter manila rope or equivalent, free from deterioration, chemical damage, flaws, or other imperfections. Rope connections shall be such that the platform cannot shift or slip. If two ropes are use with each float, they shall be arranged so as to provide four ends which are to be securely fastened to an overhead support. Each of the two supporting ropes shall be hitched around one end of bearer and pass under the platforms to the other end of the bearer where it is hitched again, leaving sufficient rope at each end for the supporting ties.

(6) Each employee shall be protected by an approved safety lifebelt and lifeline.

(x) *Form scaffolds.* (1) Form scaffolds shall be constructed of wood or other suitable materials, such as steel or aluminum members of known strength characteristics. All scaffolds shall be designed and erected with a minimum safety factor of 4, computed on the basis of the maximum rated load.

(2) All scaffold planking shall be a minimum of 2- x 10-inch nominal Scaffold Grade, as recognized by approved grading rules for the species of lumber used, or equivalent material. Maximum permissible spans shall not exceed 8 feet on centers for 2- x 10-inch nominal planking. Scaffold planks shall be either nailed or bolted to the ledgers or of such length that they overlap the ledgers at least 6 inches. Unsupported projecting ends of scaffolding planks shall be limited to a maximum overhang of 12 inches.

(3) Scaffolds shall not be loaded in excess of the working load for which they were designed.

(4) Figure-four form scaffolds: (i) Figure-four scaffolds are intended for light duty and shall not be used to support loads exceeding 25 pounds per square foot unless specifically designed for heavier loading. For minimum design criteria, see Table L-17.

Table L-17 – Minimum Design Criteria For Figure-Four Form Scaffolds

Members	Dimensions
Uprights	2 X 4 in. or 2 X 6 in.
Outriggers ledgers (two)	1 X 6 in.
Braces	1 X 6 in.
Guardrails	2 X 4 in.
Guardrail height	Approximately 42 in.
Intermediate guardrails	1 X 6 in.
Toeboards	4 in. (minimum)
Maximum length of ledgers	3 ft. 6 in. (unsupported)
Planking	2 X 10 in.
Upright spacing	8 ft. 0 in. (on centers)

(ii) Figure-four form scaffold frames shall be spaced not more than 8 feet on centers and constructed from sound lumber, as follows: The outrigger ledger shall consist of two pieces of 1- x 6-inch or heavier material nailed on opposite sides of the vertical form support. Ledgers shall project not more than 3 feet 6 inches from the outside of the form support and shall be substantially braced and secured to prevent tipping or turning. The knee or angle brace shall intersect the ledger at least 3 feet from the form at an angle of approximately 45°, and the lower end shall be nailed to a vertical support. The platform shall consist of two or more 2- x 10-inch planks, which shall be of such length that they extend at least 6 inches beyond ledgers at each end unless secured to the ledgers. When planks are secured to the ledgers (nailed or bolted), a wood filler strip shall be used between the ledgers. Unsupported projecting ends of planks shall be limited to an overhang of 12 inches.

(15) Metal bracket form scaffolds: (i) Metal brackets or scaffold jacks which are an integral part of the form shall be securely bolted or welded to the form. Folding type brackets shall be either bolted or secured with a locking-type pin when extended for use.

(ii) "Clip-on" or "hook-over" brackets may be used, provided the form walers are bolted to the form or secured by snap ties or shea-bolt extending through the form and securely anchored.

(iii) Metal brackets shall be spaced not more than 8 feet on centers.

(iv) Scaffold planks shall be either bolted to the metal brackets or of such length that they overlap the brackets at each end by at least 6 inches. Unsupported projecting ends of scaffold planks shall be limited to a maximum overhang of 12 inches.

(v) Metal bracket form scaffolds shall be equipped with wood guardrails, intermediate rails, toeboards, and scaffold planks meeting the minimum dimensions shown in Table L-18. (Metal may be substituted for wood, providing it affords equivalent or greater design strength.)

Table L-18 – Minimum Design Criteria For Metal Bracket Form Scaffolds

Members	Dimensions
Uprights	2 X 4 in.
Guardrails	2 X 4 in.
Guardrail height	Approximately 42 in.
Intermediate guardrails	1 X 6 in.
Toeboards	4 in. (minimum)
Planking	2 X 9 in.

(6) Wooden bracket form scaffolds: (i) Wooden bracket form scaffolds shall be an integral part of the form panel. The minimum design criteria set forth herein and in Table L-19 cover scaffolding intended for light duty and shall not be used to support loads exceeding 25 pounds per square foot, unless specifically designed for heavier loading.

(ii) Scaffold planks shall be either nailed or bolted to the ledgers or of such length that they overlap the ledgers at each end by at least 6 inches. Unsupported projecting ends of scaffold planks shall be limited to a maximum overhang of 12 inches.

Table L-19 – Minimum Design Criteria For Wooden Bracket Form Scaffolds

Members	Dimensions
Uprights	2 X 4 in. or 2 X 6 in.
Support ledgers	2 X 6 in.
Maximum scaffold width	3 ft. 6 in.
Braces	1 X 6 in.
Guardrails	2 X 4 in.
Guardrail height	Approximately 42 in.
Intermediate guardrails	1 X 6 in.
Toeboards	4 in. (minimum)
Upright spacing	8 ft. 0 in. (on centers)

(iii) Guardrails and toeboards shall be installed on all open sides and ends of platforms and scaffolding over 10 feet above floor or ground. Guardrails shall be made of lumber 2 x 4 inch nominal dimension (or other material providing equivalent protection), approximately 42 inches high, supported at intervals not to exceed 8 feet. Guardrails shall be equipped with midrails constructed of 1 x 6 inch nominal lumber (or other material providing equivalent protection). Toeboards shall extend not less than 4 inches above the scaffold plank.

(y) *Pump jack scaffolds.* (1) Pump jack scaffolds shall:

(i) Not carry a working load exceeding 500 pounds; and

(ii) Be capable of supporting without failure at least four times the maximum intended load.

(iii) The manufactured components shall not be loaded in excess of the manufacturer's recommended limits.

(2) Pump jack brackets, braces, and accessories shall be fabricated from metal plates and angles. Each pump jack bracket shall have two positive gripping mechanisms to prevent any failure or slippage.

(3) The platform bracket shall be fully decked and the planking secured. Planking, or equivalent, shall conform with paragraph (a) of this section.

(4) (i) When wood scaffold planks are used as platforms, poles used for pump jacks shall not be spaced more than 10 feet center to center. When fabricated platforms are used that fully comply with all other provisions of this paragraph (y), pole spacing may exceed 10 feet center to center.

(ii) Poles shall not exceed 30 feet in height.

(iii) Poles shall be secured to the work wall by rigid triangular bracing, or equivalent, at the bottom, top and other points as necessary, to provide a maximum vertical spacing of not more than 10 feet between braces. Each brace shall be capable of supporting a minimum of 225 pounds tension or compression.

(iv) For the pump jack bracket to pass bracing already installed, an extra brace shall be used approximately 4 feet above the one to be passed until the original brace is reinstalled.

(5) All poles shall bear on mud sills or other adequate firm foundations.

(6) Pole lumber shall be two 2 X 4's, of Douglas fir, or equivalent, straight-grained, clear, free of cross-grain, shakes large loose or dead knots, and other defects which might impair strength.

(7) When poles are constructed of two continuous lengths, they shall be two by fours, spiked together with the seam parallel to the bracket, and with 10d common nails, no more than 12 inches center to center, staggered uniformly from opposite outside edges.

(8) If two by fours are spliced to make up the pole, the splices shall be so constructed as to develop the full strength of the member.

(9) A ladder shall be provided for access to the platform during use.

(10) Not more than two persons shall be permitted at one time upon a pump jack scaffold between any two supports.

(11) Pump jacks scaffolds shall be provided with standard guardrails. No guardrail is required when safety belts with lifelines are provided for employees.

(12) When a work bench is used at an approximate height of 42 inches, the top guardrail may be eliminated, if the work bench is fully decked, the planking secured, and is capable of withstanding 200 pounds pressure in any direction.

(13) Employees shall not be permitted to use a work bench as a scaffold platform.

Definitions applicable to this subpart.
(a) "Ladders." (1) "Cleats" – Ladder crosspieces of rectangular cross section placed on edge which a person may step in ascending or descending.

(2) "Single cleat ladder" – One which consists of a pair of side rails, usually parallel, but with flared side rails permissible, connected together with cleats that are joined to the side rails at regular intervals.

(3) "Double cleat ladder" – One that is similar to a single cleat ladder, but is wider, with an additional center rail which will allow for two-way traffic for workmen in ascending and descending.

(b) "Scaffolding." (1) "Bearer" – A horizontal member of a scaffold upon which the platform rests and which may be supported by ledgers.

(2) "Boatswain's chair" – A seat supported by slings attached to a suspended rope, designed to accommodate one workman in a sitting position.

(3) "Brace" – A tie that holds one scaffold member to a fixed position with respect to another member.

(4) "Bricklayers' square scaffold" – A scaffold composed of framed wood squares which support a platform, limited to light and medium duty.

(5) "Carpenters' bracket scaffold" – A scaffold consisting of wood or metal brackets supporting a platform.

(6) "Coupler" – A device for locking together the component parts of a tubular metal scaffold. (The material used for the couplers shall be of a structural type, such as a drop-forged steel, malleable iron, or structural grade aluminum.)

(7) "Crawling board or chicken ladder" – A plank with cleats spaced and secured at equal intervals, for use by a worker on roofs, not designed to carry any material.

(8) "Double pole or independent pole scaffold" – A scaffold supported from the base by a double row of uprights, independent of support from the walls and constructed of uprights, ledgers, horizontal platform bearers, and diagonal bracing.

(9) "Float or ship scaffold" – A scaffold hung from overhead supports by means of ropes and consisting of a substantial platform having diagonal bracing underneath, resting upon and securely fastened to two parallel plank bearers at right angles to the span.

(10) "Guardrail" – A rail secured to uprights and erected along the exposed sides and ends of platforms.

(11) "Heavy duty scaffold" – A scaffold designed and constructed to carry a working load not to exceed 75 pounds per square foot.

(12) "Horse scaffold" – A scaffold for light or medium duty, composed of horses supporting a work platform.

(13) "Interior hung scaffold" – A scaffold suspended from the ceiling or roof structure.

(14) "Ladder jack scaffold" – A light duty scaffold supported by brackets attached to ladders.

(15) "Ledgers (stringers)" – A horizontal scaffold member which extends from post to post and which supports the putlogs or bearers forming a tie between the posts.

(16) "Light duty scaffold" – A scaffold designed and constructed to carry a working load not to exceed 25 pounds per square foot.

(17) "Manually propelled mobile scaffold" – A portable rolling scaffold supported by casters.

(18) "Masons' adjustable multiple-point suspension scaffold" – A scaffold having a continuous platform supported by bearers suspended by wire rope from overhead supports, so arranged and operated as to permit the raising or lowering of the platform to desired working positions.

(19) "Maximum rated load" – The total of all loads including the working load, the weight of the scaffold, and such other loads as may be reasonably anticipated.

(20) "Medium duty scaffold" – A scaffold designed and constructed to carry a working load not to exceed 50 pounds per square foot.

(21) "Midrail" – A rail approximately midway between the guardrail and platform, secured to the uprights erected along the exposed sides and ends of platforms.

(22) "Needle beam scaffold" – A light duty scaffold consisting of needle beams supporting a platform.

(23) "Outrigger scaffold" – A scaffold supported by outriggers or thrustouts projecting beyond the wall or face of the building or structure, the inboard ends of which are secured inside of such building or structure.

(24) "Putlog" – A scaffold member upon which the platform rests.

(25) "Roofing or bearer bracket" – A bracket used in slope roof construction, having provisions for fastening to the roof or supported by ropes fastened over the ridge and secured to some suitable object.

(26) "Runner" – The lengthwise horizontal bracing or bearing member or both.

(27) "Scaffold" – Any temporary elevated platform and its supporting structure used for supporting both workmen or materials, or both.

(28) "Single-point adjustable suspension scaffold" – A manually or power-operated unit designed for light duty use, supported by a single wire rope from an overhead support so arranged and operated as to permit the raising or lowering of platform to desired working positions.

(29) "Single-pole scaffold" – Platforms resting on putlogs or cross beams, the outside ends of which are supported on ledgers secured to a single row of posts or uprights, and the inner ends of which are supported on or in a wall.

(30) "Stone setters" adjustable multiple-point suspension scaffold" – A swinging type scaffold having a platform supported by hangers suspended at four points so as to permit the raising or lowering of the platform to the desired working position by the use of hoisting machines.

(31) "Toeboard" – A barrier secured along the sides and ends of a platform to guard against the falling of material.

(32) "Tube and coupler scaffold" – An assembly consisting of tubing which serves as posts, bearers, braces, ties and runners, a base supporting the posts, and special couplers which serve to connect the uprights and to join the various members.

(33) "Tubular welded frame scaffold" – A sectional panel or frame metal scaffold substantially built up of prefabricated welded sections which consists of posts and horizontal bearer with intermediate members.

(34) "Two-point suspension scaffold (swinging scaffold)" – A scaffold, the platform of which is supported by hangers (stirrups) at two points, suspended from overhead supports so as to permit the raising or lowering of the platform to the desired working position by tackle or hoisting machines.

(35) "Window jack scaffold" – A scaffold, the platform of which is supported by a bracket or jack which projects through a window opening.

(36) "Working load" – Load imposed by men, materials, and equipment.

Floor and Wall Openings, and Stairways
Guardrails, handrails, and covers.

(a) *General provision.* This subpart shall apply to temporary or emergency conditions where there is danger of employees or materials falling through floor, roof, or wall openings, or from stairways or runways.

(b) *Guarding of floor openings and floor holes.* (1) Floor openings shall be guarded by a standard railing and toeboards or cover, as specified in paragraph (f) of this section. In general, the railing shall be provided on all exposed sides, except at entrances to stairways.

(2) Ladderway floor openings or platforms shall be guarded by standard railings with standard toeboards on all exposed sides, except at entrance to opening, with the passage through the railing either provided with a swinging gate or so offset that a person cannot walk directly into the opening.

(3) Hatchways and chute floor openings shall be guarded by one of the following:

(i) Hinged covers of standard strength and construction and a

standard railing with only one exposed side. When the opening is not in use, the cover shall be closed or the exposed side shall be guarded at both top and intermediate positions by removable standard railings;

(ii) A removable standard railing with toe board on not more than two sides of the opening and fixed standard railings with toeboards on all other exposed sides. The removable railing shall be kept in place when the opening is not in use and should preferably be hinged or otherwise mounted so as to be conveniently replaceable.

(4) Wherever there is danger of falling through a skylight opening, it shall be guarded by a fixed standard railing on all exposed sides or a cover capable of sustaining the weight of a 200-pound person.

(5) Pits and trap-door floor openings shall be guarded by floor opening covers of standard strength and construction. While the cover is not in place, the pit or trap openings shall be protected on all exposed sides by removable standard railings.

(6) Manhole floor openings shall be guarded by standard covers which need not be hinged in place. While the cover is not in place, the manhole opening shall be protected by standard railings.

(7) Temporary floor openings shall have standard railings.

(8) Floor holes, into which persons can accidentally walk, shall be guarded by either a standard railing with standard toeboard on all exposed sides, or a floor hole cover of standard strength and construction that is secured against accidental displacement. While the cover is not in place, the floor hole shall be protected by a standard railing.

(9) Where doors or gates open directly on a stairway, a platform shall be provided, and the swing of the door shall not reduce the effective width of the platform to less than 20 inches.

(c) *Guarding of wall openings.* (1) Wall openings, from which there is a drop of more than 4 feet, and the bottom of the opening is less than 3 feet above the working surface, shall be guarded as follows:

(i) When the height and placement of the opening in relation to the working surface is such that either a standard rail or intermediate rail will effectively reduce the danger of falling, one or both shall be provided;

(ii) The bottom of a wall opening, which is less than 4 inches above the working surface, regardless of width, shall be protected by a standard toeboard or an enclosing screen either of solid construction or as specified in paragraph (f)(7)(ii) of this section.

(2) An extension platform outside a wall opening onto which materials can be hoisted for handling shall have side rails or equivalent guards of standard specifications. One side of an extension platform may have removable railings in order to facilitate handling materials.

(3) When a chute is attached to an opening the provisions of paragraph (c)(1) of this section shall apply, except that a toeboard is not required.

(d) *Guarding of open-sided floors, platforms, and runways.* (1) Every open-sided floor or platform 6 feet or more above adjacent floor or ground level shall be guarded by a standard railing, or the equivalent, as specified in paragraph (f)(1)(i) of this section, on all open sides, except where there is entrance to a ramp, stairway, or fixed ladder. The railing shall be provided with a standard toeboard wherever, beneath the open sides, persons can pass, or there is moving machinery, or there is equipment with which falling materials could create a hazard.

(2) Runways shall be guarded by a standard railing, or the equivalent, as specified in paragraph (f) of this section, on all open sides, 4 feet or more above floor or ground level. Wherever tools, machine parts, or materials are likely to be used on the runway, a toeboard shall also be provided on each exposed side.

(3) Runways used exclusively for special purposes may have the railing on one side omitted where operating conditions necessitate such omission, providing the falling hazard is minimized by using a runway not less than 18 inches wide.

(4) Where employees entering upon runways become thereby exposed to machinery, electrical equipment, or other danger not a falling hazard, additional guarding shall be provided.

(5) Regardless of height, open-sided floors, walkways, platforms, or runways above or adjacent to dangerous equipment, pickling or galvanizing tanks, degreasing units, and similar hazards shall be guarded with a standard railing and toeboard.

(e) *Stairway railings and guards.* (1) Every flight of stairs having four or more risers shall be equipped with standard stair railings or standard handrails as specified below, the width of the stair to be measured clear of all obstructions except handrails:

(i) On stairways less than 44 inches wide having both sides enclosed, at least one handrail, preferably on the right side descending;

(ii) On stairways less than 44 inches wide having one side open, at least one stair railing on the open side;

(iii) On stairways less than 44 inches wide having both sides open, one stair railing on each side;

(iv) On stairways more than 44 inches wide but less than 88 inches wide, one handrail on each enclosed side and one stair railing on each open side;

(v) On stairways 88 or more inches wide, one handrail on each enclosed side, one stair railing on each open side, and one intermediate stair railing located approximately midway of the width.

(2) Winding stairs shall be equipped with a handrail offset to prevent walking on all portions of the treads having width less than 6 inches.

(f) *Standard specifications.* (1) A standard railing shall consist of top rail, intermediate rail, toeboard, and posts, and shall have a vertical height of approximately 42 inches from upper surface of top rail to floor, platform, runway, or ramp level. The top rail shall be smooth-surfaced throughout the length of the railing. The intermediate rail shall be halfway between the top rail and the floor, platform, runway or ramp. The ends of the rails shall not overhang the terminal posts except where such overhand does not constitute a projection hazard. Minumum requirements for standard railings under various types of construction are specified in the following paragraphs:

(i) For wood railings, the posts shall be of at least 2-inch by 4-inch stock spaced not to exceed 8 feet; the toprail shall be at least 2-inch by 4-inch stock; the intermediate rail shall be of at least 1-inch by 6-inch stock.

(ii) For pipe railings, posts and top and intermediate railings shall be at least 1½ inches nominal diameter with post spaced not more than 8 feet on centers.

(iii) For structural steel railings, posts and top and intermediate rails shall be of 2-inch by 2-inch by ⅜-inch angles or other metal shapes of equivalent bending strength, with posts spaced not more than 8 feet on centers.

(iv) The anchoring of posts and framing of members for railings of all types shall be of such construction that the completed structure shall be capable of withstanding a load of at least 200 pounds applied in any direction at any point on the top rail, with a minimum of deflection.

(v) Railings receiving heavy stresses from employees trucking or handling materials shall be provided additional strength by the use of heavier stock, closer spacing of posts, bracing, or by other means.

(vi) Other types, sizes, and arrangements of railing construction are acceptable, provided they meet the following conditions:

(a) A smooth-surfaced top rail at a height above floor, platform, runway, or ramp level of approximately 42 inches;

(b) A strength to withstand at least the minimum requirement of 200 pounds top rail pressure with a minimum of deflection;

(c) Protection between top rail and floor, platform, runway, ramp or stair treads, equivalent at least to that afforded by a standard intermediate rail;

(d) Elimination of overhang of rail ends unless such overhang does not constitute a hazard.

(2) A stair railing shall be of construction similar to a standard railing, but the vertical height shall be not more than 34 inches nor less than 30 inches from upper surface of top rail to surface of tread in line with face of riser at forward edge of tread.

(3) (i) A standard toeboard shall be 4 inches minimum in vertical height from its top edge to the level of the floor, platform, runway, or ramp. It shall be securely fastened in place and have not more than ¼-inch clearance above floor level. It may be made of any substantial material, either solid, or with openings not over 1 inch in greatest dimension.

(ii) Where material is piled to such height that a standard toeboard does not provide protection, paneling or screening from floor to intermediate rail or to top rail shall be provided.

(4) (i) A standard handrail shall be of construction similar to a standard railing except that it is mounted on a wall or partition, and does not include an intermediate rail. It shall have a smooth surface along the top and both sides of the handrail. The handrail shall have an adequate handhold for any one grasping it to avoid falling. Ends of the handrail shall be constructed so as not to constitute a projection hazard.

(ii) The height of handrails shall be not more than 34 inches nor less than 30 inches from upper surface of handrail to surface of tread, in line with face of riser or to surface of ramp.

(iii) All handrails and railings shall be provided with a clearance of approximately 3 inches between the handrail or railing and any other object.

(5) Floor opening covers shall be of any material that meets the following strength requirements:

(i) Conduits, trenches, and manhole covers and their supports, when located in roadways, and vehicular aisles, shall be designed to carry a truck rear-axle load of at least 2 times the maximum intended load;

(ii) The floor opening cover shall be capable of supporting the maximum intended load and so installed as to prevent accidental displacement.

(6) Skylight openings that create a falling hazard shall be guarded with standard railing, or covered in accordance with paragraph (f)(5)(ii) of this section.

(7) Wall opening protection shall meet the following requirements:

(i) Barriers shall be of such construction and mounting that, when in place at the opening, the barrier is capable of withstanding a load of at least 200 pounds applied in any direction (except upward), with a minimum of deflection at any point on the top rail or corresponding member.

(ii) Screens shall be of such construction and mounting that they are capable of withstanding a load of at least 200 pounds applied horizontally at any point on the near side of the screen. They may be of solid construction, of grill work with openings not more than 8 inches long, or of slat work with openings not more than 4 inches wide with length unrestricted.

(g) *Guarding of low-pitched roof perimeters during the performance of built-up roofing work* – (1) *General provisions.* During the performance of built-up roofing work on low-pitched roof with a ground to eave height greater than 16 feet (4.9 meters), employees engaged in such work shall be protected from falling from all unprotected sides and edges of the roof as follows:

(i) By the use of a motion-stopping safety system (MSS system); or

(ii) By the use of a warning line system erected and maintained as provided in paragraph (g)(3) of this section and supplemented for employees working between the warning line and the roof edge by the use of either an MSS system or, where mechanical equipment is not being used or stored, by the use of a safety monitoring system; or

(iii) By the use of a safety monitoring system on roofs fifty feet (15.25 meters) or less in width where mechanical equipment is not being used or stored.

(2) *Exception.* The provisions of paragraph (g)(1) of this section do not apply at points of access such as stairways, ladders, and ramps, or when employees are on the roof only to inspect, investigate, or estimate roof level conditions. Roof edge materials handling areas and materials storage areas shall be guarded as provided in paragraph (g)(5) of this section.

(3) *Warning lines.* (i) Warning lines shall be erected around all sides of the work area.

(a) When mechanical equipment is being used, the warning line shall be erected not less than six feet (1.8 meters) from the roof edge.

(b) When mechanical equipment is being used, the warning line shall be erected not less that six feet (1.8 meters) from the roof edge which is parallel to the direction of mechanical equipment operation, and not less than 10 feet (3.1 meters) from the roof edge which is perpendicular to the direction of mechanical equipment operation.

(ii) The warning line shall consist of a rope, wire, or chain, and supporting stanchions erected as follows:

(a) The rope, wire, or chain shall be flagged at not more than six foot (1.8 meters) intervals with high-visibility material;

(b) The rope, wire, or chain shall be rigged and supported in such a way that its lowest point (include sag) is no less than 34 inches (.86 meters) from the roof surface and its highest point is no more than 39 inches (1 meter) from the roof surface;

(c) After being erected, with the rope, wire, or chain attached, stanchions shall be capable of resisting, without tipping over, a force of at least 16 pounds (71 Newtons) applied horizontally against the stanchion, 30 inches (0.76 meters) above the roof surface, perpendicular to the warning line, and in the direction of the roof edge;

(d) The rope, wire, or chain shall have a minimum tensile strength of 500 pounds (227 Kilograms), and after being attached to the stanchions, shall be capable of supporting, without breaking, the loads applied to the stanchions as prescribed in paragraph (g)(3)(ii)(c) of this section; and

(e) The line shall be attached at each stanchion in such a way that pulling on one section of the line between stanchions will not result in slack being taken up in adjacent sections before the stanchion tips over.

(iii) Access paths shall be erected as follows:

(a) Points of access, materials handling areas and storage areas shall be connected to the work area by a clear access path formed by two warning lines.

(b) When the path to a point of access is not in use, a rope, wire, or chain, equal in strength and height to the warning line, shall be placed across the path at the point where the path intersects the warning line erected around the work area.

(4) *Mechanical equipment.* Mechanical equipment may be used or stored only in areas where employees are being protected by either a warning line or an MSS system. Mechanical equipment may not be used or stored between the warning line and the roof edge unless the employees are being protected by an MSS system. Mechanical equipment may not be used or stored where the only protection provided is by a safety monitoring system.

(5) *Roof edge materials handling areas and materials storage.* Employees working in a roof edge materials handling or materials storage area located on a low-pitched roof with a ground to eave height greater than 16 feet (4.9 meters) shall be protected from falling by the use of an MSS system along all unprotected roof sides and edges of the area.

(i) When guardrails are used at hoisting areas, a minimum of four feet of guardrail shall be erected on each side of the access point through which materials are hoisted.

(ii) A chain or gate shall be placed across the opening between the guardrail sections when hoisting operations are not taking place.

(iii) When guardrails are used at bitumen pipe outlets, a minimum of four feet of guardrail shall be erected on each side of the pipe.

(iv) When safety belt systems are used, they shall not be attached to the hoist.

(v) When safety belt systems are used they shall be rigged to allow the movement of employees only as far as the roof edge.

(vi) Materials may not be stored within six feet of the roof edge unless guardrails are erected at the roof edge.

(vii) Materials which are piled, grouped, or stacked shall be stable and self-supporting.

(6) *Training.* (i) The employer shall provide a training program for all employees engaged in built-up roofing work so that they are able to recognize and deal with the hazards of falling associated with working near a roof perimeter. The employees shall also be trained in the safety procedures to be followed in order to prevent such falls.

(ii) The employer shall assure that employees engaged in built-up roofing work have been trained and instructed in the following areas:

(a) The nature of fall hazards in the work area near a roof edge;

(b) The function, use, and operation of the MSS system, warning line, and safety monitoring systems to be used;

(c) The correct procedures for erecting, maintaining, and disassembling the systems to be used;

(d) The role of each employee in the safety monitoring system when this system is used;

(e) The limitations on the use of mechanical equipment; and

(f) The correct procedures for the handling and storage of equipment and materials.

Stairways.

(a) On all structures, two or more floors (20 feet or over) in height, stairways, ladders, or ramps shall be provided for employees during the construction period.

(c) All parts of stairways shall be free of hazardous projections, such as protruding nails.

(d) Debris, and other loose materials, shall not be allowed on or under stairways.

(e) Slippery conditions on stairways shall be eliminated as soon as possible after they occur.

(f) Permanent steel or other metal stairways, and landings with hollow pan-type treads that are to be filled with concrete or other materials, when used during construction, shall be filled to the level of the nosing with solid material. The requirement shall not apply during the period of actual construction of the stairways themselves.

(g) Wooden treads for temporary service shall be full width.

(h) Metal landings shall be secured in place before filling.

(i) Temporary stairs shall have a landing not less than 30 inches in the direction of travel at every 12 feet of vertical rise.

(j) Stairs shall be installed at angles to the horizontal of between 30° and 50°.

(k) Rise height and tread width shall be uniform throughout any flight of stairs including any foundation structure used as one or more treads of the stairs.

(m) Spiral stairways shall not be permitted except for special limited usage and secondary access situations where it is not practical to provide a conventional stairway.

Cranes, Derricks, Hoists, Elevators, and Conveyors

Cranes and derricks.

(a) *General requirements.* (1) The employer shall comply with the manufacturer's specifications and limitations applicable to the operation of any and all cranes and derricks. Where manufacturer's specifications are not available, the limitations assigned to the equipment shall be based on the determination of a qualified engineer competent in this field and such determinations will be appropriately documented and recorded.

Attachments used with cranes shall not exceed the capacity, rating, or scope recommended by the manufacturer.

(2) Rated load capacities, and recommended operating speeds, special hazard warnings, or instructions, shall be conspicuously posted on all equipment. Instructions or warnings shall be visible to the operator while he is at his control station.

(4) Hand signals to crane and derrick operators shall be those prescribed by the applicable ANSI standard for the type of crane in use. An illustration of the signals shall be posted at the job site.

(5) The employer shall designate a competent person who shall inspect all machinery and equipment prior to each use, and during use, to make sure it is in safe operating condition. Any deficiencies shall be repaired, or defective parts replaced, before continued use.

(6) A thorough, annual inspection of the hoisting machinery shall be made by a competent person, or be a government or private agency recognized by the U.S. Department of Labor. The employer shall maintain a record of the dates and results of inspections for each hoisting machine and piece of equipment.

(7) Wire rope shall be taken out of service when any of the following conditions exist:

(i) In running ropes, six randomly distributed broken wires in one lay or three broken wires in one strand in one lay;

(ii) Wear of one-third the original diameter of outside individual wires. Kinking, crushing, bird caging, or any other damage resulting in distortion of the rope structure;

(iii) Evidence of any heat damage from any cause;

(iv) Reductions from nominal diameter of more than one-sixty-fourth inch for diameters up to and including five-sixteenths inch, one-thirty-second inch for diameters three-eighths inch to and including one-half inch, three-sixty-fourths inch for diameters nine-sixteenths inch to and including three-fourths inch, one-sixteenth inch to 1⅛ inches inclusive, three-thirty-seconds inch for diameters 1¼ to 1½ inclusive;

(v) In standing ropes, more than two broken wires in one lay in sections beyond end connections or more than one broken wire at an end connection.

(vi) Wire rope safety factors shall be in accordance with American National Standards Institute B 30.5-1968 or SAE J959-1966.

(8) Belts, gears, shafts, pulleys, sprockets, spindles, drums, fly wheels, chains, or other reciprocating, rotating, or other moving parts or equipment shall be guarded if such parts are exposed to contact by employees, or otherwise create a hazard. Guarding shall meet the requirements of the American National Standards Institute B 15.1-1958 Rev., Safety Code for Mechanical Power Transmission Apparatus.

(9) Accessible areas within the swing radius of the rear of the rotating superstructure of the crane, either permanently or temporarily mounted, shall be barricaded in such a manner as to prevent an employee from being struck or crushed by the crane.

(10) All exhaust pipes shall be guarded or insulated in areas where contact by employees is possible in the performance of normal duties.

(11) Whenever internal combustion engine powered equipment exhausts in enclosed spaces, tests shall be made and recorded to see that employees are not exposed to unsafe concentrations of toxic gases or oxygen deficient atmospheres.

(12) All windows in cabs shall be of safety glass, or equivalent, that introduces no visible distortion that will interfere with the safe operation of the machine.

(13) (i) Where necessary for rigging or service requirements, a ladder, or steps, shall be provided to give access to a cab roof.

(ii) Guardrails, handholds, and steps shall be provided on cranes for easy access to the car and cab, conforming to American National Standards Institute B30.5

(iii) Platforms and walkways shall have anti-skid surfaces.

(14) Fuel tank filler pipe shall be located in such a position, or protected in such manner, as to not allow spill or overflow to run onto the engine, exhaust, or electrical equipment of any machine being fueled.

(i) An accessible fire extinguisher of 5BC rating, or higher, shall be available at all operator stations or cabs of equipment.

(ii) When fuel is transported by vehicles on public highways, Department of Transportation rules contained in 49 CFR Parts 177 and 393 concerning such vehicular transportation are considered applicable.

(15) Except where electrical distribution and transmission lines have been deenergized and visibly grounded at point of work or where insulating barriers, not a part of or an attachment to the equipment or machinery, have been erected to prevent physical contact with the lines, equipment or machines shall be operated proximate to power lines only in accordance with the following:

(i) For lines rated 50 kV. or below, minimum clearance between the lines and any part of the crane or load shall be 10 feet;

(ii) For lines rated over 50 kV., minimum clearance between the lines and any part of the crane or load shall be 10 feet plus 0.4 inch for each 1 kV. over 50 kV., or twice the length of the line insulator, but never less than 10 feet;

(iii) In transit with no load and boom lowered, the equipment clearance shall be a minimum of 4 feet for voltages less than 50 kV., and 10 feet for voltages over 50 kV., up to and including 345 kV., and 16 feet for voltages up to and including 750 kV.

(iv) A person shall be designated to observe clearance of the equipment and give timely warning for all operations where it is difficult for the operator to maintain the desired clearance by visual means;

(v) Cage-type boom guards, insulating links, or proximity warning devices may be used on cranes, but the use of such devices shall not alter the requirements of any other regulation of this part even if such device is required by law or regulation;

(vi) Any overhead wire shall be considered to be an energized line unless and until the person owning such line or the electrical utility authorities indicate that it is not an energized line and it has been visibly grounded;

(vii) Prior to work near transmitter towers where an electrical charge can be induced in the equipment or materials being handled, the transmitter shall be de-energized or tests shall be made to determine if electrical charge is induced on the crane. The following precautions shall be taken when necessary to dissipate induced voltages:

(a) The equipment shall be provided with an electrical ground directly to the upper rotating structure supporting the boom; and

(b) Ground jumper cables shall be attached to materials being handled by boom equipment when electrical charge is induced while working near energized transmitters. Crews shall be provided with nonconductive poles having large alligator clips or other similar protection to attach the ground cable to the load.

(c) Combustible and flammable materials shall be removed from the immediate area prior to operations.

(16) No modifications or additions which affect the capacity or safe operation of the equipment shall be made by the employer without the manufacturer's written approval. If such modifications or changes are made, the capacity, operation, and maintenance instruction plates, tags, or decals, shall be changed accordingly. In no case shall the original safety factor of the equipment be reduced.

(17) The employer shall comply with Power Crane and Shovel Association Mobile Hydraulic Crane Standard No. 2.

(18) Sideboom cranes mounted on wheel or crawler tractors shall meet the requirements of SAE J743a-1964.

(b) *Crawler, locomotive, and truck cranes.* (1) All jibs shall have positive stops to prevent their movement of more than 5° above the straight line of the jib and boom on conventional type crane booms. The use of cable type belly slings does not constitute compliance with this rule.

(2) All crawler, truck, or locomotive cranes in use shall meet the applicable requirements for design, inspection, construction, testing, maintenance and operation as prescribed in the ANSI B30.5-1968, Safety Code for Crawler, Locomotive and Truck Cranes.

(c) *Hammerhead tower cranes.* (1) Adequate clearance shall be maintained between moving and rotating structures of the crane and fixed objects to allow the passage of employees without harm.

(2) Employees required to perform duties on the horizontal boom of hammerhead tower cranes shall be protected against falling by guardrails or by safety belts and lanyards attached to lifelines.

(3) Buffers shall be provided at both ends of travel of the trolley.

(4) Cranes mounted on rail tracks shall be equipped with limit switches limiting the travel of the crane on the track and stops or buffers at each end of the tracks.

(5) All hammerhead tower cranes in use shall meet the applicable requirements for design, construction, installation, testing, maintenance, inspection, and operation as prescribed by the manufacturer.

(d) *Overhead and gantry cranes.* (1) The rated load of the crane shall be plainly marked on each side of the crane, and if the crane has more than one hoisting unit, each hoist shall have its rated load marked on it or its load block, and this marking shall be clearly legible from the ground or floor.

(2) Bridge trucks shall be equipped with sweeps which extend below the top of the rail and project in front of the truck wheels.

(3) Except for floor-operated cranes, a gong or other effective audible warning signal shall be provided for each crane equipped with a power traveling mechanism.

(4) All overhead and gantry cranes in use shall meet the applicable requirements for design, construction, installation, testing, maintenance, inspection, and operation as prescribed in the ANSI B30.2.0-1967, Safety Code for Overhead and Gantry Cranes.

(e) *Derricks.* All derricks in use shall meet the applicable requirements for design, construction, installation, inspection, testing, maintenance, and operation as prescribed in American National Standards Institute B30.6-1969, Safety Code for Derricks.

(f) *Floating cranes and derricks* — (1) *Mobile cranes mounted on barges.* (i) When a mobile crane is mounted on a barge, the rated load of the crane shall not exceed the original capacity specified by the manufacturer.

(ii) A load rating chart, with clearly legible letters and figures, shall be provided with each crane, and securely fixed at a location easily visible to the operator.

(iii) When load ratings are reduced to stay within the limits for list of the barge with a crane mounted on it, a new load rating chart shall be provided.

(iv) Mobile cranes on barges shall be positively secured.

(2) *Permanently mounted floating cranes and derricks.* (i) When cranes and derricks are permanently installed on a barge, the capacity and limitations of use shall be based on competent design criteria.

(ii) A load rating chart with clearly legible letters and figures shall be provided and securely fixed at a location easily visible to the operator.

(iii) Floating cranes and floating derricks in use shall meet the applicable requirements for design, construction, installation, testing, maintenance, and operation as prescribed by the manufacturer.

Helicopters

(a) *Helicopter regulations.* Helicopter cranes shall be expected to comply with any applicable regulations of the Federal Aviation Administration.

(b) *Briefing.* Prior to each day's operation a briefing shall be conducted. This briefing shall set forth the plan of operation for the pilot and ground personnel.

(c) *Slings and tag lines.* Load shall be properly slung. Tag lines shall be of a length that will not permit their being drawn up into rotors. Pressed sleeve, swedged eyes, or equivalent means shall be used for all freely suspended loads to prevent hand splices from spinning open or cable clamps from loosening.

(d) *Cargo hooks.* All electrically operated cargo hooks shall have the electrical activating device so designed and installed as to prevent inadvertent operation. In addition, these cargo hooks shall be equipped with an emergency mechanical control for releasing the load. The hooks shall be tested prior to each day's operation to determine that the release functions properly, both electrically and mechanically.

(e) *Personal protective equipment.* (1) Personal protective equipment for employees receiving the load shall consist of complete eye protection and hard hats secured by chinstraps.

(2) Loose-fitting clothing likely to flap in the downwash, and thus be snagged on hoist line, shall not be worn.

(f) *Loose gear and objects.* Every practical precaution shall be taken to provide for the protection of the employees from fly-ing objects in the rotor downwash. All loose gear within 100 feet of the place of lifting the load, depositing the load, and all other areas susceptible to rotor downwash shall be secured or removed.

(g) *Housekeeping.* Good housekeeping shall be maintained in all helicopter loading and unloading areas.

(h) *Operator responsibility.* The helicopter operator shall be responsible for size, weight, and manner in which loads are connected to the helicopter. If, for any reason, the helicopter operator believes the lift cannot be made safely, the lift shall not be made.

(i) *Hooking and unhooking loads.* When employees are required to perform work under hovering craft, a safe means of access shall be provided for employees to reach the hoist line hook and engage or disengage cargo slings. Employees shall not perform work under hovering craft except when necessary to hook or unhook loads.

(j) *Static charge.* Static charge on the suspended load shall be dissipated with a grounding device before ground personnel touch the suspended load, or protective rubber gloves shall be worn by all ground personnel touching the suspended load.

(k) *Weight limitation.* The weight of an external load shall not exceed the manufacturer's rating.

(l) *Ground lines.* Hoist wires or other gear, except for pulling lines or conductors that are allowed to "pay out" from a container or roll off a reel, shall not be attached to any fixed ground structure, or allowed to foul on any fixed structure.

(m) *Visibility.* When visibility is reduced by dust or other conditions, ground personnel shall exercise special caution to keep clear of main and stabilizing rotors. Precautions shall also be taken by the employer to eliminate as far as practical reduced visibility.

(n) *Signal systems.* Signal systems between aircrew and ground personnel shall be understood and checked in advance of hoisting the load. This applies to either radio or hand signal systems.

(o) *Approach distance.* No unauthorized person shall be allowed to approach within 50 feet of the helicopter when the rotor blades are turning.

(p) *Approaching helicopter.* Whenever approaching or leaving a helicopter with blades rotating, all employees shall remain in full view of the pilot and keep in a crouched position. Employees shall avoid the area from the cockpit or cabin rearward unless authorized by the helicopter operator to work there.

(q) *Personnel.* Sufficient ground personnel shall be provided when required for safe helicopter loading and unloading operations.

(r) *Communications.* There shall be constant reliable communication between the pilot, and a designated employee of the ground crew who acts as a signalman during the period of loading and unloading. This signalman shall be distinctly recognizable from other ground personnel.

(s) *Fires.* Open fires shall not be permitted in an area that could result in such fires being spread by the rotor downwash.

Material hoists, personnel hoists, and elevators.

(a) *General requirements.* (1) The employer shall comply with the manufacturer's specifications and limitations applicable to the operation of all hoists and elevators. Where manufacturer's specifications are not available, the limitations assigned to the equipment shall be based on the determinations of a professional engineer competent in the field.

(2) Rated load capacities, recommended operating speeds, and special hazard warnings or instructions shall be posted on cars and platforms.

(3) Wire rope shall be removed from service when any of the following conditions exists:

(i) In hoisting ropes, six randomly distributed broken wires in one rope lay or three broken wires in one strand in one rope lay;

(ii) Abrasion, scrubbing, flattening, or peening, causing loss of more than one-third of the original diameter of the outside wires;

(iii) Evidence of any heat damage resulting from a torch or any damage caused by contact with electrical wires;

(iv) Reduction from nominal diameter of more than three sixty-fourths inch for diameters up to and including three-fourths inch; one-sixteenth inch for diameters seven-eighths to 1⅛ inches; and three thirty-seconds inch for diameters 1¼ to 1½ inches.

(4) Hoisting ropes shall be installed in accordance with the wire rope manufacturers' recommendations.

(5) The installation of live booms on hoists is prohibited.

(6) The use of endless belt-type manlifts on construction shall be prohibited.

(b) *Material hoists.* (1) (i) Operating rules shall be established and posted at the operator's station of the hoist. Such rules shall include signal system and allowable line speed for various loads. Rules and notices shall be posted on the car frame or crosshead in a conspicuous location, including the statement "No Riders Allowed."

(ii) No person shall be allowed to ride on material hoists except for the purposes of inspection and maintenance.

(2) All entrances of the hoistways shall be protected by substantial gates or bars which shall guard the full width of the landing entrance. All hoistway entrance bars and gates shall be painted with diagonal contrasting colors, such as black and yellow stripes.

(i) Bars shall be not less than 2- by 4-inch wooden bars or the equivalent, located 2 feet from the hoistway line. Bars shall be located not less than 36 inches nor more than 42 inches above the floor.

(ii) Gates or bars protecting the entrances to hoistways shall be equipped with a latching device.

(3) Overhead protective covering of 2-inch planking, ¾-inch plywood, or other solid material of equivalent strength, shall be provided on the top of every material hoist cage or platform.

(4) The operator's station of a hoisting machine shall be provided with overhead protection equivalent to tight planking not less than 2 inches thick. The support for the overhead protection shall be of equal strength.

(5) Hoist towers may be used with or without an enclosure on all sides. However, whichever alternative is chosen, the following applicable conditions shall be met:

(i) When a hoist tower is enclosed, it shall be enclosed on all sides for its entire height with a screen enclosure of ½-inch mesh, No. 18 U.S. gauge wire or equivalent, except for landing access.

(ii) When a hoist tower is not enclosed, the hoist platform or car shall be totally enclosed (caged) on all sides for the full height between the floor and the overhead protective covering with ½-inch mesh of No. 14 U.S. gauge wire or equivalent. The hoist platform enclosure shall include the required gates for loading and unloading. A 6-foot high enclosure shall be provided on the unused sides of the hoist tower at ground level.

(6) Car arresting devices shall be installed to function in case of rope failure.

(7) All material hoist towers shall be designed by a licensed professional engineer.

(8) All material hoists shall conform to the requirements of ANSI A10.5-1969, Safety Requirements for Material Hoists.

(c) *Personnel hoists.* (1) Hoist towers outside the structure shall be enclosed for the full height on the side or sides used for entrance and exit to the structure. At the lowest landing, the enclosure on the sides not used for exit or entrance to the structure shall be enclosed to a height of at least 10 feet. Other sides of the tower adjacent to floors or scaffold platforms shall be enclosed to a height of 10 feet above the level of such floors or scaffolds.

(2) Towers inside of structures shall be enclosed on all four sides throughout the full height.

(3) Towers shall be anchored to the structure at intervals not exceeding 25 feet. In addition to tie-ins, a series of guys shall be installed. Where tie-ins are not practical the tower shall be anchored by means of guys made of wire rope at least one-half inch in diameter, securely fastened to anchorage to ensure stability.

(4) Hoistway doors or gates shall be not less than 6 feet 6 inches high and shall be provided with mechanical locks which cannot be operated from the landing side, and shall be accessible only to persons on the car.

(5) Cars shall be permanently enclosed on all sides and the top, except sides used for entrance and exit which have car gates or doors.

(6) A door or gate shall be provided at each entrance to the car which shall protect the full width and height of the car entrance opening.

(7) Overhead protective covering of 2-inch planking, ¾-inch plywood or other solid material or equivalent strength shall be provided on the top of every personnel hoist.

(8) Doors or gates shall be provided with electric contacts which do not allow movement of the hoist when door or gate is open.

(9) Safeties shall be capable of stopping and holding the car and rated load when traveling at governor tripping speed.

(10) Cars shall be provided with a capacity and data plate secured in a conspicuous place on the car or crosshead.

(11) Internal combustion engines shall not be permitted for direct drive.

(12) Normal and final terminal stopping devices shall be provided.

(13) An emergency stop switch shall be provided in the car and marked "Stop."

(14) Ropes: (i) The minimum number of hoisting ropes used shall be three for traction hoists and two for drum-type hoists.

(ii) The minimum diameter of hoisting and counterweight wire ropes shall be ½-inch.

(iii) Safety factors:

Minimum Factors of Safety for Suspension Wire Ropes

Rope speed in feet per minute	Minimum factor of safety
50	7.60
75	7.75
100	7.95
125	8.10
150	8.25
175	8.40
200	8.60
225	8.75
250	8.90
300	9.20
350	9.50
400	9.75
450	10.00
500	10.25
550	10.45
600	10.70

(15) Following assembly and erection of hoists, and before being put in service, an inspection and test of all functions and safety devices shall be made under the supervision of a competent person. A similar inspection and test is required following major alteration of an existing installation. All hoists shall be inspected and tested at not more than 3-month intervals. Records shall be maintained and kept on file for the duration of the job.

(16) All personnel hoists used by employees shall be constructed of materials and components which meet the specifications for materials, construction, safety devices, assembly, and structural integrity as stated in the American National Standard A10.4-1963, Safety Requirements for Workmens' Hoists. The requirements of this paragraph (c) (16) do not apply to cantilever type personnel hoists.

(17) (i) Pesonnel hoists used in bridge tower construction shall be approved by a registered professional engineer and erected under the supervision of a qualified engineer competent in this field.

(ii) When a hoist tower is not enclosed, the hoist platform or car shall be totally enclosed (caged) on all sides for the full height between the floor and the overhead protective covering with ¾-inch mesh of No. 14 U.S. gauge wire or equivalent. The hoist platform enclosure shall include the required gates for loading and unloading.

(iii) These hoists shall be inspected and maintained on a weekly basis. Whenever the hoisting equipment is exposed to winds exceeding 35 miles per hour it shall be inspected and put in operable condition before reuse.

(iv) Wire rope shall be taken out of service when any of the following conditions exist:

(a) In running ropes, six randomly distributed broken wires in one lay or three broken wires in one strand in one lay;

(b) Wear of one-third the original diameter of outside individual wires. Kinking, crushing, bird caging, or any other damage resulting in distortion of the rope structure;

(c) Evidence of any heat damage from any cause;

(d) Reductions from nominal diameter of more than three-sixty-fourths inch for diameters to and including three-fourths inch, one-sixteenth inch for diameters seven-eighths inch to 1⅛ inches inclusive, three-thirty-seconds inch for diameters 1¼ to 1½ inches inclusive;

(e) In standing ropes, more than two broken wires in one lay in sections beyond end connections or more than one broken wire at an end connection.

(d) Permanent elevators under the care and custody of the employer and used by employees for work covered by this Act shall comply with the requirements of American National Standards Institute A17.1-1965 with addenda A17.1a-1967, A17.1b-1968, A17.1c-1969, A17.1d-1970, and inspected in accordance with A17.2-1960 with addenda A17.2a-1965, A17.2b-1967.

Base-mounted drum hoists.

(a) General requirements. (1) Exposed moving parts such as gears, projecting screws, setscrews, chain, cables, chain sprockets, and reciprocating or rotating parts, which constitute a hazard, shall be guarded.

(2) All controls used during the normal operation cycle shall be located within easy reach of the operator's station.

(3) Electric motor operated hoists shall be provided with:

(i) A device to disconnect all motors from the line upon power failure and not permit any motor to be restarted until the controller handle is brought to the "off" position;

(ii) Where applicable, an overspeed preventive device;

(iii) A means whereby remotely operated hoists stop when any control is ineffective.

(4) All base-mounted drum hoists in use shall meet the applicable requirements for design, construction, installation, testing, inspection, maintenance, and operations, as prescribed by the manufacturer.

Overhead hoists.

(a) General requirements. (1) The safe working load of the overhead hoist, as determined by the manufacturer, shall be indicated on the hoist, and this safe working load shall not be exceeded.

(2) The supporting structure to which the hoist is attached shall have a safe working load equal to that of the hoist.

(3) The support shall be arranged so as to provide for free movement of the hoist and shall not restrict the hoist from lining itself up with the load.

(4) The hoist shall be installed only in locations that will permit the operator to stand clear of the load at all times.

(5) Air hoists shall be connected to an air supply of sufficient capacity and pressure to safely operate the hoist. All air hoses supplying air shall be positively connected to prevent their becoming disconnected during use.

(6) All overhead hoists in use shall meet the applicable requirements for construction, design, installation, testing, inspection, maintenance, and operation, as prescribed by the manufacturer.

Conveyors.

(a) *General requirements.* (1) Means for stopping the motor or engine shall be provided at the operator's station. Conveyor systems shall be equipped with an audible warning signal to be sounded immediately before starting up the conveyor.

(2) If the operator's station is at a remote point, similar provisions for stopping the motor or engine shall be provided at the motor or engine location.

(3) Emergency stop switches shall be arranged so that the conveyor cannot be started again until the actuating stop switch has been reset to running or "on" position.

(4) Screw conveyors shall be guarded to prevent employee contact with turning flights.

(5) Where a conveyor passes over work areas, aisles, or thoroughfares, suitable guards shall be provided to protect employees required to work below the conveyors.

(6) All crossovers, aisles, and passageways shall be conspicuously marked by suitable signs.

(7) Conveyors shall be locked out or otherwise rendered inoperable, and tagged out with a "Do Not Operate" tag during repairs and when operation is hazardous to employees performing maintenance work.

(8) All conveyors in use shall meet the applicable requirements for design, construction, inspection, testing, maintenance, and operation, as prescribed in the ANSI B20.1-1957, Safety Code for Conveyors, Cableways and Related Equipment.

Aerial lifts.

(a) *General requirements.* (1) Unless otherwise provided in this section, aerial lifts acquired for use on or after the effective date of this section shall be designed and constructed in conformance with the applicable requirements of the American National Standards for "Vehicle Mounted Elevating and Rotating Work Platforms," ANSI A92.2-1969, including appendix. Aerial lifts acquired before the effective date of this section, which do not meet the requirements of ANSI A92.2-1969, may not be used after January 1, 1976, unless they shall have been modified so as to conform with the applicable design and construction requirements of ANSI A92.2-1969. Aerial lifts include the following types of vehicle-mounted aerial devices used to elevate personnel to job-sites above ground; (i) Extensible boom platforms, (ii) aerial ladders, (iii) articulating boom platforms, (iv) vertical towers, and (v) a combination of any of the above. Aerial equipment may be made of metal, wood, fiberglass reinforced plastic (FRP), or other material; may be powered or manually operated; and are deemed to be aerial lifts whether or not they are capable of rotating about a substantially vertical axis.

(2) Aerial lifts may be "field modified" for uses other than those intended by the manufacturer provided the modification has been certified in writing by the manufacturer or by any other equivalent entity, such as a nationally recognized testing laboratory, to be in conformity with all applicable provisions of ANSI A92.2-1969 and this section and to be at least as safe as the equipment was before modification.

(b) *Specific requirements* — (1) *Ladder trucks and tower trucks.* Aerial ladders shall be secured in the lower traveling position by the locking device on top of the truck cab, and the manually operated device at the base of the ladder before the truck is moved for highway travel.

(2) *Extensible and articulating boom platforms.* (i) Lift controls shall be tested each day prior to use to determine that such controls are in safe working condition.

(ii) Only authorized persons shall operate an aerial life.

(iii) Belting off to an adjacent pole, structure, or equipment while working from an aerial life shall not be permitted.

(iv) Employees shall always stand firmly on the floor of the basket, and shall not sit or climb on the edge of the basket or use planks, ladders, or other devices for a work position.

(v) A body belt shall be worn and a lanyard attached to the boom or basket when working from an aerial lift.

(vi) Boom and basket load limits specified by the manufacturer shall not be exceeded.

(vii) The brakes shall be set and when outriggers are used, they shall be positioned on pads or a solid surface. Wheel chocks shall be installed before using an aerial lift on an incline, provided they can be safely installed.

(viii) An aerial lift truck shall not be moved when the boom is elevated in a working position with men in the basket, except for equipment which is specifically designed for this type of operation in accordance with the provisions of paragraphs (a) (1) and (2) of this section.

(ix) Articulating boom and extensible boom platforms, primarily designed as personnel carriers, shall have both platform (upper) and lower controls. Upper controls shall be in or beside the platform within easy reach of the operator. Lower controls shall provide for overriding the upper controls. Controls shall be plainly marked as to their function. Lower level controls shall not be operated unless permission has been obtained form the employee in the lift, except in case of emergency.

(x) Climbers shall not be worn while performing work from an aerial lift.

(xi) The insulated portion of an aerial lift shall not be altered in any manner that might reduce its insulating value.

(xii) Before moving an aerial lift for travel, the boom(s) shall be inspected to see that it is properly cradled and outriggers are in stowed position except as provided in paragraph (b) (2) (viii) of this section.

(3) *Electrical tests.* All electrical tests shall conform to the requirements of ANSI A92.2-1969 section 5. However equivalent d.c voltage tests may be used in lieu of the a.c. voltage specified in A92.2-1969; d.c. voltage tests which are approved by the equipment manufacturer or equivalent entity shall be considered an equivalent test for the purpose of this paragraph (b) (3).

(4) *Bursting safety factor.* The provisions of the American National Standards Institute standard ANSI A92.2-1969, section 4.9 Bursting Safety Factor shall apply to all critical hydraulic and pneumatic components. Critical components are those in which a failure would result in a free fall or free rotation of the boom. All noncritical components shall have a bursting safety factor of at least 2 to 1.

(5) *Welding standards.* All welding shall conform to the following standards as applicable:

(i) Standard Qualification Procedure, AWS B3.0-41.

(ii) Recommended Practices for Automotive Welding Design, AWS D8.4-61.

(iii) Standard Qualification of Welding Procedures and Welders for Piping and Tubing, AWS D10.9-69.

(iv) Specifications for Welding Highway and Railway Bridges, AWS D2.0-69.

Pile driving equipment.

(a) *General requirements.* (1) Boilers and piping systems which are a part of, or used with, pile driving equipment shall meet the applicable requirements of the American Society of Mechanical Engineers, Power Boilers (section I).

(2) All pressure vessels which are a part of, or used with, pile driving equipment shall meet the applicable requirements of the American Society of Mechanical Engineers, Pressure Vessels (section VIII).

(3) Overhead protection, which will not obscure the vision of the operator shall be provided. Protection shall be the equivalent of 2-inch planking or other solid material of equivalent strength.

(4) Stop blocks shall be provided for the leads to prevent the hammer from being raised against the head block.

(5) A blocking device, capable of safely supporting the weight of the hammer, shall be provided for placement in the leads under the hammer at all times while employees are working under the hammer.

(6) Guards shall be provided across the top of the head block to prevent the cable from jumping out of the sheaves.

(7) When the leads must be inclined in the driving of batter piles, provisions shall be made to stabilize the leads.

(8) Fixed leads shall be provided with ladder, and adequate rings, or similar attachment points, so that the loft worker may engage his safety belt lanyard to the leads. If the leads are provided with loft platform(s), such platform(s) shall be protected by standard guardrails.

(9) Steam hose leading to a steam hammer or jet pipe shall be securely attached to the hammer with an adequate length of at least ¼-inch diameter chain or cable to prevent whipping in the event the joint at the hammer is broken. Air hammer hoses shall be provided with the same protection as required for steam lines.

(10) Safety chains, or equivalent means, shall be provided for each hose connection to prevent the line from thrashing around in case the coupling becomes disconnected.

(11) Steam line controls shall consist of two shutoff valves, one of which shall be a quick-acting lever type within easy reach of the hammer operator.

(12) Guys, outriggers, thrustouts, or counterbalances shall be provided as necessary to maintain stability of pile driver rigs.

(c) *Pile driving equipment.* (1) Engineers and winchmen shall accept signals only from the designated signalmen.

(2) All employees shall be kept clear when piling is being hoisted into the leads.

(3) When piles are being driven in an excavated pit, the walls of the pit shall be sloped to the angle of repose or sheet-piled and braced.

(4) When steel tube piles are being "blown-out", employees shall be kept well beyond the range of falling materials.

(5) When it is necessary to cut off the tops of driven piles, pile driving operations shall be suspended except where the cutting operations are located at least twice the length of the longest pile from the driver.

(6) When driving jacked piles, all access pits shall be provided with ladders and bulkheaded curbs to prevent material from falling into the pit.

Site clearing.

(a) *General requirements.* (1) Employees engaged in site clearing shall be protected from hazards of irritant and toxic plants and suitably instructed in the first aid treatment available.

(2) All equipment used in site clearing operations shall be equipped with rollover guards. In addition, rider-operated equipment shall be equipped with an overhead and rear canopy guard meeting the following requirements:

(i) The overhead covering on this canopy structure shall be of not less than ⅛-inch steel plate or ¼-inch woven wire mesh with openings no greater than 1 inch, or equivalent.

(ii) The opening in the rear of the canopy structure shall be covered with not less than ¼-inch woven wire mesh with openings no greater than 1 inch.

(a) *Material handling operations.* (1) Operations fitting the definition of "material handling" shall be performed in conformance with applicable requirements of Part 1918, "Safety and Health Regulations for Longshoring" of this chapter. The term "longshoring operations" means the loading, unloading, moving, or handling of construction materials, equipment and supplies, etc. into, in, on, or out of any vessel from a fixed structure or shore-to-vessel, vessel-to-shore or fixed structure or vessel-to-vessel.

(b) *Access to barges.* (1) Ramps for access of vehicles to or between barges shall be of adequate strength, provided with side boards, well maintained, and properly secured.

(2) Unless employees can step safely to or from the wharf, float, barge, or river towboat, either a ramp, meeting the requirements of paragraph (b)(1) of this section, or a safe walkway, shall be provided.

(3) Jacob's ladders shall be of the double rung or flat tread type. They shall be well maintained and properly secured.

(4) A Jacob's ladder shall either hang without slack from its lashings or be pulled up entirely.

(5) When the upper end of the means of access rests on or is flush with the top of the bulwark, substantial steps properly secured and equipped with at least one substantial hand rail approximately 33 inches in height, shall be provided between the top of the bulwark and the deck.

(6) Obstructions shall not be laid on or across the gangway.

(7) The means of access shall be adequately illuminated for its full length.

(8) Unless the structure makes it impossible, the means of access shall be so located that the load will not pass over employees.

(c) *Working surfaces of barges.* (1) Employees shall not be permitted to walk along the sides of covered lighters or barges with coamings more than 5 feet high, unless there is a 3-foot clear walkway, or a grab rail, or a taut handline is provided.

Excavations, Trenching, and Shoring

General protection requirements.

(a) Walkways, runways, and sidewalks shall be kept clear of excavated material or other obstructions and no sidewalks shall be undermined unless shored to carry a minimum live load of one hundred and twenty-five (125) pounds per square foot.

(b) If planks are used for raised walkways, runways, or sidewalks, they shall be laid parallel to the length of the walk and fastened together against displacement.

(c) Planks shall be uniform in thickness and all exposed ends shall be provided with beveled cleats to prevent tripping.

(d) Raised walkways, runways, and sidewalks shall be provided with plank steps on strong stringers. Ramps, used in lieu of steps, shall be provided with cleats to insure a safe walking surface.

(e) All employees shall be protected with personal protective equipment for the protection of the head, eyes, respiratory organs, hands, feet, and other parts of the body.

(f) Employees exposed to vehicular traffic shall be provided with and shall be instructed to wear warning vests marked with or made of reflectorized or high visibility material.

(g) Employees subjected to hazardous dusts, gases, fumes, mists, or atmospheres deficient in oxygen, shall be protected with approved respiratory protection.

(h) No person shall be permitted under loads handled by power shovels, derricks, or hoists. To avoid any spillage employees shall be required to stand away from any vehicle being loaded.

(i) Daily inspections of excavations shall be made by a competent person. If evidence of possible cave-ins or slides is apparent, all work in the excavation shall cease until the necessary precautions have been taken to safeguard the employees.

Specific excavation requirements.

(a) Prior to opening an excavation, effort shall be made to determine whether underground installations; i.e., sewer, telephone, water, fuel, electric lines, etc., will be encountered, and if so, where such underground installations are located. When the excavation approaches the estimated location of such an installation, the exact location shall be determined and when it is uncovered, proper supports shall be provided for the existing installation. Utility companies shall be contacted and advised of proposed work prior to the start of actual excavation.

(b) Trees, boulders, and other surface encumbrances, located so as to create a hazard to employees involved in excavation work or in the vicinity thereof at any time during operations, shall be removed or made safe before excavating is begun.

(c) The walls and faces of all excavations in which employees are exposed to danger from moving ground shall be guarded by a shoring system, sloping of the ground, or some other equivalent means.

(d) Excavations shall be inspected by a competent person after every rainstorm or other hazard-increasing occurrence, and the protection against slides and cave-ins shall be increased if necessary.

(e) The determination of the angle of repose and design of the supporting system shall be based on careful evaluation of perti-

nent factors such as: Depth of cut; possible variation in water content of the material while the excavation is open; anticipated changes in materials from exposure to air, sun, water, or freezing; loading imposed by structures, equipment, overlying material, or stored material; and vibration from equipment, blasting, traffic, or other sources.

(f) Supporting systems; i.e., piling, cribbing, shoring, etc., shall be designed by a qualified person and meet accepted engineering requirements. When tie rods are used to restrain the top of sheeting or other retaining systems, the rods shall be securely anchored well back of the angle of repose. When tight sheeting or sheet piling is used, full loading due to ground water table shall be assumed, unless prevented by weep holes or drains or other means. Additional stringers, ties, and bracing shall be provided to allow for any necessary temporary removal of individual supports.

(g) All slopes shall be excavated to at least the angle of repose except for areas where solid rock allows for line drilling or presplitting.

(h) The angle of repose shall be flattened when an excavation has water conditions, silty materials, loose boulders, and areas where erosion, deep frost action, and slide planes appear.

(i)(1) In excavations which employees may be required to enter, excavated or other material shall be effectively stored and retained at least 2 feet or more from the edge of the excavation.

(2) As an alternative to the clearance prescribed in paragraph (i)(1) of this section, the employer may use effective barriers or other effective retaining devices in lieu thereof in order to prevent excavated or other materials from falling into the excavation.

(j) Sides, slopes, and faces of all excavations shall meet accepted engineering requirements by scaling, benching, barricading, rock bolting, wire meshing, or other equally effective means. Special attention shall be given to slopes which may be adversely affected by weather or moisture content.

(k) Support systems shall be planned and designed by a qualified person when excavation is in excess of 20 feet in depth, adjacent to structures or improvements, or subject to vibration or ground water.

(l) Materials used for sheeting, sheet piling, cribbing, bracing, shoring, and underpinning shall be in good serviceable condition, and timbers shall be sound, free from large or loose knots, and of proper dimensions.

(m) Special precautions shall be taken in sloping or shoring the sides of excavations adjacent to a previously backfilled excavation or a fill, particularly when the separation is less than the depth of the excavation. Particular attention also shall be paid to joints and seams of material comprising a face and the slope of such seams and joints.

(n) Except in hard rock, excavations below the level of the base of footing of any foundation or retaining wall shall not be permitted, unless the wall is underpinned and all other precautions taken to insure the stability of the adjacent walls for the protection of employees involved in excavation work or in the vicinity thereof.

(o) If the stability of adjoining buildings or walls is endangered by excavations, shoring, bracing, or underpinning shall be provided as necessary to insure their safety. Such shoring, bracing, or underpinning shall be inspected daily or more often, as conditions warrant, by a competent person and the protection effectively maintained.

(p) Diversion ditches, dikes, or other suitable means shall be used to prevent surface water from entering an excavation and to provide adequate drainage of the area adjacent to the excavation. Water shall not be allowed to accumulate in an excavation.

(q) If it is necessary to place or operate power shovels, derricks, trucks, materials, or other heavy objects on a level above and near an excavation, the side of the excavation shall be sheet-piled, shored, and braced as necessary to resist the extra pressure due to such superimposed loads.

(s) When mobile equipment is utilized or allowed adjacent to excavations, substantial stop logs or barricades shall be installed. If possible, the grade should be away from the excavation.

(t) Adequate barrier physical protection shall be provided at all remotely located excavations. All wells, pits, shafts, etc., shall be barricaded or covered. Upon completion of exploration and similar operations, temporary wells, pits, shafts, etc., shall be backfilled.

(u) If possible, dust conditions shall be kept to a minimum by the use of water, salt, calcium chloride, oil, or other means.

(v) In locations where oxygen deficiency or gaseous conditions are possible, air in the excavation shall be tested. Controls shall be established to assure acceptable atmospheric conditions. When flammable gases are present, adequate ventilation shall be provided or sources of ignition shall be eliminated. Attended emergency rescue equipment, such as breathing apparatus,

a safety harness and line, basket stretcher, etc., shall be readily available where adverse atmospheric conditions may exist or develop in an excavation.

(w) Where employees or equipment are required or permitted to cross over excavations, walkways or bridges with standard guardrails shall be provided.

(x) Where ramps are used for employees or equipment, they shall be designed and constructed by qualified persons in accordance with accepted engineering requirements.

Specific trenching requirements.

(a) Banks more than 5 feet high shall be shored, laid back to a stable slope, or some other equivalent means of protection shall be provided where employees may be exposed to moving ground or cave-ins. Trenches less than 5 feet in depth shall also be effectively protected when examination of the ground indicates hazardous ground movement may be expected.

(b) Sides of trenches in unstable or soft material, 5 feet or more in depth, shall be shored, sheeted, braced, sloped, or otherwise supported by means of sufficient strength to protect the employees working within them.

(c) Sides of trenches in hard or compact soil, including embankments, shall be shored or otherwise supported when the trench is more than 5 feet in depth and 8 feet or more in length. In lieu of shoring, the sides of the trench above the 5-foot level may be sloped to preclude collapse, but shall not be steeper than a 1-foot rise to each ½-foot horizontal. When the outside diameter of a pipe is greater than 6 feet, a bench of 4-foot minimum shall be provided at the toe of the sloped portion.

Table P-1

Approximate Angle of Repose
For Sloping of Sides of Excavations

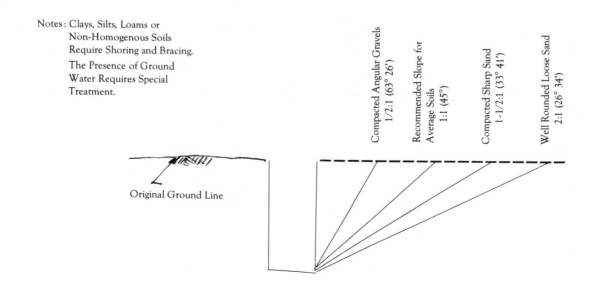

Notes: Clays, Silts, Loams or
Non-Homogenous Soils
Require Shoring and Bracing.

The Presence of Ground
Water Requires Special
Treatment.

Compacted Angular Gravels
1/2:1 (63° 26')

Recommended Slope for
Average Soils
1:1 (45°)

Compacted Sharp Sand
1-1/2:1 (33° 41')

Well Rounded Loose Sand
2:1 (26° 34')

Original Ground Line

Table P-2 — Trench Shoring — Minimum Requirements

Depth of trench	Kind or condition of earth	Uprights		Stringers		Cross braces [1], Width of trench					Maximum spacing	
		Minimum dimension	Maximum spacing	Minimum dimension	Maximum spacing	Up to 3 feet	3 to 6 feet	6 to 9 feet	9 to 12 feet	12 to 15 feet	Vertical	Horizontal
Feet		Inches	Feet	Inches	Feet	Inches	Inches	Inches	Inches	Inches	Feet	Feet
5 to 10	Hard, compact	3 x 4 or 2 x 6	6			2 x 6	4 x 4	4 x 6	6 x 6	6 x 8	4	6
	Likely to crack	3 x 4 or 2 x 6	3	4 x 6	4	2 x 6	4 x 4	4 x 6	6 x 6	6 x 8	4	6
	Soft, sandy, or filled	3 x 4 or 2 x 6	Close sheeting	4 x 6	4	4 x 4	4 x 6	6 x 6	6 x 8	8 x 8	4	6
	Hydrostatic pressure	3 x 4 or 2 x 6	Close sheeting	6 x 8	4	4 x 4	4 x 6	6 x 6	6 x 8	8 x 8	4	6
10 to 15	Hard	3 x 4 or 2 x 6	4	4 x 6	4	4 x 4	4 x 6	6 x 6	6 x 8	8 x 8	4	6
	Likely to crack	3 x 4 or 2 x 6	2	4 x 6	4	4 x 4	4 x 6	6 x 6	6 x 8	8 x 8		6
	Soft, sandy, or filled	3 x 4 or 2 x 6	Close sheeting	4 x 6	4	4 x 6	6 x 6	6 x 8	8 x 8	8 x 10	4	6
	Hydrostatic pressure	3 x 6	Close sheeting	8 x 10	4	4 x 6	6 x 6	6 x 8	8 x 8	8 x 10	4	6
15 to 20	All kinds or conditions	3 x 6	Close sheeting	4 x 12	4	4 x 12	6 x 8	8 x 8	8 x 10	10 x 10	4	6
Over 20	All kinds or conditions	3 x 6	Close sheeting	6 x 8	4	4 x 12	8 x 8	8 x 10	10 x 10	10 x 12	4	6

[1]Trench jacks may be used in lieu of, or in combination with, cross braces.
Shoring is not required in solid rock, hard shale, or hard slag.
Where desirable, steel sheet piling and bracing of equal strength may be substituted for wood.

(d) Materials used for sheeting and sheet piling, bracing, shoring, and underpinning, shall be in good serviceable condition, and timbers used shall be sound and free from large or loose knots, and shall be designed and installed so as to be effective to the bottom of the excavation.

(e) Additional precautions by way of shoring and bracing shall be taken to prevent slides or cave-ins when excavations or trenches are made in locations adjacent to backfilled excavations, or where excavations are subjected to vibrations from railroad or highway traffic, the operation of machinery, or any other source.

(f) Employees entering bell-bottom pier holes shall be protected by the installation of a removable-type casing of sufficient strength to resist shifting of the surrounding earth. Such temporary protection shall be provided for the full depth of that part of each pier hole which is above the bell. A lifeline, suitable for instant rescue and securely fastened to a shoulder harness, shall be worn by each employee entering the shafts. This lifeline shall be individually manned and separate from any line used to remove materials excavated from the bell footing.

(2) Braces and diagonal shores in a wood shoring system shall not be subjected to compressive stress in excess of values given by the following formula:

$$S = 13 - 20L/D$$
$$\text{Maximum ratio } L/D = 50$$

Where:
L = Length, unsupported, in inches
D = Least side of the timber in inches.
S = Allowable stress in pounds per square inch of cross-section.

(h) When employees are required to be in trenches 4 feet deep or more, an adequate means of exit, such as a ladder or steps, shall be provided and located so as to require no more than 25 feet of lateral travel.

(i) Bracing or shoring of trenches shall be carried along with the excavation.

(j) Cross braces or trench jacks shall be placed in true horizontal position, be spaced vertically, and be secured to prevent sliding, falling, or kickouts.

(k) Portable trench boxes or sliding trench shields may be used for the protection of personnel in lieu of a shoring system or sloping. Where such trench boxes or shields are used, they shall be designed, constructed, and maintained in a manner which will provide protection equal to or greater than the sheeting or shoring required for the trench.

(l) Backfilling and removal of trench supports shall progress together from the bottom of the trench. Jacks or braces shall be released slowly and in, unstable soil, ropes shall be used to pull out the jacks or braces from above after employees have cleared the trench.

Concrete, Concrete Forms, and Shoring

General provisions.

(a) *General.* All equipment and materials used in concrete construction and masonry work shall meet the applicable requirements for design, construction, inspection, testing, maintenance and operations as prescribed in ANSI A10..9-1970, Safety Requirements for Concrete Construction and Masonry Work.

(b) *Reinforcing steel.* (1) Employees working more than 6 feet above any adjacent working surfaces, placing and tying reinforcing steel in walls, piers, columns, etc., shall be provided with a safety belt, or equivalent device.

(2) Employees shall not be permitted to work above vertically protruding reinforcing steel unless it has been protected to eliminate the hazard of impalement.

(3) Guying: Reinforcing steel for walls, piers, columns, and similar vertical structures shall be guyed and supported to prevent collapse.

(4) Wire mesh rolls: Wire mesh rolls shall be secured at each end to prevent dangerous recoiling action.

(c) *Bulk concrete handling.* Bulk storage bins, containers, or silos shall have conical or tapered bottoms with mechanical or pneumatic means of starting the flow of material.

(d) *Concrete placement* — (1) *Concrete mixers.* Concrete mixers equipped with 1-yard or larger loading skips shall be equipped with a mechanical device to clear the skip of material.

(2) *Guardrails.* Mixers of 1-yard capacity or greater shall be equipped with protective guardrails installed on each side of the skip.

(3) *Bull floats.* Handles on bull floats, used where they may contact energized electrical conductors, shall be constructed of nonconductive material, or insulated with a nonconductive sheath whose electrical and mechanical characteristics provide the equivalent protection of a handle constructed of non-conductive material.

(4) *Powered concrete trowels.* Powered and rotating-type concrete troweling machines that are manually guided shall be equipped with a control switch that will automatically shut off the power whenever the operator removes his hands from the equipment handles.

(5) *Concrete buggies.* Handles of buggies shall not extend beyond the wheels on either side of the buggy. Installation of knuckle guards on buggy handles is recommended.

(6) *Pumpcrete systems.* Pumpcrete or similar systems using discharge pipes shall be provided with pipe supports designed for 100 percent overload. Compressed air hose in such systems shall be provided with positive failsafe joint connectors to prevent separation of sections when pressurized.

(7) *Concrete buckets.* (i) Concrete buckets equipped with hydraulic or pneumatically operated gates shall have positive safety latches or similar safety devices installed to prevent aggregate and loose material from accumulating on the top and sides of the bucket.

(ii) Riding of concrete buckets for any purpose shall be prohibited, and vibrator crews shall be kept out from under concrete buckets suspended from cranes or cableways.

(8) When discharging on a slope, the wheels of ready-mix trucks shall be blocked and the brakes set to prevent movement.

(9) Nozzlemen applying a cement, sand, and water mixture through a pneumatic hose shall be required to wear protective head and face equipment.

(e) *Vertical shoring* — (1) *General requirements.* (i) When temporary storage of reinforcing rods, material, or equipment on top of formwork becomes necessary, these areas shall be strengthened to meet the intended loads.

510

(ii) The sills for shoring shall be sound, rigid, and capable of carrying the maximum intended load.

(iii) All shoring equipment shall be inspected prior to erection to determine that it is as specified in the shoring layout. Any equipment found to be damaged shall not be used for shoring.

(iv) Erected shoring equipment shall be inspected immediately piror to, during, and immediately after the placement of concrete. Any shoring equipment that is found to be damaged or weakened shall be immediately reinforced or reshored.

(v) Reshoring shall be provided when necessary to safely support slabs and beams after stripping, or where such members are subjected to superimposed loads due to construction work done.

(2) *Tubular welded frame shoring.* (i) Metal tubular frames used for shoring shall not be loaded beyond the safe working load recommended by the manufacturer.

(ii) All locking devices on frames and braces shall be in good working order; coupling pins shall align the frame or panel legs; pivoted cross braces shall have their center pivot in place; and all components shall be in a condition similar to that of original manufacture.

(iii) When checking the erected shoring frames with the shoring layout, the spacing between towers and cross brace spacing shall not exceed that shown on the layout, and all locking devices shall be in the closed position.

(iv) Devices for attaching the external lateral stability bracing shall be securely fastened to the legs of the shoring frames.

(v) All baseplates, shore heads, extension devices, or adjustment screws shall be in firm contact with the footing sill and the form.

Forms and shoring.

(a) *General provisions.* (1) Formwork and shoring shall be designed, erected, supported, braced, and maintained so that it will safely support all vertical and lateral loads that may be imposed on it during placement of concrete.

(2) Drawings or plans showing the jack layout, formwork, shoring, working decks, and scaffolding, shall be available at the jobsite.

(3) Stripped forms and shoring shall be removed and stockpiled promptly after stripping, in all areas in which persons are required to work or pass. Protruding nails, wire ties, and other form accessories not necessary to subsequent work shall be pulled, cut, or other means taken to eliminate the hazard.

(4) Imposition of any construction loads on the partially completed structure shall not be permitted unless such loading has been considered in the design and approved by the engineer-architect.

(b) *Vertical slip forms.* (1) The steel rods or pipe on which the jacks climb or by which the forms are lifted shall be specifically designed for the purpose. Such rods shall be adequately braced where not encased in concrete.

(2) Jacks and vertical supports shall be positioned in such a manner that the vertical loads are distributed equally and do not exeed the capacity of the jacks.

(3) The jacks or other lifting devices shall be provided with mechanical dogs or other automatic holding devices to provide protection in case of failure of the power supply or the lifting mechanism.

(4) Lifting shall proceed steadily and uniformly and shall not exceed the predetermined safe rate of lift.

(5) Lateral and diagonal bracing of the forms shall be provided to prevent excessive distortion of the structure during the jacking operation.

(6) During jacking operations, the form structure shall be maintained in line and plumb.

(7) All vertical lift forms shall be provided with scaffolding or work platforms completely encircling the area of placement.

(c) *Tube and coupler shoring.* (1) Couplers (clamps) shall not be used if they are deformed, broken, or have defective or missing threads on bolts, or other defects.

(2) The material used for the couplers (clamps) shall be of a structural type such as drop-forged steel, malleable iron, or structural grade aluminum. Gray cast iron shall not be used.

(3) When checking the erected shoring towers with the shoring layout, the spacing between posts shall not exceed that shown on the layout and all interlocking of tubular members and tightness of couples shall be checked.

(4) All baseplates, shore heads, extention devices, or adjustment screws shall in in firm contact with the footing sill and the form material and shall be snug against the posts.

(d) *Single post shores.* (1) For stability, single post shores shall be horizontally braced in both the longitudinal and transverse directions, and diagonal bracing shall also be installed. Such bracing shall be installed as the shores are being erected.

(2) All baseplates or shore heads of single post shores shall be in firm contact with the footing sill and the form materials.

(3) Whenever single post shores are used in more than one tier, the layout shall be designed and inspected by a structural engineer.

(4) When formwork is at an angle, or sloping, or when the surface shored is sloping, the shoring shall be designed for such loading.

(5) Adjustment of single post shores to raise formwork shall not be made after concrete is in place.

(6) Fabricated single post shores shall not be used if heavily rusted, bent, dented, rewelded, or having broken weldments or other defects. If they contain timber, they shall not be used if timber is split, cut, has sections removed, is rotted, or otherwise structurally damaged.

(7) All timber and adjusting devices to be used for adjustable timber single post shores shall be inspected before erection.

(8) Timber shall not be used if it is split, cut, has sections removed, is rotted, or is otherwise structurally damaged.

(9) Adjusting devices shall not be used if heavily rusted, bent, dented, rewelded, or having broken weldments or other defects.

(10) All nails used to secure bracing or adjustable timber single post shores shall be driven home and the point of the nail bent over if possible.

Definitions applicable to this subpart.

(a) "Bull float" — A tool used to spread out and smooth the concrete.

(b) "Formwork" or "falsework" — The total system of support for freshly placed concrete, including the mold or sheathing which contacts the concrete as well as all supporting members, hardware, and necessary bracing.

(c) "Guy" — A line that steadies a high piece or structure by pulling against an off-center load.

(d) "Shore" — A supporting member that resists a compressive force imposed by a load.

(e) "Vertical slip forms" — Forms which are jacked vertically and continuously during placing of the concrete.

Steel Erection

Flooring requirements.

(a) *Permanent flooring — skeleton steel construction in tiered buildings.*

(1) The permanent floors shall be installed as the erection of structural members progresses, and there shall be not more than eight stories between the erection floor and the uppermost permanent floor, except where the structural integrity is maintained as a result of the design.

(2) At no time shall there be more than four floors or 48 feet of unfinished bolting or welding above the foundation or uppermost permanently secured floor.

(b) *Temporary flooring — skeleton steel construction in tiered buildings.* (1) (i) The derrick or erection floor shall be solidly planked or decked over its entire surface except for access openings. Planking or decking of equivalent strength, shall be of proper thickness to carry the working load. Planking shall be not less than 2 inches thick full size undressed, and shall be laid tight and secured to prevent movement.

(ii) On buildings or structures not adaptable to temporary floors, and where scaffolds are not used, safety nets shall be installed and maintained whenever the potential fall distance exceeds two stories or 25 feet. The nets shall be hung with sufficient clearance to prevent contacts with the surface of structures below.

(iii) Floor periphery — safety railing. A safety railing of ½-inch wire rope or equal shall be installed, approximately 42 inches high, around the periphery of all temporary-planked or temporary metal-decked floors of tier buildings and other multifloored structures during structural steel assembly.

(2) (i) Where skeleton steel erection is being done, a tightly planked and substantial floor shall be maintained within two stories or 30 feet, whichever is less, below and directly under that portion of each tier of beams on which any work is being performed, except when gathering and stacking temporary floor planks on a lower floor, in preparation for transferring such planks for use on an upper floor. Where such a floor is not practicable, paragraph (b) (1) (ii) of this section applies.

(ii) When gathering and stacking temporary floor planks, the planks shall be removed successively, working toward the last panel of the temporary floor so that the work is always done from the planked floor.

(iii) When gathering and stacking temporary floor planks from the last panel, the employees assigned to such work shall be protected by safety belts with safety lines attached to a catenary line or other substantial anchorage.

(c) *Flooring — other construction.* (1) In the erection of a building having double wood floor construction, the rough flooring shall be completed as the building progresses, including the tier below the one on which floor joists are being installed.

(2) For single wood floor or other flooring systems, the floor immediately below the story where the floor joists are being installed shall be kept planked or decked over.

Structural steel assembly.

(a) During the final placing of solid web structural members, the load shall not be released from the hoisting line until the members are secured with not less than two bolts, or the equivalent at each connection and drawn up wrench tight.

(b) Open web steel joists shall not be placed on any structural steel framework unless such framework is safely bolted or welded.

(c) (1) In steel framing, where bar joists are utilized, and columns are not framed in at least two directions with structural steel members, a bar joist shall be field-bolted at columns to provide lateral stability during construction.

(2) Where longspan joists or trusses, 40 feet or longer, are used, a center row of bolted bridging shall be installed to provide lateral stability during construction prior to slacking of hoisting line.

(3) No load shall be placed on open web steel joists until these security requirements are met.

(d) Tag lines shall be used for controlling loads.

Bolting, riveting, fitting-up, and plumbing-up.

(a) *General requirements.* (1) Containers shall be provided for storing or carrying rivets, bolts, and drift pins, and secured against accidental displacement when aloft.

(2) Pneumatic hand tools shall be disconnected from the power source, and pressure in hose lines shall be released, before any adjustments or repairs are made.

(3) Air line hose sections shall be tied together except when quick disconnect couplers are used to join sections.

(b) *Bolting.* (1) When bolts or drift pins are being knocked out, means shall be provided to keep them from falling.

(2) Impact wrenches shall be provided with a locking device for retaining the socket.

(c) *Riveting.* (1) Riveting shall not be done in the vicinity of combustible material unless precautions are taken to prevent fire.

(2) When rivet heads are knocked off, or backed out, means shall be provided to keep them from falling.

(3) A safety wire shall be properly installed on the snap and on the handle of the pneumatic riveting hammer and shall be used at all times. The wire size shall be not less than No. 9 (B&S gauge), leaving the handle and annealed No. 14 on the snap, or equivalent.

(d) *Plumbing-up.* (1) Connections of the equipment used in plumbing-up shall be properly secured.

(2) The turnbuckles shall be secured to prevent unwinding while under stress.

(3) Plumbing-up guys related equipment shall be placed so that employees can get at the connection points.

(4) Plumbing-up guys shall be removed only under the supervision of a competent person.

(e) Wood planking shall be of proper thickness to carry the working load, but shall be not less than 2 inches thick full size undressed, exterior grade plywood, at least ¾-inch thick, or equivalent material.

(f) Metal decking of sufficient strength shall be laid tight and secured to prevent movement.

(g) Planks shall overlap the bearing on each end by a minimum of 12 inches.

(h) Wire mesh, exterior plywood, or equivalent, shall be used around columns where planks do not fit tightly.

(i) Provisions shall be made to secure temporary flooring against displacement.

(j) All unused openings in floors, temporary or permanent, shall be completely planked over or guarded.

(k) Employees shall be provided with safety belts when they are working on float scaffolds.

Demolition

Preparatory operations.

(a) Prior to permitting employees to start demolition operations, an engineering survey shall be made, by a competent person, of the structure to determine the condition of the framing, floors, and walls, and possibility of unplanned collapse of any portion of the structure. Any adjacent structure where employees may be exposed shall also be similarly checked. The employer shall have in writing evidence that such a survey has been performed.

(b) When employees are required to work within a structure to be demolished which has been damaged by fire, flood, explosion, or other cause, the walls or floor shall be shored or braced.

(c) All electric, gas, water, steam, sewer, and other service lines shall be shut off, capped, or otherwise controlled, outside the building line before demolition work is started. In each case, any utility company which is involved shall be notified in advance.

(d) If it is necessary to maintain any power, water or other utilities during demolition, such lines shall be temporarily relocated, as necessary, and protected.

(e) It shall also be determined if any type of hazadous chemicals, gases, explosives, flammable materials, or similarly dangerous substances have been used in any pipes, tanks, or other equipment on the property. When the presence of any such substances is apparent or suspected, testing and purging shall be performed and the hazard eliminated before demolition is started.

(f) Where a hazard exists from fragmentation of glass, such hazards shall be removed.

(g) Where a hazard exists to employees falling through wall openings, the opening shall be protected to a height of approximately 42 inches.

(h) When debris is dropped through holes in the floor without the use of chutes, the area onto which the material is dropped shall be completely enclosed with barricades not less than 42 inches high and not less than 6 feet back from the projected edge of the opening above. Signs, warning of the hazard of falling materials, shall be posted at each level. Removal shall not be permitted in this lower area until debris handling ceases above.

(i) All floor openings, not used as material drops, shall be covered over with material substantial enough to support the weight of any load which may be imposed. Such material shall be properly secured to prevent its accidental movement.

(j) Except for the cutting of holes in floors for chutes, holes through which to drop materials, preparation of storage space, and similar necessary preparatory work, the demolition of exterior walls and floor construction shall begin at the top of the structure and proceed downward. Each story of exterior wall and floor construction shall be removed and dropped into the storage space before commencing the removal of exterior walls and floors in the story next below.

(k) Employee entrances to multistory structures being demolished shall be completely protected by sidewalk sheds or canopies, or both, providing protection from the face of the building for a minimum of 8 feet. All such canopies shall be at least 2 feet wider than the building entrances or openings (1 foot wider on each side thereof), and shall be capable of sustaining a load of 150 pounds per square foot.

Stairs, passageways, and ladders.

(a) Only those stairways, passageways, and ladders, designated as means of access to the structure of a building, shall be used. Other access ways shall be entirely closed at all times.

(b) All stairs, passageways, ladders and incidental equipment thereto, which are covered by this section, shall be periodically inspected and maintained in a clean safe condition.

(c) In a multistory building, when a stairwell is being used, it shall be properly illuminated by either natural or artificial means, and completely and substantially covered over at a point not less than two floors below the floor on which work is being performed, and access to the floor where the work is in progress shall be through a properly lighted, protected, and separate passageway.

Chutes.

(a) No material shall be dropped to any point lying outside the exterior walls of the structure unless the area is effectively protected.

(b) All materials chutes, or sections thereof, at an angle of more than 45° from the horizontal, shall be entirely enclosed, except for openings, equipped with closures at or about floor level for the insertion of materials. The openings shall not exceed 48 inches in height measured along the wall of the chute. At all stories below the top floor, such openings shall be kept closed when not in use.

(c) A substantial gate shall be installed in each chute at or near the discharge end. A competent employee shall be assigned to control the operation of the gate, and the backing and loading of trucks.

(d) When operations are not in progress, the area surrounding the discharge end of a chute shall be securely closed off.

(e) Any chute opening, into which workmen dump debris, shall be protected by a substantial guardrail approximately 42 inches above the floor or other surface on which the men stand to dump the material. Any space between the chute and the edge of openings in the floors through which it passes shall be solidly covered over.

(f) Where the material is dumped from mechanical equipment or wheelbarrows, a securely attached toeboard or bumper, not less than 4 inches thick and 6 inches high, shall be provided at each chute opening.

(g) Chutes shall be designed and constructed of such strength as to eliminate failure due to impact of materials or debris loaded therein.

Removal of materials through floor openings.

Any openings cut in a floor for the disposal of materials shall be no larger in size than 25 percent of the aggregate of the total floor area, unless the lateral supports of the removed flooring remain in place. Floors weakened or otherwise made unsafe by demolition operations shall be shored to carry safely the intended imposed load from demolition operations.

Removal of walls, masonry sections, and chimneys.

(a) Masonry walls, or other sections of masonry, shall not be permitted to fall upon the floors of the building in such masses as to exceed the safe carrying capacities of the floors.

(b) No wall section, which is more than one story in height, shall be permitted to stand alone without lateral bracing, unless such wall was originally designed and constructed to stand without such lateral support, and is in a condition safe enough to be self-supporting. All walls shall be left in a stable condition at the end of each shift.

(c) Employees shall not be permitted to work on the top of a wall when weather conditions constitute a hazard.

(d) Structural or load-supporting members on any floor shall not be cut or removed until all stories above such a floor have been demolished and removed. This provision shall not prohibit the cutting of floor beams for the disposal of materials or for the installation of equipment.

(e) Floor openings within 10 feet of any wall being demolished shall be planked solid, except when employees are kept out of the area below.

(f) In buildings of "skeleton-steel" construction, the steel framing may be left in place during the demolition of masonry. Where this is done, all steel beams, girders, and similar structural supports shall be cleared of all loose material as the masonry demolition progresses downward.

(g) Walkways or ladders shall be provided to enable employees to safely reach or leave any scaffold or wall.

(h) Walls, which serve as retaining walls to support earth or adjoining structures, shall not be demolished until such earth has been properly braced or adjoining structures have been properly underpinned.

(i) Walls, which are to serve as retaining walls against which debris will be piled, shall not be so used unless capable of safely supporting the imposed load.

Manual removal of floors.

(a) Openings cut in a floor shall extend the full span of the arch between supports.

(b) Before demolishing any floor arch, debris and other material shall be removed from such arch and other adjacent floor area. Planks not less than 2 inches by 10 inches in cross section, full size undressed, shall be provided for, and shall be used by employees to stand on while breaking down floor arches between beams. Such planks shall be so located as to provide a safe support for the workmen should the arch between the beams collapse. The open space between planks shall not exceed 16 inches.

(c) Safe walkways, not less than 18 inches wide, formed of planks not less than 2 inches thick if wood, or of equivalent strength if metal, shall be provided and used by workmen when necessary to enable them to reach any point without walking upon exposed beams.

(d) Stringers of ample strength shall be installed to support the flooring planks, and the ends of such stringers shall be supported by floor beams or girders, and not by floor arches alone.

(e) Planks shall be laid together over solid bearings with the ends overlapping at least 1 foot.

(f) When floor arches are being removed, employees shall not be allowed in the area directly underneath, and such an area shall be barricaded to prevent access to it.

(g) Demolition of floor arches shall not be started until they, and the surrounding floor area for a distance of 20 feet, have been cleared of debris and any other unnecessary materials.

Removal of walls, floors, and material with equipment.

(a) Mechanical equipment shall not be used on floors or working surfaces unless such floors or surfaces are of sufficient strength to support the imposed load.

(b) Floor openings shall have curbs or stop-logs to prevent equipment from running over the edge.

Storage.

(a) The storage of waste material and debris on any floor shall not exceed the allowable floor loads.

(b) In buildings having wooden floor construction, the flooring boards may be removed from not more than one floor above grade to provide storage space for debris, provided falling material is not permitted to endanger the stability of the structure.

(c) When wood floor beams serve to brace interior walls or free-standing exterior walls, such beams shall be left in place until other equivalent support can be installed to replace them.

(d) Floor arches, to an elevation of not more than 25 feet above grade, may be removed to provide storage area for debris: *Provided*, That such removal does not endanger the stability of the structure.

(e) Storage space into which material is dumped shall be blocked off, except for openings necessary for the removal of material. Such openings shall be kept closed at all times when material is not being removed.

Removal of steel construction.

(a) When floor arches have been removed, planking shall be provided for the workers engaged in razing the steel framing.

(c) Steel construction shall be dismantled column length by column length, and tier by tier (columns may be in two-story lengths).

(d) Any structural member being dismembered shall not be overstressed.

Mechanical demolition.

(a) No workers shall be permitted in any area, which can be adversely affected by demolition operations, when balling or clamming is being performed. Only those workers necessary for the performance of the operations shall be permitted in this area at any other time.

(b) The weight of the demolition ball shall not exceed 50 percent of the crane's rated load, based on the length of the boom and the maximum angle of operation at which the demolition ball will be used, or it shall not exceed 25 percent of the nominal breaking strength of the line by which it is suspended, whichever results in a lesser value.

(c) The crane boom and loadline shall be as short as possible.

(d) The ball shall be attached to the loadline with a swivel-type connection to prevent twisting of the loadline, and shall be attached by positive means in such manner that the weight cannot become accidentally disconnected.

(e) When pulling over walls or portions thereof, all steel members affected shall have been previously cut free.

(f) All roof cornices or other such ornamental stonework shall be removed prior to pulling walls over.

(g) During demolition, continuing inspections by a competent person shall be made as the work progresses to detect hazards resulting from weakened or deteriorated floors, or walls, or loosened material. No employee shall be permitted to work where such hazards exist until they are corrected by shoring, bracing, or other effective means.

INDEX